Library of
Davidson College

Electronic Properties of Crystalline Solids

AN INTRODUCTION TO FUNDAMENTALS

Electronic Properties of Crystalline Solids

AN INTRODUCTION TO FUNDAMENTALS

RICHARD H. BUBE

Department of Materials Science and Engineering
Stanford University
Stanford, California

ACADEMIC PRESS   New York and London   1974
A Subsidiary of Harcourt Brace Jovanovich, Publishers

COPYRIGHT © 1974, BY ACADEMIC PRESS, INC.
ALL RIGHTS RESERVED.
NO PART OF THIS PUBLICATION MAY BE REPRODUCED OR
TRANSMITTED IN ANY FORM OR BY ANY MEANS, ELECTRONIC
OR MECHANICAL, INCLUDING PHOTOCOPY, RECORDING, OR ANY
INFORMATION STORAGE AND RETRIEVAL SYSTEM, WITHOUT
PERMISSION IN WRITING FROM THE PUBLISHER.

ACADEMIC PRESS, INC.
111 Fifth Avenue, New York, New York 10003

United Kingdom Edition published by
ACADEMIC PRESS, INC. (LONDON) LTD.
24/28 Oval Road, London NW1

Library of Congress Cataloging in Publication Data

Bube, Richard H   DATE
    Electronic properties of crystalline solids.

    Bibliography: p.
    1.   Energy-band theory of solids.   2.   Wave
mechanics.   I.   Title.
QC176.8.E4B8   530.4'1   72–9322
ISBN 0–12–138550–7

PRINTED IN THE UNITED STATES OF AMERICA

By faith
we understand
that the world was created
by the Word of God,
so that what is seen
was made
out of things which do not appear.

Hebrews 11:3 (RSV)

Contents

PREFACE xi

Chapter 1 **Classical Waves: A Review** 1

1.1 General Properties of Waves 1
1.2 General Approach to Wave Problems 4
1.3 Long-Wavelength Waves in Strings and Rods 5
1.4 Lattice Waves in a One-Dimensional Crystal 5
1.5 Electromagnetic Waves 15
1.6 Summary 17
    Problems 18

Chapter 2 **Wave Approach to Quantum Mechanics** 20

2.1 Simple Applications of a Wave Analogy 21
2.2 The Schroedinger Equation 23
2.3 Basic Postulates of Quantum Mechanics 26
2.4 Interpretation of the Wavefunction 31
2.5 Orthogonality 34
2.6 Expectation Values 38
2.7 Dirac Notation 42
2.8 Summary 43
    Problems 45

Chapter 3 **Quantum Mechanical Treatment of Simple Systems** . . . 47

3.1 A Free Particle 48
3.2 A Particle in a One-Dimensional Potential Well 49
3.3 A Linear Harmonic Oscillator 53

3.4	A Hydrogenic Atom	58
3.5	Summary	68
	Problems	70

Chapter 4 Free-Electron Model of Metals   73

4.1	Atomic Energy Levels and the Periodic Table	75
4.2	The Sommerfeld Free-Electron Model	83
4.3	Traveling Waves and Periodic Boundary Conditions	90
4.4	Hartree Model for Free Electrons in a Metal	93
4.5	Occupancy of Allowed Energy Levels for Free Electrons in a Metal	98
4.6	Examples of Applications of the Free-Electron Model for Metals	106
4.7	Summary	117
	Problems	118

Chapter 5 Origin of Energy Bands in Solids   121

5.1	Wavefunction for an Electron in a Periodic Potential	122
5.2	The Cellular Method	124
5.3	Geometrical Considerations: Reciprocal Lattice and Brillouin Zones	128
5.4	Energy Bands in a Perturbed Nearly Free Electron System	141
5.5	Energy Bands in the Tight-Binding Approximation	153
5.6	Summary	161
	Problems	163

Chapter 6 Properties of Energy Bands   166

6.1	Energy-Band Calculations	167
6.2	Density of States in Energy Bands	171
6.3	Electron Velocity and Effective Mass	174
6.4	The Band Model and Electrical Properties	179
6.5	Energy Bands in Real Crystals	184
6.6	Excitons and Polarons	199
6.7	Bands and Bonds	200
6.8	Summary	206
	Problems	207

Chapter 7 Carrier Transport   211

7.1	Wave Packets	212
7.2	Description of Particle Motion Using Wave Packets	217
7.3	The Boltzmann Equation	220
7.4	Solution of the Boltzmann Equation	222
7.5	Relaxation-Time Solution of the Boltzmann Equation	223

Contents

7.6	Electrical Conductivity in the Relaxation-Time Approximation	229
7.7	Electrical Conductivity in Semiconductors and Metals	231
7.8	Thermal Conductivity Due to Electrons	239
7.9	Thermoelectric Effect	244
7.10	Summary	246
	Problems	248

Chapter 8 Scattering Processes 250

8.1	Scattering by Acoustic-Mode Lattice Waves: Simple Model of Wave Reflection	251
8.2	Scattering by Acoustic-Mode Lattice Waves: Perturbation Calculation	257
8.3	Charged-Imperfection Scattering	273
8.4	Other Scattering Mechanisms	285
8.5	High-Electric-Field Effects in Semiconductors	288
8.6	Summary	299
	Problems	302

Chapter 9 Localized Energy Levels 305

9.1	Energy Levels in an Imperfect Crystal	306
9.2	Imperfection Terminology	310
9.3	Description of Imperfection Incorporation	310
9.4	Description of Electronic Behavior	316
9.5	Theory of Shallow Imperfection Energy Levels	318
9.6	Thermal-Equilibrium Fermi Level in Semiconductors and Insulators	324
9.7	Fermi-Level Description of Electrical Conductivity	329
9.8	Imperfection Interactions	343
9.9	Device Applications Describable by the Band Picture of Imperfect Semiconductors	346
9.10	Summary	353
	Problems	355

Chapter 10 Magnetic-Field Effects 357

10.1	Low Magnetic Fields in the Linear Approximation	359
10.2	Types of Mobility	367
10.3	General Treatment of the Low-Magnetic-Field Range	369
10.4	Effects of Scattering Mechanisms	377
10.5	Effects of Band Structure	383
10.6	Magnetothermal Effects	386
10.7	High-Magnetic-Field Effects	387
10.8	Summary	394
	Problems	396

Chapter 11 Optical Absorption 401

11.1 Free-Carrier Absorption 403
11.2 Optical Transitions between Bands 413
11.3 Direct Intrinsic Transition 414
11.4 Indirect Intrinsic Transition 427
11.5 Exciton Absorption 442
11.6 Summary 444
Problems 445

Chapter 12 Photoelectronic Effects 449

12.1 General Concepts 450
12.2 Electrical Contacts 460
12.3 Analytical Approaches 471
12.4 Models of Photoconductivity 477
12.5 Recombination Mechanisms 485
12.6 Recombination Kinetics 490
12.7 Related Photoelectronic Effects 506
12.8 Summary 509
Problems 510

Appendix Units and Conversion Factors 513

Bibliography 516

Index 519

Preface

The material presented in this book is the result of courses in the electronic properties of solids taught over the past decade in the Department of Materials Science and Engineering at Stanford University. It is intended for upper-level undergraduates in a science major, or for first- or second-year graduate students with an interest in the scientific basis for our understanding of the properties of materials. It assumes only an elementary previous acquaintance with quantum mechanics and develops in a self-consistent way all the background necessary for the subject matter, beyond that expected in a usual undergraduate science or engineering program. Students taking these courses have come from a wide diversity of backgrounds including metallurgy, materials science, electrical engineering, applied physics, chemistry, chemical engineering, and various other engineering disciplines. As usual in such cases, a special word of appreciation must be spoken for all those students whose patience and questions helped to shape the final text; it was written with them in mind.

The book starts with a brief review of classical wave mechanics, since the concept of waves and their properties plays such a unifying role in the interactions of electrons, phonons, and photons. An introduction to quantum mechanics by the way of its interpretation as wave mechanics is illustrated for a number of simple systems of noninteracting particles. These results are then applied to the free electron model for metals, and subsequently to the origin, derivation, and properties of allowed and forbidden energy bands for electrons in crystalline materials. The remainder of the book deals with transport phenomena and optical effects in crystalline materials, including electrical conductivity, scattering phenomena, thermal conductivity, Hall and thermoelectric effects, magnetoresistance, optical absorption, photoconductivity, and other photoelectronic effects in both ideal (intrinsic or defect-free) and real (containing localized imperfections) materials. In the course of this development such basic concepts as quantum statistics, Fermi level analysis, wave packets and their properties, and time-independent and time-dependent perturbation theory are introduced. Frequent examples of typical problems with the indicated method of treatment are included at appropriate places in the text. Each chapter also contains at its end an assortment of unsolved (but solvable) problems, on which the student may test his ability to apply the principles of the chapter. The present text seeks to utilize the best features of previously published works in this field, to which every succeeding author must be indebted, and to integrate them into an overall context which is broader than any currently available with a basic emphasis on student-oriented pedagogical readability.

Limitations on length, desire for comprehensive treatment, resolution of technical roadblocks for the student—all of these are in a sense mutually exclusive. The final book can be viewed best as a compromise—but a compromise which I hope has some claim to uniqueness. The subtitle of the book, An Introduction to Fundamentals, was chosen to emphasize that one such compromise involves at least a partial limitation on the breadth of the subject matter covered. I have tried to provide the student with the basic concepts and ideas needed for understanding current literature and treatments of a variety of phenomena of relatively recent interest, rather than to give only sketchy summaries of effects judged primarily for their relevance. It is expected that this course will only be preliminary to other courses on quantum electronics, detailed techniques for the calculation of energy bands, details of electronic devices, and modern advances in the physics of metals and noncrystalline materials. Experience seems to confirm that the choice and treatment of material presented here does provide the kind of desired and needed bridge into the field of electronic properties of materials, particularly for students without a strong background in classical physics and quantum mechanics.

At least three ingredients are required for successful learning in a scientific or engineering discipline. There must be a growth of understanding of (1) the broad underlying physical concepts, (2) the mathematical techniques necessary to analyze and predict, and (3) the types of problems that are encountered. This book attempts to incorporate all three of these ingredients. What helps the student has been of prime concern in the choice of material and organization. Thus, within the limitations of space and breadth of subject matter, it has been considered more useful to treat a few examples in considerable depth than to treat many examples superficially. Although mathematics as the language of science must of necessity

Preface

make up a large proportion of the text, it is intended that the mathematics per se will ultimately be secondary to the understanding of the physical concepts involved. Elegance (not necessarily mathematical difficulty, but mathematical formalism which may obscure the physical relationships for a relative beginner) is sacrificed to comprehensibility. Paradoxically, to prevent mathematical details from hindering the student's progress often requires that particular attention to these details be paid so as to remove the aura of mystery from the situation.

I wish to thank the following journals and publishers for permission to reproduce previously published figures, as noted individually in each case in the place where the figure appears in the text: *The Physical Review*, *The Journal of Applied Physics*, *The Journal of Chemical Physics*, *The Journal of Physics and Chemistry of Solids*, The Royal Society of London, The Physical Society of London, the *Journal of Electronics and Control*, North-Holland Publishing Company, *RCA Review*, McGraw-Hill Book Company, Prentice-Hall Publishers, John Wiley and Sons, Publishers, Addison-Wesley Publishing Company, and Academic Press.

I thank heartily all who have made this book possible and have encouraged me during its preparation, I am grateful for those students who may find these pages useful, and trust that any possible errors in concept or copy may be forgiven and overcome by the reader's own persistence.

Electronic Properties of Crystalline Solids

AN INTRODUCTION TO FUNDAMENTALS

Chapter 1

Classical Waves: A Review

One of the unifying themes that runs throughout physical phenomena of many different kinds is that of wave motion. Sound waves and light waves are part of the subject matter of classical physics, and a classical type of wave mechanics is sufficient for at least a partial treatment of the interaction of these kinds of wave with crystalline solids. Many of the properties of these classical wave systems are found also in a variety of forms in quantum wave mechanics. It is primarily for this reason that we begin with a brief review of the classical treatment of waves suitable for a partial description of sound and light.

1.1 General Properties of Waves

A wave is any periodic disturbance in time and position, characterized by a *velocity*, a *wavelength*, and a *frequency*. Such a disturbance may have quite a general distribution in space as long as these characteristics are definable. The relationship between velocity, wavelength, and frequency is

$$v = \lambda \nu \qquad (1.1)$$

Values of these parameters vary widely for different types of wave. Representative values are indicated in Table 1.1.

One of the simplest and most analytically useful wave forms is that of a harmonic wave which can be represented in terms of sine and cosine functions. Any general wave form can be expressed in terms of a Fourier

TABLE 1.1 TYPICAL VALUES OF WAVE PARAMETERS

	Velocity, v (cm/sec)	Wavelength, λ (cm)	Frequency, ν (sec^{-1})
Sound (in air)	3×10^4	1 to 3×10^3	10 to 2×10^4
Visible light (in vacuum)	3×10^{10}	4×10^{-5} to 7×10^{-5}	10^{15}
Free electron (300°K)	5×10^6	8×10^{-7}	6×10^{12}

expansion of harmonic waves. A harmonic wave can be expressed mathematically in the following kinds of relationship between the *displacement* ξ, the *angular frequency* $\omega = 2\pi\nu$, and the *wavenumber* $k = 2\pi/\lambda$.

$$\xi = A \cos(\omega t - kx) \quad \text{or} \quad \xi = B \sin(\omega t - kx) \quad (1.2)$$

or

$$\xi = C\, e^{i(kx-\omega t)} \quad (1.3)$$

All of these representations indicate a harmonic wave moving in the $+x$ direction. A wave moving in the $-x$ direction is indicated by changing the sign of the kx term in Eq. (1.2), or choosing the signs so that the kx and ωt terms in Eq. (1.3) have the same sign.

Equation (1.1) may be rewritten in terms of ω and k to give

$$\omega = vk \quad (1.4)$$

The relationship between ω and k is in general known as the *dispersion* relationship. If v is a constant, independent of the values of ω and k, the wave motion is said to be *nondispersive*; if v varies with ω and k, the wave motion is said to be *dispersive*.

Equations (1.2) and (1.3) represent *traveling waves*. The sum of two traveling waves moving in opposite directions but with equal frequency is a *standing wave*, i.e., a wave for which the dependence of the displacement on position is independent of time. Such a standing wave is obtained from

$$e^{i(kx-\omega t)} + e^{-i(kx+\omega t)} = 2 \cos kx\, e^{-i\omega t} \quad (1.5)$$

Wave motion in a vibrating string provides one of the most familiar examples of such behavior. If one end of the string is held and the other end is free, then a wave started at the held end travels down the string until it

1.1 General Properties of Waves

is dissipated. But if wave motion is induced in a string with both ends held, a standing displacement pattern is produced as the result of reflection and subsequent superposition of traveling waves moving in opposite directions along the string.

Waves may be either *transverse* waves or *longitudinal* waves, depending on the relationship between the direction of the disturbance displacement and the direction of the wave motion. Figure 1.1 pictures both a transverse and a longitudinal wave in terms of the displacement of particles from an equilibrium position, as is useful for considering wave motion in crystals. In the transverse case, the wave is a mathematical curve drawn through the various displaced particles; in the longitudinal case, the wave can be visualized in terms of alternating regions of condensation and rarefaction. In a transverse wave, the displacement is in a direction perpendicular to the direction of wave motion; in a longitudinal wave, the displacement is in the same direction as that of the wave motion. Light waves are transverse waves; sound waves are longitudinal waves. Atoms in crystals can be displaced in ways that can be described in terms of both transverse and longitudinal waves. When a wave form is used to describe the motion of particles, the motion of the disturbance, which is the actual wave motion, must be distinguished from the motion of the individual particles; the wave may progress throughout the whole crystal, whereas the displacement of any

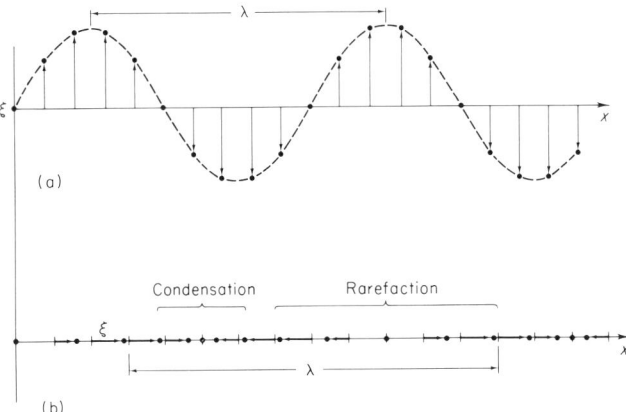

Fig. 1.1 Transverse and longitudinal waves moving in the x direction. The arrows indicate the direction of the displacement. For a transverse wave (a), the displacement is perpendicular to the direction of wave motion. For a longitudinal wave (b), the displacement, indicated by the arrows, is in the direction of wave motion. The occurrence of a longitudinal wave leads to regions of condensation (smaller-than-normal particle separation) and of rarefaction (larger-than-normal particle separation).

particular particle from its equilibrium position may be exceedingly small.

The velocity v defined by Eqs. (1.1) and (1.4) is the *phase velocity* of the wave. It represents the actual velocity with which waves of frequency and wavenumber k propagate. In a nondispersive system, this velocity is constant, but in a dispersive system it varies with frequency and wavelength. In treating wave motion in a dispersive system, it is necessary also to define a *group velocity* as

$$v_g = \frac{d\omega}{dk} \qquad (1.6)$$

Consider, for example, the propagation of a pulse of arbitrary waveform (i.e., composed of many harmonic waves with different ω and k) through a dispersive medium. As the pulse progresses through the material, its waveform changes, and it becomes necessary to distinguish between two velocities: (a) the velocity with which the "boundaries" of the pulse propagate—the group velocity; and (b) the velocity with which the waves within the pulse propagate—the phase velocity. A simple example of the effects of such behavior can be seen by considering the sum of two harmonic waves with different frequency and wavenumber, but equal amplitudes

$$\xi_1 + \xi_2 = A \cos(\omega_1 t - k_1 x) + A \cos(\omega_2 t - k_2 x) \qquad (1.7)$$

Using trigonometric identities, Eq. (1.7) can be transformed to

$$\xi_1 + \xi_2 = 2A \cos \tfrac{1}{2}(t\, \Delta\omega - x\, \Delta k) \cos(\bar{\omega} t - \bar{k} x) \qquad (1.8)$$

where $\bar{\omega} = (\omega_1 + \omega_2)/2$, $\bar{k} = (k_1 + k_2)/2$, $\Delta\omega = \omega_2 - \omega_1$, and $\Delta k = k_2 - k_1$. Equation (1.8) describes a wave system in which the waves have a velocity $v = \bar{\omega}/\bar{k}$, these waves being modulated by an envelope function moving with a velocity $v_g = \Delta\omega/\Delta k$. Now if we choose frequencies and wavenumbers such that $\omega_2 \approx \omega_1$, and $k_2 \approx k_1$, these two velocities correspond respectively to the phase and group velocities of the wave system. We conclude that the phase velocity and the group velocity are identical in a nondispersive system, but not in a dispersive system.

1.2 General Approach to Wave Problems

Certain general principles can be set forth that apply to a variety of different wave problems.

1. First we must consider *the nature of the medium*, as this determines the form that waves propagating in this medium take on. In the case of

mechanical waves involving matter, considerations related to $\mathbf{F} = m\mathbf{a}$ dominate. For electromagnetic waves, we must consider the relations between electric and magnetic fields. For electron waves, we must consider the corresponding appropriate formalisms.

2. We must formulate a *wave equation* that is some kind of differential equation expressing the space and time dependence of the displacement, $\xi(x, t)$. The wave equation will be the result of considerations of the medium.

3. The *general solution* of this differential wave equation is found, giving explicitly the general function $\xi(x, t)$.

4. Possible restrictions on the general solution may arise out of the requirements that the specific solution *must correspond to physical reality*. Mathematically allowed solutions that do not conform to physical conditions must be discarded.

5. Possible restrictions on the general solution may also arise out of *boundary conditions* which define the particular problem. The effect of such restrictions is to limit the allowed frequencies that the wave may have to a set of discrete values.

6. Standing waves will be used to describe situations in which there are rigid boundary conditions; traveling waves will be used to describe other situations, such as those involved in reflection, absorption, and interference of waves.

1.3 Long-Wavelength Waves in Strings and Rods

The properties of transverse waves in a string of linear density ϱ and under a tension $\mathscr{E}$, and of longitudinal waves in a rod of density ϱ and under a stress X, are summarized in Table 1.2. The approach used is that of continuum mechanics and is appropriate for long-wavelength vibrations. Under these conditions, these two systems are nondispersive. Applying boundary conditions of fixed ends results in the requirement that allowed waves must have a wavelength such that an integral number of half-wavelengths fits into the length of the string or rod. The waves produced by these boundary conditions are standing waves.

1.4 Lattice Waves in a One-Dimensional Crystal

The properties of transverse waves in a one-dimensional monatomic crystal with atoms with mass m and lattice spacing a, of longitudinal waves in a similar crystal, and of transverse waves in a one-dimensional diatomic

TABLE 1.2 PROPERTIES OF SIMPLE WAVE SYSTEMS (LONG-WAVELENGTH WAVES)

	Transverse waves in a string of linear density ϱ and under tension $\mathscr{C}$	Longitudinal waves in a rod of density ϱ and under a stress X
Equation of motion	$\varrho\,dx\,\dfrac{d^2\xi}{dt^2} = \mathscr{C}\,\dfrac{d^2\xi}{dx^2}\,dx$	$\varrho S\,dx\,\dfrac{d^2\xi}{dt^2} = S\,\dfrac{dX}{dx}\,dx$
Wave equation	$\dfrac{d^2\xi}{dt^2} = \dfrac{\mathscr{C}}{\varrho}\,\dfrac{d^2\xi}{dx^2}$	$\dfrac{d^2\xi}{dt^2} = \dfrac{Y}{\varrho}\,\dfrac{d^2\xi}{dx^2}, \quad Y = \dfrac{X}{d\xi/dx}$
General solution	$\xi = A\,e^{i(kx-\omega t)} + B\,e^{-i(kx-\omega t)}$	$\xi = A\,e^{i(kx-\omega t)} + B\,e^{-i(kx+\omega t)}$
Phase velocity = group velocity	$v = \left(\dfrac{\mathscr{C}}{\varrho}\right)^{1/2}$	$v = \left(\dfrac{Y}{\varrho}\right)^{1/2}$
Effect of fixed-ends boundary condition $0 = \xi(0) = \xi(L)$	$k = \dfrac{n\pi}{L}$	$k = \dfrac{n\pi}{L}$
Allowed frequencies for fixed-ends boundary condition	$\omega_n = \dfrac{n\pi}{L}\left(\dfrac{\mathscr{C}}{\varrho}\right)^{1/2}$	$\omega_n = \dfrac{n\pi}{L}\left(\dfrac{Y}{\varrho}\right)^{1/2}$
Displacements of standing wave for fixed-ends boundary condition	$\xi_n = A_n \sin\left(\dfrac{n\pi x}{L}\right)$	$\xi_n = A_n \sin\left(\dfrac{n\pi x}{L}\right)$

TABLE 1.3 Properties of Lattice Waves in a One-Dimensional Crystal

	Transverse waves in a monatomic crystal with lattice spacing a	Longitudinal waves in a monatomic crystal with lattice spacing a	Transverse waves in a diatomic crystal (masses m and M, $M > m$) with lattice spacing a
Equation of motion	$\dfrac{d^2\xi_r}{dt^2} - \eta\xi_{r-1} + 2\eta\xi_r - \eta\xi_{r+1} = 0$ $\eta \equiv \dfrac{F}{ma}$	$\dfrac{d^2\zeta_r}{dt^2} - \eta'\zeta_{r-1} + 2\eta'\zeta_r - \eta'\zeta_{r+1} = 0$ $\eta' \equiv \dfrac{1}{m}\left(\dfrac{dF}{d\zeta}\right)_a$	$\dfrac{d^2\xi_r}{dt^2} + 2\eta_M\xi_r - \eta_M\zeta_r - \eta_M\zeta_{r+1} = 0$ $\dfrac{d^2\zeta_r}{dt^2} + 2\eta_m\zeta_r - \eta_m\xi_r - \eta_m\xi_{r-1} = 0$ $\eta_M \equiv \dfrac{F}{Ma}, \quad \eta_m \equiv \dfrac{F}{ma}$
Harmonic waves	$\xi_r = A\,e^{i(kra-\omega t)}$	$\zeta_r = B\,e^{i(kra-\omega t)}$	$\xi_r = C\,e^{i(k(2r+1)a-\omega t)}$ $\zeta_r = D\,e^{i(2kra-\omega t)}$
Dispersion relation	$\omega = 2\eta^{1/2}\left|\sin\left(\dfrac{ka}{2}\right)\right|$	$\omega = 2\eta'^{1/2}\left|\sin\left(\dfrac{ka}{2}\right)\right|$	$\omega_\pm^2 = (\eta_M + \eta_m)$ $\pm\,[(\eta_M + \eta_m)^2 - 4\eta_M\eta_m \sin^2 ka]^{1/2}$
Normal modes for ends of crystal fixed	$\omega_n = 2\eta^{1/2}\left|\sin\left(\dfrac{n\pi a}{2L}\right)\right|$	$\omega_n = 2\eta'^{1/2}\left|\sin\left(\dfrac{n\pi a}{2L}\right)\right|$	
Normal displacements of standing wave with ends of crystal fixed	$\xi_{rn} = A_n \sin\left(\dfrac{n\pi ra}{L}\right)$	$\zeta_{rn} = B_n \sin\left(\dfrac{n\pi ra}{L}\right)$	

crystal with atoms with mass m and M ($M > m$) equally spaced with lattice spacing a, are summarized in Table 1.3. The appropriate definitions of displacements are given in Fig. 1.2 for each of these three cases. Three basic assumptions are involved in the calculation: (a) only forces between nearest neighbors are considered, (b) a restoring force results from an attractive force F between atoms, and (c) for small displacements compared to the lattice spacing a, it is assumed that F is both constant and in the direction of the nearest neighbor.

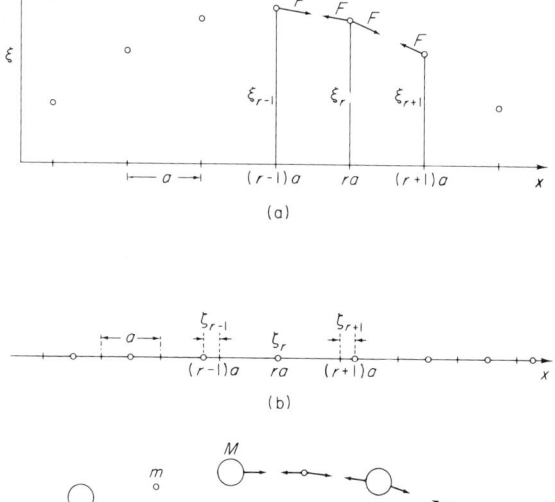

Fig. 1.2 (a) Displacements and active forces assumed for a monatomic one-dimensional lattice with lattice spacing a for transverse waves. (b) Displacements for a monatomic one-dimensional lattice with lattice spacing a for longitudinal waves. (c) Displacements and active forces assumed for a diatomic one-dimensional lattice with atoms of two different masses and equal lattice spacing for transverse waves.

In the monatomic crystal a single type of vibrational branch is found, corresponding to acoustic modes of vibration. Except for very long wavelengths ($k \approx 0$), the system is dispersive. Similar behavior is found for both transverse and longitudinal vibrations. Figure 1.3a shows the dispersion relationship predicted for the corresponding three-dimensional case, where we need to specify two transverse and one longitudinal acoustic mode

1.4 Lattice Waves in a One-Dimensional Crystal

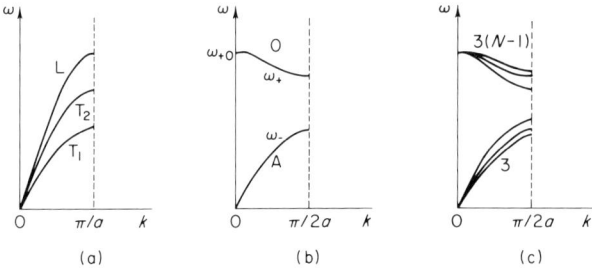

Fig. 1.3 (a) Dispersion relationships for two transverse and one longitudinal branch of acoustic modes. (b) Dispersion relationships for an acoustic branch and an optical branch. (c) Dispersion relationships for a crystal with N atoms per unit cell, showing three acoustic branches and $3(N-1)$ optical branches. The drawing is specifically for $N = 2$.

branches. The two transverse branches are taken in orthogonal directions, and are identical if the material is isotropic. All of the information in the dispersion relationship can be portrayed by showing the frequency ω as a function of k for values of k lying between $-\pi/a$ and $+\pi/a$. This is because the displacement of the individual atoms given by a value of $k' = k + 2\pi n/a$ is identical with the displacement of the atoms given by k itself.

In the diatomic crystal two types of vibrational branches are found, one corresponding to the acoustic modes described above for the monatomic crystal, and the other corresponding to optical modes. Figure 1.3b shows the dispersion relationship predicted, comparing a single acoustic branch with a single optical branch. Physically, long-wavelength acoustic modes correspond to vibrations in which nearby light and heavy atoms displace in the same direction; long-wavelength optical modes correspond to vibrations in which nearby light and heavy atoms displace in opposite directions. The origin of optical modes, here derived in terms of two different-mass atoms equally spaced, may also arise from two equal-mass atoms with two different lattice spacings, as when there are two atoms per unit cell. The expected vibrational spectrum for the case of N atoms per unit cell in three dimensions is indicated in Fig. 1.3c.

Example 1.1 Describe the vibrations in a system of six equally spaced atoms of equal mass, the two end atoms being fixed in position.

The equation of motion for each of the four free atoms is

$$\frac{d^2\xi_r}{dt^2} + 2\eta\xi_r - \eta\xi_{r+1} - \eta\xi_{r-1} = 0$$

Four such equations exist, for $r = 1, 2, 3,$ and 4. Furthermore $\xi_0 = \xi_5 = 0$. The boundary conditions imposed by the two fixed end atoms lead to

$$k = \frac{n\pi}{5a}, \quad n = 1, 2, 3, 4, 5$$

The allowed wavelengths for the vibrations are $\lambda = 10a/n$, giving a maximum wavelength of $10a$ and a minimum wavelength of $2a$. The normal modes are

$$\omega_n = \omega_{\max} \sin\left(\frac{n\pi}{10}\right)$$

and the corresponding normal displacements are

$$\xi_{nr} = A_n \sin\left(\frac{n\pi r}{5}\right) e^{-i\omega_n t}$$

Consider that particular mode for which $\lambda = 10a/3$, i.e., the mode corresponding to $n = 3$. The wave describing this mode has *nodes* ($\xi = 0$) at $r = 5/3, 10/3$, as well as at $r = 0$ and 5. This wave has a maximum at $r = 5/6$, a minimum at $r = 5/2$, and another maximum at $r = 25/6$. None of these nodes or extrema correspond to the actual location of the atoms. In this mode, the displacement of the second atom is $+0.95$ A, the displacement of the third atom is -0.59 A, the displacement of the fourth atom is -0.59 A, and the displacement of the fifth atom is $+0.95$ A. The results are shown in Fig. 1.4.

To test the remarks in the text above about the displacements for $k' = k + 2n\pi/a$ being the same as for k, consider the wave corresponding to the above value of $k = 3\pi/5a$ plus $2\pi/a$, i.e., for $k' = 13\pi/5a$.

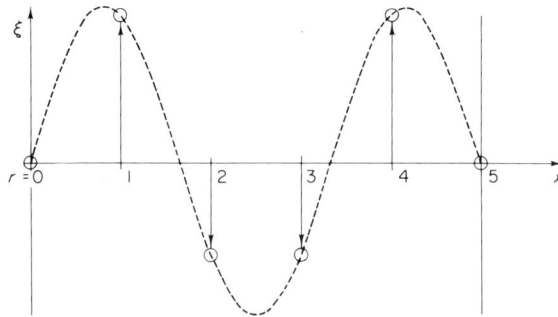

Fig. 1.4 The normal mode corresponding to $\lambda = 10a/3$ in a system of six equally spaced atoms of equal mass, the two end atoms being fixed in position.

1.4 Lattice Waves in a One-Dimensional Crystal

Now the wavelength is $10a/13$ and the wave has nodes at twelve points corresponding to $r = 5m/13$ with $m = 0, 1, \ldots, 13$. But if we examine the displacement of the atoms corresponding to k', we find that they are identical with those described above. For example,

$$\xi_{31}(k') = A\sin\left(\frac{13\pi}{5}\right) = \xi_{31}(k) = A\sin\left(\frac{3\pi}{5}\right)$$

Example 1.2 Figure 1.5 is a plot of a transverse optical mode branch and a transverse acoustical mode branch for one of the II–VI crystals (ZnS, ZnSe, ZnTe, CdS, CdSe, CdTe) as would be predicted from a one-dimensional lattice model of two different atoms all spaced equally. What is the compound involved?

There are four points of possible reference that can be derived from the data of Fig. 1.5.

$$\omega_{+0} = 2^{1/2}\left[\frac{F}{a}\left(\frac{1}{M} + \frac{1}{m}\right)\right]^{1/2} = 3.2\times 10^{13} \text{ Hz at } k = 0$$

$$\omega_{+} = 2^{1/2}\left(\frac{F}{a}\frac{1}{m}\right)^{1/2} = 2.6\times 10^{13} \text{ Hz at } k = \pi/2a$$

$$\omega_{-} = 2^{1/2}\left(\frac{F}{a}\frac{1}{M}\right)^{1/2} = 1.8\times 10^{13} \text{ Hz at } k = \pi/2a$$

$$\frac{\omega}{k} = \left(\frac{2Fa}{M+m}\right)^{1/2} = 1.5\times 10^{13}a \text{ cm/sec at } k = 0$$

The second and third of these can be used to obtain a ratio $(M/m) = 1.97$. The first and fourth add no new information. We may compare this value of (M/m) with calculated values for the compounds: ZnS (2.04), ZnSe (1.21), ZnTe (1.95), CdS (3.50), CdSe (1.42), and CdTe (1.13). The compound is therefore ZnTe.

The acoustic branches derive their name from the fact that the long-wavelength longitudinal acoustical modes correspond to sound waves; the optical branches derive their name from the fact that the long-wavelength transverse optical modes in ionic crystals can be directly excited by light waves. The strongest interaction of the latter type corresponds to $\omega_{\text{light}} = \omega_{+0}$, and has historically been given the name *Reststrahlen* (residual ray) absorption, since the high reflection associated with this infrared absorption was used to separate optical wavelengths in the Reststrahlen absorption band from other wavelengths in a polychromatic black-body

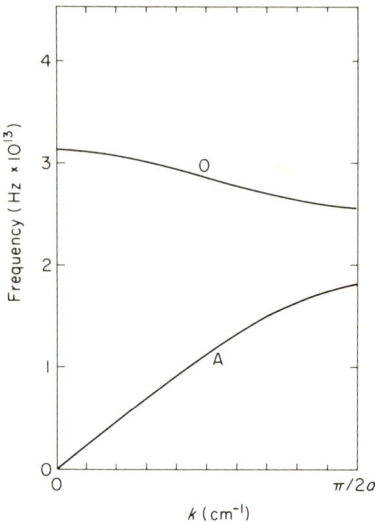

Fig. 1.5 Plot of a transverse optical branch and a transverse acoustic branch for one of the II–VI crystals as would be predicted from a simple one-dimensional lattice model of two different atoms all spaced equally. See Example 1.2 for further details.

source. The Reststrahlen frequency corresponds to the highest frequency of crystal vibrations that exists in a particular crystal. Another characteristic parameter of materials, the *Debye temperature* Θ_D, is related to the highest acoustical frequency,

$$k\Theta_D \approx \hbar\omega_{\max}^- \qquad (1.9)$$

where k is Boltzmann's constant (1.38×10^{-16} erg/degree) and $\hbar$ is Planck's constant (6.63×10^{-27} erg sec) divided by 2π.

Dispersion curves determined by experiment are shown in Fig. 1.6 for three different elemental materials: graphite, sodium, and lead. The longitudinal acoustic branch for graphite has a shape very close to that of the ideal sine wave indicated in Table 1.3. The results for sodium can be described reasonably well theoretically if effects from first and second nearest neighbors are included, although indications of effects out to fifth nearest neighbors are obtained. It is evident that the results for lead depart appreciably from the predictions of the simple theory; analysis indicates that very long-range forces beyond fifth nearest neighbors are active. Dispersion curves for two elemental semiconductors are given in Fig. 1.7; although the materials are covalent, both germanium and silicon have two atoms per unit cell and therefore have both acoustic and optical branches.

1.4 Lattice Waves in a One-Dimensional Crystal

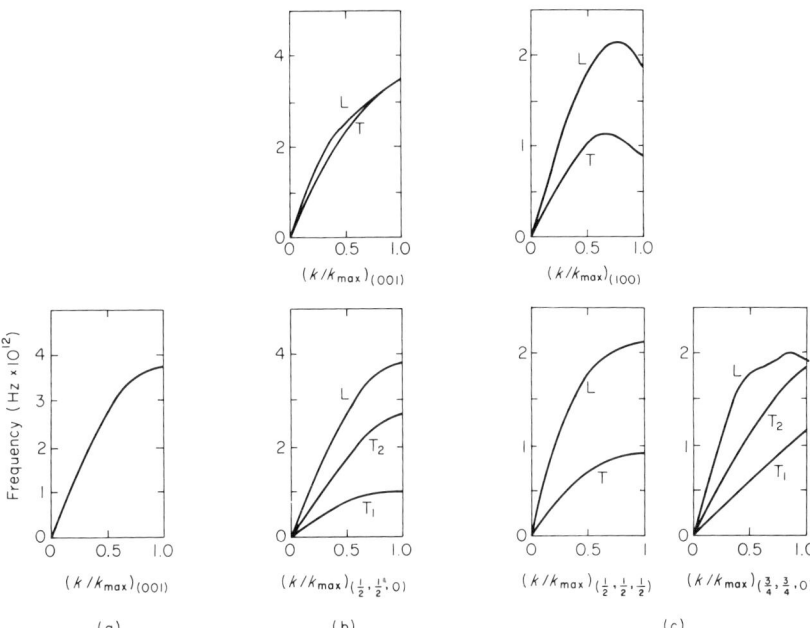

Fig. 1.6 Measured dispersion curves for lattice waves in three different materials. (a) Longitudinal acoustic mode in graphite. [After G. Dolling and B. N. Brockhouse, *Phys. Rev.* **128**, 1120 (1962).] (b) Acoustic modes in sodium for wave vectors in two different crystalline directions. [After A. D. B. Woods, B. N. Brockhouse, R. H. March, A. T. Stewart, and R. Bowers, *Phys. Rev:* **128**, 1112 (1962).] (c) Acoustic mode branches in lead for wave vectors in three different crystalline directions. [After B. N. Brockhouse, T. Arase, G. Caglioti, K. R. Rao, and A. D. B. Woods, *Phys. Rev.* **128**, 1099 (1962).]

Example 1.3 The wavelengths of light involved in Reststrahlen absorption for several alkali halides are as follows: 61 μm in NaCl, 71 μm in KCl, 88 μm in KBr, and 102 μm in KI. The wavelength for ZnS is 33 μm. Are these values consistent with known properties of these crystals?

As a working assumption, treat these materials as ionic compounds, i.e., a complete transfer of valence electrons is assumed to have occurred in the formation of the crystals. This is reasonably accurate for the alkali halides, somewhat less so for ZnS. A knowledge of the Reststrahlen wavelengths gives information about the force between atoms from the simple picture described above. We can also calculate the simple electrostatic attraction between the ions and compare this force with the value derived from the Reststrahlen data.

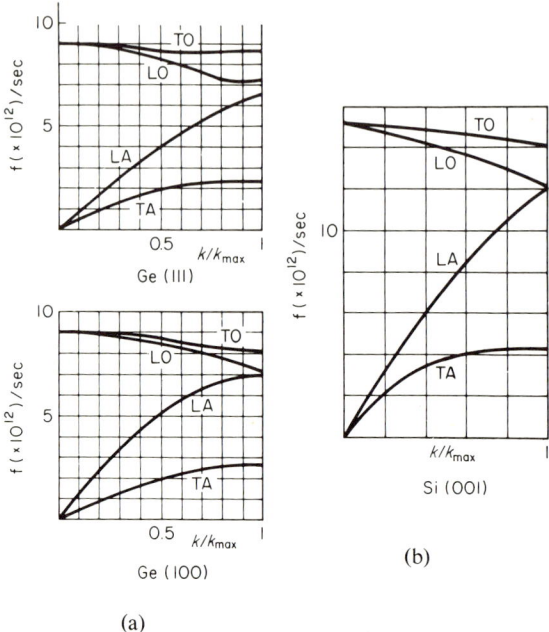

Fig. 1.7 (a) Acoustic and optical modes for germanium with two atoms per unit cell for two different crystalline directions. [After B. N. Brockhouse and P. K. Iyengar, *Phys. Rev.* **111**, 747 (1958).] (b) Acoustic and optical modes for silicon with two atoms per unit cell. [After B. N. Brockhouse, *Phys. Rev. Letters* **2**, 256 (1959).]

The relation between ω_{light} and ω_{+0} for Reststrahlen absorption can be rewritten to give

$$F = \frac{2\pi^2 ac^2}{\lambda_{\text{light}}^2}\left(\frac{mM}{m+M}\right)$$

Here a is the sum of the ionic radii, e.g., in the case of NaCl, the sum of the Na^+ radius and the Cl^- radius. For the ratio-of-masses factor, the same ratio of the atomic masses can be multiplied by the mass of a proton $(1.67 \times 10^{-24}\text{ g})$. To calculate the electrostatic attraction between two atoms directly, take

$$F = \frac{Z^2 e^2}{a^2}$$

where a has the same definition, and Z is the number of charges, i.e., $Z = 1$ for alkali halides and $Z = 2$ for ZnS. The results are summarized in Table 1.4. The agreement between the values of F calculated in the two different ways, in view of the approximations involved, is

1.5 Electromagnetic Waves

TABLE 1.4 Comparison of Estimates of the Force between Atoms

Compound	Cation radius (Å)	Anion radius (Å)	a (Å)	$Mm/(M+m)$ ($\times 1.7 \times 10^{-24}$ g)	Reststrahlen F (dynes $\times 10^{-4}$)	Electrostatic F (dynes $\times 10^{-4}$)
NaCl	0.95	1.81	2.76	14	3.1	3.0
KCl	1.33	1.81	3.14	18.6	3.4	2.3
KBr	1.33	1.95	3.28	25.6	3.2	2.2
KI	1.33	2.16	3.49	29.9	3.0	1.9
ZnS	0.74	1.84	2.58	21.5	15	14

quite good. The range of experimentally measured values for the Reststrahlen wavelengths for the alkali halides is seen to be the direct result of differences in atomic spacing and atomic masses, the basic force between atoms remaining the same throughout the series. The appreciably lower value of Reststrahlen wavelength for ZnS is the result of a larger charge on the ions, an effective charge of 2 fitting these particular data very well.

1.5 Electromagnetic Waves

The approximate values of wavelength associated with the different parts of the electromagnetic spectrum are summarized in Table 1.5. The electromagnetic spectrum encompasses a wavelength range that covers over twenty orders of magnitude; yet all radiation of this type propagates with the same velocity.

There are many different kinds of interaction between electromagnetic radiation and solids. Some of the most common are *reflection*, *refraction*, and *absorption*. For a given material, these phenomena are usually described quantitatively in terms of phenomenological parameters empirically assigned to the material: *index of refraction* and *absorption index*. These two optical quantities of refraction and absorption index can be correlated with two electrical quantities of the solid: the *dielectric constant* (a measure of polarizability), and *electrical conductivity* (a measure of free-carrier density).

The basic relationships needed to describe electromagnetic waves in a homogeneous isotropic material with dielectric constant ε, permeability μ, and electrical conductivity σ, are summarized in Table 1.6. The starting

TABLE 1.5 Characteristic Ranges of the Electromagnetic Spectrum

	Wavelength (Å)
Cosmic rays	10^{-4}
Gamma rays	10^{-2}
X rays	1
Ultraviolet	10^3
Visible light	$4\text{–}7 \times 10^3$
Infrared	10^5
Radar	10^9
Television	10^{10}
Radio	10^{12}
60-Hz current from a generator	10^{17}

point for the definition of the properties of the medium is the set of relationships between electric field $\mathscr{E}$, magnetic field **H**, and charge density ϱ, known as Maxwell's equations. In a material without electrical conductivity, electromagnetic waves are nondispersive and propagate with a velocity

TABLE 1.6 Properties of Electromagnetic Waves for Homogeneous and Isotropic Media

Basic relationships for defining waves	$\nabla \cdot \mathscr{E} = \dfrac{4\pi}{\varepsilon} \varrho \qquad \nabla \times \mathscr{E} = -\dfrac{\mu}{c} \dfrac{\partial \mathbf{H}}{\partial t}$
	$\nabla \cdot \mathbf{H} = 0 \qquad \nabla \times \mathbf{H} = \dfrac{\varepsilon}{c} \dfrac{\partial \mathscr{E}}{\partial t} + \dfrac{4\pi\sigma}{c} \mathscr{E}$
General wave equation for $\mathscr{E}$	$\dfrac{4\pi}{\varepsilon} \nabla \varrho - \nabla^2 \mathscr{E} = -\dfrac{\varepsilon\mu}{c^2} \dfrac{\partial^2 \mathscr{E}}{\partial t^2} - \dfrac{4\pi\mu\sigma}{c^2} \dfrac{\partial \mathscr{E}}{\partial t}$
Wave equation for $\mathscr{E}$ if $\sigma = 0$	$\nabla^2 \mathscr{E} = \dfrac{\varepsilon\mu}{c^2} \dfrac{\partial^2 \mathscr{E}}{\partial t^2}, \qquad \left[\varrho = \varrho_0 \, e^{-t/\tau_r}; \quad \tau_r = \dfrac{\varepsilon}{4\pi\sigma} \right]$
Harmonic wave for $\mathscr{E}$	$\mathscr{E} = \mathscr{E}_0 \, e^{i(\mathbf{k}\cdot\mathbf{r} - \omega t)}$
Dispersion relation for $\sigma = 0$	$\omega = \dfrac{c}{\sqrt{\varepsilon\mu}} k$
Dispersion relation for $\sigma \neq 0$	$\dfrac{1}{v^2} = \dfrac{k^2}{\omega^2} = \dfrac{\varepsilon\mu}{c^2} + i \dfrac{4\pi\mu\sigma}{c^2 \omega}$

$v = c/(\varepsilon\mu)^{1/2}$, where $c = 3 \times 10^{10}$ cm/sec, the velocity of light in vacuum. In a material with electrical conductivity, the phase velocity becomes complex, implying an attenuation of the wave, which corresponds to the phenomenon of optical absorption. We consider some of these relations further in later portions of this book where they are needed.

1.6 Summary

The properties of waves and of wave motion are important for the understanding of a variety of topics relevant to the electronic processes that occur in solids. In this chapter we have briefly summarized certain basic characteristics of wave motion, and have given a few examples of classical wave systems.

The analysis of wave problems starts with the formulation of a wave equation based on a knowledge of the system involved. This wave equation involves space and time derivatives of a suitable displacement. The general solution of this wave equation is obtained, and specific solutions are generated by consideration of boundary conditions or other restrictions imposed by physical reality. The specific solutions correspond to certain allowed frequencies that are consistent with the boundary conditions and other possible restrictions.

The major requirement imposed by boundary conditions that terminate the displacement at the ends of the region of interest is that an integral number of half-wavelengths must be contained within the region.

The variation of the phase and group velocities of a wave with wavenumber are given by the dispersion relationship between frequency and wavenumber. For long-wavelength vibrations in a string or rod, and for electromagnetic waves in a nonabsorbing medium, the velocity is constant and the phase velocity is equal to the group velocity. For lattice vibrations in a one-dimensional crystal or for electromagnetic waves in an absorbing medium, however, the velocity is not constant and the phase and group velocities are not equal to one another.

In crystals with more than one kind of atom, or with more than one interatomic lattice distance (more than one atom per unit cell), two branches of the vibration spectrum are found, called the acoustic branch (because long-wavelength longitudinal waves of this branch are sound waves), and the optical branch (because absorption of energy from light with wavelength in the Reststrahlen absorption range acts as a driving force for these crystal vibrations in ionic or partially ionic crystals).

Deriving the properties of electromagnetic waves from Maxwell's convenient summary of electrical and magnetic relationships shows that these waves are transverse waves that travel with the velocity of light. The identification of light with an electromagnetic wave allows the derivation of the relationship between such optical properties as index of refraction, absorption constant, and reflection coefficient, and such electrical properties as dielectric constant (caused by polarization) and electrical conductivity (caused by free charge carriers).

In subsequent chapters we correlate the ideas reviewed here for lattice and electromagnetic waves with similar views for electrons and their behavior in crystals.

Problems

1.1 The flow of heat into the earth is described by the following equation:

$$\frac{\partial T}{\partial t} = b \frac{\partial^2 T}{\partial x^2}$$

where T is the temperature, x is the distance normal to the surface of the earth, and b is a constant involving thermal conductivity, density, and specific heat of the earth. Assuming that a harmonic wave is a suitable solution of this equation because of the periodic variation of temperature with day and night at the surface of the earth, calculate $T(x, t)$. Interpret the physical significance of the solution in view of its mathematical form. (Remember that $(i)^{1/2} = (1 + i)/(2)^{1/2}$.)

1.2 Consider the following wave equation (actually that suitable for electron waves):

$$-\frac{\partial^2 \phi}{\partial x^2} = ib \frac{\partial \phi}{\partial t}$$

where b is a real constant. Find the dispersion relation required if harmonic waves are to be a solution of this equation. Under the boundary conditions that $\phi = 0$ at $x = 0$ and at $x = L$, what are the allowed frequency values?

1.3 Consider a one-dimensional crystal in which the atoms all have the same mass m, but have different spacings, a and b, as indicated in Fig. 1.8. Such a situation is similar to that encountered in an elemental crystal with two atoms per unit cell. Treat the problem by analogy with the problem of different masses and equal spacings, to derive the properties of the acoustical and optical vibration branches.

Problems

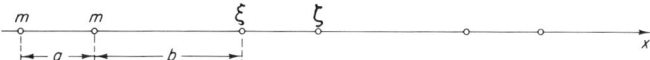

Fig. 1.8 A linear one-dimensional crystal with all masses equal but with two different spacings between atoms as in a crystal with two atoms per unit cell. See Problem 1.3.

1.4 Figure 1.9 gives representative acoustical and optical branches of the vibrational frequency spectrum for a crystal like GaAs. (a) What is the velocity of sound? (b) What wavelength of light would be involved in Reststrahlen absorption? (c) Is the general shape of the spectrum consistent with the masses of Ga and As?

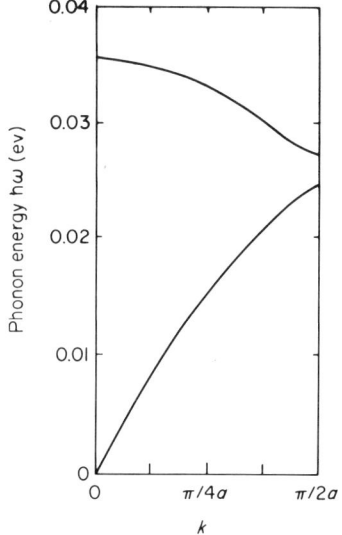

Fig. 1.9 Typical vibrational spectrum for analysis according to Problem 1.4.

1.5 Plot the ratio of group velocity to phase velocity as a function of wave number for the acoustic and for the optical lattice-vibration branches in a one-dimensional crystal with two kinds of atoms and equal spacing between atoms.

1.6 Assume that CdTe is an ionic compound, $Cd^{+2}Te^{-2}$, and estimate the wavelength of light that is involved in Reststrahlen absorption on the basis of a one-dimensional model ($a = 3.18$ Å).

Chapter 2

Wave Approach to Quantum Mechanics

One of the great revolutions in modern physics occurred when it was realized that many phenomena related to the atomic and subatomic range cannot be adequately described in terms of a classical framework. When electrons are diffracted from the surface of a crystal, the direction of maximum intensity is found to vary with the velocity of the incident electrons, as though the effective electron wavelength in diffraction were related to the particle momentum. Although α-particles have an energy less than that required to escape from the potential well of a suitable nucleus, nevertheless radioactive decay does occur with the α-particles "tunneling through" the barrier. Such phenomena require that entities previously conceived as discrete particles be reinterpreted to include extension in space with wavelike properties.

The new effects also required that previously continuous quantities, such as the energy of an atomic system, be reinterpreted in terms of a quantized series of discrete allowed energies. When energy is supplied to a monatomic gas, for example, it is found that *all* of the energy supplied thermally is converted into kinetic energy of the gas atoms, and *none* of the energy is used to increase the internal energy of electrons in the atom. It is concluded that there is at least a certain minimum energy which is required before it is possible to increase the internal energy of the atoms, and that the value of this minimum energy is much larger than the available thermal energy kT at or near room temperature. Electron-scattering experiments in gases showed that either (a) electrons lose no energy and come through the gas unscattered, or (b) electrons are strongly scattered if their energy is equal

2.1 Simple Applications of a Wave Analogy

to one of a set of discrete values increasing in magnitude up to the ionization energy of the gas atoms. This result implies that only energies corresponding to specific energy differences inside the atom can be absorbed from the impinging electrons.

Similar evidence exists for the quantization of light energy. A hot furnace does not emit x rays; thermal energy available to the atoms of the furnace is not great enough to provide that minimum energy necessary for the emission of x rays. When light is absorbed by a material, the kinetic energy of ejected electrons is given by

$$E_{kin} \leq \hbar\omega - \phi \tag{2.1}$$

where $\hbar = h/2\pi$, is Planck's constant ($h = 6.626 \times 10^{-27}$ erg sec), and $\hbar\omega$ is interpreted as the energy of a quantum of light energy (a *photon*) corresponding to light of frequency ω. The quantity ϕ is a characteristic threshold energy of the material, called the work function. Light with frequency $\omega < \phi/\hbar$ can irradiate the material forever without the emission of a single electron. When electrons produce x rays by striking a suitable target, the maximum frequency of the emitted x rays is related to the accelerating voltage V of the electrons by

$$\hbar\omega_{max} = eV \tag{2.2}$$

Once again the existence of x-ray quanta is indicated.

Both the wave properties of classical particles and the quantization of classically continuous energies can be treated in terms of a wave approach to quantum mechanics. This approach is not unique; an approach known as matrix mechanics yields identical results and in some areas provides a more powerful analytical framework. Since the concepts of quantum mechanics represent in general, however, a gross abstraction compared to the more familiar concepts of classical mechanics, we develop our treatment of the subject here in terms of the wave interpretation to establish continuity with our previous discussion of waves and to emphasize the overlap of ideas between classical and quantum mechanics.

2.1 Simple Applications of a Wave Analogy

Consider the following problem raised by the experimental findings cited above. The energy of *free* electrons is continuous and not quantized, but the energy of electrons *in* an atom is quantized. It appears that electron energies are quantized when the electron is *confined* (as in an atom), but are not quantized when the electron is free. Casting about for a model that will permit only a set of discrete allowed energy values when confining boundary conditions are imposed, but that is free from these restrictions in the absence of these boundary conditions, we think of the analogy with waves in a string with fixed ends and with free ends. When the string has fixed ends, the frequency spectrum of waves is quantized, i.e., only certain normal modes are allowed; when the ends are not fixed, these restrictions are removed. If the electron in an atom were like a wave in a string with fixed ends, then these properties of quantization might be directly expected! From our previous experience with standing waves, let us take as a necessary condition for the wave-mechanical treatment of phenomena that an integral number of half-wavelengths must fit into the confined system.

Before this analogy can be tested out on some simple systems, however, a correlation between the momentum of the classical particle and the wavelength of the corresponding wave in the wave-mechanical approach is needed. Since the wavevector **k** for a wave is a vector in the direction of wave propagation, a proportionality between the particle momentum **p** and the wave vector **k** is expected. The specific relationship proposed by de Broglie is

$$\mathbf{p} = \hbar \mathbf{k} \quad \text{or} \quad p = \frac{h}{\lambda} \tag{2.3}$$

where the proportionality constant between **p** and **k** is simply $\hbar$.

Consider first the problem of a particle in a one-dimensional box of length a. Applying our wave analogy produces the quantization condition

$$n\left(\frac{\lambda}{2}\right) = a \tag{2.4}$$

The corresponding allowed values of energy are

$$E = \frac{p^2}{2m} = \frac{\hbar^2 k^2}{2m} = \frac{h^2}{2m\lambda^2} = \frac{n^2 h^2}{8ma^2} \tag{2.5}$$

Since the application of the complete wave-mechanical treatment to a particle in a box leads to the same quantization condition, the predicted values for allowed energies are exactly the same as those in Eq. (2.5).

As a second example, consider a simple harmonic oscillator in the form of a mass m on a spring with restoring force $-gx$. If the maximum displacement of the mass from its rest position is x_0, the quantization condition is that an integral number of half-wavelengths must fit in the distance $2x_0$,

2.2 The Schroedinger Equation

since this is the total length within which the oscillator is confined. The total energy of the oscillator is $W = gx_0^2/2$, which is all potential energy for $x = \pm x_0$, and is all kinetic energy for $x = 0$. Defining the wavelength in terms of the maximum kinetic energy,

$$\lambda = \frac{h}{(2mW)^{1/2}} \qquad (2.6)$$

Since the quantization condition is that

$$n(\lambda/2) = 2x_0 \qquad (2.7)$$

combining Eqs. (2.6) and (2.7) gives

$$W = \frac{n^2 h^2}{32 m x_0^2} \qquad (2.8)$$

Substituting $x_0^2 = 2W/g$ and recalling that for an oscillator, $\omega = (g/m)^{1/2}$, we obtain

$$W = n\hbar\omega\left(\frac{2\pi}{8}\right) \qquad (2.9)$$

We conclude that the allowed energy values for a one-dimensional oscillator are integral multiples of the basic quantum $\hbar\omega$. The form of Eq. (2.9) is similar to that of the exact solution with the following exceptions: n should be replaced by $(n + \frac{1}{2})$, and $(2\pi/8)$ should be replaced by unity.

These analogical calculations indicate in a crude way the kind of effect that is involved in the wave approach to quantum mechanics. We now return to the basic question: what is the appropriate wave equation for electrons and other atomic particles?

2.2 The Schroedinger Equation

Wave equations for strings, rods, lattices, and electromagnetic fields were derived from a knowledge of the medium in which the waves were to propagate and of relationships between the quantities involved. The velocity of the wave was closely linked to such physical parameters as tension, density, force, spatial distribution, dielectric constant, and permeability. This kind of approach to constructing the wave equation for electrons and other atomic particles is not very fruitful; we are embarrassed by our ignorance of the medium and its properties. We have really only one guide:

the wave equation must be valid and applicable in describing the wavelike properties of a "particle," regardless of the specific energy or momentum of the particle.

There are some other guides as to the *form* of the wave equation. Based on previous experience with wave phenomena, we expect that (a) the solution of the wave equation has the familiar form

$$\Psi = A\, e^{i(kx-\omega t)} \tag{2.10}$$

where Ψ plays the role of some kind of appropriate "displacement," and (b) the wave equation contains appropriate derivatives of this displacement Ψ with respect to coordinates and time.

To require that the wave equation be valid independent of the specific energy or momentum of the particle means that the wave equation must contain those derivatives of Ψ that yield only E and p^2, since for the particle these are related by

$$E = \frac{p^2}{2m} + V \tag{2.11}$$

where V is the potential energy and E is the total energy. Inspection of Eq. (2.10) shows that the first derivative of Ψ with respect to time contains $E = \hbar\omega$, and that the second derivative with respect to coordinates contains $p^2 = (\hbar k)^2$. The wave equation must then have the following form in one dimension:

$$\frac{\partial \Psi}{\partial t} = a \frac{\partial^2 \Psi}{\partial x^2} + bV\Psi \tag{2.12}$$

where a and b are constants to be determined by the requirement that substitution of Ψ from Eq. (2.10) into the wave equation (2.12) must be consistent with the relationship between E, p, and V given in Eq. (2.11). Upon substitution for Ψ from Eq. (2.10), the exponential term cancels out and the following equation for a and b results:

$$-\frac{i}{\hbar}E = -\frac{p^2}{\hbar^2}a + bV \tag{2.13}$$

Comparison with Eq. (2.11) gives

$$a = \frac{i\hbar}{2m} \tag{2.14}$$

and

$$b = -\frac{i}{\hbar} \tag{2.15}$$

2.2 The Schroedinger Equation

When these values are substituted into Eq. (2.12) the result is the desired wave equation,

$$-\frac{\hbar^2}{2m}\frac{\partial^2 \Psi}{\partial x^2} + V\Psi = i\hbar \frac{\partial \Psi}{\partial t} \qquad (2.16)$$

The generalization to three dimensions is direct,

$$-\frac{\hbar^2}{2m}\nabla^2 \Psi + V\Psi = i\hbar \frac{\partial \Psi}{\partial t} \qquad (2.17)$$

Equation (2.17) is known as the *time-dependent Schroedinger equation*, since the solution Ψ and the energy are functions of time. It is the full equation that must be used to describe processes in which dynamic transitions involving a change in energy of the particle occur.

In the special case where the energy is not a function of time, i.e., when we describe only what energy states are available, not how the energy changes with time, a *stationary state* is said to exist and a simplification of Eq. (2.17) is possible. The description of a stationary state requires the use of a wave for which the dependence on time can be separated from the dependence on coordinates. Such a wave can be formulated as a solution of Eq. (2.17),

$$\Psi(x, y, z, t) = \psi(x, y, z)\, e^{-i\omega t} \qquad (2.18)$$

If this particular expression for Ψ is substituted into Eq. (2.17), the Laplacian affects only the $\psi(x, y, z)$ portion of Ψ, and the time-dependent exponential term cancels out. The resulting equation is the *time-independent Schroedinger equation*,

$$-\frac{\hbar^2}{2m}\nabla^2 \psi(x, y, z) + V(x, y, z)\, \psi(x, y, z) = E\, \psi(x, y, z) \qquad (2.19)$$

In this equation ψ and V are functions only of the coordinates, and E is a constant, giving the value of allowed energy for a particular ψ and V. If we wish to find the allowed values of energy for an electron or an atomic particle, therefore, the problem is in principle quite simple. All that is needed is to obtain the appropriate solutions of the wave equation, Eq. (2.19), for the potential energy V that characterizes a given problem of interest. The elegance of this formulation is that the same equation can deal with *any* problem, the only difference from one problem to the next lying in the specific form given to the potential function V. Of course the practicality of this approach depends crucially on the ease of solution of the wave equation for different forms of V. Unfortunately, exact solutions are limited to only a few elementary problems and most of the applications of Eq. (2.19) require some kind of approximate method.

We should not leave this "derivation" of the Schroedinger wave equations without explicitly commenting on what has probably been obvious to the reader. We have not derived these equations in any real sense at all. All we have done is to supply a kind of reasonable analogy by which the existence and form of the wave equations can be better appreciated. Actually, like other basic relationships in science, the Schroedinger wave equations cannot be rigorously derived from more fundamental principles. The value and acceptance of these wave equations are established by their pragmatic success in describing and predicting experimental observations.

And what of this "displacement" Ψ, which we have introduced but never given physical substance? Its interpretation is also guided by empirical study, and we return to this subject a few sections later. But first, in order to understand the traditional terminology of quantum mechanics, we must recognize that equations like Eq. (2.19), arrived at from a wave perspective, can also be viewed in a somewhat different framework. Since this latter framework is used in discussions of the results of quantum mechanics as much, or probably more, than the purely wave interpretation, we must understand this approach also.

2.3 Basic Postulates of Quantum Mechanics

The theoretical structure of a discipline can often be codified conveniently and succinctly in a set of basic postulates, which are not themselves derivable from more fundamental understanding but which receive their authority from their pragmatic ability to lead to results confirmable by experience and experiment. In a sense, Maxwell's equations given in Chapter 1 form such a set, summarizing in mathematical equations the properties of much empirical research on electrical and magnetic fields. For our present purposes it is most convenient to gain an introduction to quantum mechanics via this device of a set of postulates.

The postulates we use are three in number.

I. A mathematical operator is to be associated with every observable.

II. If **A** is the operator associated with the observable a, and **B** is the operator associated with the observable b, the operator associated with the observable $[a, b]$, where the brackets imply the Poisson bracket, defined as

$$[a, b] = \frac{\partial a}{\partial x} \frac{\partial b}{\partial p} - \frac{\partial a}{\partial p} \frac{\partial b}{\partial x} \qquad (2.20)$$

2.3. Basic Postulates of Quantum Mechanics

is **C** and is related to **A** and **B** by

$$\mathbf{AB} - \mathbf{BA} = i\hbar\mathbf{C} \tag{2.21}$$

Here a and b are assumed to be expressible as functions of x and p.

III. The numerical values of an observable that can be obtained by experimental measurement are the eigenvalues belonging to the eigenfunctions of the operator associated with the observable.

Let us relate these apparently abstract postulates to the previous wave equation, and then go back and fill in a few gaps.

FROM CLASSICAL HAMILTONIAN TO SCHROEDINGER EQUATION

The Schroedinger equation of Eq. (2.19) can be obtained by transforming the classical Hamiltonian to a Hamiltonian operator. If the energy is not a function of time, the classical Hamiltonian function of the coordinate x and the momentum p in one dimension is

$$H(x, p) = \frac{p^2}{2m} + V(x) = E \tag{2.22}$$

Hamilton's canonical equations of motion follow directly,

$$\frac{\partial H}{\partial p} = \frac{p}{m} = v = \dot{x} \tag{2.23}$$

$$\frac{\partial H}{\partial x} = -\frac{\partial V(x)}{\partial x} = m\dot{v} = \dot{p} \tag{2.24}$$

If we now make an ad hoc replacement, to be partially justified later, of the observable x by a mathematical operator involving simply multiplication by x, and of the observable p by the mathematical operator $(-i\hbar\,\partial/\partial x)$ (to satisfy Postulate II), the result is the Hamiltonian operator,

$$\mathbf{H(X, P)} = \mathbf{H}\left(x, -i\hbar\frac{\partial}{\partial x}\right) = -\frac{\hbar^2}{2m}\frac{\partial^2}{\partial x^2} + V(x) \tag{2.25}$$

Generalizing this result to three dimensions, operating with the Hamiltonian operator on some function ψ, and invoking the relationship of Eq. (2.22) leads to the equation

$$\mathbf{H}\psi = -\frac{\hbar^2}{2m}\nabla^2\psi + V(x, y, z)\psi = E\psi \tag{2.26}$$

This is an operator equation which can be interpreted in terms of Postulate III. The values of energy E that are allowed (those that can be measured in an experiment) are the eigenvalues corresponding to the eigenfunctions ψ of this operator equation. The equation, its solution, and its interpretation from this point on are the same as from the wave perspective that led to Eq. (2.19), which is identical to Eq. (2.26).

When the energy is not independent of time, Eq. (2.22) must be replaced by

$$\mathbf{H}(x, p, t) = \mathbf{E}(t) \tag{2.27}$$

By properly assigning an operator to E, we may proceed in this case to the time-dependent Schroedinger Eq. (2.17). Comparison with Eq. (2.17) suggests that the association of the operator $(i\hbar \, \partial/\partial t)$ with the observable $E(t)$ is appropriate.

POSTULATES I AND II

The meaning of postulates I and II for the observables of position and momentum is that an operator $\mathbf{X}$ must be associated with the observable x, and that an operator $\mathbf{P}$ must be associated with the observable p, in such a way that

$$\mathbf{XP} - \mathbf{PX} = i\hbar \mathbf{C} \tag{2.28}$$

where $\mathbf{C}$ is the operator associated with

$$[x, p] = \frac{\partial x}{\partial x} \frac{\partial p}{\partial p} - \frac{\partial x}{\partial p} \frac{\partial p}{\partial x} = 1 \tag{2.29}$$

Since the operator associated with a constant is simply multiplication by that constant, Eq. (2.28) becomes

$$\mathbf{XP} - \mathbf{PX} = i\hbar \tag{2.30}$$

Several choices might be made for appropriate operators to be associated with the observables x and p. Equation (2.30) does not uniquely determine this choice. The simplest choices would be to associate the operator $\mathbf{X}$ with x, or to associate the operator $\mathbf{P}$ with p. In each case an appropriate form can be found from Eq. (2.30) for the operator to be associated with the other variable. For example, if we choose to associate the operator $\mathbf{X}$ with the observable x, then Eq. (2.30) leads to the choice of the operator $(-i\hbar \, \partial/\partial x)$ to be associated with the observable p. To see that this choice

2.3 Basic Postulates of Quantum Mechanics

satisfies Eq. (2.30), let the operators in this equation operate on some dummy function ϕ,

$$(\mathbf{XP} - \mathbf{PX})\phi = x\left(-i\hbar \frac{\partial}{\partial x}\right)\phi - \left(-i\hbar \frac{\partial}{\partial x}\right)x\phi$$

$$= -i\hbar x \frac{\partial \phi}{\partial x} + i\hbar \frac{\partial}{\partial x}(x\phi)$$

$$= i\hbar \phi$$

On the other hand, if we choose to associate the operator $\mathbf{P}$ with the observable p, then Eq. (2.30) leads to the choice of the operator $(i\hbar \, \partial/\partial p)$ to be associated with the observable x. This also satisfies Eq. (2.30). It is usually conventional to make the first of these operator-pair choices, although for certain problems the second choice may prove more convenient mathematically.

Operators associated with a function of an observable are formed as the same function of the operator. Thus the operator associated with x^2 is the operator $\mathbf{X}^2$, the operator associated with $V(x, y, z)$ is $V(\mathbf{X}, \mathbf{Y}, \mathbf{Z})$, and the operator associated with p_x^2 is, as we have seen in the last section, $(-\hbar^2 \, \partial^2/\partial x^2)$.

Example 2.1 Derive the operator to be associated with the observable xp_x.

First we must determine what Poisson bracket has the value xp_x. The desired Poisson bracket is $[x^2/2, p_x^2/2]$. Then from Eq. (2.21),

$$\left(\frac{x^2}{2}\right)\left(-\frac{\hbar^2}{2} \frac{\partial^2}{\partial x^2}\right) - \left(-\frac{\hbar^2}{2} \frac{\partial^2}{\partial x^2}\right)\left(\frac{x^2}{2}\right) = i\hbar \mathbf{O}_{xp_x}$$

This is an operator equation, and $\mathbf{O}_{xp_x}$ is the desired operator. The quantity on the left side reduces to the operator

$$\frac{\hbar^2}{4}\left(\frac{\partial^2}{\partial x^2} x^2 - x^2 \frac{\partial^2}{\partial x^2}\right)$$

By operating on a dummy function ϕ with this operator, we may reduce it to a simpler form. First we find that

$$\frac{\hbar^2}{4}\left(\frac{\partial^2}{\partial x^2} x^2 - x^2 \frac{\partial^2}{\partial x^2}\right) = \frac{\hbar^2}{2}\left(1 + 2x \frac{\partial}{\partial x}\right)$$

and then by a similar process, we can reduce still further to

$$\frac{\hbar^2}{2}\left(1 + 2x \frac{\partial}{\partial x}\right) = \frac{\hbar^2}{2}\left(x \frac{\partial}{\partial x} + \frac{\partial}{\partial x} x\right)$$

Thus it is found finally that the operator to be associated with xp_x is

$$-\frac{i\hbar}{2}\left(x\frac{\partial}{\partial x} + \frac{\partial}{\partial x}x\right) = \frac{1}{2}\left[x\left(-i\hbar\frac{\partial}{\partial x}\right) + \left(-i\hbar\frac{\partial}{\partial x}\right)x\right]$$

The operator to be associated with the observable xp_x is therefore a kind of "average" of the operators $\mathbf{XP}_x$ and $\mathbf{P}_x\mathbf{X}$.

POSTULATE III

The third postulate of quantum mechanics associates the numerical values of an observable to be expected from an experimental measurement with the eigenvalues of the operator associated with that observable. If the operator $\mathbf{W}$ is associated with an observable w, then if

$$\mathbf{W}u_i = w_i u_i \tag{2.31}$$

for a particular set of *well behaved* functions u_i satisfying the proper boundary conditions of the problem, the numbers w_i are called the *eigenvalues* belonging to the *eigenfunctions* u_i of the operator $\mathbf{W}$. If the system is in the state described by the eigenfunction u_j, then a measurement of w will yield the value w_j exactly.

The Schroedinger equation is one example of such an operator equation, where the Hamiltonian operator $\mathbf{H}$ corresponds to the observable values of energy E,

$$\mathbf{H}\psi_i = E_i \psi_i \tag{2.32}$$

If we find the eigenfunctions ψ_i of the operator $\mathbf{H}$, then the corresponding eigenvalues E_i are the values of energy that are experimentally measured. If the system is in the state given by ψ_j, then the value of the energy is exactly E_j. Our principal concern in most of the time-independent problems is to find the eigenfunctions ψ_i and their corresponding energy eigenvalues E_i for Hamiltonian operators with various forms of the potential energy V.

The transformation of the time-dependent to the time-independent Schroedinger equation and the statement of Eq. (2.18) can be viewed in terms of these concepts. The time-dependent equation is

$$\mathbf{H}\Psi = i\hbar\frac{\partial \Psi}{\partial t}$$

For a conservative system, we can write the operator equation for the energy operator $i\hbar\, \partial/\partial t$,

$$i\hbar\frac{\partial \Psi}{\partial t} = E\Psi \tag{2.33}$$

where E, not a function of time, is an eigenvalue of the energy operator. The eigenfunctions of Eq. (2.33) are

$$\Psi = \psi(x, y, z)\, e^{-iEt/\hbar} \tag{2.34}$$

which are identical with Eq. (2.18).

2.4 Interpretation of the Wavefunction

Because the ψ eigenfunctions of the Schroedinger equation represent in some way the behavior of waves, they are commonly called *wavefunctions*. An eigenfunction must be mathematically well-behaved. This is a mathematical restriction imposed to give physical interpretability to the function. It requires that the function be (a) single valued, not taking on more than one value for a specific set of coordinate values; (b) not identically zero, which represents a trivial solution; (c) continuous and with continuous derivatives everywhere; and (d) finite everywhere.

But what is the physical meaning of a "ψ displacement" in the wave picture? Perhaps our brief excursion into the operator version of the Schroedinger equation has prepared us for the conclusion that there is *no* simple physical or visualizable connection between the ψ function *itself* (usually a complex function) and the particular "particle–wave" being treated. It is this departure from the conventional conceptual model building that characterizes classical physics, which makes the quantum-mechanical view so strange at first acquaintance.

Even if ψ is a complex function, however, the product $\psi^*\psi = |\psi|^2$ is a real quantity. It is to this real quantity that physical significance can be associated. Suppose that a plot of $|\psi(x)|^2$ versus x representing a "particle–wave" is like that shown in Fig. 2.1. As drawn, $|\psi(x)|^2$ has a maximum value at $x = x_0$, a finite but decreasing value for values of x larger or smaller than x_0, and an effectively negligible value for ranges of x very far removed from x_0. Since the location of the "particle" is described by $\psi(x)$, we might suppose that in regions of space where $|\psi(x)|^2 \approx 0$, there is very small probability of the particle being found in an experimental measurement, whereas in regions of space where $|\psi(x)|^2 > 0$, there is a finite probability of finding the particle.

We can make this more specific by defining the magnitude

$$\psi^*\psi\, dx = |\psi|^2\, dx$$

as the probability of finding the particle between x and $x + dx$.

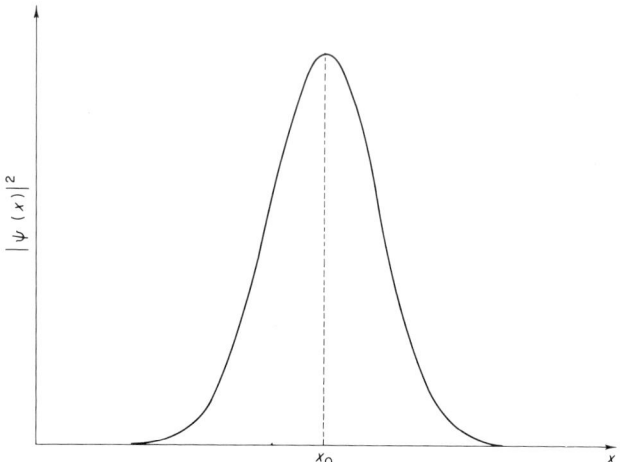

Fig. 2.1 Representative variation of the square of the amplitude of the wavefunction in one dimension with the value of x. The corresponding particle is most likely to be found at $x = x_0$ by an appropriate measurement.

There are two conceptual approaches to this definition of $|\psi|^2$:

(a) The particle *has* a position, i.e., the particle can be considered as a mathematical point corresponding to a specific set of coordinates. The value of $|\psi(x)|^2$ is then the probability that the measured position of this particle lies within dx of x as determined by a suitable experiment.

(b) The particle does *not* have a position, but is distributed in space according to the value of $|\psi(x)|^2$. For example, for an electron, $-e|\psi|^2$ represents the spatial distribution of charge corresponding to a single electron. The particle is no longer considered to be identifiable as a point with a particular position; the whole probability "smear" *is* the particle. Measurement of course does give a particular position for the particle, but this is a result of the measurement process and the probability of measuring a particular value is proportional to the magnitude of the "density" function $|\psi|^2$ at that point.

Because of the interpretation of $|\psi^2|$ as a probability, one further condition must be added to the previous list applying to well-behaved functions: the wavefunction ψ must also be *square-integrable*, i.e., if the wavefunction is defined over the range from $x = a$ to $x = b$,

$$\int_a^b \psi^*\psi \, dx = \text{constant} \tag{2.35}$$

2.4 Interpretation of the Wavefunction

When the wavefunction ψ is multiplied by an appropriate constant such that

$$A^2 \int_a^b \psi^* \psi \, dx = 1 \tag{2.36}$$

the wavefunction is said to be *normalized*, and A is called the *normalization constant*. In order for $|\psi|^2$ to be used as a probability, it is evident that the wavefunction must be normalized. Also if, for example, $-e|\psi|^2$ is to be interpretable as the spatial distribution of an electron, it is clear that

$$\int_{-\infty}^{+\infty} -e|\psi|^2 \, dx = -e \tag{2.37}$$

and that the wavefunction entering Eq. (2.37) must contain the necessary normalization constant.

The normalization condition written above in one dimension can be readily generalized to three dimensions. A normalized wavefunction in three dimensions satisfies

$$\int_{-\infty}^{+\infty} |\psi(x, y, z)|^2 \, dx \, dy \, dz = 1 \tag{2.38}$$

in Cartesian coordinates, or

$$\int_0^{2\pi} \int_0^{\pi} \int_0^{\infty} |\psi(r, \theta, \phi)|^2 r^2 \sin\theta \, dr \, d\theta \, d\phi = 1 \tag{2.39}$$

in spherical coordinates.

Example 2.2 Show that the eigenfunctions of the operator $-i\,\partial/\partial x$ are well-behaved for any real-valued eigenvalue, but are not well-behaved for an imaginary eigenvalue.

The operator equation,

$$-i\frac{\partial \phi}{\partial x} = c\phi$$

has the solution

$$\phi = A\,e^{icx}$$

This function is well behaved if c is real, but if c is imaginary then $\phi = A\,e^{\pm ax}$, where a is real, which becomes infinite as x goes to ∞ or $-\infty$.

Example 2.3 What is the normalization constant for the functions

$$\phi_n = \sin n\pi x/L$$

defined over the interval between $x = 0$ and $x = L$?

It is desired that

$$A^2 \int_0^L \sin^2 \frac{n\pi x}{L} \, dx = 1$$

Since

$$\int_0^L \sin^2 \frac{n\pi x}{L} \, dx = \frac{L}{n\pi} \left[\frac{n\pi x}{2L} - \frac{1}{4} \sin \frac{2n\pi x}{L} \right]_0^L = \frac{L}{2}$$

the normalization constant is

$$A = \left(\frac{2}{L}\right)^{1/2}$$

2.5 Orthogonality

In the Cartesian coordinate system, the unit vectors in the three coordinate directions, $\mathbf{e}_1$, $\mathbf{e}_2$, and $\mathbf{e}_3$, are said to be *orthogonal*. This means that, for example, $\mathbf{e}_1 \cdot \mathbf{e}_2 = 0$, or in general

$$\mathbf{e}_i \cdot \mathbf{e}_j = \delta_{ij} \tag{2.40}$$

where the symbol δ_{ij} is the *Kronecker delta*,

$$\delta_{ij} \equiv \begin{Bmatrix} 1, & i = j \\ 0, & i \neq j \end{Bmatrix} \tag{2.41}$$

Because the unit vectors in the Cartesian coordinate system obey Eq. (2.40), they are said to constitute an *orthonormal set*. Because every vector in three-dimensional space can be constructed as a linear combination of the three unit vectors, they are said to be the *basis* vectors for the space and to constitute a *complete* orthonormal set. This same terminology can be applied to the solutions of the Schroedinger equation in the following way.

The Schroedinger equation

$$\mathbf{H}\psi_i = E_i \psi_i$$

has in general many different eigenfunctions, $\psi_1, \psi_2, \ldots, \psi_n$, corresponding to energy eigenvalues of $E_1, E_2, \ldots, E_n$ respectively. If each of these eigenfunctions is normalized, they form a *complete orthonormal set* which

2.5 Orthogonality

can serve as the *basis* functions for the construction of any other well-behaved function as a linear combination of these basis eigenfunctions.

Two wavefunctions ψ_i and ψ_j are said to be *orthogonal* if

$$\int \psi_i^* \psi_j \, d\mathbf{r} = 0 \tag{2.42}$$

The integration over coordinates takes the place in this system of the scalar product in the unit Cartesian vector system. The symbol $d\mathbf{r}$ represents the general integration over coordinates, i.e., in one dimension $d\mathbf{r}$ reduces to dx, in three dimensions $d\mathbf{r} = dx\, dy\, dz$ or $d\mathbf{r} = r^2 \sin\theta\, dr\, d\theta\, d\phi$. If ψ_i and ψ_j are normalized, and if

$$\int \psi_i^* \psi_j \, d\mathbf{r} = \delta_{ij} \tag{2.43}$$

in analogy with Eq. (2.40), then ψ_i and ψ_j constitute an *orthonormal set*.

The normalized eigenfunctions of the Hamiltonian operator form an orthonormal set. Before showing this, it is useful to look briefly at the properties of the Hamiltonian operator.

A *linear* operator with exclusively *real* eigenvalues is called an *Hermitian* operator. Because every Schroedinger operator has eigenvalues that correspond to physically measurable quantities, i.e., real eigenvalues, it follows that every Schroedinger operator, including of course the Hamiltonian operator, is a Hermitian operator.

If **O** is a Hermitian operator, and

$$\mathbf{O}\phi_i = a\phi_i \tag{2.44}$$

then a is a real number, i.e.,

$$a = a^* \tag{2.45}$$

From this requirement we derive a particular relationship obeyed by Hermitian operators that is useful in present and future calculations. Multiply both sides of Eq. (2.44) by ϕ_j^* and integrate over coordinates:

$$a \int \phi_j^* \phi_i \, d\mathbf{r} = \int \phi_j^* \mathbf{O} \phi_i \, d\mathbf{r} \tag{2.46}$$

If Eq. (2.44) holds, then also

$$\mathbf{O}^* \phi_j^* = a^* \phi_j^* \tag{2.47}$$

Multiply both sides of Eq. (2.47) by ϕ_i and integrate over coordinates:

$$a^* \int \phi_j^* \phi_i \, d\mathbf{r} = \int \phi_i (\mathbf{O}\phi_j)^* \, d\mathbf{r} \qquad (2.48)$$

Comparing Eqs. (2.45), (2.46), and (2.48) shows that a Hermitian operator **O** obeys the following relationship:

$$\int \phi_j^* \mathbf{O} \phi_i \, d\mathbf{r} = \int \phi_i (\mathbf{O}\phi_j)^* \, d\mathbf{r} \qquad (2.49)$$

Now we return to the claim that the eigenfunctions of the Hamiltonian operator form a complete orthonormal set. Consider two such eigenfunctions ψ_1 and ψ_2 and the following two equations satisfied by them

$$(\mathbf{H}\psi_1)^* = \mathbf{H}\psi_1^* = E_1 \psi_1^* = (E_1 \psi_1)^* \qquad (2.50)$$

$$\mathbf{H}\psi_2 = E_2 \psi_2 \qquad (2.51)$$

Multiply Eq. (2.51) by ψ_1^* and subtract from the center two terms of Eq. (2.50) multiplied by ψ_2; then integrate over coordinates to obtain

$$\int \psi_2 \mathbf{H} \psi_1^* \, d\mathbf{r} - \int \psi_1^* \mathbf{H} \psi_2 \, d\mathbf{r} = (E_1 - E_2) \int \psi_1^* \psi_2 \, d\mathbf{r} \qquad (2.52)$$

Since **H** is a Hermitian operator satisfying Eq. (2.49), it follows that the left-hand side of Eq. (2.52) is zero. This means that unless $E_1 = E_2$,

$$\int \psi_1^* \psi_2 \, d\mathbf{r} = 0 \qquad (2.53)$$

and ψ_1 and ψ_2 are orthogonal. States for which $E_1 \neq E_2$ are called *nondegenerate* states. We have therefore shown that the eigenfunctions of the Hamiltonian operator corresponding to nondegenerate states are orthogonal. If $E_1 = E_2$, we have a pair of *degenerate* states and it is not necessarily true that eigenfunctions corresponding to degenerate states are orthogonal. If, however, such a situation exists, it is possible to construct two orthogonal wavefunctions describing these states in the following way.

Suppose that ψ_1 and ψ_2 are eigenfunctions associated with degenerate states, that they are normalized but not necessarily orthogonal. Then

$$\int \psi_1^* \psi_2 \, d\mathbf{r} = c$$

2.5 Orthogonality

where c is some nonzero constant. Construct the wavefunction ψ_3 as a linear combination of ψ_1 and ψ_2,

$$\psi_3 = c\psi_1 - \psi_2 \tag{2.54}$$

ψ_1 and ψ_3 form an orthogonal pair of degenerate wavefunctions with eigenvalue E_1. Their orthogonality is seen by considering that

$$\int \psi_1^* \psi_3 \, d\mathbf{r} = c \int \psi_1^* \psi_1 \, d\mathbf{r} - \int \psi_1^* \psi_2 \, d\mathbf{r} = 0 \tag{2.55}$$

Thus we have shown that the eigenfunctions of the Hamiltonian operator form an orthonormal set. That they form a *complete* orthonormal set, i.e., that any well behaved function ϕ can be expressed as

$$\phi = \sum_n a_n \psi_n \tag{2.56}$$

where a_n are constant coefficients, is difficult to prove in general. It has been proved specifically for the operators of position and linear and angular momentum, and we shall accept it as a reasonable postulate.

Example 2.4 Show that the Hamiltonian operator is a Hermitian operator.

If the Hamiltonian operator is a Hermitian operator, then

$$\int \psi_i^* \left[-\frac{\partial^2}{\partial x^2} + V(x) \right] \psi_j \, dx = \int \psi_j \left[-\frac{\partial^2}{\partial x^2} + V(x) \right]^* \psi_i^* \, dx$$

setting up the problem in one dimension for convenience and neglecting the $-\hbar^2/2m$ multiplying constant of the $\partial^2/\partial x^2$ term for the same reason. If we take the above expression term by term we note that the terms involving $V(x)$ are the same and cancel out. We are left with the question

$$\int \psi_i^* \frac{\partial^2 \psi_j}{\partial x^2} \, dx \stackrel{?}{=} \int \psi_j \frac{\partial^2 \psi_i^*}{\partial x^2} \, dx$$

Proceed by integrating the left-hand side of this equation by parts to obtain

$$\int \psi_i^* \frac{\partial^2 \psi_j}{\partial x^2} \, dx = \psi_i^* \frac{\partial \psi_j}{\partial x} \bigg|_{-\infty}^{+\infty} - \int \frac{\partial \psi_i^*}{\partial x} \frac{\partial \psi_j}{\partial x} \, dx$$

The first term on the right of this equation is zero, since to be square-integrable ψ must vanish at infinity. The second term on the right can be integrated a second time by parts to obtain

$$-\int \frac{\partial \psi_i^*}{\partial x} \frac{\partial \psi_j}{\partial x} dx = -\psi_j \frac{\partial \psi_i^*}{\partial x}\bigg|_{-\infty}^{+\infty} + \int \psi_j \frac{\partial^2 \psi_i^*}{\partial x^2} dx$$

Comparison with our initial question above shows that we may write QED.

Example 2.5 Show that the wavefunctions $e^{i\phi}$, 1, $e^{-i\phi}$ form an orthogonal set in the interval $(0, 2\pi)$ of ϕ. What is the corresponding orthonormal set?

Consider first the orthogonality of the first and second of these wavefunctions

$$\int_0^{2\pi} e^{-i\phi} d\phi = \int_0^{2\pi} (\cos\phi - i\sin\phi) d\phi$$

$$= \sin\phi \bigg|_0^{2\pi} + i\cos\phi \bigg|_0^{2\pi} = 0$$

Similarly

$$\int_0^{2\pi} e^{i\phi} d\phi = \int_0^{2\pi} e^{2i\phi} d\phi = 0$$

Thus these functions do form an orthogonal set.

The normalizing constant can be calculated as

$$A^2 \int_0^{2\pi} d\phi = 1$$

and the orthonormal set is therefore

$$\frac{1}{(2\pi)^{1/2}} e^{i\phi}, \qquad \frac{1}{(2\pi)^{1/2}}, \qquad \frac{1}{(2\pi)^{1/2}} e^{-j\phi}$$

2.6 Expectation Values

In a system such that

$$\mathbf{H}\psi_j = E_j \psi_j \tag{2.57}$$

the energy is E_j when the system is in the state ψ_j. We may rewrite Eq. (2.57) to get an explicit expression for E_j by multiplying both sides by ψ_j^* and integrating over coordinates:

$$E_j = \frac{\int \psi_j^* \mathbf{H} \psi_j \, d\mathbf{r}}{\int \psi_j^* \psi_j \, d\mathbf{r}} \tag{2.58}$$

2.6 Expectation Values

If the ψ_j are normalized, then

$$E_j = \int \psi_j^* \mathbf{H} \psi_j \, d\mathbf{r} \tag{2.59}$$

Suppose we are interested in the measured values of some other observable w. We define an "average" or "*expectation*" value (the value to be expected from a measurement) of the observable w as

$$\langle w \rangle \equiv \int \psi_j^* \mathbf{W} \psi_j \, d\mathbf{r} \tag{2.60}$$

where **W** is the operator associated with the observable w. For an observable that is a function only of the coordinates, the expression given in Eq. (2.60) corresponds to the normal way of calculating the average value,

$$\langle f(x) \rangle = \int f(x) \, | \, \psi(x) \, |^2 \, dx \tag{2.61}$$

and we shall assume that it holds in other cases as well.

The expectation value of the energy when the system is in the state ψ_j (when $\mathbf{H}\psi_j = E_j\psi_j$) is just exactly E_j. The expectation value of an observable w for the system in the state ψ_j will be the exact value w_j *only* if

$$\mathbf{W}\psi_j = w_j\psi_j \tag{2.62}$$

i.e., only if ψ_j is also an eigenfunction of the operator **W** as well as of the Hamiltonian operator **H**. If ψ_j is not an eigenfunction of the operator **W**, then the expectation value $\langle w \rangle$ calculated from Eq. (2.73) corresponds to a statistical expectation in a series of measurements, and *not* to an exact value characterizing the observable w in the state ψ_j.

Measurement determines the exact value of an observable in the state ψ_j (where $\mathbf{H}\psi_j = E_j\psi_j$) only if the operator associated with that observable has *simultaneous eigenfunctions* with the Hamiltonian operator. What is the condition for such an occurrence? Consider the two operator equations,

$$\mathbf{H}\psi_j = E_j\psi_j \tag{2.63}$$

$$\mathbf{W}\psi_j = w_j\psi_j \tag{2.64}$$

Operate on Eq. (2.63) by **W** and on Eq. (2.64) by **H** as follows:

$$\mathbf{WH}\psi_j = E_j\mathbf{W}\psi_j = E_jw_j\psi_j \tag{2.65}$$

$$\mathbf{HW}\psi_j = w_j\mathbf{H}\psi_j = w_jE_j\psi_j \tag{2.66}$$

Subtracting Eq. (2.66) from Eq. (2.65) yields

$$(\mathbf{WH} - \mathbf{HW})\psi_j = 0 \tag{2.67}$$

The quantity $(\mathbf{WH} - \mathbf{HW})$ is called the *commutator* of the two operators $\mathbf{W}$ and $\mathbf{H}$. An operator $\mathbf{W}$ has simultaneous eigenfunctions with the Hamiltonian operator if their commutator is zero, i.e., if they *commute*. Therefore an observable whose operator commutes with the Hamiltonian has an exact value in a state defined by ψ_j, an eigenfunction of the Hamiltonian operator.

Consider some of the properties of a free electron in the light of these conclusions. First note that the operator associated with the momentum commutes with the Hamiltonian,

$$\left(-i\hbar \frac{\partial}{\partial x}\right)\left(-\frac{\hbar^2}{2m} \frac{\partial^2}{\partial x^2}\right)\phi = \left(-\frac{\hbar^2}{2m} \frac{\partial^2}{\partial x^2}\right)\left(-i\hbar \frac{\partial}{\partial x}\right)\phi \tag{2.68}$$

The momentum has an exact value for the state ψ_j just as the energy does. Note also, however, that the operator associated with the position does not commute with the Hamiltonian,

$$x\left(-\frac{\hbar^2}{2m} \frac{\partial^2}{\partial x^2}\right)\phi \neq \left(-\frac{\hbar^2}{2m} \frac{\partial^2}{\partial x^2}\right)x\phi \tag{2.69}$$

Therefore the position of the electron does not have an exact value in the state ψ_j.

These considerations are one way of approaching one of the most interesting aspects of quantum mechanics: the realization that there are certain classical quantities, such as position and momentum, that cannot both be simultaneously known exactly. The conclusion is summarized in the *Heisenberg indeterminacy principle*, which states that the product of the indeterminacies in position and momentum must be larger than some finite quantity. Specifically,

$$\Delta x \, \Delta p_x \geq \frac{\hbar}{2} \tag{2.70}$$

Quantities such as position and momentum are known as *complementary* quantities. Although the basis for Eq. (2.70) can be presented from the viewpoint of interference by the experimenter on the system in the process of experimentation, it also appears to be a direct result of the present state of the quantum mechanical formulation. Indeed it can be viewed really as another statement of the particle–wave duality that characterizes the

2.6 Expectation Values

quantum picture. If the particle nature is stressed by a close determination of the position, then the wave character is obscured, i.e., the wavelength and the momentum become indeterminate. If the wave character is stressed by an accurate determination of wavelength (momentum), then the position and hence the particle character are obscured. Since it is not possible to simultaneously present a full particle-like and a full wavelike representation, the indeterminacy principle follows.

In order to investigate how the indeterminacies of Eq. (2.70) can be actually calculated, we define the indeterminacy in an observable as the root-mean-square deviation from the mean,

$$
\begin{aligned}
(\Delta x)^2 &= \langle (x - \langle x \rangle)^2 \rangle \\
&= \langle x^2 - 2x\langle x \rangle + \langle x \rangle^2 \rangle \\
&= \langle x^2 \rangle - \langle x \rangle^2
\end{aligned}
\quad (2.71)
$$

since $\langle x \langle x \rangle \rangle = \langle x \rangle^2$. Similarly,

$$(\Delta p_x)^2 = \langle p_x^2 \rangle - \langle p_x \rangle^2 \quad (2.72)$$

Given a state function ψ_j, the various quantities entering into Eqs. (2.71) and (2.72) can be directly calculated and the indeterminacy relation confirmed. If the function ψ_j is normalized,

$$\langle x \rangle = \int \psi_j^* x \psi_j \, dx \quad (2.73)$$

$$\langle x^2 \rangle = \int \psi_j^* x^2 \psi_j \, dx \quad (2.74)$$

$$\langle p_x \rangle = -i\hbar \int \psi_j^* \frac{\partial \psi_j}{\partial x} \, dx \quad (2.75)$$

$$\langle p_x^2 \rangle = -\hbar^2 \int \psi_j^* \frac{\partial^2 \psi_j}{\partial x^2} \, dx \quad (2.76)$$

Regardless of the particular function ψ_j used in these calculations, the indeterminacy relation of Eq. (2.70) holds.

Example 2.6 We find in Chapter 3 that the wavefunction corresponding to the lowest energy state of a particle in a one-dimensional potential box extending from $x = 0$ to $x = L$ is

$$\psi_i = \left(\frac{2}{L}\right)^{1/2} \sin \frac{\pi x}{L}$$

In Example 2.3 we calculated the normalization constant for this wavefunction. Test the indeterminacy principle for this state.

We need to apply Eqs. (2.73) through (2.76) to the specific given wavefunction. The results are as follows:

$$\langle x \rangle = \frac{2}{L} \int_0^L x \sin^2 \frac{\pi x}{L} dx = \frac{L}{2}$$

$$\langle x^2 \rangle = \frac{2}{L} \int_0^L x^2 \sin^2 \frac{\pi x}{L} dx = L^2 \left[\frac{1}{3} - \frac{1}{2\pi^2} \right]$$

$$\langle p_x \rangle = -i\hbar \frac{2}{L} \int_0^L \left(\sin \frac{\pi x}{L} \right) \frac{\pi}{L} \cos \frac{\pi x}{L} dx = 0$$

$$\langle p_x^2 \rangle = \hbar^2 \frac{2}{L} \int_0^L \frac{\pi^2}{L^2} \sin^2 \frac{\pi x}{L} dx = \frac{\hbar^2 \pi^2}{L^2}$$

Note that because $\langle p_x \rangle = 0$ it does not follow that $\langle p_x^2 \rangle$ is zero. Using the results of the above calculations, we find

$$(\Delta x)^2 = L^2 \left(\frac{1}{3} - \frac{1}{2\pi^2} - \frac{1}{4} \right)$$

or

$$\Delta x = 0.17 L$$

and

$$\Delta p_x = \frac{\hbar \pi}{L}$$

so that

$$\Delta x \, \Delta p_x = 0.55 \hbar > 0.50 \hbar.$$

2.7. Dirac Notation

In many places it proves convenient to develop the language of quantum mechanics in terms of a notation introduced by Dirac. Although we do not find it necessary to use this shorthand at any great length in this book, it is appropriate to outline its main characteristics so that other literature using this notation will be understandable.

A function ψ_j is represented by the *ket* symbol $|j\rangle$, and its complex conjugate ψ_j^* is represented by the *bra* symbol $\langle j|$. The *product* of two such functions is defined as the integral of the mathematical product over all coordinates and is written as follows:

$$\int \psi_i^* \psi_j \, d\mathbf{r} = \langle i | j \rangle \tag{2.77}$$

Thus the product of a *bra* function and a *ket* function yields a bra(c)ket. Using this Dirac notation, we rewrite Eq. (2.59), for example, as

$$E_j = \langle j | \mathbf{H} | j \rangle \tag{2.78}$$

and the time-independent Schroedinger equation becomes

$$\mathbf{H} | j \rangle = E_j | j \rangle \tag{2.79}$$

2.8 Summary

A classical description of matter requires rather radical changes when atomic and subatomic proportion are of principle interest. Entities such as electrons, previously conceived as particles, must be given a wavelike character through the de Broglie relation to account for many of their properties. Wave phenomena such as lattice waves and light must experience quantization of the energy and hence acquire a particle-like character. Neither particle nor wave picture is wholly adequate by itself; the essential particle–wave duality manifests itself in the Heisenberg indeterminacy principle.

Quantized energy states for particle–waves can be considered to result from the existence of the constraints of a confined system. Energies for a free electron are not quantized; energies for an electron in a box, a crystal, or an atom are quantized.

The appropriate wave equation for particle–waves appears to be the Schroedinger equation. This equation can be produced either by an analogical procedure starting with previous wave equations, or by transforming to an operator interpretation of the classical Hamiltonian function.

The Schroedinger operator approach to quantum mechanics involves three basic postulates. The essential point is that the measured values of an observable are the eigenvalues belonging to the eigenfunctions of an operator associated with the observable according to certain rules. This operator eigenvalue equation for the Hamiltonian operator is identical with the time-independent Schroedinger wave equation. The calculation of allowed energy states proceeds from a solution of the Schroedinger equation in accordance with the physical boundary conditions and the particular potential function of the system under investigation.

Although the wavefunction itself, the eigenfunction of the Hamiltonian operator or the solution of the Schroedinger wave equation, is frequently a complex quantity and does not have a physical interpretation, the square of the magnitude of the wavefunction is interpreted as giving the probability distribution in space of the corresponding particle. All the properties of the system can in principle be calculated once these eigenfunctions are known.

Relevant concepts of normalization, orthogonality, orthonormal sets, basis functions, and complete sets were introduced and applied to the properties of wavefunctions. It was seen how these concepts, commonly applied to such systems as the Cartesian coordinate system (or Fourier analysis, or expansion in terms of spherical harmonics), could be extended and applied to the wavefunction space of interest in quantum mechanics. In particular, the eigenfunctions of a particular Hamiltonian operator form a complete orthonormal set that can serve as the basis functions for the construction of any other well-behaved function of physical significance.

Because all Schroedinger operators must have real eigenvalues to be physically significant, all Schroedinger operators fall into the class of Hermitian operators.

The difference was developed between an exact value found in a measurement of an observable when the system is in an eigenstate of the operator associated with that observable, and the statistically determined expectation value when the system is not in such an eigenstate. The requirement that a particular observable have an exact value in a particular eigenstate of the Hamiltonian operator is that the operator associated with the observable commute with the Hamiltonian operator. As an example it was shown that the linear momentum of a free electron has an exact value for the system in an eigenstate of the Hamiltonian operator, but that the position does not have an exact value under the same conditions. This was viewed as another way of looking at the indeterminacy principle.

Books have been written on the philosophical significance of the indeterminacy principle. Is there an ultimate indeterminacy in the structure of nature, or is this apparent result only the sign of our present, and possibly future, incapabilities? Is there really an indeterminacy in position and momentum as valid concepts, or are the concepts of position and momentum themselves as inappropriate for an electron as are the concepts of color and temperature? Is there any connection between indeterminacy in the physical atomic and subatomic realm with personal and historical indeterminacy in a completely different area of life? The greatest minds in modern science and philosophy have tangled with these questions, and still it seems that it is not yet decided whether these questions have substance or are semantically misphrased. The reader will do himself an injustice as a human being, if, in his pursuit of an understanding of quantum theory, he limits himself only to the pragmatic aspects of mathematical utility and neglects to think at least a little about the broad meaning for science, nature, and himself as a constituent of nature.

Problems

2.1 The (111) planes of hexagonal CdS are separated by 6.72 Å. For what angle of incidence will diffraction be observed for electrons that have passed through a potential difference of 10^4 V?

2.2 A collection of radioactively decaying atoms decays on the average according to a well-defined exponential decay constant, i.e., if N^* is the number of undecayed atoms at time t,

$$\frac{dN^*}{dt} = -bN^*$$

What causes a particular atom to decay at a particular time t?

2.3 Derive the Bohr relation for the quantization of the angular momentum of a particle executing circular motion, in multiples of $\hbar$, by applying the wave analogy to this particular problem.

2.4 (a) Show that the Schroedinger operator associated with the observable xp_x^2 is

$$\frac{1}{2}\left[x\left(-\hbar^2\frac{\partial^2}{\partial x^2}\right) + \left(-\hbar^2\frac{\partial^2}{\partial x^2}\right)x\right]$$

by starting with Postulate II.
(b) Does the observable xp_x^2 have exact values in an eigenstate of the Hamiltonian operator?
(c) Calculate the expectation value of xp_x^2 in the ground state of a particle in a one-dimensional box, using the wavefunction for this state given in Example 2.6.

2.5 (a) Show that the Poisson bracket of the classical observables H and x is $-v_x$.

(b) Use this result to show that the Schroedinger operator to be associated with the observable v_x is given by

$$\frac{1}{m}\left(-i\hbar \frac{\partial}{\partial x}\right)$$

2.6 Can the operator $\partial/\partial x$ be used to represent a physical observable?

2.7 What restriction is imposed on the potential energy for both energy and momentum to have exact values in the same state?

2.8 Starting with an ideal simple wave equation

$$\nabla^2 \Psi - \frac{1}{v^2} \frac{\partial^2 \Psi}{\partial t^2} = 0$$

show that the time-independent Schroedinger equation can be derived by assuming only the de Broglie wavelength relationship and that

$$\Psi(x, y, z, t) = \psi(x, y, z)\, e^{-i\omega t}.$$

2.9 Consider the functions

$$\phi_n = x^n\, e^{-ax^2}$$

defined over all x. Normalize ϕ_0 and ϕ_1 and discuss their orthogonality.

2.10 Calculate the expectation value of the radius r for the ground state of the hydrogen atom for which the normalized wavefunction is

$$\psi = \frac{1}{(\pi a_0^3)^{1/2}}\, e^{-r/a_0}$$

where a_0 is a constant, the Bohr radius.

2.11 What is the indeterminacy in r in the ground state defined in Problem 2.10?

Chapter 3

Quantum Mechanical Treatment of Simple Systems

In principle the allowed energy levels for any system can be calculated from the Schroedinger equation. For a given potential energy and set of boundary conditions, the well-behaved eigenfunctions of the Schroedinger equation define the allowed energy states for the system. In most practical situations of interest, however, it is not possible to follow this procedure in any but approximate ways. The possibility of obtaining exact solutions is limited to those special cases where only a single particle is involved. Fortunately there are several simple systems involving only one particle that illustrate the type of solution characteristic of a number of physical systems. It is to these simple systems that we turn in this chapter.

We consider here the following systems:

1. A free particle. $V = 0$ throughout all space.
2. A particle in a one-dimensional rectangular potential well (a "box"). For $x > 0$ and $x < L$, the potential V is infinite. For $0 \leq x \leq L$, $V = 0$.
3. A particle executing simple harmonic motion, i.e., the linear harmonic oscillator. The particle moves under a restoring force proportional to the magnitude of the displacement from a rest position. For a restoring force $F = -gx$, the associated potential energy is $V = gx^2/2$.
4. An electron in a hydrogenic atom. The potential $V = -Ze^2/r$, is a function only of the radial distance of the particle from the center of the force field.

The procedure followed in each of these cases is essentially the same as that outlined for general wave problems in Section 1.2.

3.1 A Free Particle

A system consisting of a single noninteracting particle is the simplest of all. It corresponds to an identically zero potential energy throughout all space. Yet it has interest because it forms the basis for the representation of those electrons that contribute to electrical and thermal transport in metals and semiconductors (the "free" electrons).

The time-independent Schroedinger equation for the allowed energy states in one dimension is

$$-\frac{\hbar^2}{2m}\frac{d^2\psi}{dx^2} = E\psi \qquad (3.1)$$

The solution for $\psi(x)$ is

$$\psi(x) = A\,e^{ikx} + B\,e^{-ikx} \qquad (3.2)$$

Substitution of this solution into Eq. (3.1) gives directly

$$\frac{\hbar^2 k^2}{2m} = E \qquad (3.3)$$

and the alternative form for $\psi(x)$,

$$\psi(x) = A\,e^{i(2mE)^{1/2}x/\hbar} + B\,e^{-i(2mE)^{1/2}x/\hbar} \qquad (3.4)$$

Since the de Broglie wavelength is $\lambda = h/(2mE)^{1/2}$, the wavefunction of a free particle of momentum $p = (2mE)^{1/2}$ has the de Broglie wavelength.

All values of $E \geq 0$ are allowed for a free particle, and Eq. (3.3) is the dispersion relationship for this wave system,

$$\omega = \frac{\hbar k^2}{2m} \qquad (3.5)$$

Values of $E < 0$ are not allowed, for then the exponential terms of Eq. (3.4) would have real exponents and one or the other would become infinite as x approaches infinity. Because of the absence of confinement for a free particle, no quantization of energy occurs.

Since both terms in Eq. (3.2) represent independent solutions to the free-particle wave equation, it follows that the solutions are doubly degenerate, giving the same value of E for both $+k$ and $-k$.

The eigenfunctions of the Hamiltonian for the free particle are not strictly square-integrable. The distribution of the probability density $\psi^*\psi$ is a constant throughout space, but can be normalized strictly only in a finite volume.

3.2 A Particle in a One-Dimensional Potential Well

The eigenfunctions for the free particle in three dimensions have the form

$$\psi(\mathbf{r}) = A\, e^{i\mathbf{k}\cdot\mathbf{r}} \qquad (3.6)$$

with

$$\mathbf{k} = \mathbf{e}_1 k_x + \mathbf{e}_2 k_y + \mathbf{e}_3 k_z \qquad (3.7)$$

$$= \frac{1}{\hbar}(\mathbf{e}_1 p_x + \mathbf{e}_2 p_y + \mathbf{e}_3 p_z) \qquad (3.8)$$

$$= \frac{2\pi}{\lambda}(\mathbf{e}_1 \cos a + \mathbf{e}_2 \cos b + \mathbf{e}_3 \cos c) \qquad (3.9)$$

where $\mathbf{e}_1$, $\mathbf{e}_2$, and $\mathbf{e}_3$ are unit vectors in the x, y, and z directions respectively, and $\cos a$, $\cos b$, and $\cos c$ are the direction cosines of $\mathbf{k}$. In three dimensions the wavefunction for a free particle is infinitely degenerate, since the same energy E is obtained for any direction of $\mathbf{k}$, provided only that $|\mathbf{k}|$ is the same.

3.2 A Particle in a One-Dimensional Potential Well

Consider a particle in the one-dimensional potential well with infinitely high sides, as shown in Fig. 3.1. This corresponds to what is often called the "particle in a box" problem. This particular system has some of the characteristics of a particle that is essentially free but only within a confined portion of space.

Classically we expect that all energies are allowed for such a particle, and that the probability of finding the particle at some value x between 0 and L

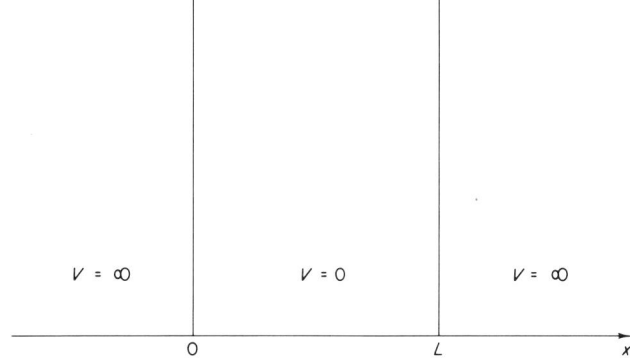

Fig. 3.1 Potential box with $V = 0$ between $x = 0$ and $x = L$, and $V = \infty$ outside the box.

is a constant equal to $1/L$. The wave picture alters both of these classical expectations. Only specific values of energy are allowed, and the probability of finding the particle varies with x in a way that is different for each of the allowed energy states.

The Schroedinger equation to be considered both inside and outside the box is

$$\frac{d^2\psi}{dx^2} + \frac{2m}{\hbar^2}(E - V)\psi = 0 \qquad (3.10)$$

Outside the box $V = \infty$ and only $\psi_{out} = 0$ will satisfy Eq. (3.10). Inside the box $V = 0$, and the solutions of Eq. (3.10) are the same as Eq. (3.2),

$$\psi_{in} = A\, e^{ikx} + B\, e^{-ikx} \qquad (3.11)$$

with $k = (2mE)^{1/2}/\hbar$ for the state with energy E. The boundary conditions are that $\psi = 0$ at $x = 0$ and $x = L$, in order to maintain continuity of the wavefunction at the boundaries of the box. Continuity of the slope of the wavefunction at the boundaries is *not* required in this special case, because of the infinite change in the potential energy at these points.

Applying these boundary conditions to Eq. (3.11) gives

$$0 = A + B \qquad (3.12)$$

and

$$0 = A\, e^{ikL} + B\, e^{-ikL} \qquad (3.13)$$

Satisfaction of the boundary conditions requires that

$$k = \frac{n\pi}{L} \qquad n = 1, 2, \ldots \qquad (3.14)$$

and that

$$E_n = \frac{n^2\hbar^2\pi^2}{2mL^2} \qquad (3.15)$$

This is the same result we found by simple analogy with waves in Eq. (2.5). The particular state with energy E_n is associated with the particular integer n. We may call n the *quantum number* associated with the state with energy E_n and wavefunction ψ_n. The appropriate wavefunctions for a particle in a one-dimensional box can be obtained by using Eq. (3.14) for k in Eq. (3.11),

$$\psi_n = A\, e^{in\pi x/L} + B\, e^{-in\pi x/L} \qquad (3.16)$$

3.2 A Particle in a One-Dimensional Potential Well

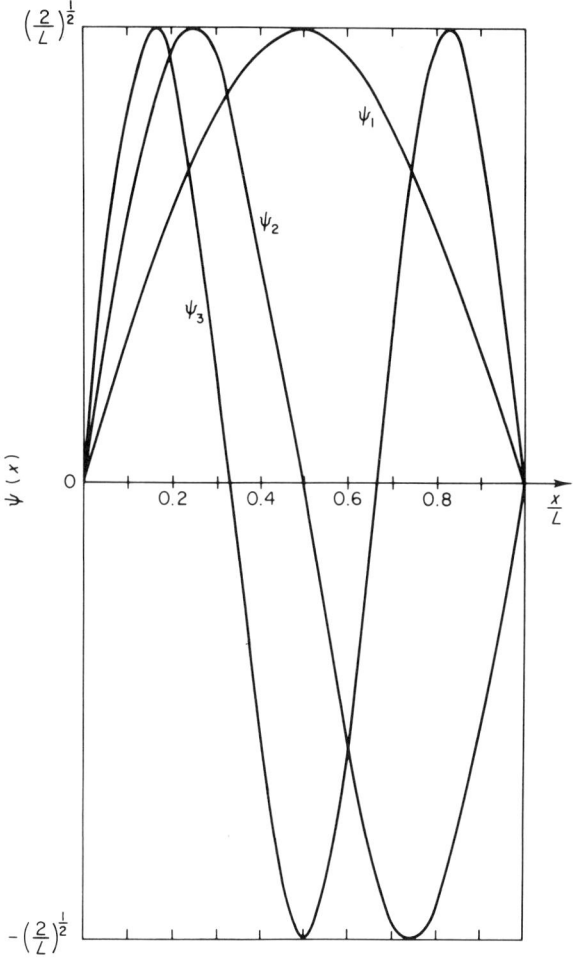

Fig. 3.2 Eigenfunctions for the ground state and first two excited states for a particle in a one-dimensional box of width L.

which, in view of Eq. (3.12) becomes

$$\psi_n = C_n \sin \frac{n\pi x}{L} \tag{3.17}$$

where C_n is a normalizing constant; $C_n = (2/L)^{1/2}$ from Example 2.3. Equation (3.17) describes a standing wave, the necessary consequence of the fact that in the system being considered the particle was restricted to a finite range of coordinate values.

A plot of ψ_1, ψ_2, and ψ_3 is given in Fig. 3.2. In the lowest energy state, $n = 1$. The value $n = 0$ is excluded to avoid having a wavefunction that is identically zero. The lowest energy state is frequently called the *ground state* of the system, while higher energy state are called *excited states*. The ground state of the particle in a one-dimensional box has an energy

$$E_1 = \frac{\hbar^2 \pi^2}{2mL^2} \qquad (3.18)$$

and a wavefunction

$$\psi_1(x) = \left(\frac{2}{L}\right)^{1/2} \sin\left(\frac{\pi x}{L}\right) \qquad (3.19)$$

The first excited state has an energy

$$E_2 = \frac{4\hbar^2 \pi^2}{2mL^2} \qquad (3.20)$$

and a wavefunction

$$\psi_2 = \left(\frac{2}{L}\right)^{1/2} \sin\left(\frac{2\pi x}{L}\right) \qquad (3.21)$$

The nth excited state has an energy given by Eq. (3.15) and a wavefunction given by Eq. (3.17). Each of the states of a particle in a one-dimensional box is nondegenerate.

We can summarize the insights gained by this calculation as follows.

a. We have an example of the meaning of obtaining solutions of the Schroedinger equation

$$\mathbf{H}\psi_n = E_n \psi_n$$

for a particular system characterized by a specific potential energy, and we have a set of eigenfunctions ψ_n and eigenvalues E_n.

b. Figure 3.2 illustrates a basic property of wavefunctions: the higher the energy of the state being described, the larger the number of nodes in the wavefunction. Here, for example, we see that the number of nodes in the wavefunction is given by $(n - 1)$.

c. It is easy to demonstrate for this system that

$$\int_0^L \psi_n^* \psi_m \, d\mathbf{r} = \delta_{mn}$$

i.e., that the eigenfunctions of this Hamiltonian form an orthonormal set.

3.3 A Linear Harmonic Oscillator

d. The classical expectation of a continuous range of allowed energies is replaced by a set of discrete values, the difference between successive energy levels increasing as the energy of the states increases.

e. The classical expectation of a uniform probability of finding the particle anywhere in the box is replaced by a probability that has a maximum at $x = L/2$ in the ground state, at $x = L/4$ and at $x = 3L/4$ in the first excited state, and so on.

f. As n increases, the probability of finding the particle at a specific point in the box approaches a uniform distribution.

g. Expectation values of other observables associated with the particle in the box can be calculated using this set of wavefunctions ψ_n, as indicated in part in Example 2.6.

3.3 A Linear Harmonic Oscillator

Properties of a linear harmonic oscillator are of wide significance for many different physical phenomena. Since both lattice vibrations and electromagnetic vibrations can be considered as the result of a collection of oscillators representing the normal modes of the system, the quantization of the energy of a harmonic oscillator means the quantization of the energy of lattice waves (where the quantum $\hbar\omega$ is called the *phonon*) and the quantization of the energy of electromagnetic waves (where the quantum $\hbar\omega$ is called the *photon*).

In a linear harmonic oscillator, the potential energy is a continuous function of the coordinate, $V = gx^2/2$, if the restoring force for the oscillator $F = -gx$. The classical equation of motion is

$$m \frac{d^2 x}{dt^2} = -gx \tag{3.22}$$

with solution

$$x = A\, e^{i\omega t} + B\, e^{-i\omega t} \tag{3.23}$$

where the classical frequency ω is given by

$$\omega = \left(\frac{g}{m}\right)^{1/2} \tag{3.24}$$

All values of energy greater than or equal to zero are classically allowed for the oscillator. If the maximum displacement from equilibrium is $\pm x_0$, there is no probability of finding the particle further than x_0 from equilibrium, and the probability distribution for finding the particle is always a maximum at the turning points where $x = \pm x_0$. All of these classical expectations are altered by the quantum mechanical calculation.

The Schroedinger equation for the linear harmonic oscillator is

$$-\frac{\hbar^2}{2m}\frac{d^2\psi}{dx^2} + \frac{gx^2}{2}\psi = E\psi \qquad (3.25)$$

Although the particle in the oscillator is confined, so that quantized energy values are expected, there are no specific boundary conditions. Quantization of the energy must arise in such a situation from the condition that the general solution of Eq. (3.25) be mathematically well behaved so as to be able to describe physical reality. It is rather amazing that profound properties of the physical world are revealed simply by requiring that the mathematical solutions to an elementary differential equation have a mathematical form that does not exclude them from interpretation in a physically significant way.

In order to obtain the solutions of Eq. (3.25), rewrite the equation with the following defined quantities:

$$\beta \equiv \frac{2mE}{\hbar^2} \qquad (3.26)$$

$$\gamma^2 \equiv \frac{mg}{\hbar^2} \qquad (3.27)$$

Then the Schroedinger equation for the linear harmonic oscillator is

$$\frac{d^2\psi(x)}{dx^2} + (\beta - \gamma^2 x^2)\psi(x) = 0 \qquad (3.28)$$

As a trial solution for this differential equation, take

$$\psi(x) = f(x)\, e^{-\gamma x^2/2} \qquad (3.29)$$

where $f(x)$ is a polynomial in x that terminates after a finite number of terms and the exponent in the exponential term is appropriately dimensionless since the dimensions of γ are cm^{-2}. If $f(x)$ does not terminate, or if it contains infinitely high powers of x, then $\psi(x)$ will not be square-integrable. Thus the requirement that $\psi(x)$ be mathematically well behaved is expressed as the requirement that the polynomial $f(x)$ in the trial solution of Eq. (3.29) have only a finite number of terms.

By substitution of Eq. (3.29) into Eq. (3.28), we obtain an equation in $f(x)$,

$$\frac{d^2f(x)}{dx^2} - 2\gamma x\frac{df(x)}{dx} + (\beta - \gamma)f(x) = 0 \qquad (3.30)$$

3.3 A Linear Harmonic Oscillator

It is convenient now to define a dimensionless parameter,

$$\xi = \gamma^{1/2}x \tag{3.31}$$

and to define a new function $H(\xi)$ such that

$$H(\xi) \equiv f(x) \tag{3.32}$$

Upon substitution of Eqs. (3.32) and (3.31) into Eq. (3.30), we obtain an equation in $H(\xi)$,

$$\frac{d^2H(\xi)}{d\xi^2} - 2\xi\frac{dH(\xi)}{d\xi} + \left(\frac{\beta}{\gamma} - 1\right)H(\xi) = 0 \tag{3.33}$$

since

$$\frac{d^2f(x)}{dx^2} = \frac{d^2f(\xi/\gamma^{1/2})}{d(\xi/\gamma^{1/2})^2} = \gamma\frac{d^2H(\xi)}{d\xi^2} \tag{3.34}$$

and so on. To determine the solutions of Eq. (3.33), consider the trial polynomial,

$$H(\xi) = \sum_{n=0}^{l} a_n \xi^n \tag{3.35}$$

Substitution into Eq. (3.33) gives

$$\sum_{n=0}^{l}(n-1)na_n\xi^{n-2} - 2\sum_{n=0}^{l}na_n\xi^n + \left(\frac{\beta}{\gamma}-1\right)\sum_{n=0}^{l}a_n\xi^n = 0 \tag{3.36}$$

A typical term for the coefficients of ξ^n may be written,

$$\cdots + \left[(n+1)(n+2)a_{n+2} + \left(\frac{\beta}{\gamma} - 1 - 2n\right)a_n\right]\xi^n + \cdots = 0 \tag{3.37}$$

Since in such a solution the coefficient of each power of ξ must be identically zero,

$$a_{n+2} = -\frac{\beta/\gamma - 1 - 2n}{(n+1)(n+2)}a_n \tag{3.38}$$

This is a recursion formula that generates the coefficients for the odd or even powers of ξ. Now for $H(\xi)$ to terminate after a finite number of terms, the multiplying quantity in Eq. (3.38) must go to zero at some point in the series. To terminate $H(\xi)$ at the term in ξ^n,

$$\frac{\beta}{\gamma} - 1 - 2n = 0 \tag{3.39}$$

This, then, is the condition for $H(\xi)$ to be finite, hence for $f(x)$ to be finite, hence for $\psi(x)$ to be square-integrable. Equation (3.39) does represent a quantization condition that can be seen explicitly by substituting from Eqs. (3.26) and (3.27) to obtain

$$E_n = \left(n + \frac{1}{2}\right)\hbar\left(\frac{g}{m}\right)^{1/2} = \left(n + \frac{1}{2}\right)\hbar\omega \qquad (3.40)$$

This means that the allowed energy states for a linear harmonic oscillator with classical frequency ω are multiples of the basic quantum $\hbar\omega$. These energy levels are equally spaced, and the lowest energy level is not zero, as in the classical expectation, but is rather $\hbar\omega/2$, sometimes called the *zero-point energy*. Whenever energy is given to or taken from a harmonic oscillator, therefore, the energy difference involved ΔE must be an integral multiple of $\hbar\omega$,

$$\Delta E = n\hbar\omega \qquad (3.41)$$

The eigenfunctions of the Hamiltonian for the linear harmonic oscillator are the functions

$$\psi_n(x) = A_n H_n(\gamma^{1/2}x)\, e^{-\gamma x^2/2} \qquad (3.42)$$

referring back to Eq. (3.29). The functions $H_n(\gamma^{1/2}x)$ form a class of mathematical functions known as *Hermite polynomials*.

With the condition given in Eq. (3.40), Eq. (3.33) for $H(\xi)$ becomes

$$\frac{d^2 H(\xi)}{d\xi^2} - 2\xi\, \frac{d H(\xi)}{d\xi} + 2n\, H(\xi) = 0 \qquad (3.43)$$

The polynomials known as Hermite polynomials have the form

$$H_n(\xi) = (-1)^n\, e^{\xi^2}\, \frac{d^n}{d\xi^n}\, e^{-\xi^2} \qquad (3.44)$$

We may show that the Hermite polynomials are indeed the solutions of Eq. (3.43) by substituting Eq. (3.44) into Eq. (3.43) and obtaining

$$\frac{d^{n+2}}{d\xi^{n+2}} e^{-\xi^2} + 2\xi\, \frac{d^{n+1}}{d\xi^{n+1}} e^{-\xi^2} + 2(n+1)\, \frac{d^n}{d\xi^n} e^{-\xi^2} = 0 \qquad (3.45)$$

If this relationship holds for any values of n, then the Hermite polynomials are the desired solutions. We note first that differentiating Eq. (3.45) with respect to ξ is the same result as replacing n wherever it appears by $(n + 1)$. Therefore if we can show that the equation is satisfied for any particular

3.3 A Linear Harmonic Oscillator

value of n, it is satisfied for all n. It is relatively simple to show that the equation is satisfied for $n = 0$, and therefore by this method of induction we have proved that the Hermite polynomials are the desired solutions of Eq. (3.43).

Let us now examine the properties of several of these eigenfunctions for the linear harmonic oscillator. The ground state of the oscillator is given by $n = 0$,

$$\psi_0 = A_0 e^{-\gamma x^2/2}$$
$$E_0 = \frac{\hbar\omega}{2} \tag{3.46}$$

since $H_0(\gamma^{1/2} x) = 1$. For the first excited state, $n = 1$,

$$\psi_1 = A_1 2\gamma^{1/2} x \, e^{-\gamma x^2/2}$$
$$E_1 = \frac{3\hbar\omega}{2} \tag{3.47}$$

since $H_1(\gamma^{1/2} x) = 2\gamma^{1/2} x$. For the second excited state, $n = 2$,

$$\psi_2 = A_2 2(2\gamma x^2 - 1) e^{-\gamma x^2/2}$$
$$E_2 = \frac{5\hbar\omega}{2} \tag{3.48}$$

since $H_2(\gamma^{1/2} x) = 2(2\gamma x^2 - 1)$. These three eigenfunctions are plotted in Fig. 3.3. In the ground state, the maximum probability location for the particle is at the rest position, $x = 0$. In the first excited state, however, the probability of finding the particle at $x = 0$ is zero, and the positions with maximum probability are $x = \pm 1/\gamma^{1/2}$. In the second excited state the maximum probability positions are at $x = 0$, $\pm(5/2\gamma)^{1/2}$, with nodes (zero probability) at $x = \pm 1/(2\gamma)^{1/2}$. Note that once more the number of nodes is given by $(n - 1)$ and increases with the energy of the state being described. A major difference between the wave picture and the classical picture is that in any state there is a finite probability according to the wave picture of finding the particle at large distances from the origin.

Determination of the normalization constants for the eigenfunctions of the linear harmonic oscillator is obtained by use of the definite integral

$$\int_{-\infty}^{+\infty} x^{2n} e^{-ax^2} dx = \frac{1 \cdot 3 \cdot 5 \cdots (2n-1)}{2^n a^n} \left(\frac{\pi}{a}\right)^{1/2} \tag{3.49}$$

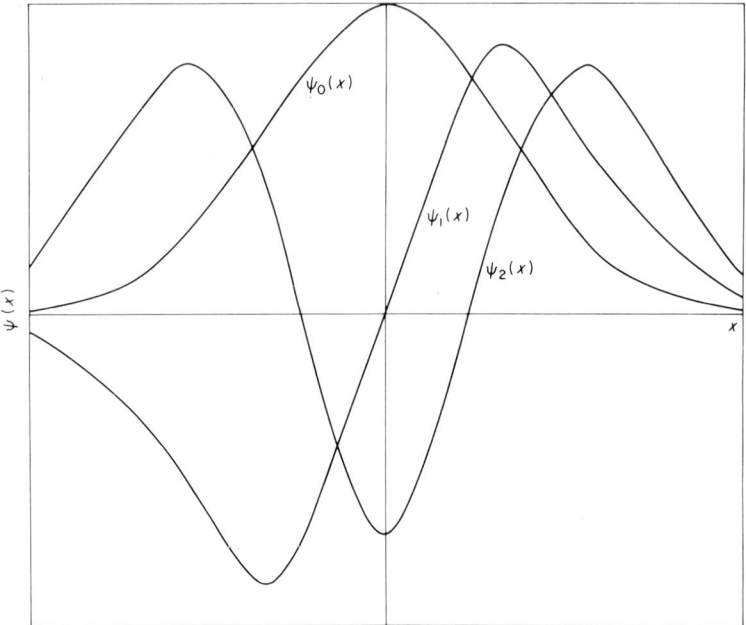

Fig. 3.3 Normalized eigenfunctions for the ground state and first two excited states of a linear harmonic oscillator.

A general expression for the normalization constant A_n is

$$A_n = \frac{1}{[2^n n! (\pi/\gamma)^{1/2}]^{1/2}} \tag{3.50}$$

As in the case of a particle in a box, quantization for the linear harmonic oscillator can be expressed via a *quantum number n*, which specifies both the eigenfunctions and the eigenvalues of the Hamiltonian. As in the case of a particle in a box the quantum probability $|\psi(x)|^2$ approaches the classical expectation of a uniform value throughout the box for large values of n, so also in the case of the harmonic oscillator the quantum probability $|\psi(x)|^2$ for large values of n approaches the classical expectation with maximum probability of being found at the classical "turn-around" points.

3.4 A Hydrogenic Atom

The problem of the allowed energy states for an electron with charge $-e$ bound to a nucleus with charge $+Ze$ is called the problem of a *hydro-*

3.4 A Hydrogenic Atom

genic atom, since the special case of $Z = 1$ corresponds directly to the hydrogen atom. It is a problem in which the electron moves in a *central force field*, i.e., one in which the potential energy has a magnitude that is a function only of the distance from the center of force (the nucleus), and is independent of the particular direction from that center. In a spherical coordinate system, such as that shown in Fig. 3.4, the potential energy is a function only of $|\mathbf{r}|$, and not of θ or ϕ.

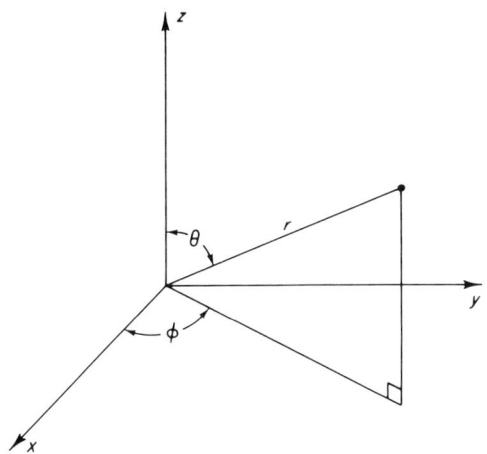

Fig. 3.4 The spherical coordinate system with $x = r \sin \theta \cos \phi$, $y = r \sin \theta \sin \phi$, and $z = r \cos \theta$. The volume element $d\mathbf{r} = r^2 \sin \theta \, dr \, d\theta \, d\phi$.

The Schroedinger equation for a hydrogenic atom is

$$\nabla^2 \psi + \frac{2m}{\hbar^2} \left(E + \frac{Ze^2}{r} \right) \psi = 0 \tag{3.51}$$

with

$$\nabla^2 = \frac{1}{r^2} \frac{\partial}{\partial r} \left(r^2 \frac{\partial}{\partial r} \right) + \frac{1}{r^2 \sin \theta} \frac{\partial}{\partial \theta} \left(\sin \theta \frac{\partial}{\partial \theta} \right) + \frac{1}{r^2 \sin^2 \theta} \frac{\partial^2}{\partial \phi^2} \tag{3.52}$$

The use of spherical coordinates makes it possible to approach the solution of the Schroedinger equation by a separation of variables,

$$\psi(r, \theta, \phi) = R(r) \, \Theta(\theta) \, \Phi(\phi) \tag{3.53}$$

If Eq. (3.53) is substituted into Eq. (3.51), it is possible to separate the terms so that one side of the resulting equation contains functions only of r and θ, and the other side contains functions only of ϕ. It follows that both sides must be equal to a constant A. The equation containing functions of r and θ in terms of A can then be separated further into functions of r only and functions of θ only, which then must both be equal to a second constant B. The separated equations are the following:

$$-\frac{1}{\Phi}\frac{d^2\Phi}{d\phi^2} = A \tag{3.54}$$

$$\frac{1}{\sin\theta}\frac{d}{d\theta}\left(\sin\theta\frac{d\Theta}{d\theta}\right) + \left(B - \frac{A}{\sin^2\theta}\right)\Theta = 0 \tag{3.55}$$

$$\frac{d^2R}{dr^2} + \frac{2}{r}\frac{dR}{dr} + \left[\frac{2m}{\hbar^2}\left(E + \frac{Ze^2}{r}\right) - \frac{B}{r^2}\right]R = 0 \tag{3.56}$$

Examination of these equations shows that the potential energy enters only the equation for $R(r)$. The solutions of the problem of a central force field as far as the $\Theta(\theta)$ and $\Phi(\phi)$ solutions are concerned, are the same for any arbitrary $V(r)$. The angular dependence of the solutions, expressible finally in what are called *spherical harmonics*, can be separated from the radial dependence of the solutions, which determines the allowed energy values and includes the potential energy term.

ANGULAR DEPENDENCE

Solution of Eq. (3.54) for $\Phi(\phi)$ is required to be single valued. This requirement is satisfied only if

$$\Phi(\phi) = C_1 e^{im_l\phi}, \qquad m_l = 0, \pm 1, \pm 2, \ldots \tag{3.57}$$

The index m_l is the so-called *magnetic quantum number*; it acquired this name because the energy states are independent of m_l in the absence of a magnetic field, but in the presence of a magnetic field, states with different m_l numbers may have different energies.

The solution of Eq. (3.55) for $\Theta(\theta)$ can be approached in a manner similar to that used for the linear harmonic oscillator; a solution in the form of a polynomial is substituted and the condition that it must terminate after a finite number of terms introduces the restrictive conditions and the related quantum number. This requirement leads to

$$B = l(l+1), \qquad l = 0, 1, 2, \ldots \tag{3.58}$$

where l is a second quantum number, called the *angular momentum quantum number*, and

$$|m_l| \leq l \tag{3.59}$$

3.4 A Hydrogenic Atom

The association of angular momentum with the number l arises from the fact that the magnitude of the total angular momentum in a given state is $\hbar[l(l+1)]^{1/2}$.

The form of the angular eigenfunctions for small values of l and m_l are simple, and are indicated in Table 3.1. The table also lists the common designation for these states as defined by their quantum number l; $l = 0$ is called an s state, $l = 1$ is called a p state, $l = 2$ is called a d state, and $l = 3$ is called an f state. These apparently arbitrary letter designations arose out of the early days of experimental spectroscopy and are related to the types of spectra associated with certain transitions: s for "sharp," p for "principal," and d for "diffuse."

TABLE 3.1 ANGULAR EIGENFUNCTIONS FOR THE HYDROGENIC ATOM

Spectroscopic designation	l	m_l	$\Theta(\theta)$ [a]	$\Phi(\phi)$ [b]
s	0	0	1	1
p	1	0	$\cos\theta$	1
p	1	± 1	$\sin\theta$	$e^{\pm i\phi}$
d	2	0	$(3\cos^2\theta - 1)/2$	1
d	2	± 1	$3\sin\theta\cos\theta$	$e^{\pm i\phi}$
d	2	± 2	$3\sin^2\theta$	$e^{\pm i2\phi}$

[a] General form: $\Theta(\theta) = \dfrac{1}{2^l l!} (1 - \cos^2\theta)^{|m_l|/2} \dfrac{d^{l+|m_l|}(\cos^2\theta - 1)^l}{d(\cos\theta)^{l+|m_l|}}$.

[b] General form given in Eq. (3.53).

Each set of angular eigenfunctions for a given set of l and m_l values corresponds to a particular angular distribution of the probability density $|\psi(r, \theta, \phi)|^2$ for the electronic states. For an s state, $l = 0$ and $m_l = 0$, and both $\Theta(\theta)$ and $\Phi(\phi)$ are constants. This means that *an s state has spherical symmetry*. The classical or quasi-quantum picture of an s state as an electron moving about the nucleus in a circular orbit is not consistent with the description of a state of zero angular momentum. Either a static, nonrotating spherical charge distribution must be pictured, or if a model involving a moving electron is insisted upon, the electron must be moving along radial directions to and from the nucleus!

For a p state, $l = 1$ and $m_l = -1, 0, +1$. Since, in the absence of a magnetic field, the energy is not dependent on the value of m_l, a p state is triply degenerate. The three eigenfunctions corresponding to these degenerate states are

$$\psi_{-1} = R(r) \sin \theta \, e^{-i\phi} \quad (3.60)$$

$$\psi_0 = R(r) \cos \theta \quad (3.61)$$

$$\psi_{+1} = R(r) \sin \theta \, e^{i\phi} \quad (3.62)$$

The angular distribution for the p state, given by $|\psi(r, \theta, \phi)|^2$, is not a function of ϕ, and is therefore symmetric about the z axis. It corresponds to a dumbell shape extending in the z-axis direction for ψ_0, as shown in

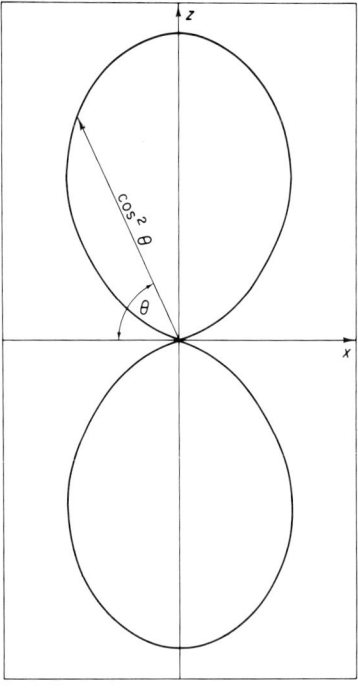

Fig. 3.5 Angular dependence of a p-state eigenfunction for $m_l = 0$ for a central force field.

Fig. 3.5, and to a toroidal shape for both ψ_{-1} and ψ_{+1} with axis as the z axis, the sense of the rotation being opposite for ψ_{-1} and ψ_{+1} as indicated by the opposite signs in the exponent of the $e^{i\phi}$ term. Unlike the spherical symmetry of the s state, therefore, the p state is characterized by lobes of

3.4 A Hydrogenic Atom

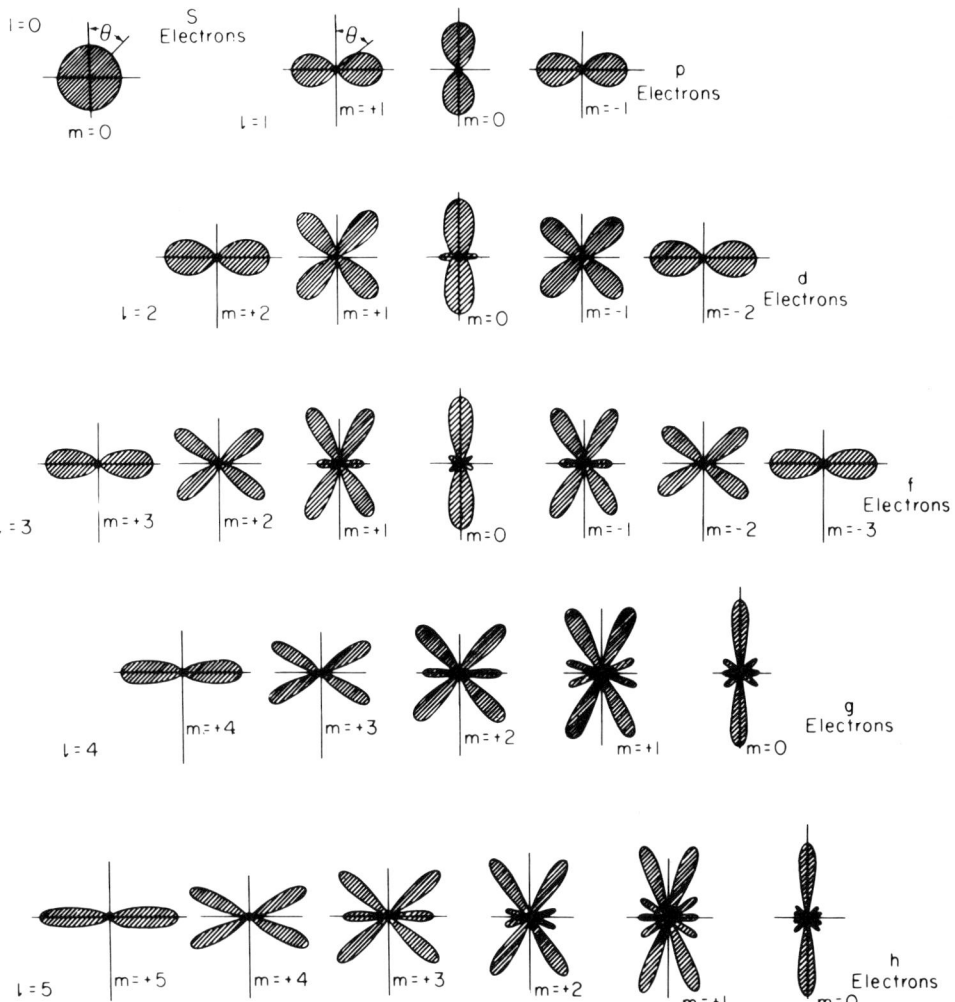

Fig. 3.6 Angular dependence of eigenfunctions for a central force field for different values of l and m_l. (From H. E. White, "Introduction to Atomic Spectra," McGraw-Hill, New York, 1934, p. 63.)

probability density extending away from the origin. Typical probability density distributions for different values of l are given in Fig. 3.6.

Sometimes in discussing these distributions, linear combinations of the independent eigenfunctions for $\Phi(\phi)$, $e^{i|m_l|\phi}$, and $e^{-i|m_l|\phi}$ are taken to form a new set of useful functions, for which, however, m_l is no longer a good quantum number. In terms of these linear combination functions, we can write for a p state,

$$\psi'_{\pm 1} = R(r) \sin \theta \cos \phi = \frac{x}{r} R(r) \tag{3.63}$$

$$\psi'_0 = R(r) \cos \theta = \frac{z}{r} R(r) \tag{3.64}$$

$$\psi'_{\mp 1} = R(r) \sin \theta \sin \phi = \frac{y}{r} R(r) \tag{3.65}$$

These functions preserve symmetry about the z axis only for ψ'_0, showing instead a dumbell distribution along the x axis for $\psi'_{\pm 1}$ and along the y axis for $\psi'_{\mp 1}$.

RADIAL DEPENDENCE

The radial equation to be solved to obtain the eigenvalues for the energy in a hydrogenic atom is Eq. (3.56) with $B = l(l + 1)$. The appropriate well behaved solutions to this equation can be obtained by the assumption of a suitable polynomial form and then deriving the quantization conditions from the requirement that this polynomial terminate after a finite number of terms. This quantization condition involves a third quantum number n, called the *principal quantum number* because in the hydrogenic atom the energy depends only on the value of n. The relationship between n and l is given by

$$l \leq (n - 1) \tag{3.66}$$

Typical values for the radial eigenfunctions and corresponding eigenvalues for $n = 1$, 2, and 3 are given in Table 3.2. The values given in the table illustrate the fact that the energy depends only on the value of n, being given in general by

$$E_n = -\frac{Z^2 m e^4}{2\hbar^2} \frac{1}{n^2} \tag{3.67}$$

The ground state of the hydrogen atom is given with $Z = 1$ and $n = 1$; it is $E_1 = -13.60$ eV. When an electron in energy state E_n undergoes a transition to the ground state E_1, the energy difference $(E_n - E_1)$ may be emitted as a photon. A series of optical emission lines are observed, corresponding to transitions from excited states to the ground state, with energies given by

$$E_n = (E_n - E_1) = \frac{Z^2 m e^4}{2\hbar^2} \left(1 - \frac{1}{n^2}\right) \tag{3.68}$$

3.4 A Hydrogenic Atom

TABLE 3.2 RADIAL EIGENFUNCTIONS AND EIGENVALUES FOR A HYDROGENIC ATOM

Spectroscopic designation	n	l	$R_{n,l}(r)$ [a]	$E_{n,l}$
1s	1	0	e^{-Zr/a_0}	$-\dfrac{Z^2 me^4}{2\hbar^2}$
2s	2	0	$2\left(2 - \dfrac{Zr}{a_0}\right) e^{-Zr/2a_0}$	$-\dfrac{Z^2 me^4}{2\hbar^2} \dfrac{1}{2^2}$
2p	2	1	$\dfrac{6Z}{a_0} r\, e^{-Zr/2a_0}$	$-\dfrac{Z^2 me^4}{2\hbar^2} \dfrac{1}{2^2}$
3s	3	0	$18\left[1 - \dfrac{2}{3}\left(\dfrac{Zr}{a_0}\right) + \dfrac{2}{27}\left(\dfrac{Zr}{a_0}\right)^2\right] e^{-Zr/3a_0}$	$-\dfrac{Z^2 me^4}{2\hbar^2} \dfrac{1}{3^2}$
3p	3	1	$64\left(\dfrac{Zr}{a_0}\right)\left[1 - \dfrac{1}{6}\left(\dfrac{Zr}{a_0}\right)\right] e^{-Zr/3a_0}$	$-\dfrac{Z^2 me^4}{2\hbar^2} \dfrac{1}{3^2}$
3d	3	2	$\dfrac{160}{3}\left(\dfrac{Zr}{a_0}\right)^2 e^{-Zr/3a_0}$	$-\dfrac{Z^2 me^4}{2\hbar^2} \dfrac{1}{3^2}$

[a] General form:

$$R_{n,l}(r) = -\varrho^l\, e^{-\varrho/2}\, \frac{d^{2l+1}}{d\varrho^{2l+1}} \left[e^\varrho\, \frac{d^{n+l}}{d\varrho^{n+l}} (\varrho^{n+l}\, e^{-\varrho}) \right]$$

with $\varrho = (2Zme^2/n\hbar^2)r$, $a_0 = \hbar^2/me^2$.

The energies predicted by this equation are in good agreement with the experimental values for the Lyman series of emission lines for the hydrogen atom.

GENERAL REMARKS

A particular energy level for the hydrogenic atom may be associated with a relatively large number of different wavefunctions because of the dependence of the energy only on n, and not on l or m_l. A $(2l + 1)$-fold degeneracy arises simply because of the independence of the energy on m_l. The total degeneracy of a given energy level is equal to n^2 (neglecting of course the twofold spin degeneracy of the electron, which we have not yet introduced).

If the mathematical form of the eigenfunctions for the hydrogenic atom is explored, it is seen that the angular functions corresponding to the quantum number l have l nodes, and that the radial functions corresponding to the quantum numbers n and l have $[n - (l + 1)]$ nodes. The total number of nodes for the whole wavefunction is therefore given by $(n - 1)$, or by the maximum value of l for a given n. Since an s state never has any angular nodes, all nodes for an s state must be radial; similarly the state corresponding to the maximum value of l for a given n has no radial nodes.

Normalized radial functions $R_{1,0}$, $R_{2,0}$, and $R_{2,1}$ are plotted in Fig. 3.7. The quantity of physical significance, however, if we wish to calculate quantities involving the most probable distance of the electron from the nucleus, is the quantity $4\pi r^2 | R_{n,l}(r) |^2$, which can have quite a different appearance when plotted as a function of r. The factor $4\pi r^2$ accounts for the greater volume of spherical shells at larger values of r. Although $R_{1,0}$ and $R_{2,0}$ as given in Fig. 3.7, for example, have a maximum value at $r = 0$, the probability of finding the electron at $r = 0$ is zero. Plots of the quantity $r^2 | R_n(r) |^2$ are given in Fig. 3.8 for several values of n and l. The most

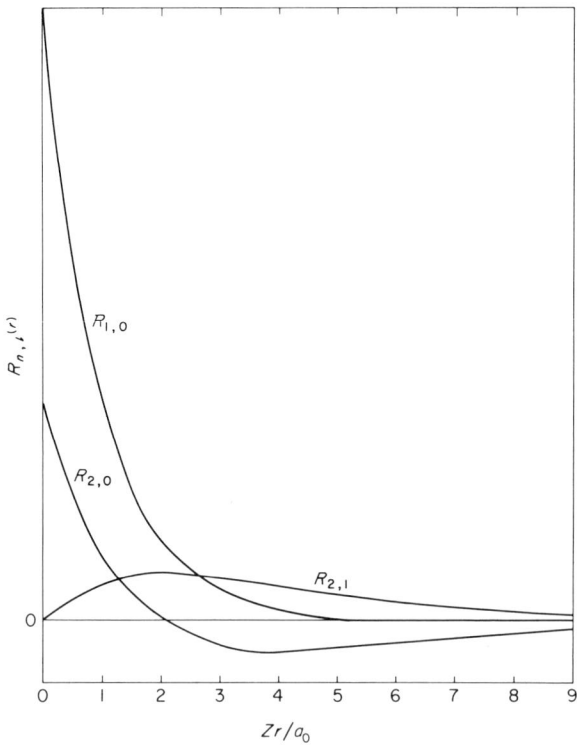

Fig. 3.7 Normalized radial eigenfunctions for the hydrogen atom for the 1s, 2s, and 2p states.

3.4 A Hydrogenic Atom

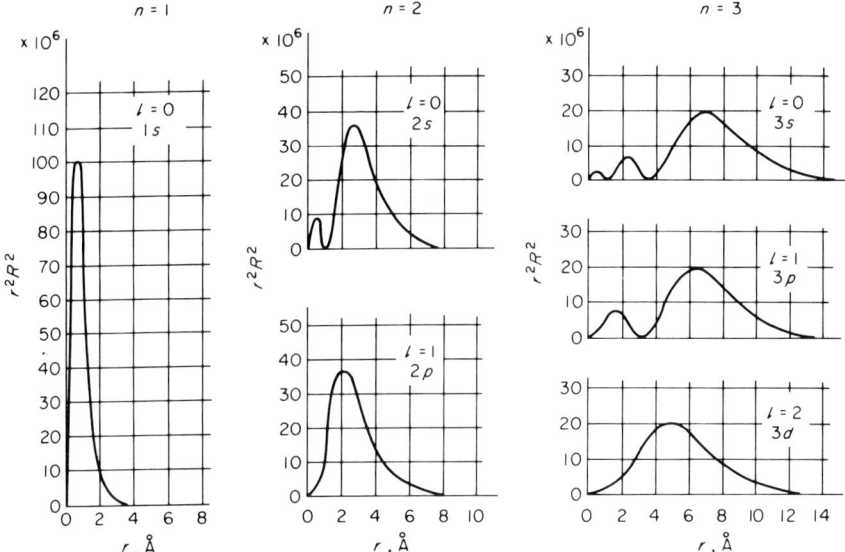

Fig. 3.8 Radial probability density distribution functions for the hydrogen atom for the 1s, 2s, 2p, 3s, 3p, and 3d states. (From W. J. Moore, "Physical Chemistry," Prentice-Hall, Englewood Cliffs, New Jersey, 1962, after G. Herzberg, "Atomic Spectra," Dover, New York, 1944, after H. A. Pohl, "Quantum Mechanics for Science and Engineering," Prentice-Hall, Englewood Cliffs, New Jersey, 1967.)

probable value of r is $a_0 = \hbar^2/me^2 = 0.5292$ Å for the 1s state and $4a_0$ for the 2p state. The probability density for the 2s state shows two maxima, the smaller at $0.75a_0$ and the larger at $5.25a_0$.

Example 3.1 What is the normalized wavefunction for the 2p state of the hydrogen atom, corresponding to $m_l = \pm 1$?

The 2p wavefunction for $m_l = \pm 1$ for a hydrogenic atom is

$$\psi_{2,1,\pm 1}(r, \theta, \phi) = A^r_{2,1} A^\theta_{1,\pm 1} A^\phi_{\pm 1} \frac{6Z}{a_0} r \, e^{-Zr/2a_0} \sin\theta \, e^{\pm i\phi}$$

The general forms for the respective normalization constants may be shown to be

$$A^\phi_{m_l} = \frac{1}{(2\pi)^{1/2}}$$

$$A^\theta_{l,m_l} = \left[\frac{(2l+1)(l-|m_l|)!}{2(l+|m_l|)!}\right]^{1/2}$$

$$A^r_{n,l} = \left[\frac{Z^3 4(n-l-1)!}{a_0^3 n^4 [(n+l)!]^3}\right]^{1/2}$$

Therefore

$$A^0_{1,\pm 1} = \frac{(3)^{1/2}}{2} \quad \text{and} \quad A^r_{2,1} = \frac{1}{12(6)^{1/2}} \left(\frac{Z}{a_0}\right)^{3/2}$$

which gives the normalized wavefunctions

$$\psi_{2,1,\pm 1}(r, \theta, \phi) = \frac{1}{8\pi^{1/2}} \left(\frac{1}{a_0}\right)^{5/2} r\, e^{-r/2a_0} \sin \theta\, e^{\pm i\phi}$$

Example 3.2 Show that the Bohr orbits given by $r_n = n^2(a_0/Z)$ correspond to the most probable values of r for maximum l and m_l in the quantum mechanical picture of the hydrogen atom.

We have already seen that this is satisfied for 1s and 2p states, for which the most probable values of r were calculated to be a_0 and $4a_0$ respectively. The radial function for maximum l and m_l is

$$R_{n,(n-1)} = -\varrho^{(n-1)} e^{-\varrho/2} \frac{d^{2n-1}}{d\varrho^{2n-1}} \left[e^{\varrho} \frac{d^{2n-1}}{d\varrho^{2n-1}} (\varrho^{2n-1} e^{-\varrho}) \right]$$

The form of this generating function is such that for maximum l and m_l, the total process of differentiation indicated above will result in only a constant that we can neglect in our calculation of the most probable value. Therefore

$$R_{n,(n-1)} \propto \left(\frac{2r}{na_0}\right)^{(n-1)} e^{-r/na_0}$$

and the most probable value of r is given by

$$\frac{d}{dr} \left[\left(\frac{2r}{na_0}\right)^{2(n-1)} e^{-2r/na_0} r^2 \right] = 0$$

Differentiation gives

$$r^{2n}\left(-\frac{2}{na_0}\right) + 2nr^{2n-1} = 0 \quad \text{or} \quad r = n^2 a_0$$

as desired.

3.5 Summary

Essentially all of the types of systems for which complete and exact solutions of the Schroedinger equation can be obtained are represented in the cases selected for consideration in this chapter. The common feature

3.5 Summary

of each is that we are able to consider the behavior of a single particle in a given potential field without the complications inherent in the interactions between more than one particle.

For each particular system considered, the allowed energy levels are determined by the appropriate boundary conditions applied to the general solutions of the wave equation, and by the requirement that the wavefunction solutions be mathematically well-behaved, i.e., that they be susceptible to the description of physically meaningful phenomena. Relatively simple exact solutions are obtained for a variety of cases in which the potential energy is assumed to be constant over a specified range of coordinates. In other cases where the potential energy is a continuous function of the coordinates, as in the harmonic oscillator problem or in the central force field, exact solutions are possible in terms of families of mathematical functions.

A comparison between the results of the quantum mechanical calculation and classical calculations on the same type of system reveal a number of distinct differences. Where the classical picture includes a continuum of allowed energies, the quantum picture contains only discrete allowed energies. Classical calculations of the probability of finding a particle in a given coordinate range are quite different from quantum calculations of the probability, particularly for the lowest energy states of the system. Such differences can be attributed directly to the wave characteristics assumed in the quantum picture as contrasted to the particle characteristics assumed in the classical picture.

Specifically we have seen the following results:

1. A free particle does not show quantization of the energy, and is characterized by a wavelength which is the de Broglie wavelength.

2. The properties of a particle in a potential box are completely analogous to the properties of waves in a string. Major deviations from the classical view occur when the corresponding wavelength is of the same order of magnitude as the dimensions of the box itself. When the wavelength is much smaller than the dimensions of the box, the results approach the classical expectations. Quantization of the energy is describable in terms of a single quantum number in one dimension.

3. The allowed energy levels of a harmonic oscillator are equally spaced by the energy of one quantum, $\hbar\omega$, so that any energy interchange between an oscillator and its surroundings must occur in integral multiples of this quantum energy. Since phenomena involving lattice waves and light waves can be described in terms of an oscillator representation, the properties of

the quanta of these systems, i.e., phonons and photons respectively, are a direct consequence of the properties of the oscillator.

4. The treatment of the central force field in three dimensions leads to three quantum numbers: m_l, the magnetic quantum number, which arises from the requirement that the function of ϕ be single valued; l, the angular momentum quantum number, which arises from the requirement that the function of θ be well behaved; and n, the principal quantum number, which arises from the requirement that the function of r be well behaved for a Coulomb force field. These quantum numbers are interrelated, since $|m_l| \leq l$ and $l \leq (n-1)$. For the hydrogenic atom, the energy of a given state depends only on the value of n, and so most energy levels are highly degenerate.

The hydrogen atom is in one sense the simplest system in the atomic realm. Yet the properties of the atom, the distributions of charge, the various energy levels available, and so on are far from simple when looked at from the perspective of classical understanding. Still we find that by taking a simple second-order differential equation containing only an abstract function ψ and the specific form of the Coulomb potential energy, all of the basic properties of the hydrogen atom result only from the requirement that the solutions of this equation be suitable for use in the description of the physically real world, i.e., that they not become infinite. What does this mean? Certainly it is a remarkable occurrence, but it does not stand alone among many instances where the mathematical imaginations of man's mind have showed fantastically appropriate correspondence with the nature of the world around him. Can it be argued that this is a case where man has impressed the rationality of his own mind upon nature? Or is it perhaps more suggestive that the rationality of man's mind and the structure of the universe have a common source, that God who is the Creator of both?

Problems

3.1 Consider the analogous problem to that discussed in Section 3.2 when the potential outside the well is not infinite but finite, $V_{\text{out}} = V_0$. Obtain the general solution of the Schroedinger equation in each of the three regions ($x < 0$, $0 \leq x \leq L$, $x > L$), eliminate those solutions that are not well-behaved in the region over which they are defined, and obtain the condition for the quantization of energy by applying the boundary conditions requiring continuous joining of the wavefunctions at $x = 0$ and $x = L$.

Problems

Show that the allowed energy states are the roots of the equation

$$\tan\left[L\frac{(2mE)^{1/2}}{\hbar}\right] = \frac{2[E(V_0 - E)]^{1/2}}{2E - V_0}$$

if $E < V_0$. Show that the number of allowed energy states within the well increases with increasing L, and that for very small L there is finally only one (but always at least one) allowed energy state.

3.2 Determine the allowed energies for a particle in a potential well of depth V_0, as in Problem 3.1, when the particle energy $E > V_0$.

3.3 Consider the problem of a traveling particle–wave moving in the $+x$ direction and approaching a step in the potential from $V = 0$ to $V = V_0$. Solve the Schroedinger equation in each of these two regions of space and identify the physical significance of each term in view of the stated problem, as incident, reflected, and transmitted waves. Calculate the reflection coefficient (square of the ratio of reflected to incident amplitudes) for a wave (a) with $E > V_0$, (b) with $E < V_0$.

3.4 If in Problem 3.3, $E > V_0$, show that the transmission coefficient (1 minus the reflection coefficient) is given by

$$T = \frac{4k_1 k_2}{(k_1 + k_2)^2}$$

where $k_1 = (2mE)^{1/2}/\hbar$ and $k_2 = [2m(E - V_0)]^{1/2}/\hbar$. What must the square of the ratio of transmitted to incident amplitudes be multiplied by to obtain this value of transmission coefficient? Why?

3.5 Consider the motion of a traveling wave in the $+x$ direction as it approaches a potential barrier of height $V_0 > E$, and thickness d. Obtain the solutions of the Schroedinger equation in each of the three regions, identify each term physically, and apply the boundary conditions requiring smooth joining of the wavefunctions at $x = 0$ and $x = d$ to permit calculation of the transmission coefficient. Show that if the reflection from the far side of the barrier is neglected at the near side of the barrier, the transmission coefficient is

$$T = 16\left(\frac{E}{V_0}\right)\left[1 - \left(\frac{E}{V_0}\right)\right]\exp\left\{-\frac{2}{\hbar}[2m(V_0 - E)]^{1/2}d\right\}$$

3.6 Calculate the transmission coefficient for the situation of Problem 3.5 if $E > V_0$.

3.7 Calculate the expectation value of the kinetic energy in the ground state of the linear harmonic oscillator.

3.8 Compare the expectation value of r with the most probable value of r in the ground state of the hydrogen atom.

3.9 Calculate the normalized wavefunction for the 4f state of a hydrogenic atom with $m_l = 0$.

3.10 Compare the uncertainty in radius r for the s states of a hydrogen atom corresponding to $n = 1$, 2, and 3.

3.11 Compare the expectation value of the potential energy of the electron in a 2p state of a hydrogen atom with that found in the ground state.

3.12 Neon (atomic number 10) in its neutral state has two 1s electrons, two 2s electrons, and six 2p electrons. The ion Ne^{+9} would consist only of the nucleus and a single electron. If this system were treated as a hydrogenic atom: (a) What is the ionization energy of this last electron? (b) What is the energy of the 3d state? (c) What is the most probable distance of the electron from the nucleus in the 4f state? (d) What would the ionization energy for this last electron be if the neon ion were incorporated in a material with dielectric constant of 10?

3.13 There are ten emission lines in the emission spectrum of the hydrogen atom which correspond to transitions involving only the five lowest energy levels. Draw an energy-level diagram for the hydrogen atom showing these levels, and identify the lines predicted with lines found in spectral tables.

3.14 That choice of assigning Schroedinger operators in which the operator involving simple multiplication by p is associated with the observable p may be called the "p method." (a) Determine the eigenfunctions and eigenvalues of the Schroedinger momentum operator for a free electron using the normal "x method." (b) Determine the eigenfunctions and eigenvalues of the free electron using the p method. (c) Apply the p method to determine the eigenfunctions and eigenvalues of the Hamiltonian operator for the linear harmonic oscillator, proceeding as far as the basic operator equation that must be solved. Given that the ground-state eigenfunction is

$$\psi = A \exp\left[-\frac{p^2}{(2m\hbar\omega)}\right],$$

calculate the ground-state energy. (d) Calculate the expectation value of the momentum in the ground state of a linear harmonic oscillator using the p method.

Chapter 4

Free-Electron Model of Metals

For free atoms with more than one electron, i.e., for all elements except hydrogen, the interaction between electrons must be taken into account, and exact solution of the Schroedinger equation for the allowed electronic energy states is not possible. A variety of approximate methods have been devised, however, which permit the calculation of the allowed energy states for electrons in free atoms. When the additional property of electron *spin* is introduced, to account for specific experimental observations, and when the restrictions of the *Pauli exclusion principle* are noted, it becomes possible to describe the electronic distribution in the elements and the observed order of the Periodic Table. In all of this work, the measurement of the emission from excited atoms as electrons return to lower energy states has played a vital role in guiding the correlation of theory with experiment.

In every atom the electrons are distributed over a set of allowed energy states. In the unexcited neutral atom, lowest energy states are filled up first, then higher states in succession. The outermost electrons are called the *valence* electrons, since it is primarily these electrons that determine the chemical properties of the atom. The valence electrons are the farthest away from the positive charge of the atomic nucleus on the average, and are shielded from that charge by the other nearer electrons in lower energy states.

Our interest in this book is not in free atoms but in atoms combined to form solids. The energy levels of electrons in crystalline solids are determined by the interactions between the many atoms (typically 10^{22} cm^{-3}) that constitute the crystal. Since these atoms are arranged in an orderly array in three-dimensional space known as the crystal *lattice*, the energy levels of electrons in crystals are determined by the existence of a potential function that is periodic in space. The interaction of the electrons with this periodic potential and with each other must be considered in order to derive the allowed energy levels for electrons in crystals.

This chapter represents an approximation to this problem. As a first step we choose simply to ignore the periodic variation of the crystal potential and treat the problem as if the potential were constant on the average. A constant-potential model is equivalent to proposing that the electrons may be treated as if they were free. The practical relevance of the approach comes in its application to metals where the large number (again typically 10^{22} cm^{-3}) of nearly free *valence electrons* makes it possible to describe the allowed energy levels for these electrons in a free-electron approximation. A metal may be pictured as made up of positively charged *ion cores* (consisting of the nucleus and all completed-shell electrons) embedded in a "sea" of nearly free valence electrons with negative charge. The large number of electrons acts to shield a given electron from the positive charges constituting the periodic potential of the crystal lattice. In some metals such as sodium, magnesium, or aluminum, the approximation is reasonably good, and more accurate theoretical treatments show that the valence electrons in these metals have properties very similar to those of free electrons. The concepts and formalism derived in terms of free-electron theory can also be applied to electrons in conducting states of semiconductors and insulators, and so these considerations have a broader overall application than to metals alone.

Consideration of free-electron theory is designed to answer three principal questions:

1. What are the allowed energy levels for free electrons in a crystal?
2. How is the density of these levels (i.e., the number of levels per unit volume of the crystal) distributed with the energy of the levels?
3. Given a distribution of allowed energy levels, what is the probability that a given level is occupied by an electron at a given temperature?

The answers to the first two of these questions can be given in terms of two models. The first of these is the *Sommerfeld model*, which is the application of the three-dimensional "particle in a box" problem, discussed in one dimension in Section 3.2. The second is a *Hartree self-consistent field* calculation using traveling waves and *periodic boundary conditions*, which are more appropriate for describing electrical transport phenomena in metals.

4.1 Atomic Energy Levels and the Periodic Table

The answer to the occupancy of allowed states is given by the *Fermi–Dirac statistics*, which may be derived from a joint consideration of the indistinguishability of electrons and the Pauli exclusion principle.

Although the free-electron approximation is inadequate to describe the properties of a metal in detail, a number of simple physical phenomena can be described to a fair approximation in terms of the free-electron model. Examples of such reasonable successes are given as illustrations of the possibilities inherent in the free-electron picture.

4.1 Atomic Energy Levels and the Periodic Table

Consider a free atom with N electrons, i.e., the element of atomic number N. The potential function for such an atom is a function of the position vectors of all the electrons, and is given by

$$V(\mathbf{r}_1, \mathbf{r}_2, \ldots, \mathbf{r}_N) = -\sum_i^N \frac{Ne^2}{r_i} + \frac{1}{2}\sum_{i \neq j}^N \sum^N \frac{e^2}{r_{ij}} \qquad (4.1)$$

The first term in the potential energy results from the Coulomb interaction between the ith electron and the positive charge of the nucleus Ne, summed over all electrons. The second term is the result of the interaction between the ith electron and the jth electron, summed over all pairs of electrons. It is the interaction term that prevents analytical solution of the Schroedinger equation.

A variety of approximate methods have been developed to treat this kind of problem. The interaction term cannot ever be neglected completely, for even in the helium atom a 40-per-cent error in the energy of the ground state results from such neglect.

1. The *variational method* for the calculation of the ground-state energy of a multielectron atom makes use of the fact that the wavefunction giving the smallest value of the energy in a calculation of the energy using that particular wavefunction is the best wavefunction, and that the corresponding energy is the closest approximation to the true ground-state energy. Operationally a trial wavefunction with arbitrary parameters is chosen, the energy is calculated using this wavefunction, and then a minimization procedure is carried out with respect to these parameters in the trial wavefunction.

2. If the interaction term is small enough, it can be treated as a *perturbation* on the system without the interaction term, and a correction calculated using the eigenfunctions of the unperturbed system.

3. The actual system can be approximated by a system in which it is assumed that a separate Schroedinger equation can be written for each of the electrons in the system, including a generalized potential function to describe the interaction between the one electron and all the others. The wavefunction for the whole system is expressed as a suitable combination of the one-electron wavefunctions. The *Hartree self-consistent field* method is one way to generate the generalized potential functions, neglecting any correlation between electron motion and electron interaction. For such a potential function, a simple product of the one-electron wavefunctions forms the wavefunction for the whole atomic system.

4. When electron spin and the Pauli exclusion principle are taken into account, the best wavefunction is not a simple product of one-electron wavefunctions like the Hartree wavefunction, but is instead a *determinantal* wavefunction constructed so that the total atomic wavefunction is *antisymmetric* when both spatial and spin coordinates are considered.

We return to perturbation calculations for a variety of purposes in later chapters of this book, and leave until then further discussion of these techniques. At this point we summarize very briefly the principal points of statements 3 and 4 above.

HARTREE SELF-CONSISTENT FIELD

In the one-electron approximation the ith electron is described by the wavefunction $\psi_i(\mathbf{r}_i)$ which is a function only of the position vector of the ith electron $\mathbf{r}_i$. For each electron in the multielectron atom, there is a Schroedinger equation,

$$\left[-\frac{\hbar^2}{2m}\nabla_i^2 - \frac{Ne^2}{r_i} + V_i(\mathbf{r}_i)\right]\psi_i(\mathbf{r}_i) = E_i\,\psi_i(\mathbf{r}_i) \qquad (4.2)$$

The generalized potential function $V_i(\mathbf{r}_i)$ is also a function only of the position vector of the ith electron and describes in some average way the average potential field due to the other electrons in the system. Normally $V_i(\mathbf{r}_i)$ is assumed to have spherical symmetry and the one-electron equations can be treated as a central force field problem with radial and angular solutions separable.

Given the eigenfunctions and eigenvalues of the one-electron equations like Eq. (4.2), the total atomic wavefunction is given by

$$\Psi(\mathbf{r}_1, \mathbf{r}_2, \ldots, \mathbf{r}_N) = \psi_1(\mathbf{r}_1)\,\psi_2(\mathbf{r}_2)\cdots\psi_N(\mathbf{r}_N) \qquad (4.3)$$

4.1 Atomic Energy Levels and the Periodic Table

The total energy of the atom is given by

$$E = \frac{\int \cdots \int \Psi^* \mathbf{H} \Psi \, d\mathbf{r}_1 \, d\mathbf{r}_2 \cdots d\mathbf{r}_N}{\int \cdots \int \Psi^* \Psi \, d\mathbf{r}_1 \, d\mathbf{r}_2 \cdots d\mathbf{r}_N} \qquad (4.4)$$

where **H** is the real Hamiltonian for the atom involving the potential function of Eq. (4.1).

The Hartree potential is one example of a choice for the generalized potential function $V_i(\mathbf{r}_i)$. Assuming an exclusively Coulomb interaction between electrons and that the motion of electrons can be treated as if they moved independently in spite of the interaction between them, the Hartree potential takes the form

$$V_i(\mathbf{r}_i) = e^2 \sum_{j \ne i}^{N} \int \frac{|\psi_j(\mathbf{r}_j)|^2}{r_{ij}} \, d\mathbf{r}_j \qquad (4.5)$$

Starting with trial wavefunctions, the procedure is to solve Eqs. (4.2) and (4.5) iteratively until a self-consistent set of wavefunctions is obtained that satisfies both equations.

The change in the total energy of the atomic system caused by removing the ith electron from the atom (i.e., the ionization energy of the ith electron from the neutral atom) is equal to the one-electron eigenvalue E_i given by solution of Eq. (4.2). The total energy of the atomic system is the sum of these one-electron eigenvalues plus a term representing the average interaction energy of the electrons.

Electron Spin

If an electron is classically considered to be moving in a circular orbit in an atom, both an angular momentum and a magnetic moment are associated with this motion. For an electron moving in a orbit of radius r with velocity v in the plane $z = 0$, the angular momentum is $L_z = mrv$. This motion of the electron in its orbit is equivalent to the flow of an electric current, which in turn is equivalent to a magnetic dipole with moment perpendicular to the plane of the loop and magnitude

$$\mu_z = -\frac{e}{2mc} L_z \qquad (4.6)$$

In the wave mechanical model L_z becomes quantized as integral multiples of $\hbar$, and Eq. (4.6) is replaced by

$$\mu_z = -\frac{e\hbar m_l}{2mc} \qquad (4.7)$$

If a magnetic field is applied in the $+z$ direction, the potential energy resulting from the interaction of the electron's orbital magnetic moment with this magnetic field is given by

$$V = -\mu_z H_z = \frac{e\hbar m_l}{2mc} H_z \quad (4.8)$$

The simple picture that takes into account only the orbital angular momentum and the associated orbital magnetic moment of an electron has serious deficiencies when compared with experiment. The fine structure of atomic spectra is unexplained, i.e., the splitting into several components of emission lines predicted to be single. It would also be expected that there would be no effect of a magnetic field on a beam of atoms with total angular momentum of zero [from Eq. (4.8)], but it is observed that beams of atoms with total orbital angular momentum of zero are still split into two beams by an inhomogeneous magnetic field. There must be some other mechanism for interaction between a magnetic field and an atom with zero angular momentum.

To meet these deficiencies in the model, it has been successfully assumed that an electron possesses an *intrinsic* angular momentum quite distinct from its orbital angular momentum. This intrinsic angular momentum exists as if the electron were thought of as "spinning about its own axis," and gives rise to a *spin* magnetic moment. The spin momentum S_z has quantized values of $\pm \hbar/2$, and the spin magnetic moment (called the Bohr magneton) is given by

$$\mu_B = \frac{e\hbar}{2mc} \quad (4.9)$$

In the presence of a magnetic field, the energy of an electron is changed because of its spin by $\pm \mu_B H_z$, the energy being reduced when the spin magnetic moment is lined up in the same direction as the magnetic field.

To incorporate the spin of an electron into its wavefunction in those cases where the Hamiltonian itself does not depend on the spin, it is necessary only to add a *spin quantum number* $m_s = \pm \frac{1}{2}$, so that $S_z = m_s \hbar$, and a *spin variable* ζ, which can be considered pictorially to be the cosine of the angle between the spin axis and the z axis, this angle having allowed values of either zero or π. When the Hamiltonian is spin-independent, the one-electron wavefunction including spin may be written as the simple product of an orbital function and a spin function,

$$\psi_{n,l,m_l,m_s}(\mathbf{r}, \theta, \phi, \zeta) = \psi_{n,l,m_l}(\mathbf{r}, \theta, \phi) \, S(\zeta) \quad (4.10)$$

4.1 Atomic Energy Levels and the Periodic Table

The orbital function is the same for both possible spin functions, and the two spin functions are $S_+(\zeta)$ and $S_-(\zeta)$. When an electron is in the $+\hbar/2$ spin state S_+, its energy is increased by $\mu_B H_z$. When an electron is in the $-\hbar/2$ spin state S_-, its energy is decreased by $\mu_B H_z$. Consistent definition of the spin variable ζ is achieved by setting

$$\begin{cases} S_+(1) = S_-(-1) = 1 \\ S_+(-1) = S_-(1) = 0 \end{cases} \quad (4.11)$$

These definitions assure that the spin functions are normalized and orthogonal.

PAULI EXCLUSION PRINCIPLE

The choice of one-electron wavefunctions in Eq. (4.3) in describing a state of the atom is not arbitrary. Familiarity with the Periodic Table of the elements has shown empirically that no two electrons in a many-electron atom may be characterized by the same set of quantum numbers n, l, m_l, and m_s, i.e., no two electrons may have identical one-electron wavefunctions. As we start with hydrogen, the element with atomic number 1, and add further electrons to obtain the elements with higher atomic number, we find that the first two electrons are 1s electrons, the next higher energy electrons are 2s electrons, then come six 2p electrons, two 3s electrons, six 3p electrons, ten 3d electrons, and so on. The number of electrons that can occupy a state with quantum number l is just the degeneracy including spin of $2(2l + 1)$. Once a state corresponding to a set of four quantum numbers n, l, m_l, and m_s is occupied by one electron, that state is full and can be occupied by no additional electrons. A schematic diagram of the Periodic Table of the elements summarizing this property is given in Table 4.1. Sometimes there are minor irregularities in the filling of the inner shells, but the overall pattern is as indicated. Consider, for example, the electronic configuration of the copper atom (atomic number 29). Experimentally it is known that this configuration is

$$1s^2 \; 2s^2 \; 2p^6 \; 3s^2 \; 3p^6 \; 3d^{10} \; 4s$$

whereas the listing of Table 4.1 might have led us to expect an occupancy of $3d^9 \, 4s^2$ for the outermost electrons. The electrons in the highest-lying energy level (those farthest from the nucleus) are called the *valence* electrons and determine most of the chemical properties of the element. In the case of Cu, the 4s electron is the valence electron, the loss of which in a chemical reaction leads to the Cu^+ valence state of the Cu ion. The

TABLE 4.1 Electronic Filling of States in the Periodic Table of Elements

Row	Column							
	I	II	III	IV	V	VI	VII	VIII
1	1 H 1s							2 He $1s^2$
2	3 Li 2s	4 Be $2s^2$	5 B	6 C	7 N	8 O	9 F	10 Ne
			←——————————— $(2p^6)$ ———————————→					
3	11 Na 3s	12 Mg $3s^2$	13 Al	14 Si	15 P	16 S	17 Cl	18 A
			←——————————— $(3p^6)$ ———————————→					
4	19 K 4s	20 Ca $(^a)$ $4s^2$	31 Ga	32 Ge	33 As	34 Se	35 Br	36 Kr
			←——————————— $(4p^6)$ ———————————→					
5	37 Rb 5s	38 Sr $(^b)$ $5s^2$	49 In	50 Sn	51 Sb	52 Te	53 I	54 Xe
			←——————————— $(5p^6)$ ———————————→					
6	55 Cs 6s	56 Ba $(^c)$ $6s^2$	81 Tl	82 Pb	83 Bi	84 Po	85 At	86 Rn
			←——————————— $(6p^6)$ ———————————→					
7	87 Fr 7s	88 Ra $(^d)$ $7s^2$						

^{a} Transition elements, First Series: Sc, Ti, V, Cr, Mn, Fe, Co, Ni, Cu, Zn. Filling $3d^{10}$.
^{b} Transition elements, Second Series: Y, Zr, Nb, Mo, Tc, Ru, Rh, Pd, Ag, Cd. Filling $4d^{10}$.
^{c} Rare Earth elements: La, Ce, Pr, Nd, Pm, Sm, Eu, Gd, Tb, Dy, Ho, Er, Tm, Yb, Lu. Filling $4f^{14}$. Transition elements, Third Series: Hf, Ta, W, Re, Os, Ir, Pt, Au, Hg. Filling $5d^{10}$.
^{d} Actinide elements: Ac, Th, Pa, U, Np, Pu, Am, Cm, Bk, Cf, Es, Fm, Md, No, Lr. Filling $5f^{14}$.

fact that Cu^{+2} is also a commonly found valence state for Cu arises from the possibility of a Cu atom also giving up a d electron in chemical binding, much as if the initial state of the atom were $3d^9$ $4s^2$ and both s electrons were given up.

The formal statement that no two one-electron wavefunctions in a single atomic system can have identical quantum numbers is known as the *Pauli exclusion principle*. This principle can also be stated in terms of a property of the wavefunction for the whole system: the total wavefunction for a system of electrons must be *antisymmetric* in the generalized coordinates including spin. It is evident that the simple product wavefunction of form

4.1 Atomic Energy Levels and the Periodic Table

of Eq. (4.3), with the spin function added as in Eq. (4.10), is a symmetric wavefunction, i.e., interchange of the coordinates of two electrons produces no change in the sign of the total wavefunction. A total wavefunction which is antisymmetric, i.e., the sign of which does change when the coordinates of two electrons are interchanged, can be constructed from the one-electron wavefunctions as

$$\Psi(\mathbf{q}_1, \mathbf{q}_2, \ldots, \mathbf{q}_N) = \begin{vmatrix} \psi_1(\mathbf{q}_1) & \psi_1(\mathbf{q}_2) & \cdots & \psi_1(\mathbf{q}_N) \\ \psi_2(\mathbf{q}_1) & \psi_2(\mathbf{q}_2) & \cdots & \psi_2(\mathbf{q}_N) \\ \vdots & \vdots & \vdots & \vdots \\ \psi_N(\mathbf{q}_1) & \psi_N(\mathbf{q}_2) & \cdots & \psi_N(\mathbf{q}_N) \end{vmatrix} \quad (4.12)$$

Interchange of two columns, equivalent to the interchange of the generalized coordinates $\mathbf{q}$ (including position and spin), results in a change in sign of the total wavefunction. If two rows of the determinant are identical, i.e., if two one-electron wavefunctions are identical, the value of the total wavefunction is zero.

There are other types of particles, which do not obey the Pauli exclusion principle, and which must be expressed, not as antisymmetric total wavefunctions, but as symmetric total wavefunctions, e.g., photons.

The best one-electron wavefunctions to use in a determinantal wavefunction like that given in Eq. (4.12) are not the one-electron wavefunctions of the Hartree potential given in Eq. (4.5), but are instead the one-electron wavefunctions for what is known as the *Hartree–Fock* potential. This potential includes some contribution of correlation between electronic interaction and motion and gives a better approximation to the total energy than that given by the Hartree potential.

ATOMIC ENERGY LEVELS

The ionization energies of the various electronic states in multielectronic atoms up to Mn (atomic number 25) are given in Fig. 4.1. Energies are plotted in Rydberg units, where 1 rydberg = 13.60 eV, the ground-state energy of the hydrogen atom. Levels plotted for each atom represent the amount of energy necessary to remove an electron from that state into the free state for the neutral atom. The levels are labeled with one-electron designations. Only those levels with binding energy less than 18 rydbergs are shown in Fig. 4.1. Inner-shell binding energies rapidly become large; for Mn, for example, the ionization energy of 1s electrons is 482 rydbergs.

The highest energy level in each atom corresponds to the lowest energy

82   4 Free-Electron Model of Metals

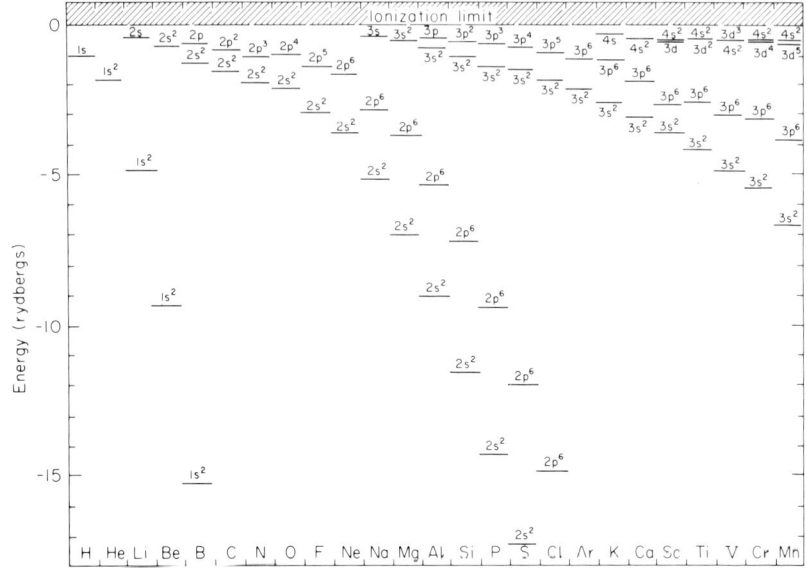

Fig. 4.1 Ionization energies for electrons in the indicated atomic levels for neutral atoms from hydrogen to manganese. (Data from J. C. Slater, "Quantum Theory of Matter," McGraw-Hill, New York, 1951, p. 145.)

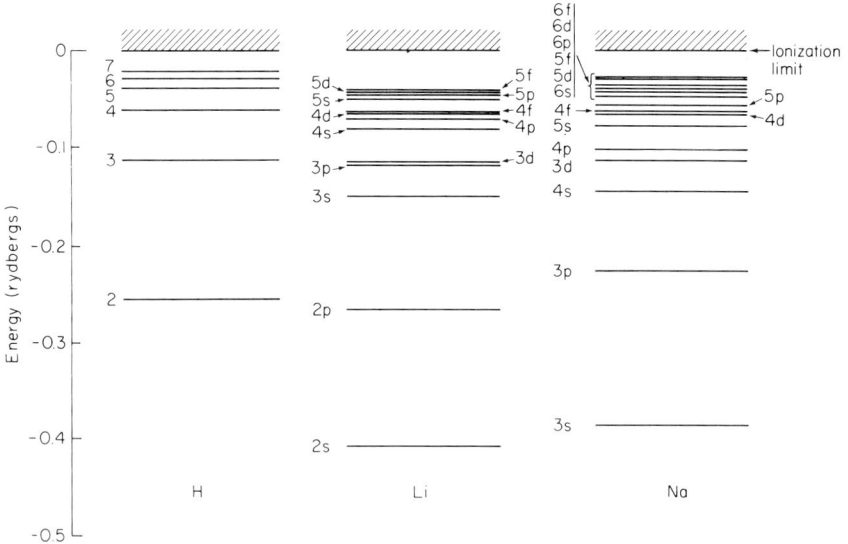

Fig. 4.2 Excited states for s valence electrons in H, Li, and Na. (After H. E. White, "Introduction to Atomic Spectra," McGraw-Hill, New York, 1934, p. 77.)

4.2 The Sommerfeld Free-Electron Model

for the valence electrons of that atom. Higher-energy excited states also exist for these valence electrons between their lowest energy and the ionization limit. Figure 4.2 shows the excited-state energies for three elements with a single s valence electron, neglecting the fine-splitting of levels.

In general, *optical* spectra involve transitions of the valence electron(s) from excited states such as are shown in Fig. 4.2 to the ground state for those electrons. Excitation producing ionization of the states shown in Fig. 4.1 requires x rays.

4.2 The Sommerfeld Free-Electron Model

The model for a metal is an orderly array of positively charged ion cores embedded in a sea of negatively charged free electrons. Any one electron is screened from the field of the ion cores by the presence of all the other electrons, causing a large reduction in potential energy. Sommerfeld (1928) assumed that the total field of electrons and ion cores at any point in the metal could be set approximately equal to zero. He applied to the problem of free electrons in a metal the results of a three-dimensional "particle in a box" calculation.

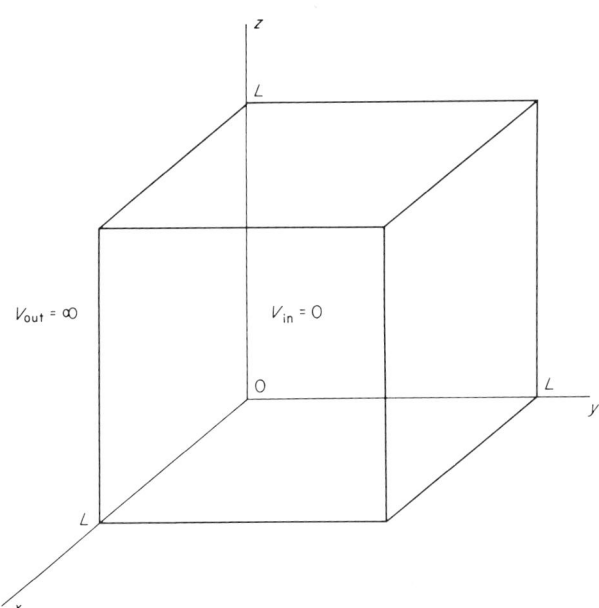

Fig. 4.3 Three-dimensional particle-in-a-box model for a cube of side L. The potential energy is zero inside the box and infinite outside the box.

Such a three-dimensional box is shown in Fig. 4.3, with $V = 0$ inside a cube of side L, and $V = \infty$ outside the box. The eigenfunctions of the Hamiltonian outside the box are identically zero. Inside the box, the eigenfunctions are solutions of the three-dimensional free-electron problem with the appropriate boundary conditions.

Inside the box the Schroedinger equation is

$$\nabla^2 \psi + \frac{2m}{\hbar^2} E\psi = 0 \qquad (4.13)$$

In order to solve this equation in three dimensions, a trial solution involving simple separation of the variables is useful,

$$\psi(x, y, z) = X(x)\, Y(y)\, Z(z) \qquad (4.14)$$

where $X(x)$ is a function only of x, $Y(y)$ is a function only of y, and $Z(z)$ is a function only of z. Substitution of Eq. (4.14) into Eq. (4.13), followed by division through by XYZ, yields

$$\frac{1}{X}\frac{d^2X}{dx^2} + \frac{1}{Y}\frac{d^2Y}{dy^2} + \frac{1}{Z}\frac{d^2Z}{dz^2} + \frac{2m}{\hbar^2}E = 0 \qquad (4.15)$$

Now since each term in this equation is a function of only one of the coordinate variables, each term must separately be equal to a constant if their sum is to be zero. Therefore, in place of Eq. (4.15), we may write three separate equations,

$$\frac{d^2X}{dx^2} + \frac{2m}{\hbar^2}E_x X = 0 \qquad (4.16)$$

$$\frac{d^2Y}{dy^2} + \frac{2m}{\hbar^2}E_y Y = 0 \qquad (4.17)$$

$$\frac{d^2Z}{dz^2} + \frac{2m}{\hbar^2}E_z Z = 0 \qquad (4.18)$$

where

$$E = E_x + E_y + E_z \qquad (4.19)$$

Each of these separated equations is of the same form as the equation for the one-dimensional box, and the solutions with the appropriate boundary conditions are already known. The solution of Eq. (4.16), for example, is

$$X(x) = \left(\frac{2}{L}\right)^{1/2} \sin\left(\frac{n_x \pi x}{L}\right) \qquad (4.20)$$

4.2 The Sommerfeld Free-Electron Model

with the energy eigenvalue

$$E_x = \frac{\hbar^2 \pi^2}{2mL^2} n_x^2 \tag{4.21}$$

and n_x an integer. Since the solutions for $Y(y)$ and $Z(z)$ are similar, the solution of Eq. (4.13) can be written as

$$\psi_{n_x n_y n_z} = \left(\frac{2}{L}\right)^{3/2} \sin\left(\frac{n_x \pi x}{L}\right) \sin\left(\frac{n_y \pi y}{L}\right) \sin\left(\frac{n_z \pi z}{L}\right) \tag{4.22}$$

and the energy given by Eq. (4.19) is

$$E_{n_x n_y n_z} = \frac{\hbar^2 \pi^2}{2mL^2} (n_x^2 + n_y^2 + n_z^2) = \frac{\hbar^2}{2m} |\mathbf{k}|^2 \tag{4.23}$$

since

$$\mathbf{k} = \mathbf{e}_1\left(\frac{n_x \pi}{L}\right) + \mathbf{e}_2\left(\frac{n_y \pi}{L}\right) + \mathbf{e}_3\left(\frac{n_z \pi}{L}\right) \tag{4.24}$$

Since n_x, n_y, and n_z must all be nonzero, the ground-state energy is given by

$$E_{111} = \frac{3\hbar^2 \pi^2}{2mL^2} \tag{4.25}$$

Some of the allowed energy states for a particle in a three-dimensional box are nondegenerate, e.g.,

$$(n_x = 1, n_y = 1, n_z = 1) \quad E_{111} = \frac{3\hbar^2 \pi^2}{2mL^2}$$

$$(n_x = 2, n_y = 2, n_z = 2) \quad E_{222} = \frac{6\hbar^2 \pi^2}{mL^2}$$

but most of the states are degenerate. For example,

$$E_{n_x n_y n_z} = 41 \frac{\hbar^2 \pi^2}{2mL^2}$$

is ninefold degenerate, corresponding to (n_x, n_y, n_z) triplets of (1, 2, 6), (1, 6, 2), (2, 1, 6), (2, 6, 1), (6, 2, 1), (6, 1, 2), (4, 3, 4), (4, 4, 3), and (3, 4, 4). Each of these triplets corresponds to a different *state*, i.e., to a different configuration in space of the probability density, but all have the same energy and correspond to a single *energy level*.

Example 4.1 Plot an energy-level diagram for all the energy states of a particle in a three-dimensional cubic box with

$$E \leq 30 \frac{\hbar^2 \pi^2}{2mL^2}$$

and list the degeneracy of each.

We need consider only the various combinations of integers that may be substituted for n_x, n_y, and n_z in Eq. (4.23). The results are shown in Fig. 4.4.

Consideration of the actual magnitude of the energy values E is instructive

$$E_{n_x n_y n_z} = \frac{3.70 \times 10^{-15}}{L^2} (n_x^2 + n_y^2 + n_z^2) \text{ eV}$$

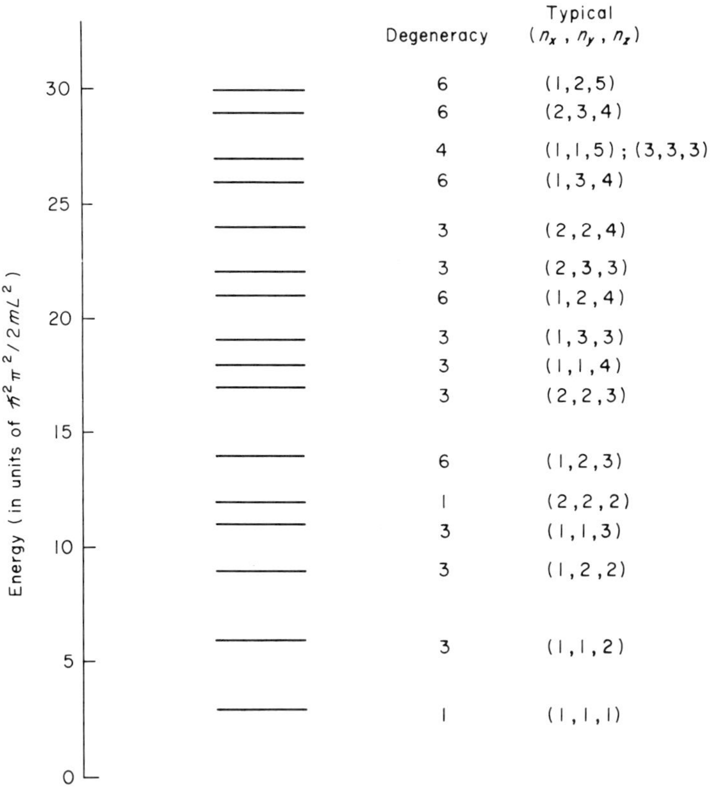

Energy (in units of $\hbar^2\pi^2/2mL^2$)	Degeneracy	Typical (n_x, n_y, n_z)
30	6	(1,2,5)
	6	(2,3,4)
	4	(1,1,5); (3,3,3)
	6	(1,3,4)
25	3	(2,2,4)
	3	(2,3,3)
	6	(1,2,4)
20	3	(1,3,3)
	3	(1,1,4)
	3	(2,2,3)
15	6	(1,2,3)
	1	(2,2,2)
	3	(1,1,3)
10	3	(1,2,2)
5	3	(1,1,2)
	1	(1,1,1)
0		

Fig. 4.4 The lowest sixteen energy levels for a particle in a three-dimensional box, together with their degeneracy and typical quantum-number sets.

4.2 The Sommerfeld Free-Electron Model

This means that for a cube 1 cm on a side, we can for all practical purposes consider the ground-state energy to be essentially zero, and we can treat the allowed energy levels as a quasi-continuous classical-like distribution of states, rather than as a series of discrete states. For a cube 10 Å on a side, however, the ground state energy $E_{111} = 1.1$ eV and the spacing between levels is still larger. Thus we have one more example of the importance of the wave properties of matter at the atomic and subatomic level, approaching the classical view of matter as the confining dimensions increase.

Upon the assumption of a quasi-continuous distribution of energy levels with energy, we proceed to the evaluation of this distribution. Define

$N(E)\,dE \equiv$ number of orbital states (half the total number of states) with energy between E and $E + dE$

$\mathcal{N}(E) \equiv$ number of orbital states with energy less than E

The density of states $N(E)$ is a quantity of basic physical significance, and enters into almost every calculation of the electronic properties of solids.

In order to calculate $N(E)$, note that

$$\mathcal{N}(E) = \int_0^E N(E)\,dE \tag{4.26}$$

so that

$$N(E) = \frac{d\mathcal{N}(E)}{dE} \tag{4.27}$$

If the quantity $\mathcal{N}(E)$ can be evaluated by counting allowed states, then $N(E)$ can also be evaluated.

$\mathcal{N}(E)$ can be evaluated by rewriting the energy given in Eq. (4.23) as

$$n_x^2 + n_y^2 + n_z^2 = \left(\frac{2mL^2}{\hbar^2\pi^2}E_{n_x n_y n_z}\right) \equiv R^2 \tag{4.28}$$

If Eq. (4.28) is plotted in n space, i.e., in three coordinate axes defined by n_x, n_y, and n_z, it is the equation of a sphere with radius R. Figure 4.5 shows the positive octant of such a sphere, which is the only portion that need be considered since only positive values of n_x, n_y, and n_z are significant in Eq. (4.28). The number of states $\mathcal{N}(E)$ with energy less than E is given by the number of representative (n_x, n_y, n_z) points lying within this positive octant with radius R corresponding to the E chosen. If the sphere is divided

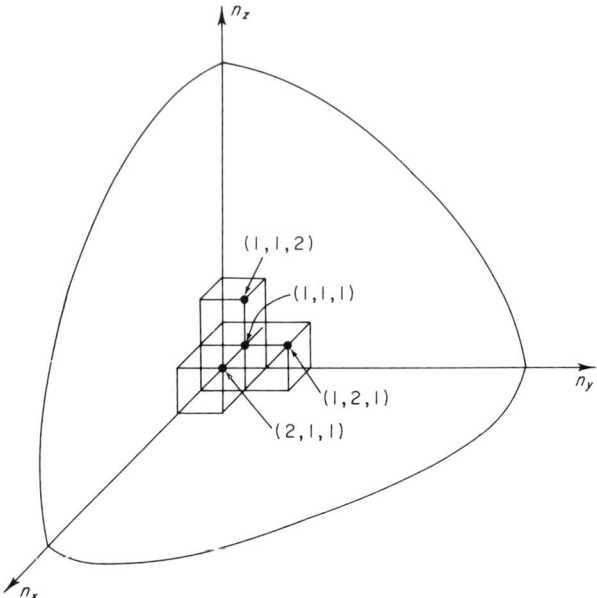

Fig. 4.5 Unit cubes in n space, illustrating the fact that there is one representative point (n_x, n_y, n_z) per unit cube.

up into unit cubic cells, each representative point (n_x, n_y, n_z) lies at the corner of one of these unit cells. Therefore there is just one representative point per unit volume, and the desired value of $\mathcal{N}(E)$ is simply the volume of the octant.

The value of $\mathcal{N}(E)$ is

$$\mathcal{N}(E) = \frac{1}{8} \frac{4\pi}{3} \left(\frac{2mL^2}{\hbar^2\pi^2} E\right)^{3/2}$$

$$= \frac{L^3}{6\pi^2} \left(\frac{2m}{\hbar^2}\right)^{3/2} E^{3/2} \qquad (4.29)$$

Therefore the density of states is given by Eq. (4.27) to be

$$N(E) = \frac{L^3}{4\pi^2} \left(\frac{2m}{\hbar^2}\right)^{3/2} E^{1/2} \qquad (4.30)$$

$N(E)$ is the density of states in a volume L^3, which is just the volume of the metal cube. The density of states per unit volume is therefore

$$N_\mathrm{v}(E) = \frac{1}{4\pi^2} \left(\frac{2m}{\hbar^2}\right)^{3/2} E^{1/2} \qquad (4.31)$$

4.2 The Sommerfeld Free-Electron Model

and the total density of states $N_v^t(E) = 2N_v(E)$ to include the effects of spin. The number of states per unit volume per energy interval increases with the energy of the state, quantitatively as the square-root of this energy.

A representative plot of $N(E)$ versus E is given in Fig. 4.6. At a temperature of absolute zero, it is reasonable to believe (to be confirmed later) that the electrons in the system fill all the lowest energy states up to some energy E_F, the *Fermi energy*. Each state is occupied by two electrons with opposite spin. The total number of free electrons in the metal determines the actual value of E_F, since

$$N_e = \int_0^{E_F} N^t(E)\, dE = \frac{\mathcal{V}}{3\pi^2} \left(\frac{2m}{\hbar^2}\right)^{3/2} E_F^{3/2} \qquad (4.32)$$

where N_e is the number of free electrons in a volume of metal $\mathcal{V}$. Solving for E_F gives

$$E_F = \frac{\hbar^2}{2m} \left(\frac{3\pi^2 N_e}{\mathcal{V}}\right)^{2/3}$$

$$= \frac{\hbar^2}{2m} (3\pi^2 n)^{2/3} \qquad (4.33)$$

where n is the density of free electrons, i.e., the number per cm^3.

The total kinetic energy of all the free electrons in the metal can be expressed directly in terms of E_F.

$$E = \int_0^{E_F} N^t(E)\, E\, dE = \frac{\mathcal{V}}{5\pi^2} \left(\frac{2m}{\hbar^2}\right)^{3/2} E_F^{5/2}$$

$$= \tfrac{3}{5} N_e E_F \qquad (4.34)$$

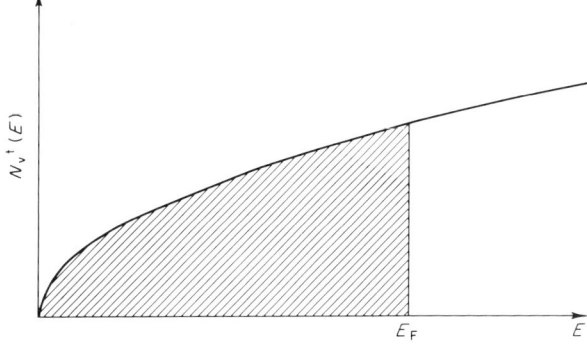

Fig. 4.6 Total density of states per unit volume as a function of energy E for the free-electron model. At absolute zero all states up to the Fermi energy E_F are occupied by electrons.

In the free-electron model the total energy is all kinetic energy. The average kinetic energy per electron is then $3E_F/5$.

Example 4.2 How many electrons per unit volume occupy energy states within 0.5 eV of the Fermi energy in metallic sodium at 0°K?

In metallic sodium each atom gives up one free electron. Since the atomic weight of sodium is 22.99 g, there are 6.023×10^{23} atoms of Na in 22.99 g, or 1 g of Na contains 2.62×10^{22} atoms. Since the density of Na is 0.97 g/cm³, there are 2.7×10^{22} atoms of Na in 1 cm³. Therefore $n = 2.7 \times 10^{22}$ cm⁻³ in Eq. (4.33), and the calculation of the Fermi energy can be completed to give $E_F = 3.34$ eV.

The number of electrons per unit volume within 0.5 eV of the Fermi energy at 0°K is therefore given by

$$N' = \int_{2.84 \text{ eV}}^{3.34 \text{ eV}} \frac{1}{2\pi^2} \left(\frac{2m}{\hbar^2}\right)^{3/2} E^{1/2} \, dE$$

$$= \frac{1}{3\pi^2} \left(\frac{2m}{\hbar^2}\right)^{3/2} E^{3/2} \Big]_{2.84 \text{ eV}}^{3.34 \text{ eV}}$$

Evaluation of this equation gives $N' = 5.8 \times 10^{21}$ cm⁻³, i.e., because of the distribution of density of states with energy, about one fifth of the total number of electrons per unit volume is found in that one seventh of the energy interval (from zero up to the Fermi energy) nearest to the Fermi energy.

4.3 Traveling Waves and Periodic Boundary Conditions

The Sommerfeld model for free electrons in a metal gives eigenfunctions that are standing waves, as indicated in Eq. (4.22). They represent the state of motion of an electron physically bouncing back and forth between the ends of the box. The expectation value for the momentum, $\langle p_x \rangle$, is zero for wavefunctions of this type.

The treatment of problems involving the transport of charge, such as electrical conductivity, requires a wavefunction description that portrays the net motion of electrons in a given direction. A *traveling wave* representation is indicated.

A traveling wave for a free particle in one dimension is represented by

$$\psi = \exp\left[\frac{i(2mE)^{1/2}x}{\hbar}\right] = e^{ikx} \quad (4.35)$$

4.3 Traveling Waves and Periodic Boundary Conditions

[see Eq. (3.4) and Section 3.1]. Can such a traveling wave be used as a solution to the Schroedinger equation for a potential box? If the wavefunction of Eq. (4.35) is substituted into

$$\frac{d^2\psi}{dx^2} + \frac{2m}{\hbar^2} E\psi = 0$$

it is found as expected that the traveling wavefunction is an eigenfunction corresponding to the eigenvalue

$$E = \frac{\hbar^2}{2m} k^2$$

But now if the boundary conditions of limitation to a box are imposed, i.e., $\psi = 0$ at $x = 0$ and $x = L$, it is evident that the traveling wave does not satisfy these.

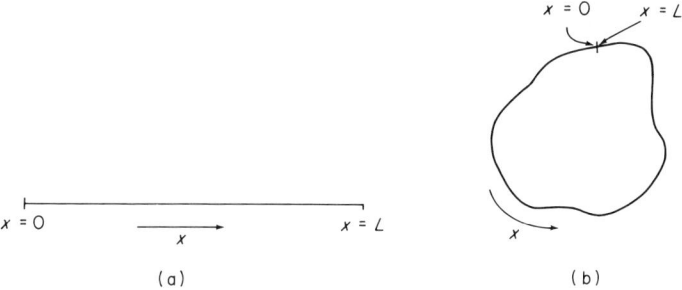

Fig. 4.7 (a) One-dimensional metal extending from $x = 0$ to $x = L$. (b) The metal of (a) bent into a loop joining the points $x = 0$ and $x = L$ in such a way as to simulate a closed electric circuit.

On the other hand, the picture of an isolated chunk of metal also does not conform to the physical situation when transport properties are normally considered. The flow of an electric current requires a closed circuit. Consider the one-dimensional metal of our discussion to be a wire, which needs to be bent into the form of a loop to obtain a closed circuit, as shown in Fig. 4.7. If we continue to measure x along the wire, the point $x = 0$ is made to coincide with the point $x = L$. In this framework, then, it is no longer required that $\psi = 0$ at $x = 0$ and $x = L$, but rather that

$$\psi(0) = \psi(L) \tag{4.36}$$

and that

$$\frac{d\psi}{dx}\bigg]_{x=0} = \frac{d\psi}{dx}\bigg]_{x=L} \tag{4.37}$$

By this procedure the infinite potential walls of the box have been effectively eliminated, and in place of the original boundary conditions we have a *periodic boundary condition,*

$$\psi(x + L) = \psi(x) \tag{4.38}$$

What justification is there for this apparently artificial and ad hoc definition of new boundary conditions? In addition to the attempt to offer a rationale already given, there is not much more to offer. The ultimate justification is the usual pragmatic one of science: "it works." Since it does "work," i.e., lead to predictions that may be confirmed by experiment, it is assumed that it satisfactorily describes the actual situation to the extent of our demands. The major test that we shall apply is whether or not it gives a density-of-states distribution consistent with experiment.

Application of the periodic boundary condition to the traveling wavefunction of Eq. (4.35) yields

$$1 = e^{ikL} \tag{4.39}$$

which requires that

$$k = \frac{2\pi n}{L}, \quad n = 0, \pm 1, \pm 2, \ldots \tag{4.40}$$

The corresponding energy is

$$E_k = \frac{\hbar^2}{2m} k^2 = \frac{\hbar^2 \pi^2}{2mL^2} (2n)^2 \tag{4.41}$$

This expression for the energy may be compared with that previously obtained from the Sommerfeld model,

$$E_n = \frac{\hbar^2 \pi^2}{2mL^2} n^2 \tag{4.42}$$

The effect of using periodic boundary conditions with a traveling wave is to reduce the total number of energy levels by a factor of 2. The density of states, however, is exactly the same for E_k and for E_n. Only positive values of n are meaningful for E_n, but there are distinguishable E_k states for both $+n$ and $-n$, corresponding to waves moving in opposite directions. The reduction in the total number of energy levels caused by using periodic boundary conditions can be effectively neglected since the density of states

4.4 Hartree Model for Free Electrons in a Metal

is so large that a quasi-continuous distribution can be assumed. Under these conditions, only the density of states is physically significant.

4.4 Hartree Model for Free Electrons in a Metal

The Sommerfeld model ignores both the electron–ion-core potential energies and the electron–electron potential energies. The Hartree model for free electrons in a metal attempts to describe the situation somewhat more realistically.

If the free electrons in a metal, with total charge $-eN_e$, are considered as essentially a free-electron gas, the effect of the mutual Coulomb electron–electron repulsion will be to disperse the gas. Physically the gas does not disperse because of the electron–ion-core attractions due to the positive charge of the ion cores. The Hartree model shows that if the total positive charge needed to prevent the electrons from dispersing, $+eN_e$, is considered to be distributed uniformly with a charge density $+eN_e/L^3$ in a metallic cube with side L, then the average electronic charge density is also uniform and given by $-eN_e/L^3$, so that the result is that the net one-electron field is simply zero.

The formal procedure of the calculation can be summarized as follows.

1. Start with a crystal having a uniform positive charge density of $+eN_e/L^3$.

2. Assume a Hartree one-electron potential that just cancels the one-electron potential due to the positive charge.

3. Solve the Hartree equation for the negative charge density in the crystal to see whether the assumption of a net field of zero is self-consistent.

The Hartree one-electron equation is

$$\left[-\frac{\hbar^2}{2m}\nabla^2 - e\left(+\frac{eN_e}{L^3}\right) - e\left(-\frac{eN_e}{L^3}\right)\right]\psi = E\psi \qquad (4.43)$$

which has the traveling-wave solution,

$$\psi_\mathbf{k}(\mathbf{r}) = A\, e^{i\mathbf{k}\cdot\mathbf{r}} \qquad (4.44)$$

the three-dimensional equivalent of Eq. (4.35). Normalization of this wavefunction gives

$$\int \psi_\mathbf{k}^*(\mathbf{r})\, \psi_\mathbf{k}(\mathbf{r})\, d\mathbf{r} = A^2 L^3 \qquad (4.45)$$

so that

$$\psi_{\mathbf{k}}(\mathbf{r}) = \frac{1}{L^{3/2}} e^{i\mathbf{k}\cdot\mathbf{r}} \tag{4.46}$$

Therefore the average negative charge density due to the electrons is

$$-eN_e |\psi_{\mathbf{k}}(\mathbf{r})|^2 = -\frac{eN_e}{L^3} \tag{4.47}$$

This is a self-consistent solution, since the acceptance of Eq. (4.47), combined with the initial assumption of a uniform positive charge density of $+eN_e/L^3$, leads to the conclusion that the net field is zero, as assumed in Eq. (4.43).

The application of periodic boundary conditions to the wavefunction of Eq. (4.44) requires that

$$\psi_{\mathbf{k}}(x+L, y, z) = \psi_{\mathbf{k}}(x, y+L, z) = \psi_{\mathbf{k}}(x, y, z+L) = \psi_{\mathbf{k}}(x, y, z) \tag{4.48}$$

which means that

$$e^{ik_x L} = e^{ik_y L} = e^{ik_z L} = 1 \tag{4.49}$$

since $\mathbf{k} = \mathbf{e}_1 k_x + \mathbf{e}_2 k_y + \mathbf{e}_3 k_z$. To satisfy Eq. (4.49),

$$k_j = \frac{2\pi n_j}{L} \quad \text{for } j = x, y, z \tag{4.50}$$

The energy levels for the Hartree model are

$$\begin{aligned} E(\mathbf{k}) &= \frac{\hbar^2}{2m} \frac{4\pi^2}{L^2} (n_x^2 + n_y^2 + n_z^2) \\ &= \frac{\hbar^2}{2m} \frac{\pi^2}{L^2} [(2n_x)^2 + (2n_y)^2 + (2n_z)^2] \end{aligned} \tag{4.51}$$

These may be compared with the energy levels predicted by the Sommerfeld model using standing wavefunctions,

$$E_{n_x n_y n_z} = \frac{\hbar^2}{2m} \frac{\pi^2}{L^2} (n_x^2 + n_y^2 + n_z^2) \tag{4.52}$$

It is seen that only one out of each eight Sommerfeld levels remains in the Hartree model, but each of these remaining levels is eight times as degenerate as in the Sommerfeld model. The result is that the density of states is the same in both models. This conclusion can be made more easily discern-

4.4 Hartree Model for Free Electrons in a Metal

ible by the following comparison for the calculation of $\mathcal{N}(E)$ in both models:

Sommerfeld model with standing waves and geometric boundary conditions

$$\mathcal{N}(E) = \left(\frac{1}{8}\right)\left[\frac{4\pi}{3}\left(\frac{2mL^2E}{\hbar^2\pi^2}\right)^{3/2}\right](1) \quad (4.53)$$

Hartree model with traveling waves and periodic boundary conditions

$$\mathcal{N}(E) = (1)\left[\frac{4\pi}{3}\left(\frac{2mL^2E}{\hbar^2\pi^2}\right)^{3/2}\right]\left(\frac{1}{8}\right) \quad (4.54)$$

The first term in the products of these two equations is the fraction of the sphere in n space to be included in the integration; for the Hartree model the total volume of the sphere is included since both $+n$ and $-n$ are significant, whereas in the Sommerfeld model only the positive octant is included. The third term in the products of these two equations is the reciprocal of the volume of a unit cube in n space; for the Hartree model the number of states with energy less than E is one eighth that of the Sommerfeld model since the unit cube in the Hartree case has a volume eight times that in the Sommerfeld.

RELATION OF k SPACE AND n SPACE

Further discussion of energy levels of electrons in crystals is done largely in the framework of **k** space. The correlation between **k** space and the n space of the Sommerfeld model is given by Eq. (4.50). Points in **k** space with coordinates $2\pi n_j/L$ represent orbital states. A cube of side $2\pi/L$ in **k** space contains one orbital state. Therefore the number of states per unit volume in **k** space is just $(2\pi/L)^{-3} = L^3/8\pi^3 = \mathcal{V}/8\pi^3$. (The number of states per unit volume in n space is simply 1.)

Surfaces of constant energy given by

$$E(\mathbf{k}) = \frac{\hbar^2}{2m} k^2$$

are spheres in **k** space with center at the origin. At $T = 0°K$, therefore, the occupied region of **k** space is a sphere, called the *Fermi sphere*, with radius k_F. The total number of electrons in this sphere is

$$N_e = 2 \frac{\mathcal{V}}{8\pi^3} \frac{4\pi}{3} k_F^3 \quad (4.55)$$

96   4   Free-Electron Model of Metals

Solving for k_F gives

$$k_F = \left(\frac{3\pi^2 N_e}{\mathscr{V}}\right)^{1/3} \tag{4.56}$$

for which the corresponding energy is

$$E(k_F) = \frac{\hbar^2}{2m} k_F^2 = \frac{\hbar^2}{2m}\left(\frac{3\pi^2 N_e}{\mathscr{V}}\right)^{2/3} = E_F \tag{4.57}$$

in agreement with Eq. (4.33), the result of the analogous calculation in terms of n space.

GENERAL EXPRESSION FOR DENSITY OF STATES FOR SPHERICAL EQUAL-ENERGY SURFACES

A general expression for the density of states for all cases of spherical equal-energy surfaces in **k** space can be derived, i.e., for all cases where the energy depends only on the magnitude of **k**. The number of orbital states with energy less than E is

$$\mathscr{N}(E) = \frac{\mathscr{V}}{8\pi^3}\frac{4\pi}{3}k^3 = \frac{\mathscr{V}}{6\pi^2}k^3 \tag{4.58}$$

The density of states $N(E)$ is therefore given by

$$N(E) = \frac{d\mathscr{N}(E)}{dE} = \frac{d\mathscr{N}(E)}{dk}\frac{dk}{dE} = \frac{\mathscr{V}k^2}{2\pi^2}\left(\frac{dE}{dk}\right)^{-1} \tag{4.59}$$

Example 4.3 Apply the general expression of Eq. (4.59) to the Hartree model.

For the Hartree model

$$E = \frac{\hbar^2}{2m}k^2 \quad \text{and} \quad \frac{dE}{dk} = \frac{\hbar^2}{m}k$$

Therefore

$$N(E) = \frac{\mathscr{V}}{2\pi^2}\frac{m}{\hbar^2}k = \frac{\mathscr{V}}{4\pi^2}\left(\frac{2m}{\hbar^2}\right)^{3/2}E^{1/2}$$

in agreement with Eq. (4.30).

Example 4.4 Calculate the energy dependence of the density of states for (a) a two-dimensional metal, and (b) a one-dimensional metal.

4.4 Hartree Model for Free Electrons in a Metal

Two-Dimensional Metal. The calculation of the density of states may be made using either the Sommerfeld approach or the Hartree approach. Both procedures should give the same density of states.

The Sommerfeld model for a two-dimensional metal gives

$$n_x^2 + n_y^2 = \frac{2mL^2}{\hbar^2\pi^2} E = R^2$$

This is the equation of a circle. The number of states with energy less than E, $\mathcal{N}(E)$, is given by the area of the positive quadrant of the circle,

$$\mathcal{N}(E) = \frac{\pi}{4} \frac{2mL^2}{\hbar^2\pi^2} E$$

so that

$$N(E) = \frac{d\mathcal{N}(E)}{dE} = \frac{mL^2}{2\pi\hbar^2}$$

The density of states in a two-dimensional metal is independent of energy. This is quite different from the result for a three-dimensional metal where the density of states increases as $E^{1/2}$.

In the generalized Hartree approach of Eq. (4.59), the number of states with energy less than E in two dimensions is

$$\mathcal{N}(E) = \frac{L^2}{4\pi^2} \pi k^2 = \frac{L^2}{4\pi} k^2$$

The density of states is therefore

$$N(E) = \frac{L^2 k}{2\pi} \frac{m}{\hbar^2 k} = \frac{mL^2}{2\pi\hbar^2}$$

which is the same as for the Sommerfeld approach, as expected.

One-Dimensional Metal. According to the Sommerfeld model,

$$n^2 = \frac{2mL^2}{\hbar^2\pi^2} E = l^2$$

The number of states with energy less than E is simply the length l,

$$\mathcal{N}(E) = (2m)^{1/2} \frac{L}{\hbar\pi} E^{1/2}$$

The density of states is

$$N(E) = \frac{d\mathcal{N}(E)}{dE} = (m/2)^{1/2} \frac{L}{\hbar\pi} E^{-1/2}$$

For a one-dimensional metal, the density of states is infinite for $E = 0$, and then decreases as $E^{-1/2}$.

In the generalized Hartree expression,

$$\mathcal{N}(E) = 2 \frac{L}{2\pi} k$$

Then

$$N(E) = \frac{L}{\pi} \frac{m}{\hbar^2 k} = \left(\frac{m}{2}\right)^{1/2} \frac{L}{\hbar\pi} E^{-1/2}$$

4.5 Occupancy of Allowed Energy Levels for Free Electrons in a Metal

In the introduction to this chapter, we posed three questions that are of principal concern. Two of these have now been answered sufficiently for us to continue on to the third. The allowed energy levels and the density-of-states distribution of these energy levels with energy have been discussed in terms of the Sommerfeld and the Hartree models.

Given a distribution of the density of available electronic energy states, and given a number of electrons N_e free in the metal, how are these electrons distributed among the available states? One particular answer to this problem has already been assumed without comment in Fig. 4.6, when it was stated that at $T = 0°K$ all available states up to some energy E_F are filled, and that no states above E_F are filled. The probability of a given state being occupied, however, would be expected to be a function of temperature, and in general it would be expected that there would be more electrons in higher-energy states above E_F the higher the temperature.

QUANTUM STATISTICS

The problem of quantum statistics is to calculate the probability distribution for the occupancy of allowed energy states by a given number of particles. In order to do this, it is *postulated* that every physically distinct distribution (as characterized by a unique wavefunction) of N particles among various energy states is equally likely to occur. The *relative probability* of any given distribution of the N particles among the various energy states is proportional to the number of distinguishable ways in which such a distribution can be constructed. The *most probable* distribution corresponds to maximizing this number of distinguishable ways subject to the constraints of a fixed total number of particles and a fixed total energy for the particles.

In order to calculate the number of distinguishable ways in which a

4.5 Occupancy of Levels for Free Electrons in a Metal

given distribution can be constructed, it is convenient to divide the entire energy range into intervals frequently called *cells*, $\Delta E_1, \Delta E_2, \ldots, \Delta E_i, \ldots$. The range of each energy cell is chosen to be small compared to the error likely in measuring the energy, but large enough so that each energy cell contains many energy states; suppose that there are G_i energy states in the ith energy cell. Finally, suppose also that there are n_i particles with energy between E_i and $E_i + \Delta E_i$.

Since the final probability distribution function depends solely on the number of distinguishable ways in which a given distribution can be constructed, the probability distribution is determined ultimately by the nature of the particles and the energy states that define what "distinguishable ways" means. There are three major categories, which can be summarized as follows.

1. Particles that are identical but distinguishable. These are classical particles for which a total system wavefunction is adequately represented by a simple product of one-particle wavefunctions. The most probable distribution is the Maxwell–Boltzmann distribution.

2. Particles that are identical but indistinguishable, that do not obey the Pauli exclusion principle. These are particles such as phonons and photons for which a total system wavefunction is a symmetric linear combination of simple products of one-particle wavefunctions. The most probable distribution is the Bose–Einstein distribution.

3. Particles that are identical and indistinguishable, but that do obey the Pauli exclusion principle. These are particles such as electrons for which a total system wavefunction is an antisymmetric linear combination of simple products of one-particle wavefunctions. The most probable distribution is the Fermi–Dirac distribution.

The setting up of the number of distinguishable ways in which a given distribution can be constructed for each of these cases is summarized in Table 4.2. Here we consider in detail only the particular case of the Fermi–Dirac distribution since it is the one appropriate for electrons.

Fermi–Dirac Statistics

The Pauli exclusion principle states that the occupancy of a one-electron energy state must be either zero or one. Consider for example the following G_i energy states in the ith energy cell:

State	a	b	c	d	e	...
Occupancy	0	0	1	0	1	...

TABLE 4.2 Summary of Probability Distributions for Occupancy of Allowed Energy States

Properties of particles	Identical and distinguishable	Identical and indistinguishable; obey Pauli principle	Identical and indistinguishable; do not obey Pauli principle
System wavefunction	$\Psi = \prod_i \psi_i(\mathbf{r}_i)$	$\Psi = \sum_P (-1)^P P \prod_i \psi_i(\mathbf{r}_i)$	$\Psi = \sum_P P \prod_i \psi_i(\mathbf{r}_i)$
Typical particles	Classical particles	Electrons, protons	Phonons, photons
Factors in counting number of distinguishable ways of constructing given distribution	Number of different ways of selecting the number of particles to be put into each cell out of the total number $= N!/n_i!$	Number of distinguishable arrangements of n_i particles among G_i states, one to a state $= G_i!/(G_i - n_i)!$	Number of ways of placing n_i distinguishable particles into G_i energy states, with any number of particles allowed in each state $= (n_i + G_i - 1)!$
	Number of different ways in which n_i particles can be arranged in G_i energy states of that cell $= G_i^{n_i}$	Permutations of n_i particles among themselves $= n_i!$	Permutations of n_i particles and ways of defining the G_i cells among themselves $= n_i!(G_i - 1)!$
Total number of distinguishable ways of constructing given distribution	$\prod_i \frac{N!}{n_i!} G_i^{n_i}$	$\prod_i \frac{G_i!}{(G_i - n_i)!} \frac{1}{n_i!}$	$\prod_i \frac{(n_i + G_i - 1)!}{n_i!(G_i - 1)!}$
Most probable distribution	$n_i = \frac{G_i}{e^{(\alpha + \beta E_i)}}$	$n_i = \frac{G_i}{e^{(\alpha + \beta E_i)} + 1}$	$n_i = \frac{G_i}{e^{(\alpha + \beta E_i)} - 1}$
Name of most probable distribution	Maxwell–Boltzmann	Fermi–Dirac	Bose–Einstein
Values of the constants	$\alpha = \ln\left[\frac{1}{2} \frac{4\pi\mathscr{V}(2m^3\pi)^{1/2}}{h^3 N}(kT)^{3/2}\right]$ $\beta = \frac{1}{kT}$	$\alpha = -\frac{E_F}{kT}$ $\beta = \frac{1}{kT}$	$\alpha = 0$ $\beta = \frac{1}{kT}$

4.5 Occupancy of Levels for Free Electrons in a Metal

There are $G_i!$ ways of permuting the G_i states, but we do not wish to count ways in which the only difference is a permutation of the n_i occupied or the $(G_i - n_i)$ unoccupied states, since electrons are indistinguishable. For example, the permutations

$$\begin{array}{ccccc} a & b & c & d & e \\ a & b & e & d & c \\ b & a & c & d & e \end{array}$$

for the first five states given above, are all indistinguishable, whereas

$$\begin{array}{ccccc} a & c & b & d & e \end{array}$$

is distinguishable from the above. Since there are $n_i!$ ways of permuting the n_i occupied states, and $(G_i - n_i)!$ ways of permuting the $(G_i - n_i)$ unoccupied states, the number of distinguishable permutations is given by

$$w_i = \frac{G_i!}{n_i!(G_i - n_i)!} \tag{4.60}$$

A similar situation holds for all the other cells covering the whole energy range, so that the total number of distinguishable permutations for the whole system is given by

$$W = \prod_i w_i = \prod_i \frac{G_i!}{n_i!(G_i - n_i)!} \tag{4.61}$$

We desire now to maximize W with respect to n_i within the limits set by the two basic constraints on the system,

$$\sum_i n_i = N_e \tag{4.62}$$

$$\sum_i n_i E_i = E \tag{4.63}$$

i.e., conservation of the total number of electrons N_e and of the total energy E. Because of the product form of W, it proves more convenient to work with the $\ln W$,

$$\ln W = \sum_i [\ln G_i! - \ln(G_i - n_i)! - \ln n_i!] \tag{4.64}$$

Since we are dealing with large numbers we may use Stirling's approximation for the factorial,

$$\ln n! = n \ln n - n \tag{4.65}$$

Substitution of Eq. (4.65) into Eq. (4.64) yields, after cancellation of terms,

$$\ln W = \sum_i [G_i \ln G_i - (G_i - n_i) \ln(G_i - n_i) - n_i \ln n_i] \quad (4.66)$$

In order to maximize $\ln W$ subject to the constraints of Eqs. (4.62) and (4.63), we use the method of Lagrangian multipliers.[†]

$$\frac{\partial}{\partial n_j} \left[\ln W + \alpha \left(N_e - \sum_i n_i \right) + \beta \left(E - \sum_i n_i E_i \right) \right] = 0 \quad (4.67)$$

Substituting $\ln W$ from Eq. (4.66) and performing the differentiation gives, after cancellation of terms,

$$\ln(G_i - n_i) - \ln n_i - \alpha - \beta E_i = 0 \quad (4.68)$$

which may be rewritten as

$$n_i = \frac{G_i}{e^{(\alpha + \beta E_i)} + 1} \quad (4.69)$$

The expression given in Eq. (4.69) is the distribution function that we set out to calculate for identical, indistinguishable particles obeying the Pauli exclusion principle. It says that n_i electrons occupying the G_i states with energy between E_i and $E_i + \Delta E_i$ are related to the energy E_i through Eq. (4.69). The ratio n_i/G_i is the *occupational probability* or *occupation index*. The distribution of Eq. (4.69) is known as the *Fermi–Dirac distribution function*.

The Lagrangian multipliers α and β must be determined so as to satisfy the constraint equations (4.62) and (4.63). Their evaluation can also be obtained by comparison with standard thermodynamic quantites. If, for example, we add a small amount of heat to the system of electrons,

$$dQ = \sum_i E_i \, dn_i \quad (4.70)$$

From Eq. (4.66) we can calculate the change in $\ln W$ due to changes in the n_i, it being assumed that E_i and G_i are unchanged by the small amount of heat being considered

$$d(\ln W) = \sum_i \ln \frac{G_i - n_i}{n_i} \, dn_i = \sum_i (\alpha + \beta E_i) \, dn_i \quad (4.71)$$

[†] See, for example, P. Franklin, "Methods of Advanced Calculus," McGraw-Hill, New York, 1944, p. 67.

4.5 Occupancy of Levels for Free Electrons in a Metal

using Eq. (4.69). Now

$$\sum_i \alpha \, dn_i = \alpha \sum_i dn_i = \alpha \, dN_e = 0 \tag{4.72}$$

since α is a constant and the total number of electrons is not changed. Also

$$\sum_i \beta E_i \, dn_i = \beta \sum_i E_i \, dn_i = \beta \, dQ \tag{4.73}$$

since β is a constant. We conclude therefore that

$$dQ = \frac{d(\ln W)}{\beta} \tag{4.74}$$

The thermodynamic relation between the temperature T and the entropy S at equilibrium is

$$dQ = T \, dS \tag{4.75}$$

These two relations agree if

$$S = k \ln W \tag{4.76}$$

$$\beta = \frac{1}{kT} \tag{4.77}$$

where k is *Boltzmann's constant*, equal to 1.38×10^{-16} ergs per degree, and must not be confused with the wavevector **k**. Generally Boltzmann's constant occurs as a multiplier of the temperature T to yield a quantity with the dimensions of energy. The entropy is seen to be a measure of disorder in the system.

The constant α may be interpreted by comparison with the thermodynamic expression for the differential energy dE in a single-phase system with one component,

$$dE = -p \, dV + T \, dS + \mu^* \, dn \tag{4.78}$$

where p is the pressure, V the volume, and μ^* a quantity called the *chemical potential*, which can be viewed as the increase in energy of a system of electrons in a metal if an additional electron is added, neglecting the effects of electronic charge. The differential energy of Eq. (4.78) is also given by

$$dE = d \sum_i n_i E_i = \sum_i n_i \, dE_i + \sum_i E_i \, dn_i \tag{4.79}$$

From our previous evaluation of β,

$$dQ = T \, dS = \sum_i (E_i + kT\alpha) \, dn_i = \sum_i E_i \, dn_i + \sum_i kT\alpha \, dn_i \tag{4.80}$$

using Eqs. (4.74) and (4.71). Substituting Eq. (4.80) into Eq. (4.78) gives

$$dE = -p\,dV + \sum_i E_i\,dn_i + \sum_i kT\alpha\,dn_i + \mu^*\,dn \tag{4.81}$$

Noting that $-p\,dV = \sum_i n_i\,dE_i$, and comparing Eq. (4.81) with Eq. (4.79) shows that

$$\sum_i kT\alpha\,dn_i = -\sum_i \mu^*\,dn_i \tag{4.82}$$

or that

$$\alpha = -\frac{\mu^*}{kT} \tag{4.83}$$

From the definition of the chemical potential μ^*, we see that it is identical with what we have called the Fermi energy for the free electrons in a metal. (In the presence of an electric field, the Fermi energy includes the electric field energy effects and is then to be identified with the *electrochemical potential*.)

Summarizing the evaluation of α and β, we may rewrite the Fermi–Dirac distribution of Eq. (4.69) as

$$f(E) = \frac{1}{\exp[(E - E_F)/kT] + 1} \tag{4.84}$$

where we have set $\mu^* \equiv E_F$. At absolute zero the Fermi–Dirac distribution has the following characteristics:

$$\text{for } E < E_F, \quad f(E) = 1$$
$$E = E_F, \quad f(E) = \tfrac{1}{2}$$
$$E > E_F, \quad f(E) = 0$$

A typical plot of Eq. (4.84) for $T = 0°K$ and for $T > 0°K$ is given in Fig. 4.8. For large values of energy, the Fermi distribution becomes identical to the Boltzmann distribution.

The density distribution of *allowed* energy states is given by Eq. (4.31). The density distribution of *occupied* allowed energy states is given by

$$n(E)\,dE = N_v^t(E)f(E)\,dE$$
$$= \frac{1}{2\pi^2}\left(\frac{2m}{\hbar^2}\right)^{3/2} \frac{E^{1/2}}{\exp[(E - E_F)/kT] + 1}\,dE \tag{4.85}$$

A typical plot of Eq. (4.85) for $T = 0°K$ is given in Fig. 4.6, and for $T > 0°K$ in Fig. 4.9.

4.5 Occupancy of Levels for Free Electrons in a Metal

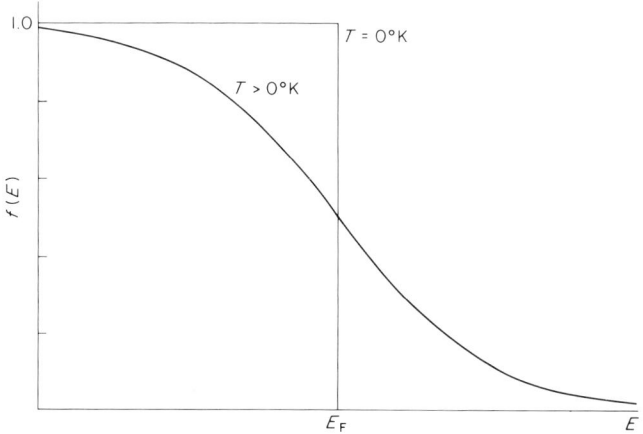

Fig. 4.8 Dependence of the Fermi–Dirac distribution function $f(E, T)$ on energy E at $T = 0°K$ and at some representative much higher temperature.

Example 4.5 At what energy E does the maximum density of occupied states occur for free electrons in a metal?

The maximum density of occupied states can be calculated from Eq. (4.85) by setting $dn(E)/dE = 0$. When this is done, the following expression is found:

$$1 + \exp\left[\frac{(E_F - E_{max})}{kT}\right] - \frac{2E_{max}}{kT} = 0$$

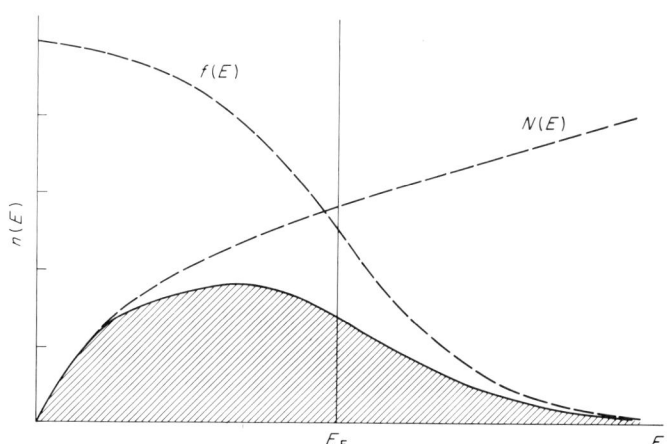

Fig. 4.9 Determination of the density of occupied states $n(E)$ at some high temperature as a function of energy E, in terms of a product between the density of allowed states $N(E)$ and the occupancy probability $f(E)$.

where $E_{\max}$ is the energy corresponding to maximum $n(E)$. Since this is a transcendental equation, it is not possible to obtain a simple expression for $E_{\max}$. It is clear that the value of $E_{\max}$ depends on the specific value of the Fermi energy E_F, as well as on the temperature T. Solution of the transcendental equation by simple graphical means, i.e., obtaining the intersections of plots of $2E_{\max}/kT$ vs. $E_{\max}$, and of $1 + \exp[(E_F - E_{\max})/kT]$ vs. $E_{\max}$, yields the following typical results, determined explicitly for 300°K.

Fermi energy, E_F (eV)	E for maximun $n(E)$
1.0	$E_F - 4.25kT$
2.0	$E_F - 5.0kT$
3.0	$E_F - 5.4kT$

At 300°K, $kT = 0.0259$ eV, so that for $E_F = 2.0$ eV, E for maximum $n(E)$ is $0.93 E_F$.

We return in later chapters to a further description of the methods of describing electronic processes in semiconductors and insulators as well as in metals in terms of the position of a Fermi level located at the Fermi energy E_F.

4.6 Examples of Applications of the Free-Electron Model for Metals

A variety of phenomena can be treated in at least a preliminary manner by the application of the free-electron model for a metal.

Spin Paramagnetism

In the absence of a magnetic field, equal numbers of free electrons in a metal have spins of $+\hbar/2$ and $-\hbar/2$. When a magnetic field is imposed, however, those electrons with spin magnetic moment in the direction of the magnetic field have their energy lowered by an amount $\mu_B H$, and those with opposite spin have their energy increased by the same amount. The Fermi energy for electrons with both kinds of spin must remain the same, however, in the presence of the magnetic field. This requires that $\mu_B H\, N(E_F)$ electrons turn their spin parallel to the magnetic field, where $N(E_F)$ is the density of states at the Fermi energy. The net result of this change in spin

4.6 Applications of Free-Electron Model for Metals

is to produce $2\mu_B H\, N(E_F)$ more electrons with spin parallel to **H** than with spin antiparallel. Thus there is an induced magnetic moment **M** due to the free electrons, and the magnetic susceptibility $\varkappa$ (**M** $= \varkappa$**H**) corresponding to this process can be calculated using the free-electron model for the value of $N(E_F)$. The result is a (paramagnetic) magnetic susceptibility due to electron spin which is independent of temperature. Experimental values for the temperature dependence of magnetic susceptibility in several metals are given in Fig. 4.10. Deviations of the experimental values from the predicted temperature independence in some cases result from contributions to the measured magnetic susceptibility from sources other than the free-electron spin, e.g., from the (diamagnetic) magnetic susceptibility of the ion cores in the metal, and from electron–electron interactions. The much larger values of susceptibility for the transition metals than for the alkali metals is the result of larger values of $N(E_F)$ for the former; this is a result of energy banding in the crystal, as we discuss in Chapter 5.

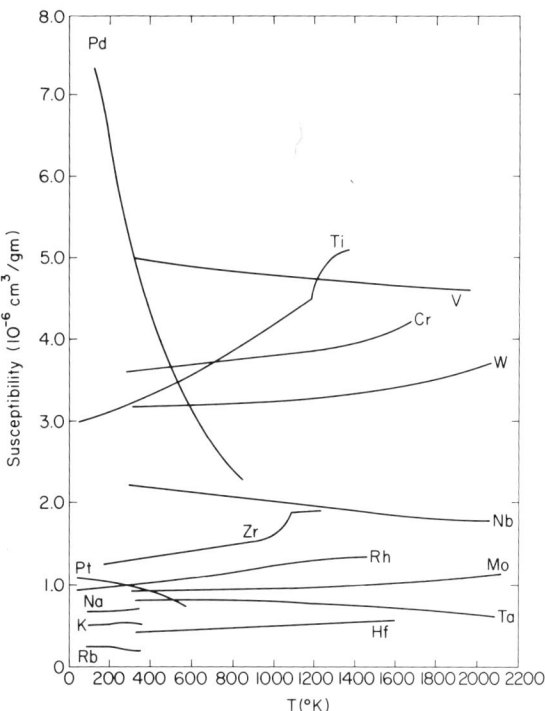

Fig. 4.10 Temperature dependence of the magnetic susceptibility of a number of different metals. (After C. Kittel, "Introduction to Solid State Physics," 3rd edition, Wiley, New York, 1966, p. 448.)

Heat Capacity

When heat is supplied to a metal, the energy of both lattice vibrations and of free electrons is increased. The rate of increase of total energy with temperature is called the *heat capacity*; it has a contribution both from lattice vibrations and from free electrons. The contribution to the total energy from lattice vibrations is obtained by integrating the energy density,

$$U_L(v)\, dv = \bar{n} h v\, N_L(v)\, dv \qquad (4.86)$$

over all frequencies v from 0 to the Debye frequency, as given in Eq. (1.9). In Eq. (4.86), $\bar{n}$ is the average number of phonons with energy hv given by the Bose–Einstein distribution of Table 4.2,

$$\bar{n} = \frac{1}{e^{hv/kT} - 1} \qquad (4.87)$$

and $N(v)$ is the density of vibrational states (standing waves) with frequency between v and $v + dv$. The contribution to the total energy from free electrons according to the free-electron model is

$$U_e = \int_0^\infty E f(E)\, N_v^t(E)\, dE \qquad (4.88)$$

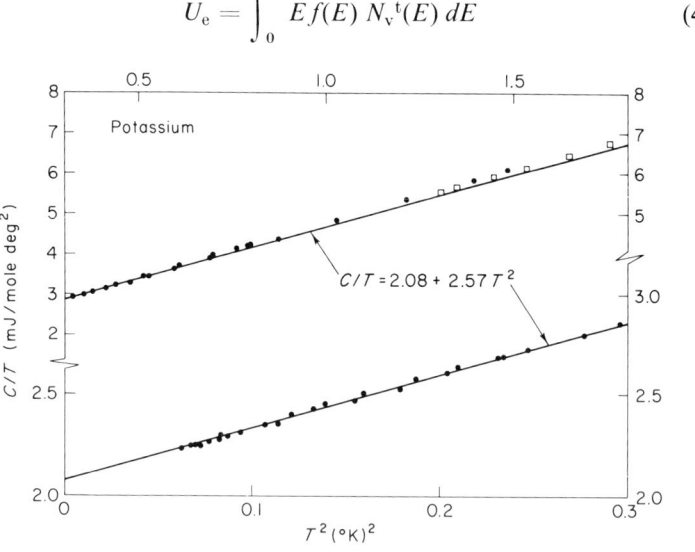

Fig. 4.11 (*Above and opposite*) Plots of measured heat capacity C as a function of temperature for potassium, rubidium, and cesium. [After W. H. Lien and N. E. Phillips, *Phys. Rev.* **133**, A1370 (1964).]

4.6 Applications of Free-Electron Model for Metals

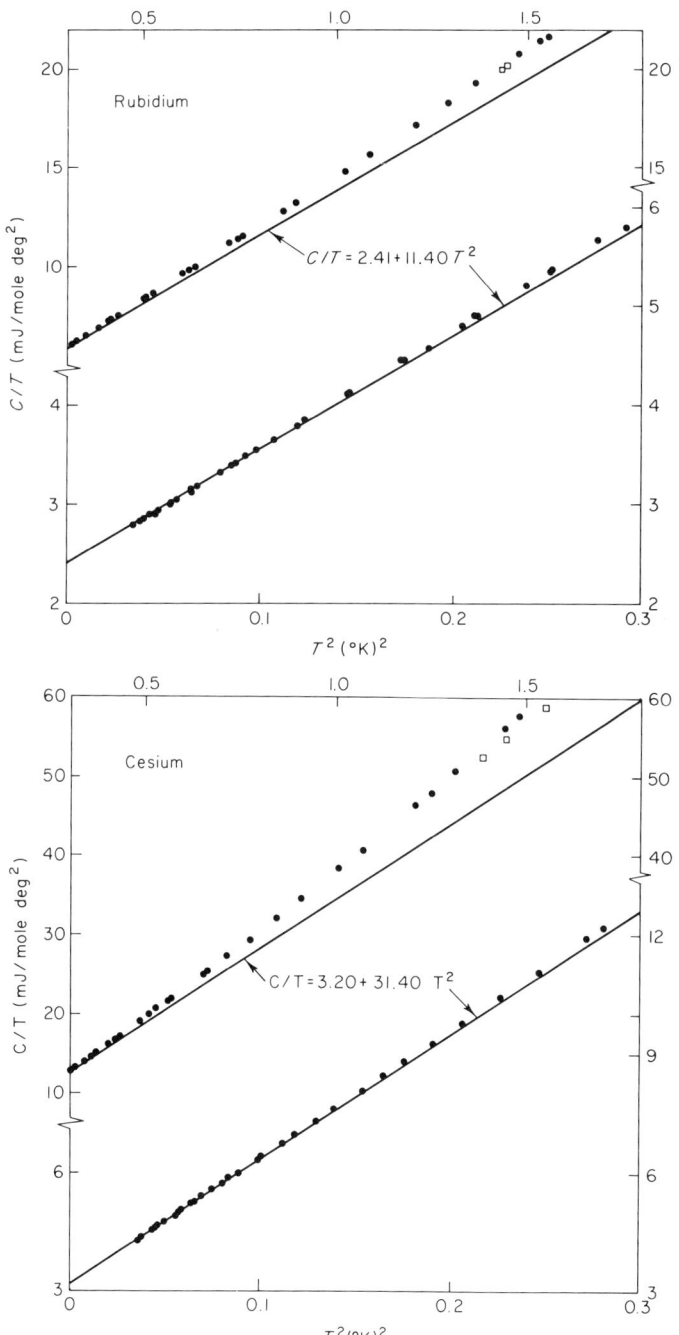

The electronic contribution to the heat capacity is therefore given by

$$C_e = \frac{\partial U_e}{\partial T} \qquad (4.89)$$

When the integral of Eq. (4.88) is evaluated by an approximation based upon $E_F \gg kT$ and the fact that the major change in the electronic distribution for $T > 0°K$ takes place close to E_F at $T = 0°K$, it is concluded that C_e is proportional to T. Evaluation of C_L leads to the conclusion that for temperatures much less than the Debye temperature, C_L is proportional to T^3. Thus the total heat capacity at low temperatures can be written as

$$C = AT + BT^3 \qquad (4.90)$$

where the first term represents C_e and the second term C_L. If measured values of C are plotted as C/T vs. T^2, a straight line is predicted with slope of B and intercept of A. Experimental data are given in Fig. 4.11, confirming this expectation. When values of the experimentally measured A are compared with calculated values from the free-electron model, they agree as far as orders of magnitude are concerned, but the measured value is frequently larger than the calculated value by some 20 to 40 per cent. A variety of energy-band, electron–electron, and electron–phonon interactions are proposed to account for the discrepancy.

THERMIONIC EMISSION

The escape of electrons from a metal under the effects of temperature can also be described to a reasonable degree of approximation in terms of the free-electron theory. All that is assumed is that the free electrons in the metal must surmount a potential barrier at the surface of the metal in order to escape into the surrounding vacuum. A schematic diagram is given in Fig. 4.12. In a sense a hybridized model is involved, for we are assuming the results of a calculation done explicitly for infinite potential barriers at the surface of the metal in order to determine the properties of a model with finite potential barriers at the surface. As long as the barrier height is much larger than kT, however, the difference between the two approximations may be neglected.

It is desired to calculate the emission current density j for electrons thermally escaping from a metal over a potential barrier defined in terms of the *work function* ϕ of the metal. The work function is defined as

$$\phi \equiv E_0 - E_F \qquad (4.91)$$

4.6 Applications of Free-Electron Model for Metals

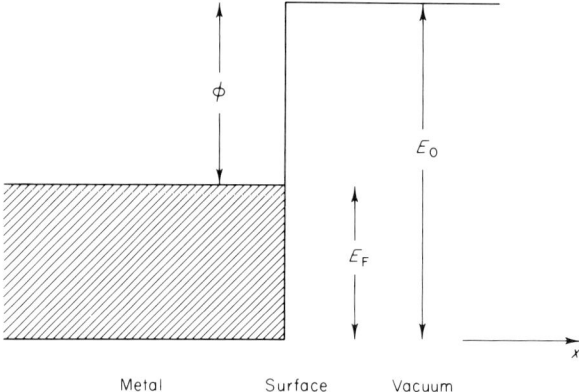

Fig. 4.12 Energy diagram for thermal escape of electrons at the surface of a metal with work function ϕ.

as shown in Fig. 4.12. The rate of emission is obtained by multiplying the density of electrons with sufficient energy to pass over the barrier by the velocity component perpendicular to the surface of the metal, a direction assumed to be that of the $+x$ axis in Fig. 4.12, and integrating over all velocities subject to the restriction that the energy $mv_x^2/2$ must be larger than E_0.

The density of occupied states in the metal is given by

$$n(E)\,dE = f(E)\,N_v^t(E) = \frac{1}{2\pi^2}\left(\frac{2m}{\hbar^2}\right)^{3/2} \frac{E^{1/2}\,dE}{\exp[(E-E_F)/kT]+1}$$

$$= \frac{8\pi m^3}{h^3} \frac{v^2\,dv}{\exp[(E-E_F)/kT]+1} \qquad (4.92)$$

where use has been made of the relationships

$$E^{1/2} = \left(\frac{m}{2}\right)^{1/2} v \qquad (4.93)$$

and

$$dE = mv\,dv \qquad (4.94)$$

In the free-electron approximation, all of the energy is kinetic energy.

To calculate the current density j, we multiply the density of occupied states by ev_x. To integrate over electron velocities, we replace the integration over $4\pi v^2\,dv$ by an integration over $dv_x\,dv_y\,dv_z$. Since only those electrons with energy greater than $(\phi + E_F)$ can escape from the surface, the lower limit for the integration over dv_x is chosen to correspond to $mv_x^2/2 \geq (\phi + E_F)$. The current density is then

$$j = \frac{2m^3 e}{h^3} \int\!\!\int_{-\infty}^{\infty} dv_y\, dv_z \int_{[\frac{2(\phi+E_F)}{m}]^{1/2}}^{\infty} \frac{v_x\, dv_x}{\exp[(E - E_F)/kT] + 1} \quad (4.95)$$

Since $(E - E_F) \gg kT$ for those electrons that can escape from the metal, we may neglect the one in the denominator of the integral over dv_x, and rewrite Eq. (4.95) as

$$j = \frac{2m^3 e}{h^3} e^{E_F/kT} \int\!\!\int_{-\infty}^{\infty} dv_y\, dv_z \int_{[\frac{2(\phi+E_F)}{m}]^{1/2}}^{\infty} v_x \exp\left[-\frac{m(v_x^2 + v_y^2 + v_z^2)}{2kT}\right] dv_x \quad (4.96)$$

The evaluation of the integrals is straightforward.

$$\int_{-\infty}^{\infty} \exp\left[-\frac{mv_y^2}{2kT}\right] dv_y = \int_{-\infty}^{\infty} \exp\left[-\frac{mv_z^2}{2kT}\right] dv_y = \left[\frac{2\pi kT}{m}\right]^{1/2} \quad (4.97)$$

$$\int_{[\frac{2(\phi+E_F)}{m}]^{1/2}}^{\infty} v_x \exp\left[-\frac{mv_x^2}{2kT}\right] dv_x = -\frac{kT}{m} \int_{[\frac{2(\phi+E_F)}{m}]^{1/2}}^{\infty} d\left(\exp\left[-\frac{mv_x^2}{2kT}\right]\right)$$

$$= \frac{kT}{m} \exp\left[\frac{-(\phi + E_F)}{kT}\right] \quad (4.98)$$

The final result for the current density is the familiar *Richardson equation*,

$$j = \frac{4\pi em}{h^3} (kT)^2 e^{-\phi/kT} \quad (4.99)$$

which shows that the current has a preexponential multiplier varying as T^2, and an effective "activation energy" given by the work function ϕ. The temperature dependence of thermionic emission will be dominated by the exponential term, and therefore depends on the energy difference between the vacuum reference energy and the Fermi energy of the free-electron system. This result is one among many in which the Fermi energy is that parameter of the free-electron system which is able to describe the particular behavior of the whole system. Experimental data on thermionic emission are given in Fig. 4.13, plotted in accordance with Eq. (4.99).

Example 4.6 Calculate the temperature dependence of thermionic emission from a one-dimensional metal.

4.6 Applications of Free-Electron Model for Metals

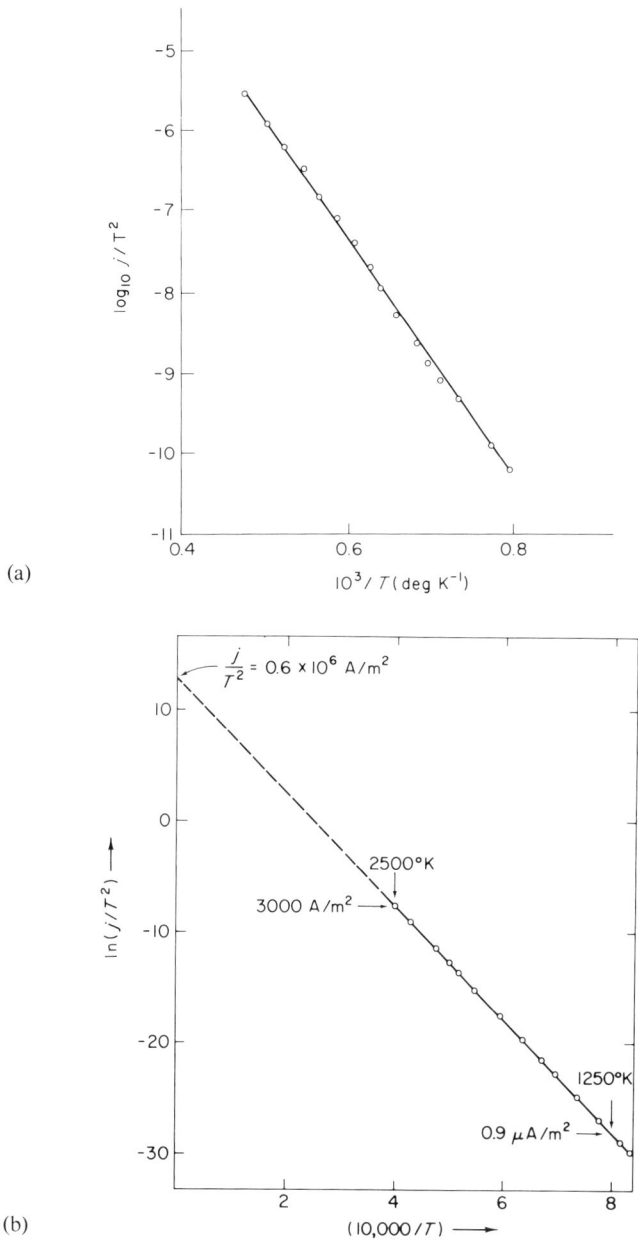

(a)

(b)

Fig. 4.13 Temperature dependence of thermionic emission from (a) uranium carbide and (b) tungsten. [After (a) L. V. Azaroff and J. J. Brophy, "Electronic Processes in Materials," McGraw-Hill, New York, 1963, p. 314; (b) R. L. Sproull, "Modern Physics," Wiley, New York, 1964, p. 441.]

According to Example 4.4, the density of states per unit length for a one-dimensional metal is given by

$$N_l^t(E) = (2m)^{1/2} \frac{1}{\hbar \pi} E^{-1/2}$$

Since

$$E^{-1/2} = \frac{1}{(m/2)^{1/2} v_x} \quad \text{and} \quad dE = m v_x \, dv_x$$

the expression for the current density analogous to Eq. (4.96) is

$$\begin{aligned} j_x &= e(2m)^{1/2} \frac{e^{E_F/kT}}{\hbar \pi} \int_{\left[\frac{2(\phi+E_F)}{m}\right]^{1/2}}^{\infty} \frac{\exp(-mv_x^2/2kT)}{(m/2)^{1/2} v_x} \, v_x (m v_x \, dv_x) \\ &= \frac{2em}{\hbar \pi} e^{E_F/kT} \int_{\left[\frac{2(\phi+E_F)}{m}\right]^{1/2}}^{\infty} v_x \exp\left(-\frac{mv_x^2}{2kT}\right) dv_x \\ &= \frac{2em}{\hbar \pi} e^{E_F/kT} \left[\frac{kT}{m} \exp\left(-\frac{\phi + E_F}{kT}\right)\right] \\ &= \frac{4e}{h} kT \, e^{-\phi/kT} \end{aligned}$$

The major temperature dependence of the thermionic emission of the one-dimensional metal is therefore the same as for a three-dimensional metal.

FIELD EMISSION

Electrons may be extracted from metals even at low temperatures, provided that a sufficiently high electric field is applied to the metal surface. The mechanism involved is tunneling of the electrons from the metal into the vacuum with the assistance of the electric field; the phenomenon is known as *field emission*.

A schematic diagram of the potential at the surface of a metal during field emission is given in Fig. 4.14. The rate of emission depends primarily on the product of the density of occupied states at a given energy and the tunneling transmission coefficient corresponding to that same energy; the total emission corresponds to an integration of this product over all energies. Most of the contribution to the field-emission current, however, comes from states close to the Fermi energy. For higher energies, the transmission coefficient of the barrier increases, but the density of occupied states decreases. For lower energies, the density of occupied states may be slightly larger, but the transmission coefficient is smaller. An approximation to the func-

4.6 Applications of Free-Electron Model for Metals

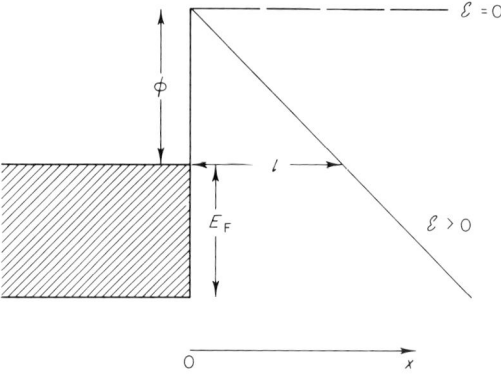

Fig. 4.14 Energy diagram for field-emission escape of electrons at the surface of a metal via tunneling through a surface barrier.

tional dependence of field-emission current, therefore, can be obtained by considering only that contribution coming from states near the Fermi energy.

If an electric field $\mathscr{E}$ is applied in the x direction to the metal surface, the potential energy varies with distance x from the surface as

$$V = -e\mathscr{E}x \tag{4.100}$$

The thickness l of the barrier at the energy corresponding to the Fermi energy is given by

$$\phi = el\mathscr{E} \tag{4.101}$$

or

$$l = \frac{\phi}{\mathscr{E}e} \tag{4.102}$$

An approximate expression for the transmission coefficient for tunneling through a barrier is given by

$$T \approx \exp\left\{-\frac{2}{\hbar} \int_0^d [2m(V_0 - E)]^{1/2}\, dx\right\} \tag{4.103}$$

where V_0 is the height of the barrier of width d. If $(V_0 - E)$ is a constant, for example, as for a square potential barrier, Eq. (4.103) predicts a transmission coefficient

$$T = \exp\left\{-\frac{2}{\hbar} [2m(V_0 - E)]^{1/2} d\right\} \tag{4.104}$$

which agrees with the dominant exponential portion of the exact calculation described in Problem 3.5. The origin of Eq. (4.103) is an approximate solution of the Schroedinger equation, known as the *WKB* (for Wentzel–Kramers–Brillouin) *approximation*. It may be shown that the general Schroedinger equation in one variable, e.g.,

$$\frac{d^2\psi}{dx^2} + \frac{2m}{\hbar^2}[E - V(x)]\psi = 0 \qquad (4.105)$$

has the following approximate solution if $V(x)$ is a slowly varying function of x,[†]

$$\psi = (2m|E - V(x)|)^{-1/4} \exp\left(\frac{i}{\hbar} \int \{2m[E - V(x)]\}^{1/2}\, dx\right) \quad (4.106)$$

When this approximation is applied to the problem of tunneling through a potential barrier the preexponential terms cancel and the net transmission

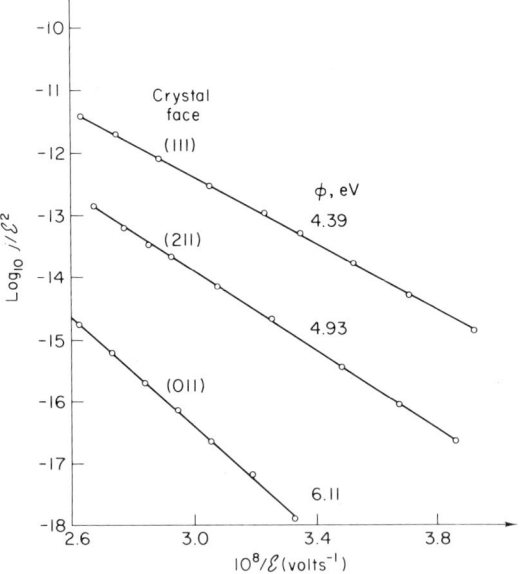

Fig. 4.15 Field emission from different crystal surfaces of tungsten. (After L. V. Azaroff and J. J. Brophy, "Electronic Processes in Materials," McGraw-Hill, New York, 1963, p. 321.)

[†] For example, see V. Rojansky, "Introductory Quantum Mechanics," Prentice-Hall, Englewood Cliffs, New Jersey, 1946, p. 186 ff.

4.7 Summary

coefficient is given by the square of the exponential term of Eq. (4.106) with integration from one side of the barrier to the other. Even when $V(x)$ is not strictly a slowly varying function of x, the approximation may still be applied to estimate orders of magnitude and functional dependence.

When Eq. (4.103) is applied to the present situation with a potential barrier described by Eq. (4.100), the transmission coefficient is

$$T(E_F) = \exp\left[-\frac{2}{\hbar}(2m\phi)^{1/2}\frac{2}{3}l\right]$$
$$= \exp\left[-\frac{4}{3\hbar e}\frac{(2m)^{1/2}\phi^{3/2}}{\mathscr{E}}\right] \quad (4.107)$$

using Eq. (4.102) to eliminate l. Thus we predict that if the major contribution to the field-emission current comes from states near the Fermi energy, a plot of the logarithm of the field-emission current as a function of $1/\mathscr{E}$ should give a straight line with a slope proportional to $\phi^{3/2}$. Experimental data for field emission from tungsten are given in Fig. 4.15. A more complete treatment of the processes involved shows that a preexponential factor of $\mathscr{E}^2$ results from integration over the electron distribution, rather than taking simply the value at the Fermi energy; the data of Fig. 4.15 are therefore plotted as $\log J/\mathscr{E}^2$ vs. $1/\mathscr{E}$.

4.7 Summary

Although the properties of electrons in solids must ultimately be treated in terms of the interactions between the electrons and the positively charged ion cores, and of the interactions between electrons themselves, it is possible to treat some of the properties of valence electrons in metals in terms of a model in which these electrons are considered to be approximately free. It is assumed that the mutual repulsion between electrons is on the average just balanced by the attraction between electrons and ion cores, so that the average potential energy for the system can be taken to be zero.

We have considered two models for treating such a system of free electrons. (1) The Sommerfeld model applies the quantum mechanical properties of noninteracting particles in a potential-free box in order to obtain wavefunctions in the form of standing waves obeying geometric boundary conditions. (2) The Hartree model includes a uniform density of positive charge and shows that a self-consistent solution can be obtained with a total potential energy for the system of zero. In order to have wavefunctions suitable for describing transport of electrons in solids, we have used traveling wavefunctions and periodic boundary conditions in the Hartree model. In treating the properties of "free electrons" in metals, we have had three goals: (1) to determine what energy states are allowed, (2) to determine the distribution of the density of these states with energy, and (3) to determine the probability that a given energy state will be occupied by an electron at a given temperature T. Both the Sommerfeld and the Hartree models give the same total density of states per unit volume,

$$N_v^t(E) = \frac{1}{2\pi^2} \left(\frac{2m}{\hbar^2}\right)^{3/2} E^{1/2}$$

Although the actual allowed energy states are not the same for the Sommerfeld and Hartree models, the fact that the states form a quasi-continuous distribution in energy leads both to give the same total density of states per unit volume. The probability that a state with energy E is occupied is given by the Fermi–Dirac distribution,

$$f(E) = \frac{1}{\exp[(E - E_F)/kT] + 1}$$

in terms of the reference Fermi energy E_F, equivalent to the chemical potential, and equal to the maximum energy of occupied energy states at $T = 0°K$. The Fermi–Dirac distribution is a direct consequence of applying quantum statistical presuppositions to a system of identical and indistinguishable particles obeying the Pauli exclusion principle.

When both the density-of-states distribution and the occupancy-probability distribution are known, the density of occupied energy states as a function of energy is given directly by

$$n(E) = N_v^t(E) f(E)$$

A number of electronic properties of free electrons in metals are associated with the quantity $n(E)$. We have considered how a simple free-electron theory can be applied to derive the essential properties of spin paramagnetism, electronic contribution to the specific heat, and thermionic and field emission from a metal.

Problems

4.1 In the ground state of the helium atom, both electrons are in the 1s state. Write the two one-electron Hartree equations for the helium atom,

Problems

and show that the ground-state energy according to the Hartree approximation is given by

$$E = 2E_i - e^2 \iint \frac{1}{r_{12}} |\psi(\mathbf{r}_1)|^2 |\psi(\mathbf{r}_2)|^2 d\mathbf{r}_1 d\mathbf{r}_2$$

where E_i is the ionization energy of a 1s electron and $\psi(\mathbf{r})$ is the Hartree wavefunction for the 1s electron.

4.2 In the ground state of the helium atom, both electrons are in the orbital 1s state, but the spins of the two electrons are opposite. Write the determinantal wavefunction for the helium ground state. Calculate the energy of the ground state and show that for this particular case, the energy has the same form as that calculated for a simple-product total wavefunction in Problem 4.1.

4.3 Calculate the spherically symmetric energy levels of a metallic sphere of radius R according to the Sommerfeld model.

4.4 Calculate the total number of electrons in a rectangular block of sodium, $3 \times 2 \times 1$ mm, with energy greater than 3.5 eV at 300°K.

4.5 Within the framework of a free-electron model for a two-dimensional metal, calculate (a) the Fermi energy and (b) the average kinetic energy per electron. Show that the spin paramagnetic susceptibility is independent of the electron density.

4.6 Calculate the temperature dependence of the thermionic emission from a two-dimensional metal.

4.7 If the energy of the nth excited state (in excess of the zero-point energy) of the mth mode of a linear harmonic oscillator is $E_{mn} = n\hbar\omega_m$, and if the probability of exciting this state is $P_{mn} = \exp(-E_{mn}/kT)$, show that the average number of phonons of energy $\hbar\omega_m$, which is given by

$$\bar{n}_m = \frac{\bar{E}_m}{\hbar\omega_m} = \frac{1}{\hbar\omega_m} \frac{\sum_{n=0}^{\infty} E_{mn} P_{mn}}{\sum_{n=0}^{\infty} P_{mn}}$$

is given by the Bose–Einstein distribution of Table 4.2 and Eq. (4.87).

4.8 Show that

$$\frac{\partial f(E,T)}{\partial k^*} = -\frac{\hbar v_g}{(kT)} f(E,T)[1 - f(E,T)]$$

where $f(E,T)$ is the Fermi–Dirac distribution function, k^* is the wave vector, k is Boltzmann's constant, and v_g is the group velocity, for a general $E(k^*)$.

4.9 (a) Calculate the ratio of $(E - E_F)$ to kT that is required in order for the Boltzmann distribution function to give no more than a 1-per-cent error if substituted for the Fermi–Dirac distribution function. (b) How much greater must the density of states $2kT$ above the Fermi level be, as compared to states $2kT$ below the Fermi level, in order for the same density of occupied levels to exist at both energies?

4.10 Show that for free electrons at $0°K$, the density of free electrons is given by

$$n = \frac{k_F^3}{3\pi^2} = \frac{8\pi}{3h^3} p_F^3$$

where k_F is the wave vector at the Fermi surface, and p_F is the momentum at the Fermi surface. In the *Thomas–Fermi* approximation, it is assumed that this relation still holds in circumstances where the electrons are not really free but where the density of electrons is large and the actual potential energy changes slowly with distance $\mathbf{r}$. In this case, show that if $-eV(\mathbf{r})$ is the potential energy of an electron due to the field of all the nuclei present and of the whole electronic charge distribution (assumed zero in the free-electron case), and if $-eV_F$ is the total energy of an electron with momentum p_F,

$$n(\mathbf{r}) = \frac{8\pi}{3h^3} \{2me[V(\mathbf{r}) - V_F]\}^{3/2}$$

Chapter 5

Origin of Energy Bands in Solids

In the previous chapter we saw that some of the properties of valence electrons in metals can be treated as if these electrons were essentially free. But this is clearly an approximation, and departures from ideal free behavior are expected because of the presence of these electrons in the periodic potential of the crystalline lattice. Appropriate eigenfunctions for the Hamiltonian of electrons in a periodic potential must manifest this periodic dependence. Such functions are the *Bloch functions*, formed by the product between a function with the periodicity of the lattice and a free-electron wavefunction.

The origin of energy bands in solids can be viewed from several complementary perspectives. It can be shown in general that a periodic potential of arbitrary form leads to a series of allowed and forbidden energy bands, but this calculation tends to obscure the physical basis for such bands. To see this physical basis, we can start with a free-electron gas that finds itself in a periodic potential, or we can start with isolated atoms that are arranged in a periodic array.

The first perspective answers the question, "Why are there energy gaps for nearly free electrons in a periodic potential?" The answer is that the free-electron waves experience Bragg reflection from the periodic planes of the crystal. The inability of an electron wave of suitable energy to undergo Bragg reflection to propagate through the crystal is equivalent to the appearance of forbidden gaps in the previously continuous energy spectrum.

The second perspective answers the question, "Why do the discrete energy levels of atoms broaden into bands when these atoms interact to form a periodic potential?" The answer is that interaction between these atoms makes it impossible to distinguish which electrons "belong" to which atoms. Application of the Pauli exclusion principle requires that individual electrons be in unique quantum states of the whole system of interacting atoms. The effect is therefore to broaden formerly discrete atomic energy levels into energy bands in the solid.

Since the shape of allowed energy bands involves the variation of the energy $E(\mathbf{k})$ with the wavevector $\mathbf{k}$, it proves convenient to describe energy bands in terms of configurations in $\mathbf{k}$ space. In one-dimensional calculations this is a trivial and easily assimilated exercise, but in three dimensions the actual spatial distribution of atoms makes the mathematical task somewhat more complicated. Not only do the various quantities change from scalars to vectors or tensors, but it is also necessary to consider explicitly the geometrical structure of $\mathbf{k}$ space in terms of the *reciprocal lattice*.

5.1 Wavefunction for an Electron in a Periodic Potential

In 1928 Bloch proposed that electrons in a periodic potential can be treated in terms of a wavefunction of the form

$$\psi_{\mathbf{k}}(\mathbf{r}) = u_{\mathbf{k}}(\mathbf{r}) \, e^{i\mathbf{k}\cdot\mathbf{r}} \qquad (5.1)$$

Here $u_{\mathbf{k}}(\mathbf{r})$ has the periodicity of the lattice, so that if $\mathbf{R}_j$ is a translation vector that takes the lattice into itself,

$$u_{\mathbf{k}}(\mathbf{r} + \mathbf{R}_j) = u_{\mathbf{k}}(\mathbf{r}) \qquad (5.2)$$

The lattice is described in terms of lattice vectors $\mathbf{R}_j$ such that

$$\mathbf{R}_j = j_1 \mathbf{a}_1 + j_2 \mathbf{a}_2 + j_3 \mathbf{a}_3 \qquad (5.3)$$

where j_i are integers, and $\mathbf{a}_i$ are the edges of the unit cell, the basic repetitive unit of the lattice. The three cubic crystal lattices are summarized in Fig. 5.3.

That the Bloch functions of Eq. (5.1) are indeed eigenfunctions of the Hamiltonian for a periodic potential can be shown in general as follows. We restrict the calculation to one dimension for simplicity. Consider the operator Γ such that $\Gamma \psi(x) = \psi(x + a)$, where a is the lattice periodic distance. We show that Γ and $\mathbf{H}$ commute, thereby showing that Γ and $\mathbf{H}$ have simultaneous eigenfunctions; we then show that the Bloch function is an eigenfunction of Γ, and conclude that it is also an eigenfunction of $\mathbf{H}$.

5.1 Wavefunction for Electron in Periodic Potential

That Γ commutes with $\mathbf{H}$ can be seen by examining

$$\Gamma \mathbf{H}\, \psi(x) = \left[-\frac{\hbar^2}{2m} \frac{d^2}{d(x+a)^2} + V(x+a) \right] \psi(x+a)$$

$$= \left[-\frac{\hbar^2}{2m} \frac{d^2}{dx^2} + V(x) \right] \psi(x+a)$$

$$= \mathbf{H}\, \psi(x+a)$$

$$= \mathbf{H}\Gamma\, \psi(x) \tag{5.4}$$

Eigenfunctions of Γ can be calculated as follows:

$$\Gamma\, \psi(x) = \lambda\, \psi(x) = \psi(x+a) \tag{5.5}$$

By successive operation with Γ we conclude that

$$\psi(x + Na) = \lambda^N\, \psi(x) \tag{5.6}$$

where N is the number of atoms in the one-dimensional crystal being considered. Periodic boundary conditions require that

$$\psi(x + Na) = \psi(x) \tag{5.7}$$

and that therefore

$$\lambda = e^{2\pi i n/N} \tag{5.8}$$

where $n = 0, \pm 1, \pm 2, \ldots$. Bloch functions are eigenfunctions of Γ, therefore, provided that

$$\psi_k(x + a) = e^{2\pi i n/N}\, \psi_k(x) \tag{5.9}$$

Now

$$\psi_k(x + a) = u_k(x + a)\, e^{ik(x+a)} = u_k(x)\, e^{ika}\, e^{ikx} = e^{ika}\, \psi_k(x)$$

Periodic boundary conditions require that $k = 2\pi n/Na$ [Eq. (4.40)], and therefore the Bloch function does satisfy Eq. (5.9).

We conclude that a Bloch function of the form of Eq. (5.1) is an appropriate eigenfunction for the Hamiltonian when the potential energy is a periodic function. The Bloch function can be considered physically to be the wavefunction for a free electron modulated by the periodic function $u_k(x)$, as shown in Fig. 5.1. The function $u_k(x)$ is the same in the vicinity of each atom, where it has a form resembling an atomic wave function, and varies smoothly between atoms where the plane wave dominates.

124   5 Origin of Energy Bands in Solids

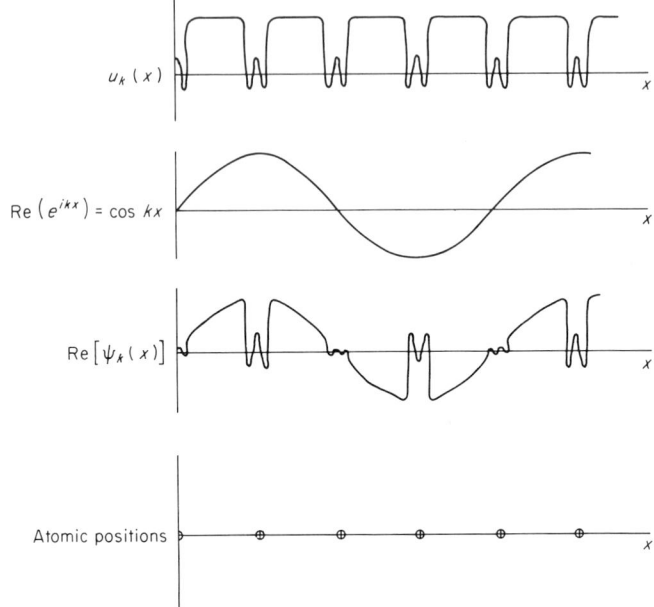

Fig. 5.1 Spatial variation of the different factors of the periodic Bloch function for a monatomic lattice.

5.2 The Cellular Method

Because $u_k(\mathbf{r})$ in the Bloch wavefunction for an electron in a crystal has the periodicity of the crystal lattice, the solution of the Schroedinger equation for the allowed electron energies needs to be carried out only within a single unit cell of the crystal. The Bloch condition,

$$\psi_\mathbf{k}(\mathbf{r} + \mathbf{R}_j) = e^{i\mathbf{k}\cdot\mathbf{R}_j}\,\psi_\mathbf{k}(\mathbf{r}) \qquad (5.10)$$

[from Eqs. (5.1) and (5.2)] contributes to the boundary conditions appropriate for the wavefunction at the boundaries of the unit cell. This approach is known as the *cellular method* and its application to the calculation of wavefunctions and energies for electrons in monovalent bcc metals such as the alkali metals dates back to Wigner and Seitz in 1933. We do not attempt here a detailed treatment of the calculation, but rather a summary of its principal features and conclusions.

In addition to the Bloch periodicity condition of Eq. (5.10), another boundary condition on the wavefunction at the boundary of the unit cell

5.2 The Cellular Method

is that of continuity. A small displacement in the direction of a lattice vector $\mathbf{R}_j$ produces from Eq. (5.10) the additional requirement that

$$\mathbf{R}_j \cdot \nabla \psi_\mathbf{k}(\mathbf{r} + \mathbf{R}_j) = e^{i\mathbf{k}\cdot\mathbf{R}_j} \mathbf{R}_j \cdot \nabla \psi_\mathbf{k}(\mathbf{r}) \qquad (5.11)$$

Finally the wavefunction must satisfy the symmetry requirements of the unit cell. In the alkali metals, the valence electrons are nearly free, and it may be assumed that the state of lowest energy $\psi_0(\mathbf{r})$ occurs at $\mathbf{k} = 0$, i.e.,

$$\psi_0(\mathbf{r}) = u_0(\mathbf{r}) \qquad (5.12)$$

In such a case $\psi_0(\mathbf{r})$ resembles an s function; since it has the full symmetry of the crystal lattice, it is reasonable to approximate the symmetry requirements on $\psi_0(\mathbf{r})$ as spherical symmetry. If such a function $\psi_0(\mathbf{r})$ is substituted into Eqs. (5.10) and (5.11), it follows that both the value of $\psi_0(\mathbf{r})$ and the value of its outward normal derivative must be the same at similarly situated points on opposite faces of the unit cell. The only way for both of these conditions to be fulfilled, i.e., the only way to satisfy the Bloch periodicity requirement with a wavefunction with spherical symmetry, is for

$$\mathbf{R}_j \cdot \nabla \psi_0(\mathbf{r}) = 0 \qquad (5.13)$$

everywhere on the surface of the unit cell.

In the approximation of spherical symmetry, the unit cell is assumed to be a sphere of radius r_s, where

$$\frac{4\pi r_s^3}{3} = \frac{\mathscr{V}}{N} \qquad (5.14)$$

with $\mathscr{V}$ the volume of the crystal and N the number of unit cells in the crystal.

The Schroedinger equation to be solved is

$$\frac{d^2\psi_0(\mathbf{r})}{dr^2} + \frac{2}{r}\frac{d\psi_0(\mathbf{r})}{dr} + \frac{2m}{\hbar^2}[E_0 - V(\mathbf{r})]\psi_0(\mathbf{r}) = 0 \qquad (5.15)$$

$V(\mathbf{r})$ is the potential energy of a valence electron in the crystal, and the eigenfunctions satisfy Eq. (5.13) for $|\mathbf{r}| = r_s$.

Values of $V(\mathbf{r})$ are approximately the same as those for the spherically symmetrical ion-core potential of a free atom, values for which have been calculated and tabulated. This approximation is reasonable for an alkali metal such as sodium for three reasons: (a) the dimensions of the ion core in sodium are much less than those of the atomic sphere, so that overlap between the core electron wavefunctions of different ions can be neglected; (b) it is a good approximation to take the electrostatic field of each cell equal to zero at points outside the cell, and hence the field acting on a valence electron is just that of the ion-core plus the self-consistent field of other electrons in that same cell; and (c) Coulomb and spin correlation effects produce a region of radius about equal to r_s about a given electron from which other electrons are excluded, so that we may consider only one electron to occupy a given cell.

Numerical solutions to Eq. (5.15) can be obtained. The solution for $\psi_0(\mathbf{r})$ for the observed value of $r_s = 3.96$ Bohr units for sodium is given in Fig. 5.2. Since the valence electron of sodium is a 3s electron, the wavefunction has two nodes for values of r much less than r_s. For most of the cell outside the ion core, $\psi_0(\mathbf{r})$ is approximately constant. Actually $\psi_0(\mathbf{r})$ is constant over about 90 per cent of the cell volume; if the electrons were completely free, then $\psi_0(\mathbf{r})$ would be exactly constant over the whole volume. The result indicates that the valence electrons in sodium are almost equivalent to free electrons.

The above calculation with eigenfunction $\psi_0(\mathbf{r})$ and eigenvalue E_0 has been for the *lowest* state of a valence electron in the metal. All but two of the electrons must lie in higher energy states, described by

$$\nabla^2 \psi_{\mathbf{k}}(\mathbf{r}) + \frac{2m}{\hbar^2} [E(\mathbf{k}) - V(\mathbf{r})] \psi_{\mathbf{k}}(r) = 0 \qquad (5.16)$$

For $\mathbf{k} \neq 0$, the eigenfunctions of Eq. (5.16) are general Bloch functions,

$$\psi_{\mathbf{k}}(\mathbf{r}) = u_{\mathbf{k}}(\mathbf{r}) \, e^{i\mathbf{k}\cdot\mathbf{r}} \qquad (5.17)$$

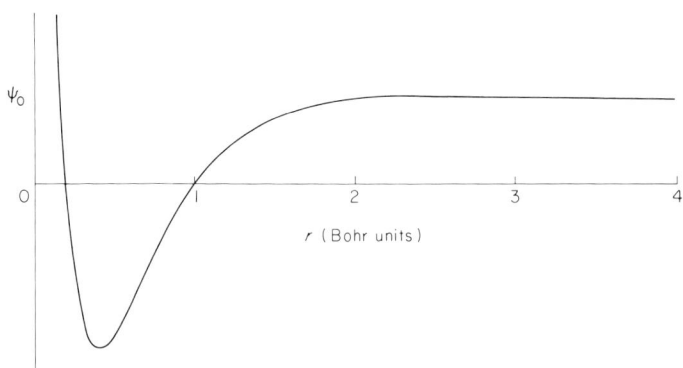

Fig. 5.2 The wavefunction for the lowest state of a valence electron in metallic sodium derived by the cellular method. (After S. Raimes, "The Wave Mechanics of Electrons in Metals," North-Holland, Amsterdam, 1961, p. 254.)

5.2 The Cellular Method

Substitution of these general Bloch functions into Eq. (5.16) gives

$$\nabla^2 u_k(\mathbf{r}) + 2i\mathbf{k} \cdot \nabla u_k(\mathbf{r}) + \frac{2m}{\hbar^2}\left[E(\mathbf{k}) - V(\mathbf{r}) - \frac{\hbar^2 k^2}{2m}\right] u_k(\mathbf{r}) = 0 \quad (5.18)$$

since

$$\nabla^2[e^{i\mathbf{k}\cdot\mathbf{r}} u_k(\mathbf{r})] = e^{i\mathbf{k}\cdot\mathbf{r}}(\nabla^2 + 2i\mathbf{k}\cdot\nabla - k^2) u_k(\mathbf{r}) \quad (5.19)$$

A first approximation replaces $u_k(\mathbf{r})$ by $u_0(\mathbf{r})$ even when $\mathbf{k} \neq 0$; this was in fact a proposal of Wigner and Seitz. This is a reasonable procedure since: (a) if $u_0(\mathbf{r})$ were exactly constant, the electrons would be free and the wavefunction of Eq. (5.17) with $u_k(\mathbf{r}) = u_0(\mathbf{r})$ would be exactly correct; (b) Eq. (5.17) with $u_k(\mathbf{r}) = u_0(\mathbf{r})$ reduces to $\psi_0(\mathbf{r})$ when $\mathbf{k} = 0$ as it should; and (c) it might be expected that $u_k(\mathbf{r})$ would be roughly the same as $u_0(\mathbf{r})$ in view of the fact that only the lower half of the first Brillouin zone is occupied in sodium. Replacing $u_k(\mathbf{r})$ by $u_0(\mathbf{r})$ in Eq. (5.18), and remembering Eq. (5.15) for $u_0(\mathbf{r})$, enables us to write

$$2i\mathbf{k}\cdot\nabla u_0(\mathbf{r}) + \left[E(\mathbf{k}) - E_0 - \frac{\hbar^2 k^2}{2m}\right]\frac{2m}{\hbar^2} u_0(\mathbf{r}) = 0 \quad (5.20)$$

If $u_0(\mathbf{r})$ is truly constant, then the term in the bracket must be equal to zero and

$$E(\mathbf{k}) = E_0 + \frac{\hbar^2 k^2}{2m} \quad (5.21)$$

The energy is given simply as the energy of the lowest state plus the energy of a free electron with wave vector $\mathbf{k}$, as expected for this free-electron approximation to the problem. A more accurate solution, however, is obtained if the term $2i\mathbf{k}\cdot\nabla u_k(\mathbf{r})$ is treated as a perturbation term for small values of $\mathbf{k}$; such a procedure is known as the $\mathbf{k}\cdot\mathbf{P}$ *approximation* (remember $\mathbf{P} = -i\hbar\nabla$) and is discussed in more detail in Chapter 6. The final result of such a calculation in the present case is

$$E(\mathbf{k}) = E_0 + \left(\frac{m}{m^*}\right)\frac{\hbar^2 k^2}{2m} \quad (5.22)$$

This result can be interpreted to mean that the allowed energies for electrons above a reference energy E_0 are like those of a free electron with mass m^* rather than mass m. The quantity m^* is therefore a kind of *effective mass* for the electron in the real periodic potential of the metal. All of the departures from a completely free-electron situation are described in the difference between m^* and m in this approximation. Values of (m/m^*) range from 0.69 for Li to 1.20 for Cs. For Na, the value of (m/m^*) is 1.02, showing that valence electrons in Na are almost completely free.

5.3 Geometrical Considerations: Reciprocal Lattice and Brillouin Zones

Geometrical considerations are not troublesome as long as discussions of the dependence of electron energy $E(\mathbf{k})$ on $\mathbf{k}$ are confined to one dimension. In three dimensions, however, the symmetry of the crystal must be considered.

THE RECIPROCAL LATTICE

The *reciprocal lattice* is a geometric construction that enables us to relate the crystal geometry directly with $\mathbf{k}$ space.

Equation (4.40) showed that the application of periodic boundary conditions to the wavefunction requires that

$$k = \frac{2\pi n}{L} = \frac{2\pi n}{Na}$$

The interatomic spacing a, the measure of distance in the one-dimensional crystal, is reciprocally related to the wave number k. If we define a reciprocal lattice where the points are separated by the distance

$$K = \frac{2\pi}{a} \tag{5.23}$$

then the wave number k is directly related to K,

$$k = \frac{n}{N} K \tag{5.24}$$

The definition of the reciprocal lattice in the three-dimensional case follows in a similar way. We define a set of vectors $\mathbf{b}_1$, $\mathbf{b}_2$, and $\mathbf{b}_3$ in terms of which the *reciprocal vector* $\mathbf{K}_m$ can be expressed

$$\mathbf{K}_m = m_1\mathbf{b}_1 + m_2\mathbf{b}_2 + m_3\mathbf{b}_3 \tag{5.25}$$

The vectors $\mathbf{b}_1$, $\mathbf{b}_2$, and $\mathbf{b}_3$ are selected perpendicular to the coordinate planes of the axes $\mathbf{a}_1$, $\mathbf{a}_2$, and $\mathbf{a}_3$ such that

$$\mathbf{a}_i \cdot \mathbf{b}_j = 2\pi \delta_{ij} \quad i, j = 1, 2, 3 \tag{5.26}$$

5.3 Reciprocal Lattice and Brillouin Zones

Explicit expressions for the reciprocal vectors are

$$b_1 = \frac{2\pi}{\Omega} a_2 \times a_3 \tag{5.27}$$

$$b_2 = \frac{2\pi}{\Omega} a_3 \times a_1 \tag{5.28}$$

$$b_3 = \frac{2\pi}{\Omega} a_1 \times a_2 \tag{5.29}$$

where $\Omega = a_1 \cdot a_2 \times a_3$ is the volume of the unit cell. Since to satisfy periodic boundary conditions in three dimensions

$$\mathbf{k} \cdot \mathbf{a}_i = \frac{2\pi n_i}{N_i} \tag{5.30}$$

where N_i is the total number of atoms in the ith direction, it follows from Eq. (5.26) that

$$\mathbf{k} = \frac{n_1}{N_1} \mathbf{b}_1 + \frac{n_2}{N_2} \mathbf{b}_2 + \frac{n_3}{N_3} \mathbf{b}_3 \tag{5.31}$$

The space of the reciprocal lattice vectors $\mathbf{b}_j$ is $\mathbf{k}$ space itself.

Example 5.1 What are the reciprocal lattice vectors for a simple cubic lattice?

For a simple cubic lattice, $\mathbf{a}_1 = a\mathbf{e}_1$; $\mathbf{a}_2 = a\mathbf{e}_2$; and $\mathbf{a}_3 = a\mathbf{e}_3$. Since $\mathbf{k} \cdot \mathbf{a}_i = 2\pi n_i/N_i$,

$$\mathbf{k} = \frac{2\pi}{a} \left(\frac{n_1}{N_1} \mathbf{e}_1 + \frac{n_2}{N_2} \mathbf{e}_2 + \frac{n_3}{N_3} \mathbf{e}_3 \right)$$

Comparison with Eq. (5.31) shows immediately that

$$\mathbf{b}_1 = \frac{2\pi}{a} \mathbf{e}_1; \quad \mathbf{b}_2 = \frac{2\pi}{a} \mathbf{e}_2; \quad \mathbf{b}_3 = \frac{2\pi}{a} \mathbf{e}_3$$

Thus the reciprocal lattice vectors are parallel to the lattice vectors of the real crystal, the points of the reciprocal lattice are separated by $2\pi/a$, and the reciprocal lattice is also a simple cubic lattice.

Example 5.2 Show that the reciprocal lattice for the face-centered cubic structure is a body-centered cubic lattice.

Description	Model
Simple Cubic $$\mathbf{a}_1 = a\mathbf{e}_1$$ $$\mathbf{a}_2 = a\mathbf{e}_2$$ $$\mathbf{a}_3 = a\mathbf{e}_3$$	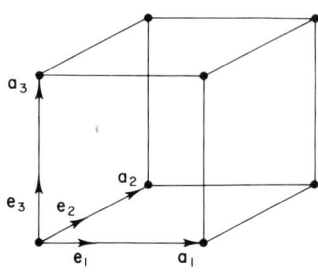
Body-Centered Cubic Two interpenetrating simple cubic lattices, the lattice points of one being at the centers of the unit cells of the second. Bravais lattice with rhombohedral unit cell: $$\mathbf{a}_1 = \tfrac{1}{2}a(-\mathbf{e}_1 + \mathbf{e}_2 + \mathbf{e}_3)$$ $$\mathbf{a}_2 = \tfrac{1}{2}a(\mathbf{e}_1 - \mathbf{e}_2 + \mathbf{e}_3)$$ $$\mathbf{a}_3 = \tfrac{1}{2}a(\mathbf{e}_1 + \mathbf{e}_2 - \mathbf{e}_3)$$	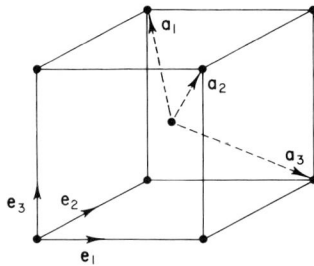
Face-Centered Cubic Four interpenetrating simple cubic lattices, the lattice points of one pair being at the centers of the faces of the second pair. Bravais lattice with rhombohedral unit cell: $$\mathbf{a}_1 = \tfrac{1}{2}a(\mathbf{e}_1 + \mathbf{e}_2)$$ $$\mathbf{a}_2 = \tfrac{1}{2}a(\mathbf{e}_2 + \mathbf{e}_3)$$ $$\mathbf{a}_3 = \tfrac{1}{2}a(\mathbf{e}_3 + \mathbf{e}_1)$$	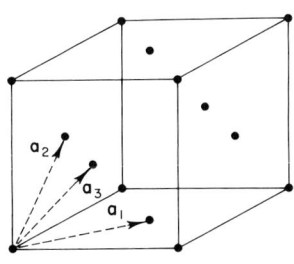

Fig. 5.3 Cubic crystal lattices and their Bravais lattice (one atom per unit cell) unit cells.

From Fig. 5.3, the lattice vectors of the face-centered cubic structure are

$$\mathbf{a}_1 = \frac{a}{2}(\mathbf{e}_1 + \mathbf{e}_2)$$

$$\mathbf{a}_2 = \frac{a}{2}(\mathbf{e}_2 + \mathbf{e}_3)$$

$$\mathbf{a}_3 = \frac{a}{2}(\mathbf{e}_3 + \mathbf{e}_1)$$

5.3 Reciprocal Lattice and Brillouin Zones

The reciprocal lattice vectors may be calculated from Eqs. (5.27)–(5.29). The volume of the unit cell $\Omega = a^3/4$. The reciprocal lattice vectors are

$$\mathbf{b}_1 = \frac{2\pi}{a}(\mathbf{e}_1 + \mathbf{e}_2 - \mathbf{e}_3)$$

$$\mathbf{b}_2 = \frac{2\pi}{a}(-\mathbf{e}_1 + \mathbf{e}_2 + \mathbf{e}_3)$$

$$\mathbf{b}_3 = \frac{2\pi}{a}(\mathbf{e}_1 - \mathbf{e}_2 + \mathbf{e}_3)$$

Similarly it may be shown that the reciprocal lattice for the body-centered cubic structure with $\Omega = a^3/2$ is a face-centered cubic lattice.

There is no simple correlation between points of the reciprocal lattice and points of the real lattice. A vector drawn between two points of the reciprocal lattice is perpendicular to a plane in the real lattice.

Many of the comparisons needed to correlate one-dimensional and three-dimensional treatments of electrons in crystals are given in Table 5.1.

Brillouin Zones

In the Bloch function in one dimension,

$$\psi_k(x) = u_k(x)\, e^{ikx} \qquad (5.32)$$

k is not uniquely specified. If we replace k by $(k + 2\pi n/a)$, i.e., by the addition of some multiple of a reciprocal lattice vector, the form of the Bloch function is unaffected

$$u_k(x)\, e^{i(k+2\pi n/a)x} = [u_k(x)\, e^{i2\pi nx/a}]\, e^{ikx}$$
$$= u_k'(x)\, e^{ikx} \qquad (5.33)$$

The term in brackets in Eq. (5.33) still has the periodicity of the crystal lattice and can be redefined as a new $u_k'(x)$. It is therefore possible, and frequently convenient, to restrict the values of k to an interval of length $2\pi/a$, including only those values of k from $-\pi/a$ to $+\pi/a$, i.e., the length of one reciprocal lattice vector long and located symmetrically about $k = 0$. Such a process is like that used in Section 1.4 to simplify the representation of transverse lattice waves in a crystal.

When a function of k [such as $E(k)$] is expressed allowing k to take on all allowed values, the representation is known as an *extended-zone*

TABLE 5.1 COMPARISON BETWEEN ONE- AND THREE-DIMENSIONAL PARAMETERS AND RESULTS

	One dimension	Three dimensions
Unit cell	Line segment of length a	Edges defined by $\mathbf{a}_1, \mathbf{a}_2, \mathbf{a}_3$
Lattice vector	ja	$\mathbf{R}_j = j_1\mathbf{a}_1 + j_2\mathbf{a}_2 + j_3\mathbf{a}_3$
Bloch function	$\psi_k(x) = u_k(x)\,e^{ikx}$ $u_k(x) = u_k(x+a)$ $\psi_k(x+a) = e^{ika}\,\psi_k(x)$	$\psi_\mathbf{k}(\mathbf{r}) = u_\mathbf{k}(\mathbf{r})\,e^{i\mathbf{k}\cdot\mathbf{r}}$ $u_\mathbf{k}(\mathbf{r}) = u_\mathbf{k}(\mathbf{r}+\mathbf{R}_j)$ $\psi_\mathbf{k}(\mathbf{r}+\mathbf{R}_j) = e^{i\mathbf{k}\cdot\mathbf{R}_j}\psi_\mathbf{k}(\mathbf{r})$
Periodic boundary condition on k	$ka = \dfrac{2\pi n}{N}$	$\mathbf{k}\cdot\mathbf{a}_i = \dfrac{2\pi n_i}{N_i};\quad i=1,2,3$
Schroedinger equation	$\dfrac{d^2\psi}{dx^2} + \dfrac{2m}{\hbar^2}[E - V(x)]\psi = 0$ $V(x) = V(x+a)$	$\nabla^2\psi + \dfrac{2m}{\hbar^2}[E - V(\mathbf{r})]\psi = 0$ $V(\mathbf{r}) = V(\mathbf{r}+\mathbf{R}_j)$
Periodic boundary condition on $\psi(x)$	$\psi(x) = \psi(x + Na)$	$\psi(\mathbf{r}) = \psi(\mathbf{r} + N_1\mathbf{a}_1)$ $\phantom{\psi(\mathbf{r})} = \psi(\mathbf{r} + N_2\mathbf{a}_2)$ $\phantom{\psi(\mathbf{r})} = \psi(\mathbf{r} + N_3\mathbf{a}_3)$
Reciprocal lattice (coordinates of $\mathbf{k}$ space)	$\dfrac{2\pi}{a}$ $a\left(\dfrac{2\pi}{a}\right) = 2\pi$	$\mathbf{b}_1, \mathbf{b}_2, \mathbf{b}_3$ $\mathbf{a}_i\cdot\mathbf{b}_j = 2\pi\,\delta_{ij}$ $i,j = 1,2,3$
Reciprocal vector	$2\pi m/a$	$\mathbf{K}_m = m_1\mathbf{b}_1 + m_2\mathbf{b}_2 + m_3\mathbf{b}_3$
Relation between $\mathbf{k}$ and reciprocal lattice from periodic boundary condition on $\mathbf{k}$	$k = \dfrac{n}{N}\left(\dfrac{2\pi}{a}\right)$	$\mathbf{k} = \dfrac{n_1}{N_1}\mathbf{b}_1 + \dfrac{n_2}{N_2}\mathbf{b}_2 + \dfrac{n_3}{N_3}\mathbf{b}_3$
$\mathbf{k}$ vector unspecified for	$k + \dfrac{2\pi n}{a}$	$\mathbf{k} + \mathbf{K}_m$
Reduced-zone wavefunction	$\psi_{kl}(x) = u_{kl}(x)\,e^{ikx}$	$\psi_{\mathbf{k}l}(\mathbf{r}) = u_{\mathbf{k}l}(\mathbf{r})\,e^{i\mathbf{k}\cdot\mathbf{r}}$
Reduced-zone definition	$-\dfrac{\pi}{a} \leq k \leq +\dfrac{\pi}{a}$	(1) Draw reciprocal lattice vectors from $\mathbf{k} = 0$ to other lattice points and construct planes that perpendicularly bisect these vectors. (2) Take smallest region bounded by these planes surrounding the origin.
Equation of zone face	$k = \pm\dfrac{\pi}{a}$	$2\mathbf{k}\cdot\mathbf{K}_m = \pm\,\lvert\mathbf{K}_m\rvert^2$

5.3 Reciprocal Lattice and Brillouin Zones

description. When the range of k values is restricted to lie between $-\pi/a$ and $+\pi/a$, the representation is known as a *reduced-zone* description. The range of k between $-\pi/a$ and $+\pi/a$ is called variously the *fundamental domain*, the *first* or *basic Brillouin zone*. In terms of a reduced zone description, the Bloch function is written with an additional subscript,

$$\psi_{kl}(x) = u_{kl}(x)\, e^{ikx} \tag{5.34}$$

The subscript $l = 1$ indicates the wavefunction suitable for values of k in the first Brillouin zone, i.e., $-\pi/a \leq k \leq +\pi/a$. The subscript $l = 2$ indicates the wavefunction suitable for values of k in the second Brillouin zone, i.e., $-2\pi/a \leq k \leq -\pi/a$, and $+\pi/a \leq k \leq +2\pi/a$. The various Brillouin zones in one dimension are illustrated in Fig. 5.4. Every Brillouin zone has the same total width.

Example 5.3 Describe E vs. k in one dimension for a free electron in the extended- and reduced-zone representations.

In the extended-zone representation,

$$E = \frac{\hbar^2}{2m} k^2$$

and a typical plot is given in Fig. 5.5a.

To treat a free electron in a reduced-zone representation is like considering the effects of a periodic potential with zero magnitude on the

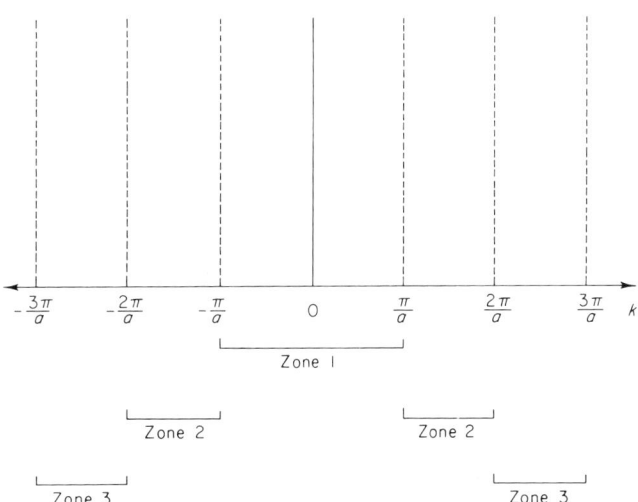

Fig. 5.4 Brillouin zones in one dimension.

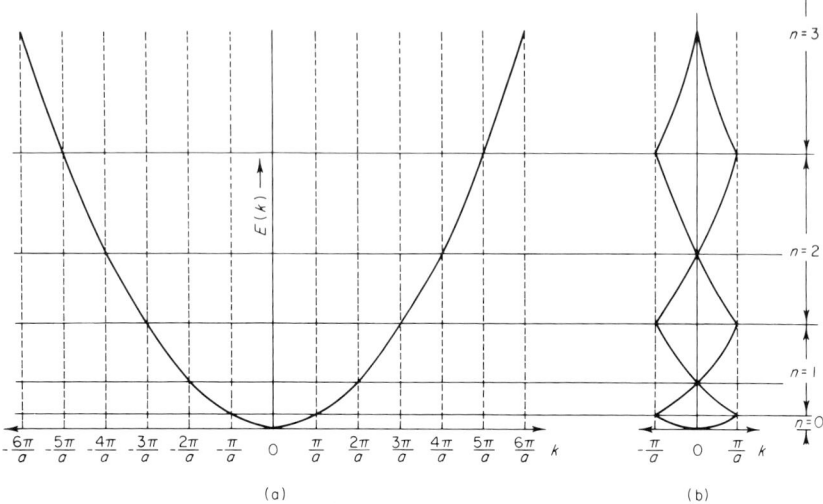

Fig. 5.5 Energy as a function of wavenumber for a one-dimensional free-electron model in both (a) the extended zone and (b) the reduced zone.

free electrons. Because it is like considering the effects of a periodic lattice on the electrons without the lattice being occupied by charged ion-cores, it is sometimes referred to as the *empty-lattice* representation.

In a periodic lattice k is unaltered by a translation of $\pm n2\pi/a$. Therefore the wavefunction for a free electron is

$$\psi_{kn}(x) = e^{i(k \pm n2\pi/a)x}$$
$$= (e^{\pm in2\pi x/a}) e^{ikx}$$

The term in parentheses is the $u_{kn}(x)$ function for the free electron. The solution of the Schroedinger equation

$$\frac{d^2\psi_{kn}}{dx^2} + \frac{2m}{\hbar^2} E\psi_{kn} = 0$$

gives energy eigenvalues of

$$E = \frac{\hbar^2}{2m} \left(k \pm \frac{n2\pi}{a}\right)$$

The reduced-zone representation for free electrons is given in Fig. 5.5b. Values of n are integers starting with $n = 0$ for the lowest energy curve.

In the three-dimensional case, **k** is also not uniquely specified in the Bloch function since replacement of **k** by (**k** + **K**$_m$) leaves the Bloch func-

5.3 Reciprocal Lattice and Brillouin Zones

tion unaffected. We may see this by considering

$$u_\mathbf{k}(\mathbf{r}) \, e^{i(\mathbf{k}+\mathbf{K}_m)\cdot\mathbf{r}} = [u_\mathbf{k}(\mathbf{r}) \, e^{i\mathbf{K}_m\cdot\mathbf{r}}] \, e^{i\mathbf{k}\cdot\mathbf{r}}$$
$$= u_\mathbf{k}'(\mathbf{r}) \, e^{i\mathbf{k}\cdot\mathbf{r}} \quad (5.35)$$

Since

$$e^{i\mathbf{K}_m\cdot(\mathbf{r}+\mathbf{R}_j)} = (e^{i\mathbf{K}_m\cdot\mathbf{r}})(e^{i\mathbf{K}_m\cdot\mathbf{R}_j}) = e^{i\mathbf{K}_m\cdot\mathbf{r}} \quad (5.36)$$

because

$$\mathbf{K}_m \cdot \mathbf{R}_j = 2\pi(\text{integer}) \quad (5.37)$$

the term in the brackets in Eq. (5.35) still has the periodicity of the lattice and the Bloch function has not been effectively changed. Because of this undetermined nature of $\mathbf{k}$, we may restrict $\mathbf{k}$ to a *single cell* of the reciprocal lattice, the *fundamental domain* of $\mathbf{k}$. In exact analogy with the one-dimensional case, the wavefunction is given by

$$\psi_{\mathbf{k}l}(\mathbf{r}) = u_{\mathbf{k}l}(\mathbf{r}) \, e^{i\mathbf{k}\cdot\mathbf{r}} \quad (5.38)$$

where the subscript l is the index indicating to which Brillouin zone the values of k relate.

The choice of the fundamental domain or first Brillouin zone of k in one dimension was a relatively obvious one: one reciprocal lattice length long and centered on $k = 0$. In order to choose the corresponding fundamental domain in the three-dimensional case, the following directions are followed:

1. Draw the reciprocal lattice vectors joining the origin at $\mathbf{k} = 0$ to the other lattice points, and construct the planes that perpendicularly bisect these vectors.

2. Take the fundamental domain of $\mathbf{k}$ to be the smallest region surrounding the origin and bounded by these planes.

It is evident that this prescription can be applied to the one-dimensional case to produce the results already stated.

The equation for the faces of the fundamental domain in one dimension was simply $k = \pm\pi/a$. The equation for the faces of the fundamental domain in three dimensions can be derived from the typical geometrical situation pictured in Fig. 5.6:

$$\overline{OP} = \frac{1}{2}|\mathbf{K}_m| = k\cos\theta = \frac{\mathbf{k}\cdot\mathbf{K}_m}{|\mathbf{K}_m|} \quad (5.39)$$

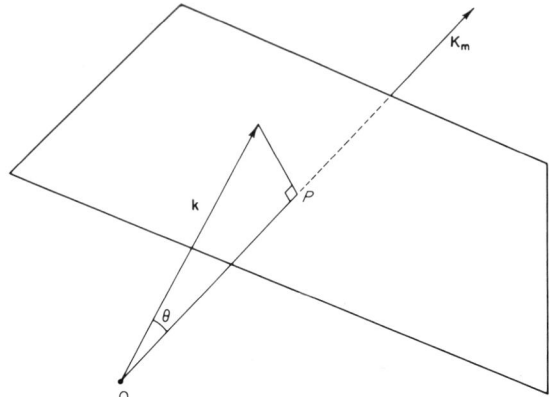

Fig. 5.6 Construction to determine the equation of a face of the Brillouin zone.

Therefore the equation for the faces of the fundamental domain has the form

$$2\mathbf{k} \cdot \mathbf{K}_m = |\mathbf{K}_m|^2 \qquad (5.40)$$

For each such face there is also a parallel face, separated by a perpendicular distance of $\mathbf{K}_m$, with the equation,

$$2\mathbf{k} \cdot \mathbf{K}_m = -|\mathbf{K}_m|^2$$

It is also evident that these more general relations can be applied to the one-dimensional case with $\mathbf{K}_m = 2\pi/a$ to obtain the previously stated results.

Example 5.4 What is the fundamental domain for a simple cubic crystal?

From the determination of the reciprocal lattice vectors in Example 5.1, we see that the fundamental domain is a simple cube with side $2\pi/a$. The vectors $\mathbf{K}_m$ that are bisected by the faces of this cube are $\pm(2\pi/a)\mathbf{e}_1$, $\pm(2\pi/a)\mathbf{e}_2$, and $\pm(2\pi/a)\mathbf{e}_3$. If we write

$$\mathbf{k} = k_1\mathbf{e}_1 + k_2\mathbf{e}_2 + k_3\mathbf{e}_3$$

then the fundamental domain is bounded by

$$k_1 = k_2 = k_3 = \pm\frac{\pi}{a}$$

5.3 Reciprocal Lattice and Brillouin Zones

Just as in one dimension, Brillouin zones in three dimensions can be represented in an extended-zone as well as in a reduced-zone scheme. As in one dimension all Brillouin zones have the same width, so in three dimensions all Brillouin zones have the same volume. Regardless of the shape of the fundamental domain, the volume is always given by

$$\mathbf{b}_1 \cdot \mathbf{b}_2 \times \mathbf{b}_3 = \frac{8\pi^3}{\Omega^3} (\mathbf{a}_2 \times \mathbf{a}_3) \cdot (\mathbf{a}_3 \times \mathbf{a}_1) \times (\mathbf{a}_1 \times \mathbf{a}_2)$$

$$= \frac{8\pi^3}{\Omega} \quad (5.41)$$

The volume of the fundamental domain of $\mathbf{k}$ is $8\pi^3$ times the reciprocal of the volume of a unit cell in the crystal lattice.

Example 5.5 Determine the extended-zone and reduced-zone representations for a two-dimensional simple square lattice.

For the two-dimensional simple square lattice, the reciprocal lattice vectors are

$$\mathbf{K}_m = m_1 \frac{2\pi}{a} \mathbf{e}_1 + m_2 \frac{2\pi}{a} \mathbf{e}_2$$

The faces of the Brillouin zones are determined by the relation of Eq. (5.40),

$$2k_1 \frac{2\pi}{a} m_1 + 2k_2 \frac{2\pi}{a} m_2 = \frac{4\pi^2}{a^2} (m_1^2 + m_2^2)$$

or

$$k_1 m_1 + k_2 m_2 = \frac{\pi}{a} (m_1^2 + m_2^2)$$

The first zone is defined by $m_1 = \pm 1$ and $m_2 = 0$, giving $k_1 = \pm \pi/a$, and by $m_1 = 0$ and $m_2 = \pm 1$, giving $k_2 = \pm \pi/a$. As expected, therefore, the first zone is a square with center at $k_1 = k_2 = 0$, and with each side of length $2\pi/a$. This is the reduced zone.

The second Brillouin zone is defined by $m_1 = \pm 1$ and $m_2 = \pm 1$, giving the four possible permutations of

$$\pm k_1 \pm k_2 = \frac{2\pi}{a}$$

Thus the second zone lies *between* the first zone and these four 45° lines.

138   5 Origin of Energy Bands in Solids

The third zone is defined by $m_1 = \pm 2, \pm 1$ and $m_2 = 0$, giving $k_1 = \pm 2\pi/a, \pm \pi/a$, and by $m_1 = 0$ and $m_2 = \pm 2, \pm 1$, giving $k_2 = \pm 2\pi/a, \pm \pi/a$. The third zone is then the *smallest* area *outside* the second zone and bounded by these eight straight lines.

Going to higher zones becomes complicated. An alternative procedure is to construct the zones geometrically, as shown in Fig. 5.7,

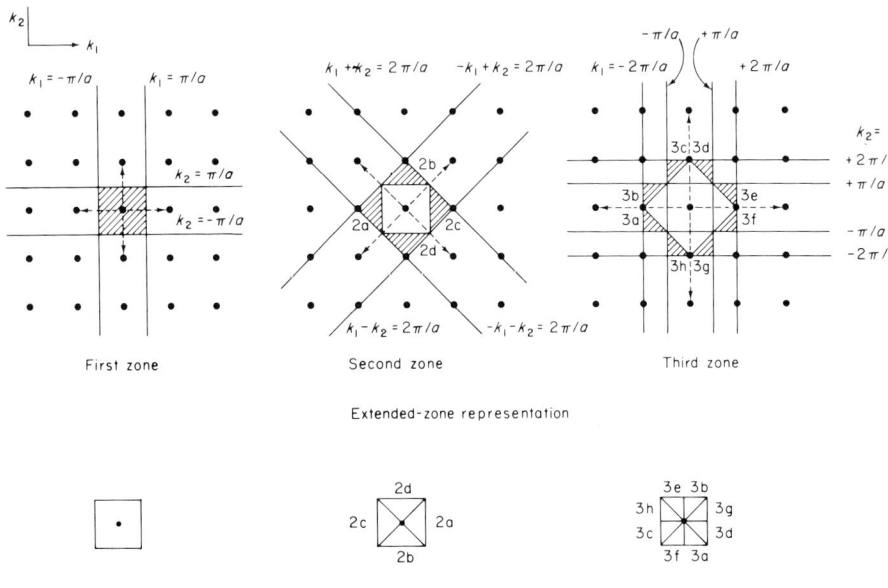

Fig. 5.7 The first three Brillouin zones for a simple square lattice, shown both in the extended-zone and in the reduced-zone representation. The dashed lines in the extended-zone representation are the $\mathbf{K}_m$ vectors.

using the prescription given earlier in this section. For the first zone, reciprocal lattice vectors are drawn from the origin to the nearest reciprocal lattice points; these vectors are bisected and the smallest area including the origin to be formed by the bisecting lines forms the first Brillouin zone. For successively higher zones, reciprocal lattice vectors are drawn to successively further reciprocal lattice points, and the same procedure of bisection is followed, care being taken to enclose the smallest area.

The way in which the extended zone is fit into the reduced zone by translation by reciprocal lattice vectors is also illustrated in Fig. 5.7.

5.3 Reciprocal Lattice and Brillouin Zones

The variation of energy E with wave vector $\mathbf{k}$ can be represented for a free electron in the empty-lattice picture in three dimensions as was done previously in Fig. 5.5 for one dimension. Since a three-dimensional variation is difficult to reproduce on a two-dimensional page, it is frequently sufficient to describe the variation of E with $\mathbf{k}$ in certain selected *directions* in the fundamental domain.

The fundamental domain (first Brillouin zone) for a face-centered cubic crystal is shown in Fig. 5.8. Certain standard notation has been adopted to describe directions and points in this fundamental domain. The 100 axis is labeled the Δ axis, the 111 axis the Λ axis, and the 110 axis the Σ axis. The center of the zone is the Γ point. If one proceeds from the Γ point along the Δ axis, the intersection with the zone surface is labelled the X point. Similarly the intersections with the zone surface along the Λ and Σ axes are respectively labeled the L and K points.

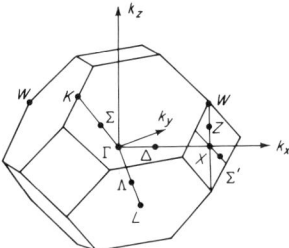

Fig. 5.8 The reduced zone for the fcc crystal with standard symmetry notation for directions and intersections.

The variation of energy E with $\mathbf{k}$ in a three-dimensional empty lattice is given by

$$E = \frac{\hbar^2}{2m} |\mathbf{k} + \mathbf{K}_m|^2 \tag{5.42}$$

which corresponds to the one-dimensional expression given in Example 5.3. Since $\mathbf{k} = k_1 \mathbf{e}_1 + k_2 \mathbf{e}_2 + k_3 \mathbf{e}_3$ and $\mathbf{K}_m$ is given by Eq. (5.25) with the reciprocal vectors of Example 5.4, Eq. (5.42) becomes

$$E = \frac{\hbar^2}{2m} \left\{ \left[k_1 + \frac{2\pi}{a}(m_1 - m_2 + m_3) \right]^2 + \left[k_2 + \frac{2\pi}{a}(m_1 + m_2 - m_3) \right]^2 \right.$$
$$\left. + \left[k_3 + \frac{2\pi}{a}(-m_1 + m_2 + m_3) \right]^2 \right\} \tag{5.43}$$

By properly adjusting the system of units used, this expression can be reduced to one of the form

$$E = (k_1 + M_1)^2 + (k_2 + M_2)^2 + (k_3 + M_3)^2 \tag{5.44}$$

A plot of E vs. $\mathbf{k}$ is given in Fig. 5.9 along the 100 axis and along the 111

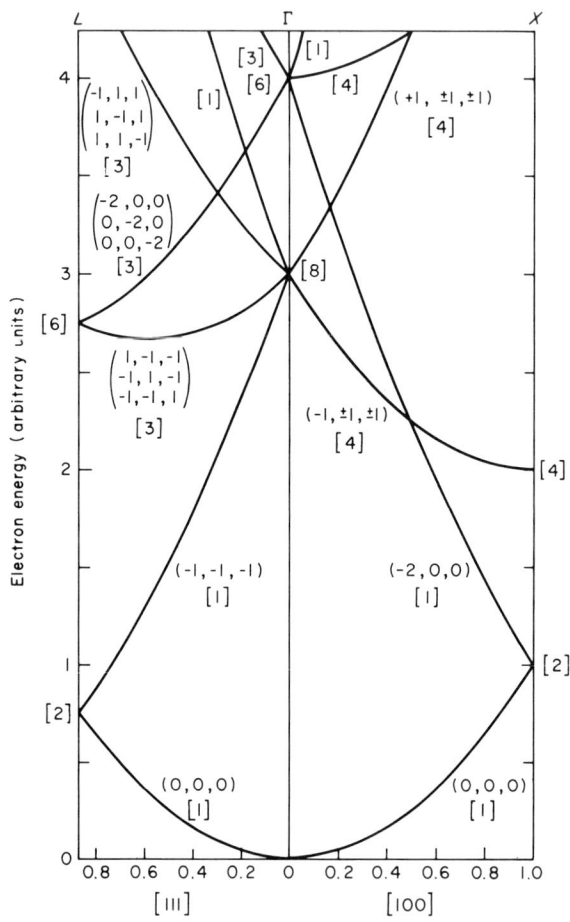

Fig. 5.9 Energy as a function of wavevector for a three-dimensional free-electron model in a cubic crystal. Values of k are plotted in the (100) and (111) directions in units of $k_{\max}$ in the (100) direction. The numbers in parentheses refers to the integers of Eq. (5.44), and the numbers in brackets indicate the degeneracy corresponding to a particular curve. (After D. Long, "Energy Bands in Semiconductors," Wiley-Interscience, New York, 1968 p. 19.)

5.4 Perturbed Nearly Free Electron System

axis in the reduced-zone representation. Along the 100 axis, for example, $k_2 = k_3 = 0$. The lowest energy band is given by $M_1 = M_2 = M_3 = 0$, i.e., by $E = k_1^2$. The next lowest energy band is given by $(M_1, M_2, M_3) = (-2, 0, 0)$, i.e., by $E = (k_1 - 2)^2$. The next lowest energy band corresponds to the four values $(M_1, M_2, M_3) = (-1, \pm 1, \pm 1)$, i.e., to $E = (k_1 - 1)^2 + 2$, and is fourfold degenerate. The appropriate values of (M_1, M_2, M_3) and the corresponding degeneracy are given for each curve in Fig. 5.9. E vs. **k** in the 111 direction is calculable in a similar way with $k_1 = k_2 = k_3$.

5.4 Energy Bands in a Perturbed Nearly Free Electron System

The allowed energy spectrum for free electrons is continuous and any positive or zero value is allowed for the energy. What happens to the allowed energies of such a free-electron system if a weak periodic potential is suddenly "turned on"? The answer to this question can be found by treating the weak periodic potential as a perturbation to the Hamiltonian for the free electrons. A brief summary of perturbation theory is necessary before this calculation can be made.

TIME INDEPENDENT PERTURBATION THEORY

Suppose that the problem for an unperturbed system has been completely solved, that

$$\mathbf{H}_0 \psi_{i0} = E_{i0} \psi_{i0} \tag{5.45}$$

and that the exact equation for the perturbed system is

$$(\mathbf{H}_0 + \mathbf{H}')\psi_i = (E_{i0} + E_i')\psi_i \tag{5.46}$$

where the ψ_i are the eigenfunctions for the perturbed system. Although these are unknown, still we can formally write them *exactly* as linear combinations of the eigenfunctions of the unperturbed system

$$\psi_i = \sum_j^\infty a_{ij} \psi_{j0} \tag{5.47}$$

If we substitute this wavefunction into Eq. (5.46) we obtain

$$\sum_j^\infty a_{ij}(\mathbf{H}_0 + \mathbf{H}')\psi_{j0} = (E_{i0} + E') \sum_j^\infty a_{ij} \psi_{j0} \tag{5.48}$$

Now multiply through by ψ_{p0}^* and integrate over all coordinates to obtain

$$\sum_j^\infty a_{ij} \int \psi_{p0}^* \mathbf{H} \psi_{j0}\, d\tau = E_i \sum_j^\infty a_{ij} \int \psi_{p0}^* \psi_{j0}\, d\tau \quad (5.49)$$

with $\mathbf{H} = (\mathbf{H}_0 + \mathbf{H}')$ and $E_i = (E_{i0} + E')$. Eq. (5.49) can be rewritten as

$$\sum_j^\infty a_{ij}(H_{pj} - E_i \delta_{pj}) = 0 \quad (5.50)$$

where $H_{pj} \equiv \int \psi_{p0}^* \mathbf{H} \psi_{j0}\, d\tau$, and is called a *matrix element*. Thus we have an infinite series of equations. For this series of equations to have solutions for nonzero values of the a_{ij}, it is necessary that the determinant of the coefficients vanish, i.e.,

$$\begin{vmatrix} H_{11} - E_i & H_{12} & H_{13} & \cdots \\ H_{21} & H_{22} - E_i & H_{23} & \cdots \\ H_{31} & H_{32} & H_{33} - E_i & \cdots \\ \vdots & \vdots & \vdots & \vdots \end{vmatrix} = 0 \quad (5.51)$$

This infinite-order secular equation has an infinite number of roots giving values for the allowed E_i of the perturbed system. These solutions are the *exact* values for the infinite number of energy states in the perturbed system. No approximation is involved through Eq. (5.51).

Examination of Eq. (5.51) shows that the effect of the perturbation is to make the off-diagonal matrix elements not equal to zero. If the perturbation term were zero, then for $p \neq j$,

$$H_{pj} = \int \psi_{p0}^* H \psi_{j0}\, d\mathbf{r} = E_{j0} \int \psi_{p0}^* \psi_{j0}\, d\mathbf{r} = 0 \quad (5.52)$$

and the roots of the equation would be simply $E_i = H_{11}, H_{22}, H_{33}, \ldots$.

For a nondegenerate state, first-order perturbation theory assumes that these relationships still hold.

For a degenerate state, first-order perturbation theory considers all off-diagonal matrix elements to be zero *except* for those involving two of the degenerate wavefunctions. Suppose for example that we have a doubly degenerate state with $E_{10} = E_{20}$. Then in addition to the diagonal elements, we leave also H_{12} and H_{21} in the secular determinant to obtain

$$\begin{vmatrix} H_{11} - E_i & H_{12} & 0 & 0 & \cdots \\ H_{21} & H_{22} - E_i & 0 & 0 & \cdots \\ 0 & 0 & H_{33} - E_i & 0 & \cdots \\ \vdots & \vdots & \vdots & \vdots & \vdots \end{vmatrix} = 0 \quad (5.53)$$

5.4 Perturbed Nearly Free Electron System

The roots now are given by

$$\begin{vmatrix} H_{11} - E_i & H_{12} \\ H_{21} & H_{22} - E_i \end{vmatrix} (H_{33} - E_i)(H_{44} - E_i) \cdots = 0 \qquad (5.54)$$

Higher-order approximations are achieved by retaining more of the off-diagonal matrix elements in the calculation.

Example 5.6 An electron in a hydrogenic atom is subjected to a small magnetic field in the $+z$ direction. What is the effect of the perturbing magnetic field on a 2p state of the electron?

The potential energy resulting from the interaction of the electron's orbital magnetic moment with the magnetic field is given by

$$V = -\mu_z H = \frac{e}{2mc} L_z H$$

where μ_z is the orbital magnetic moment of the electron and L_z is the angular momentum of the electron. The operator to be associated with L_z is $-i\hbar \, \partial/\partial\phi$. Therefore the perturbation Hamiltonian is

$$\mathbf{H}' = -i\hbar \frac{eH}{2mc} \frac{\partial}{\partial\phi}$$

Four states have the same energy in the unperturbed system. First there is the 2s state with wavefunction

$$\psi^{2s} = A_s(2 - Zr/a_0) \, e^{-Zr/2a_0}$$

and then there are the three 2p states with wavefunctions

$$\psi^{2p}_{+1} = A_{+1} r \, e^{-Zr/2a_0} \sin\theta \, e^{i\phi}$$
$$\psi^{2p}_{0} = A_0 r \, e^{-Zr/2a_0} \cos\theta$$
$$\psi^{2p}_{-1} = A_{-1} r \, e^{-Zr/2a_0} \sin\theta \, e^{-i\phi}$$

where the subscript on the 2p wavefunctions indicates the value of m_l. The secular determinant for this perturbation calculation to first order is a fourth-order determinant. We shall write the two subscripts on the matrix elements involved as "s" if an s state is involved, or as the specific value of m_l if a p state is involved. For example,

$$H'_{0,s} \equiv \int \psi^{2p*}_0 \mathbf{H}' \psi^{2s} \, d\mathbf{r}$$

The secular determinant is therefore

$$\begin{vmatrix} H'_{s,s} - E' & H'_{s,1} & H'_{s,0} & H'_{s,-1} \\ H'_{1,s} & H'_{1,1} - E' & H'_{1,0} & H'_{1,-1} \\ H'_{0,s} & H'_{0,1} & H'_{0,0} - E' & H'_{0,-1} \\ H'_{-1,s} & H'_{-1,1} & H'_{-1,0} & H'_{-1,-1} - E' \end{vmatrix}$$

There are sixteen matrix elements to be evaluated, but a quick inspection shows that this number can be greatly reduced immediately. Since **H'** involves only $\partial/\partial\phi$ and since $H'_{ij} = H'^*_{ji}$,

$$H'_{s,s} = H'_{1,s} = H'_{s,1} = H'_{0,s} = H'_{s,0} = H'_{-1,s} = H'_{s,-1} = H'_{1,0} = H'_{0,1}$$
$$= H'_{0,0} = H'_{-1,0} = H'_{0,-1} = 0$$

The four remaining matrix elements are evaluated as follows:

$$H'_{1,1} = -H'_{-1,-1} = \frac{e\hbar H}{2mc} A^2_{\pm 1} \int_0^\infty r^4 e^{-Zr/a_0} dr \int_0^\pi \sin^3\theta\, d\theta \int_0^{2\pi} d\phi$$
$$= \frac{e\hbar H}{2mc}$$

$$H'_{1,-1} = -\frac{e\hbar H}{2mc} A_{+1}A_{-1} \int_0^\infty r^4 e^{-Zr/a_0} dr \int_0^\pi \sin^3\theta\, d\theta \int_0^{2\pi} e^{-2i\phi} d\phi = 0$$

$$H'_{-1,1} = \frac{e\hbar H}{2mc} A_{-1}A_{+1} \int_0^\infty r^4 e^{-Zr/a_0} dr \int_0^\pi \sin^3\theta\, d\theta \int_0^{2\pi} e^{2i\phi} d\phi = 0$$

Thus the secular equation reduces to

$$\begin{vmatrix} -E' & 0 & 0 & 0 \\ 0 & \frac{e\hbar H}{2mc} - E' & 0 & 0 \\ 0 & 0 & -E' & 0 \\ 0 & 0 & 0 & -\frac{e\hbar H}{2mc} - E' \end{vmatrix} = 0$$

Solution for E' shows that

$$E' = 0,\ 0,\ +\frac{e\hbar H}{2mc},\ -\frac{e\hbar H}{2mc}$$

We conclude that the s state is unaffected by the perturbation, as is one of the p states. The other two p states have been altered in energy

5.4 Perturbed Nearly Free Electron System

by the perturbation, however, one being increased by $e\hbar H/(2mc)$ and one being decreased by $e\hbar H/(2mc)$. The total energies of these four degenerate (in the absence of the perturbation) states with the perturbation are given by $E = E_0 + E'$:

$$E_1 = -\frac{Z^2 me^4}{8\hbar^2}$$

$$E_2 = -\frac{Z^2 me^4}{8\hbar^2} + \frac{e\hbar H}{2mc}$$

$$E_3 = -\frac{Z^2 me^4}{8\hbar^2}$$

$$E_4 = -\frac{Z^2 me^4}{8\hbar^2} - \frac{e\hbar H}{2mc}$$

This splitting of energy levels by a magnetic field is known as the *Zeeman effect*.

ENERGY BANDS IN ONE DIMENSION

The wavefunction for a free electron in a one-dimensional periodic lattice is doubly degenerate since the same energy is indicated regardless of whether $+k$ or $-k$ is used for the wave number. The energy $E = \hbar^2 k^2/2m$ corresponds to both $\psi_{+k}(x) = (L)^{-1/2} e^{+ikx}$, and to $\psi_{-k}(x) = (L)^{-1/2} e^{-ikx}$. If we are going to treat the periodic potential as a perturbation on the free-electron system, we need to apply first-order perturbation theory for a doubly degenerate level. The secular equation for the calculation is

$$\begin{vmatrix} H'_{++} - E' & H'_{+-} \\ H'_{-+} & H'_{--} - E' \end{vmatrix} = 0 \qquad (5.55)$$

The matrix elements are

$$H'_{++} = \int_0^L \psi^*_{+k}(x)\, V(x)\, \psi_{+k}(x)\, dx \qquad (5.56)$$

$$H'_{+-} = \int_0^L \psi^*_{+k}(x)\, V(x)\, \psi_{-k}(x)\, dx \qquad (5.57)$$

$$H'_{-+} = \int_0^L \psi^*_{-k}(x)\, V(x)\, \psi_{+k}(x)\, dx \qquad (5.58)$$

$$H'_{--} = \int_0^L \psi^*_{-k}(x)\, V(x)\, \psi_{-k}(x)\, dx \qquad (5.59)$$

where $V(x) = V(x + a)$ is the small periodic potential.

Considerable simplification in the matrix elements can be achieved by realizing that

$$H'_{++} = H'_{--} = \frac{1}{L}\int_0^L V(x)\,dx = 0 \qquad (5.60)$$

since the integral of Eq. (5.60) represents an average of $V(x)$ over the whole crystal, which may be set equal to zero because of the periodic character of $V(x)$ by the appropriate choice of the reference zero of energy.

It is also seen that

$$H'_{+-} = \frac{1}{L}\int_0^L e^{-2ikx}\,V(x)\,dx = H'^{*}_{-+} \qquad (5.61)$$

With the incorporation of these simplifications, the secular equation becomes

$$E'^2 = |H'_{+-}|^2 \qquad (5.62)$$

All that remains is to evaluate the integral of Eq. (5.61).

To accomplish this we need to consider the general properties of the integral

$$\int_0^L e^{ikx}\,V(x)\,dx \qquad (5.63)$$

when $V(x) = V(x+a)$ is a periodic potential.

1. The integration of Eq. (5.63) over the whole crystal can be reduced to a sum of integrals, each over a unit cell, i.e., each over an interatomic spacing in the one-dimensional crystal

$$\int_0^L e^{ikx}\,V(x)\,dx = \sum_{m=0}^{N-1}\int_{ma}^{(m+1)a} e^{ikx}\,V(x)\,dx \qquad (5.64)$$

where m is an integer, a is the interatomic spacing, and N is the number of atoms such that $L = Na$. This reduction is made possible by the fact that we are dealing with a periodic potential that takes on the same variation with x in each unit cell.

2. The integral in Eq. (5.64) over the mth unit cell can be reduced to an integral over the first unit cell, $0 \leq x \leq a$, multiplied by an appropriate phase factor. Again periodicity provides the result. Let $x = X + ma$.

5.4 Perturbed Nearly Free Electron System

Then,

$$\int_{ma}^{(m+1)a} e^{ikx} V(x) \, dx = \int_0^a e^{ik(X+ma)} V(X+ma) \, dX$$

$$= e^{ikma} \int_0^a e^{ikX} V(X) \, dX$$

$$= e^{ikma} \int_0^a e^{ikx} V(x) \, dx \qquad (5.65)$$

since $V(X + ma) = V(X)$, and since the definite integral expressed with X as the variable is the same as the definite integral expressed with x as the variable.

To summarize, at this point we have the result

$$\int_0^L e^{ikx} V(x) \, dx = \sum_{m=0}^{N-1} e^{ikma} \int_0^a e^{ikx} V(x) \, dx \qquad (5.66)$$

The integral represents the contribution from a unit cell, these contributions then to be summed with each multiplied by the appropriate phase factor.

3. The sum over the phase factors in Eq. (5.66) shows that the integral over the whole crystal is zero except for particular values of the wavenumber k.

For a geometric progression, with A for the first term, G for the number of terms, and R for the common ratio, the sum S is given by

$$S = A \frac{R^G - 1}{R - 1} \qquad (5.67)$$

Therefore, with $A = 1$, $G = N$, and $R = e^{ika}$,

$$\sum_{m=0}^{N-1} e^{ikma} = \frac{e^{iNka} - 1}{e^{ika} - 1} \qquad (5.68)$$

Periodic boundary conditions also put certain restrictions on the value of k in the wavefunction $\psi_k(x)$; according to Eq. (4.40), $k = 2\pi n/Na$. Substituting this value into Eq. (5.68) gives

$$\sum_{m=0}^{N-1} e^{ikma} = \frac{e^{2n\pi i} - 1}{e^{2n\pi i/N} - 1} \qquad (5.69)$$

Examination of this expression shows that its value depends on the value of the ratio n/N. If n/N is not an integer, then the sum of Eq. (5.69) is zero since the numerator is zero. However, if n/N is an integer (corresponding to the interatomic spacing being equal to an integral number of wavelengths), each term of the sum is simply unity and

$$\sum_{m=0}^{N-1} e^{ikma} = N \tag{5.70}$$

whenever

$$k = q\frac{2\pi}{a} \tag{5.71}$$

where q is an integer.

We are now in a position to return to the calculation of allowed energies for nearly free electrons in the presence of a small perturbing periodic potential. The correction to the energy E' because of the perturbing potential is equal to zero for all values of k except for $k = q\pi/a$. [The factor-of-2 difference with Eq. (5.71) occurs because of the factor-of-2 difference between Eqs. (5.61) and (5.63).] According to first-order perturbation theory there is no effect on the energies of free electrons because of a periodic potential except at $k = q\pi/a$.

If the *Bragg reflection rule* ($n\lambda = 2d \sin \theta$) is written in an appropriate one-dimensional form ($n\lambda = 2d$), the condition $k = q\pi/a$ is exactly the same as that set forth by the Bragg reflection rule. This means that a wave with $k = q\pi/a$ undergoes Bragg reflection upon attempting to propagate through the crystal, and that therefore such a wave cannot propagate through the crystal. The existence of a range of forbidden energies for electrons in a crystal can be correlated with reflection of electron waves with particular values of k by the atom planes of the crystal. Furthermore, the condition $k = q\pi/a$ defines those values of k that are *on the faces* of the Brillouin zones for the one-dimensional crystal. The values of k corresponding to the faces of the various Brillouin zones are the values of k that satisfy the Bragg reflection condition, and hence are the values of k corresponding to reflected nonpropagating waves and to regions of forbidden energy.

At the values of $k = q\pi/a$, the energy levels are split into

$$E = \frac{\hbar^2}{2m}\left(\frac{q\pi}{a}\right)^2 \pm |H_q'| \tag{5.72}$$

where

$$H_q' = \frac{1}{L}\int_0^L e^{-2\pi qix/a} V(x)\,dx$$

$$= \frac{1}{a}\int_0^a e^{-2\pi qix/a} V(x)\,dx \tag{5.73}$$

In a consideration of the variation of E with k, therefore, there will be an energy gap at $k = q\pi/a$ of $2|H_q'|$. This effect is shown in Fig. 5.10 in

5.4 Perturbed Nearly Free Electron System

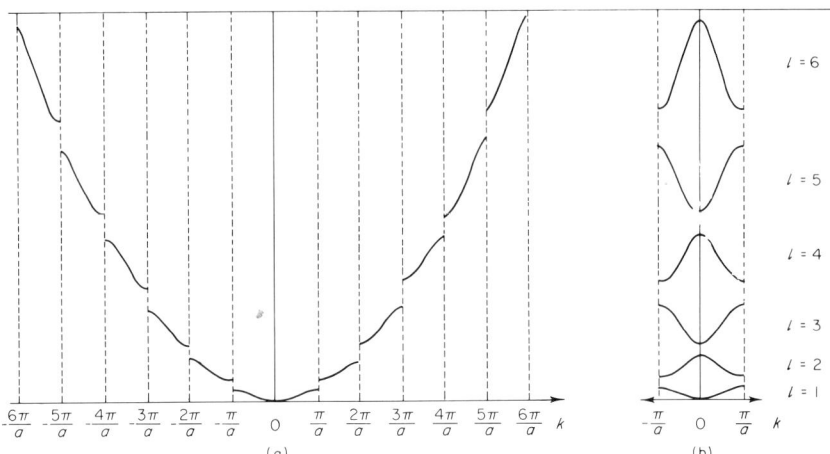

Fig. 5.10 Effect of a small periodic potential on the allowed energies for nearly free electrons in one dimension, in both (a) the extended zone and (b) the reduced zone. Compare with Fig. 5.5.

both the extended-zone and reduced-zone representation. These curves should be compared with those given in Fig. 5.5 for the empty-lattice case. Since there is a gap at $k = q\pi/a$, the actual E vs. k curves must be affected to a lesser extent near $k = q\pi/a$, as shown schematically in Fig. 5.10, but these effects are not given by a first-order perturbation calculation.

The periodic potential has opened up forbidden energy gaps in the continuous E vs. k curves of the empty-lattice case, forming a series of allowed bands separated by forbidden bands. In a Brillouin zone there are N values of k; each Brillouin zone (i.e., each allowed energy band) therefore contains N orbital states. We may see this by remembering that the restrictions of periodic boundary conditions on k give $k = 2\pi n/Na$, so that there is an interval of $2\pi/Na$ between one k value and the next. Since the width of a Brillouin zone is $2\pi/a$, it follows that there are N values of k in this zone.

The wavefunctions to be used in the presence of the perturbing potential at $k = q\pi/a$ are linear combinations of the unperturbed wavefunctions, and therefore are of the form

$$\Psi = A_1 e^{q\pi i x/a} + A_2 e^{-q\pi i x/a} \tag{5.74}$$

To evaluate A_1 and A_2, refer back to the equations leading to the secular determinant of Eq. (5.55),

$$A_1(H'_{++} - E') + A_2 H'_{+-} = 0 \tag{5.75}$$

$$A_1 H'_{-+} + A_2(H'_{--} - E') = 0 \tag{5.76}$$

which become, by Eq. (5.60),

$$-A_1 E' + A_2 H'_{+-} = 0 \tag{5.77}$$

$$A_1 H'_{-+} - A_2 E' = 0 \tag{5.78}$$

Since we know that $E' = \pm \mid H'_{+-} \mid$, it follows that $A_1 = \pm A_2$ and that the wavefunction of Eq. (5.74) is proportional to either $\sin q\pi x/a$ or $\cos q\pi x/a$, i.e., they are standing waves at $k = q\pi/a$. Therefore an electron with $k = q\pi/a$ cannot be propagated through the crystal, but undergoes reflection. This is another way of looking at the Bragg reflection effects, and says that an electron in a given band is confined to that band unless sufficient external energy is supplied to enable it to "jump" across the forbidden gap to the next-higher-energy unoccupied allowed levels.

Consider now qualitatively the following interesting properties of an electron in an otherwise unoccupied energy band. If the energy of the electron is increased gradually from zero, the electron acquires increasing velocity in a particular direction, say the $+x$ direction. This process will continue *until* the energy of the electron is such that Bragg reflection of it by the crystal planes occurs. That is, the electron continues to increase in energy until it acquires that energy corresponding to the discontinuous jump in energy at $k = q\pi/a$. At that point the attempt to provide the electron with higher energy in the $+x$ direction results in a reflection and the motion of the electron in the $-x$ direction! The acceleration of the electron in the crystal is in the opposite direction to the external force applied, this reversal being the result of the periodic potential of the crystal. If we treat the electron in the crystal as if it were a free electron with an effective mass responding only to the external force, then the observation of acceleration and force in opposite directions forces us to ascribe a *negative effective mass* to the electron, since by definition mass is the proportionality constant between force and acceleration. We develop these ideas further in Chapter 6.

ENERGY BANDS IN THREE DIMENSIONS

When we go from one to three dimensions in a consideration of the effect of a small periodic potential on a nearly free electron system, we encounter differences primarily only of degree, although a few new possibilities arise with the availability of more dimensions.

In the one-dimensional case, the wavefunction is doubly degenerate and energy gaps occur only when H'_{+-} is not zero. In three dimensions the

5.4 Perturbed Nearly Free Electron System

degeneracy is much greater since *all* of the unperturbed states lying on a sphere of radius k in $\mathbf{k}$ space have the same energy, $E = \hbar^2 |\mathbf{k}|^2/2m$. It is still true in three dimensions, however, that energy gaps occur only when some off-diagonal elements of the secular determinant are not zero, i.e., when $H'_{\mathbf{k}\mathbf{k}'} \neq 0$ for $|\mathbf{k}| = |\mathbf{k}'|$. The value of $H'_{\mathbf{k}\mathbf{k}'}$ is given by

$$H'_{\mathbf{k}\mathbf{k}'} = \int \psi_{\mathbf{k}}^*(\mathbf{r})\, V(\mathbf{r})\, \psi_{\mathbf{k}'}(\mathbf{r})\, d\mathbf{r}$$

$$= \frac{1}{\mathscr{V}} \int e^{i(\mathbf{k}'-\mathbf{k})\cdot\mathbf{r}}\, V(\mathbf{r})\, d\mathbf{r} \tag{5.79}$$

where $\mathscr{V}$ is the volume of the crystal. The integral over the whole crystal in Eq. (5.79) can be reduced to the integral over a single unit cell multiplied by a sum over the phase factors appropriate for each of the unit cells in the crystal.

To transform the integral from over the whole crystal to over a unit cell, use is made of the periodicity of $V(\mathbf{r})$. Let $\mathbf{r} = \mathbf{r}' + \mathbf{R}_j$; then

$$H'_{\mathbf{k}\mathbf{k}'} = \frac{1}{\mathscr{V}} \sum_n \int_{n\text{th cell}} e^{i(\mathbf{k}'-\mathbf{k})\cdot(\mathbf{r}'+\mathbf{R}_j)}\, V(\mathbf{r}')\, d\mathbf{r}'$$

$$= \frac{1}{\mathscr{V}} \sum_n e^{i(\mathbf{k}'-\mathbf{k})\cdot\mathbf{R}_j} \int_{\text{cell}} e^{i(\mathbf{k}'-\mathbf{k})\cdot\mathbf{r}}\, V(\mathbf{r})\, d\mathbf{r} \tag{5.80}$$

The integral represents a quantity determinable for a unit cell of the crystal. The contribution of each unit cell must then be summed when multiplied by the appropriate phase factor. Since $\mathbf{R}_j = j_1\mathbf{a}_1 + j_2\mathbf{a}_2 + j_3\mathbf{a}_3$, and $\mathbf{k} \cdot \mathbf{a}_i = 2\pi n_i/N$,

$$(\mathbf{k}' - \mathbf{k}) \cdot \mathbf{R}_j = 2\pi \left[\frac{(n_1' - n_1)}{N_1} j_1 + \frac{(n_2' - n_2)}{N_2} j_2 + \frac{(n_3' - n_3)}{N_3} j_3 \right] \tag{5.81}$$

The sum in Eq. (5.80) can be written as a triple sum, over n_1, n_2, and n_3. By considering each of these sums separately, and applying the same type of reasoning used to arrive at the result in the one-dimensional case, we conclude that the sums indicated in Eq. (5.80) are identically zero except in the special case when the relation between n_i', n_i, j_i, and N_i is such that each term in each of the sums is identically unity and the value of the triple sum of Eq. (5.80) is simply $N_1 N_2 N_3$. The required relationships are that $(n_i' - n_i)/N_i$ be integers, i.e.,

$$n_i' = n_i \pm q_i N_i \tag{5.82}$$

where q_i is the corresponding integer. By comparing Eqs. (5.25) and (5.31) with Eq. (5.82), it follows that the requirement of Eq. (5.82) is equivalent to the requirement that

$$\mathbf{k}' = \mathbf{k} \pm \mathbf{K}_m \qquad (5.83)$$

Only at those points for which $\mathbf{k}' = \mathbf{k} \pm \mathbf{K}_m$ are gaps found in the free-electron dependence of E on $\mathbf{k}$, since it is only at these points that $H'_{\mathbf{k}\mathbf{k}'} \neq 0$. Since $|\mathbf{k}| = |\mathbf{k}'|$, we square both sides of Eq. (5.83) to obtain

$$2\mathbf{k} \cdot \mathbf{K}_m = \pm |\mathbf{K}_m|^2 \qquad (5.84)$$

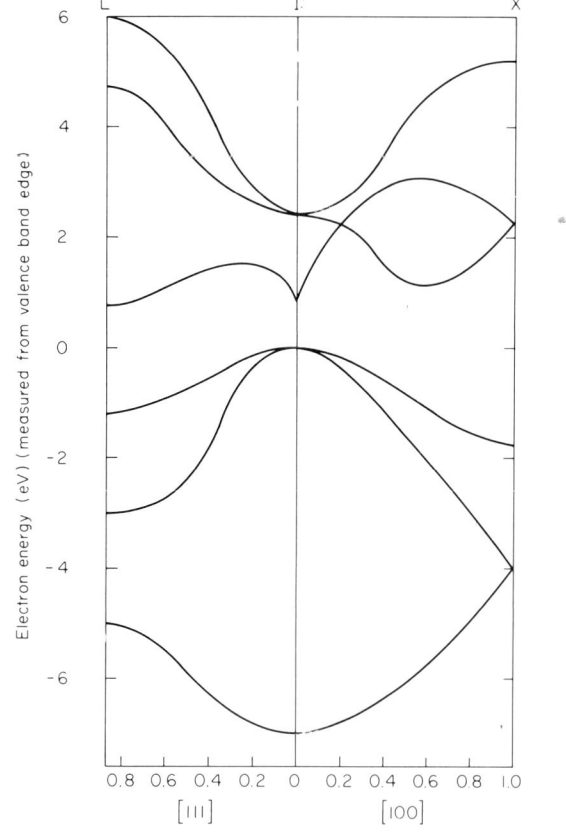

Fig. 5.11 Typical energy bands for a crystal with diamond-like crystal structure. Compare with Fig. 5.9, plotted in a similar way for free electrons. (After D. Long, "Energy Bands in Semiconductors," Wiley-Interscience, New York, 1968 p. 33.)

as the condition for the existence of a band gap. We have already seen in Eq. (5.40) that this is the equation for the faces of the Brillouin zone, and so we reconfirm once more that the energy gaps predicted for a perturbed nearly free electron system occur at the Brillouin zone boundaries only. The relation of Eq. (5.84) is also equivalent to the Bragg reflection condition in three dimensions.

Although there may be an energy gap in a given **k** direction, there is not necessarily an energy gap for an electron moving through a three-dimensional crystal. A gap in one **k** direction may be bridged by an allowed band in another **k** direction. This possibility of the three-dimensional system gives rise to the existence of *overlapping bands*, not possible in the one-dimensional system.

Typical variations of E vs. **k** in the 100 and 111 directions for a diamond structure crystal are given in Fig. 5.11. These curves may be compared with the empty lattice curves given in Fig. 5.9.

The number of **k** vectors in the fundamental domain of a three-dimensional crystal is $N_1 N_2 N_3$ (recall that the number was N in one dimension). The number of **k** vectors with endpoints in a volume $d\mathbf{k}$ of **k** space is

$$\frac{N_1 N_2 N_3}{8\pi^3/\Omega} d\mathbf{k} = \frac{\mathscr{V}}{8\pi^3} d\mathbf{k} \qquad (5.85)$$

The product $N_1 N_2 N_3$ is the number of unit cells in the crystal and the number of orbital states in a Brillouin zone. A Brillouin zone can accommodate a number of electrons equal to two times (because of spin) the number of unit cells in a crystal. If each unit cell contains one atom (as in a simple *Bravais lattice*), a Brillouin zone can accommodate two electrons per atom of the crystal. All Brillouin zones have the same volume, and hence each zone can accommodate two electrons per atom in a simple Bravais lattice.

5.5 Energy Bands in the Tight-Binding Approximation

The other extreme to considering electrons in a crystal effectively free except for a small periodic potential, is to consider electrons as tightly bound to individual atoms that interact with one another in the formation of a crystal lattice. We construct a wavefunction for the interacting system of atoms as a linear combination of the atomic wavefunctions for the state of interest, and then determine the energy values for the corresponding periodic potential.

Energy Bands in One Dimension

Suppose that $\phi(x)$ is a particular eigenfunction corresponding to a bound nondegenerate state of an electron in an isolated atom. Then $\phi(x)$ satisfies the atomic Schroedinger equation,

$$\frac{d^2 \phi(x)}{dx^2} + \frac{2m}{\hbar^2} [E_0 - V(x)] \phi(x) = 0 \tag{5.86}$$

Typical atomic $\phi(x)$ and $V(x)$ functions are shown in Fig. 5.12.

Now consider the effects of taking N of these atoms and arranging them on a line in a one-dimensional crystal lattice with the interatomic spacing a. The potential of the array of atoms is then a periodic potential,

$$U(x) = \sum_n V(x - na) \qquad n = 0, 1, \ldots, N \tag{5.87}$$

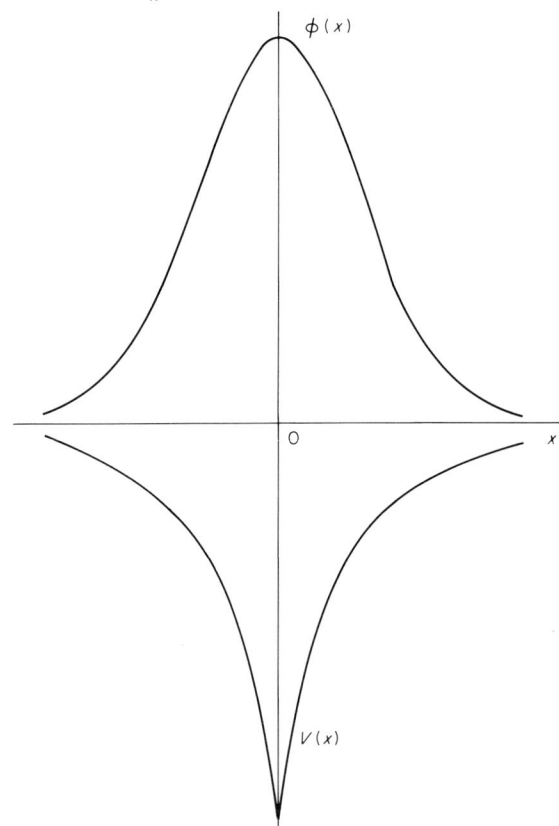

Fig. 5.12 Typical electronic eigenfunction $\phi(x)$ and potential $V(x)$ for an isolated atom.

5.5 Energy Bands in Tight-Binding Approximation

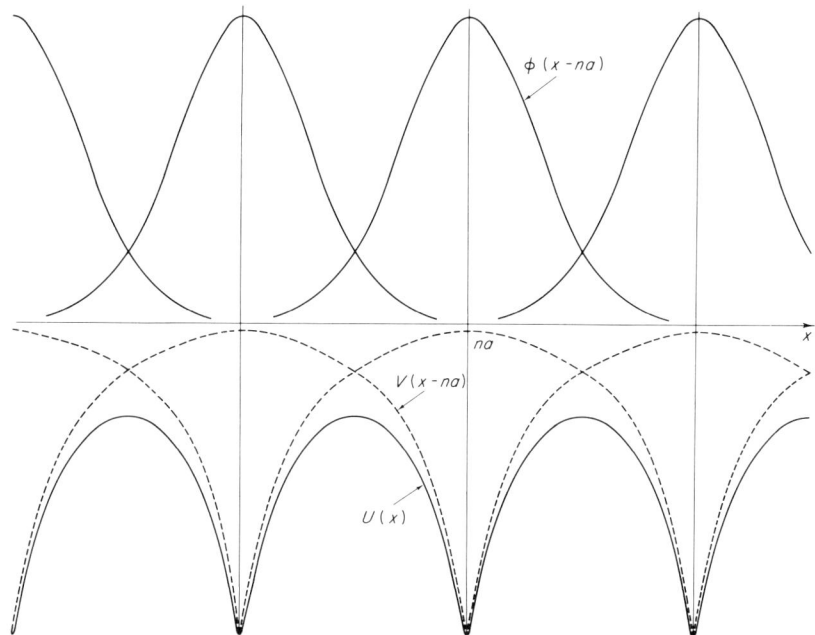

Fig. 5.13 The effect of an array of atoms in a one-dimensional crystal, producing a periodic potential $U(x)$ and an overlap between the isolated-atom wavefunctions.

where $V(x - na)$ represents the potential energy due to the atom at the position na. The potential $U(x)$ is shown schematically in Fig. 5.13.

The Schroedinger equation to be solved with the periodic potential $U(x)$ is

$$\frac{d^2\psi}{dx^2} + \frac{2m}{\hbar^2}\left[E - \sum_n V(x - na)\right]\psi = 0 \qquad (5.88)$$

For the isolated atom with potential $V(x - na)$ there is an eigenfunction $\phi(x - na)$ according to Eq. (5.86). If there is only a small interaction between the atoms of the crystal, $\phi(x - na)$ remains approximately an eigenfunction corresponding to the potential $V(x - na)$ with the energy E_0 even for the atoms in the crystal. In order to construct a wavefunction with which to calculate the energy in the periodic potential, we choose a linear combination of the atomic eigenfunctions,

$$\psi = \sum_n C_n \phi(x - na) \qquad (5.89)$$

The constants C_n are not arbitrary but must represent the appropriate phase factors to be used in summing the individual atomic wavefunctions. The

156   5   Origin of Energy Bands in Solids

Bloch periodicity condition requires that $\psi_k(x + a) = e^{ika} \psi_k(x)$. The Bloch condition is satisfied by

$$\psi_k(x) = \sum_n e^{ikna} \phi(x - na) \tag{5.90}$$

This function may be normalized as follows

$$\int \psi_k^*(x) \psi_k(x) \, dx = \sum_n \sum_m e^{ik(n-m)a} \int \phi^*(x - ma) \phi(x - na) \, dx$$

$$= \sum_n \sum_m e^{ik(n-m)a} \delta_{mn}$$

$$= N \tag{5.91}$$

if the $\phi(x)$ functions are orthonormal.

The energy is given by

$$E = \int \psi_k^*(x) \, \mathbf{H} \psi_k(x) \, dx \tag{5.92}$$

As a first step in evaluating the energy, consider the determination of $\mathbf{H} \psi_k(x)$.

$$\mathbf{H} \psi_k(x) = N^{-1/2} \sum_n e^{ikna} \left[-\frac{\hbar^2}{2m} \frac{d^2}{dx^2} + U(x) \right] \phi(x - na) \tag{5.93}$$

It is convenient to express the energy $E(k)$ in terms of the energy of the atomic state E_0 plus a correction term due to the interaction. In order to make this possible, we add and subtract $V(x - na)$ in Eq. (5.93),

$$\mathbf{H} \psi_k(x) = N^{-1/2} \sum_n e^{ikna} \left[-\frac{\hbar^2}{2m} \frac{d^2}{dx^2} + V(x - na) \right] \phi(x - na)$$

$$+ N^{-1/2} \sum_n e^{ikna} [U(x) - V(x - na)] \phi(x - na)$$

$$= E_0 \psi_k(x) + N^{-1/2} \sum_n e^{ikna} [U(x) - V(x - na)] \phi(x - na) \tag{5.94}$$

The energy $E(k)$ is given by

$$E(k) = \int \psi_k^*(x) E_0 \psi_k(x) \, dx$$

$$+ N^{-1} \sum_n \sum_m e^{ik(n-m)a} \int \phi^*(x - ma)[U(x) - V(x - na)]$$

$$\times \phi(x - na) \, dx \tag{5.95}$$

5.5 Energy Bands in Tight-Binding Approximation

We use the periodicity of $U(x)$ to eliminate the $(x - na)$ notation by introducing a transformation $X = x - na$. With this transformation, the integral of Eq. (5.95) becomes

$$\int \phi^*[X + (n - m)a][U(X + na) - V(x)]\phi(X)\, dX$$

$$= \int \phi^*(X + pa)[U(X) - V(X)]\phi(X)\, dX$$

$$= \int \phi^*(x + pa)[U(x) - V(x)]\phi(x)\, dx \quad (5.96)$$

with $p = (n - m)$. Actually in making this transformation of variables from x to X, we ought to have changed the limits of integration so that with integration with respect to X, the limits are different for each value of n. For a very long crystal, however, this effect may be neglected and all integrals taken over the same range. The energy is given by

$$E(k) = E_0 + N^{-1} \sum_n \sum_m e^{ikpa} \int \phi^*(x + pa)[U(x) - V(x)]\phi(x)\, dx \quad (5.97)$$

The double sum in Eq. (5.97) is equivalent to N times a single sum over p. This is because we take the sum over the neighbors of each atom through the difference p, and we take such a sum for each of the N atoms of the crystal. Of course if we include many neighbors for each atom, this equivalency will break down for the end atoms of the crystal, but once again if the crystal is sufficiently long only a very small error is introduced. Since, as a matter of fact, we are considering a tight-binding approximation, we consider interactions only between nearest neighbors; we restrict the value of $p = (n - m)$ to zero or ± 1. The energy is given by

$$E(k) = E_0 - E_0' - E_+'\, e^{ika} - E_-'\, e^{-ika} \quad (5.98)$$

where the E' terms represent corrections due to interaction and correspond to

$$p = 0, \quad E_0' = -\int \phi^*(x)[U(x) - V(x)]\phi(x)\, dx \quad (5.99)$$

$$p = +1, \quad E_+' = -\int \phi^*(x + a)[U(x) - V(x)]\phi(x)\, dx \quad (5.100)$$

$$p = -1, \quad E_-' = -\int \phi^*(x - a)[U(x) - V(x)]\phi(x)\, dx \quad (5.101)$$

Since $E_+' = E_-'$,

$$E(k) = E_0 - E_0' - 2E_\pm' \cos ka \tag{5.102}$$

The energy given by Eq. (5.102) is the energy of the isolated atom E_0 with an additional contribution coming from the effects of interaction in the periodic potential of the crystal. The magnitude of this interaction term depends both on the extent of overlap of wavefunctions of different atoms and on the difference between the periodic potential and the atomic potential. The electronic level with energy E_0 in the isolated atom has been shifted by an energy E_0' and split into a band with width $4E_\pm'$ in the one-dimensional crystal. In principle every atomic energy level broadens into such a band of levels under interaction. Plots of $E(k)$ vs. k for two possibilities of signs of the contribution from interaction are given in Fig. 5.14. The broadening of discrete atomic levels to form energy bands in a crystal due to the overlap of atomic wavefunctions can be thought of physically as the consequence of the Pauli exclusion principle.

For small values of k, the $\cos ka$ term in Eq. (5.102) can be expanded,

$$\begin{aligned}E(k) &= E_0 - E_0' - 2E_\pm' + E_\pm' k^2 a^2 \\ &= E_\text{ref} + E_\pm' k^2 a^2\end{aligned} \tag{5.103}$$

where the first three energy terms are treated as a reference energy for the description of the variation with k given by the last term. This expression for $E(k)$ for small k can be compared with the expression for a free electron,

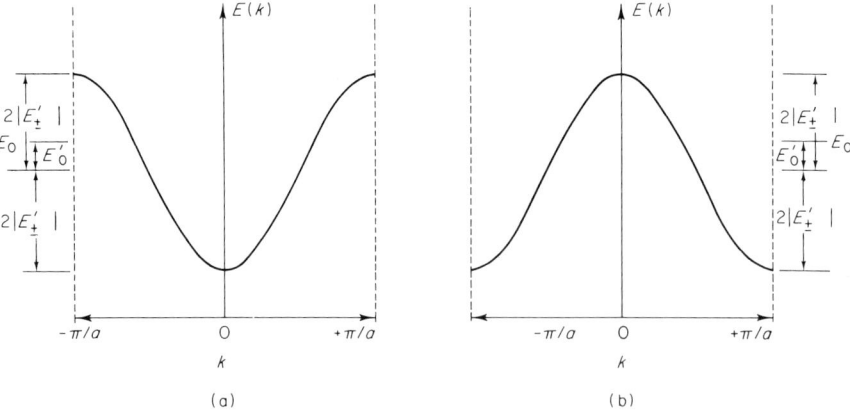

Fig. 5.14 Form of the energy bands produced by interaction between atoms in a one-dimensional crystal: (a) if E_0' is positive and $E_\pm'$ is positive; (b) if E_0' is positive and $E_\pm'$ is negative. See Eq. (5.102).

5.5 Energy Bands in Tight-Binding Approximation

$E(k) = \hbar^2 k^2/2m$. For small values of k, the electron in the crystal has an energy that varies with k above a reference energy *like* a free electron if we assign to the electron in the crystal an *effective mass* m^* given by

$$m^* = \frac{\hbar^2}{2a^2 E_{\pm}'} \tag{5.104}$$

This is another way of arriving at this concept introduced previously in Sections 5.2 and 5.4. The formal definition of effective mass is considered in Chapter 6.

ENERGY BANDS IN THREE DIMENSIONS

The extension of the one-dimensional tight-binding derivation of energy bands to three dimensions can be done almost directly without additional difficulties, provided that we keep things simple by considering a simple Bravais lattice and those energy bands that arise from a nondegenerate s state. If the potential energy of electrons in the free atom is $V(\mathbf{r})$, and if E_0 is the energy of an atomic s state with eigenfunction $\phi(\mathbf{r})$, the atomic Schroedinger equation is

$$\nabla^2 \phi(\mathbf{r}) + \frac{2m}{\hbar^2}[E_0 - V(\mathbf{r})]\phi(\mathbf{r}) = 0 \tag{5.105}$$

The periodic potential in a lattice of atoms is

$$U(\mathbf{r}) = \sum_j V(\mathbf{r} - \mathbf{R}_j) \tag{5.106}$$

where the sum is over all lattice points. We desire a solution of the general Schroedinger equation,

$$\nabla^2 \psi + \frac{2m}{\hbar^2}[E - U(\mathbf{r})]\psi = 0 \tag{5.107}$$

An approximate wavefunction is constructed by taking a linear combination of atomic orbitals with appropriate phase factors to preserve the Bloch periodicity requirement,

$$\psi_{\mathbf{k}}(\mathbf{r}) = \sum_l e^{i\mathbf{k}\cdot\mathbf{R}_l} \phi(\mathbf{r} - \mathbf{R}_l) \tag{5.108}$$

The normalizing factor for the wavefunction of Eq. (5.108) is $N^{-1/2}$ where N is the number of atoms in the crystal.

The approximate energy for the system of interacting atoms is calculated by a procedure exactly analogous to that used in the one-dimensional case.

The equivalent result to Eq. (5.97) is

$$E(\mathbf{k}) = E_0 + \sum_j e^{i\mathbf{k}\cdot\mathbf{R}_j} \int \phi^*(\mathbf{r} + \mathbf{R}_j)[U(\mathbf{r}) - V(\mathbf{r})]\phi(\mathbf{r})\,d\mathbf{r} \quad (5.109)$$

Neglecting overlap except between nearest neighbors,

$$E(\mathbf{k}) = E_0 - E_0' - E' \sum_j e^{i\mathbf{k}\cdot\mathbf{R}_j} \quad (5.110)$$

where the sum is over nearest neighbors only, and

$$E_0' = -\int \phi^*(\mathbf{r})\,[U(\mathbf{r}) - V(\mathbf{r})]\,\phi(\mathbf{r})\,d\mathbf{r} \quad (5.111)$$

$$E' = -\int \phi^*(\mathbf{r} + \mathbf{R}_j)\,[U(\mathbf{r}) - V(\mathbf{r})]\,\phi(\mathbf{r})\,d\mathbf{r} \quad (5.112)$$

Example 5.7 Calculate the allowed energy levels in a simple cubic crystal corresponding to atomic s states in the isolated atoms.

Since for a simple cubic crystal

$$\mathbf{R}_j = \pm a\mathbf{e}_1,\ \pm a\mathbf{e}_2,\ \pm a\mathbf{e}_3$$

and since

$$\mathbf{k} = k_1\mathbf{e}_1 + k_2\mathbf{e}_2 + k_3\mathbf{e}_3$$

we have from Eq. (6.121),

$$E(\mathbf{k}) = E_0 - E_0' - E'(e^{ik_1 a} + e^{-ik_1 a} + e^{ik_2 a} + e^{-ik_2 a} + e^{ik_3 a} + e^{-ik_3 a})$$
$$= E_0 - E_0' - 2E'(\cos k_1 a + \cos k_2 a + \cos k_3 a)$$

The allowed energy levels are spread over a band with width of $12E'$. All states are accounted for if $\mathbf{k}$ is restricted to lie within a cube bounded by $k_1 = k_2 = k_3 = \pm\pi/a$, the fundamental domain of $\mathbf{k}$.

For small values of $\mathbf{k}$, $\cos k_i a$ can be approximated by $(1 - k_i^2 a^2/2)$, giving

$$E(\mathbf{k}) = E_0 - E_0' - 6E' + E'a^2\,|\,\mathbf{k}\,|^2$$

Then the corresponding effective mass near $\mathbf{k} = 0$ is

$$m^* = \frac{\hbar^2}{2E'a^2}$$

In the variation of $E(\mathbf{k})$ as a function of $\mathbf{k}$ in $\mathbf{k}$ space, there are

5.6 Summary

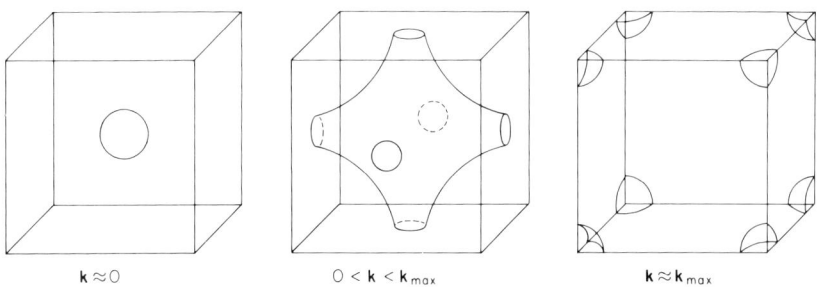

Fig. 5.15 Typical equal-energy surfaces for the simple cubic crystal.

three typical equal-energy surfaces, shown in Fig. 5.15. Near $\mathbf{k} = 0$, the equal-energy surfaces are spherical. As they expand toward the zone boundary, they become distorted since the equal-energy surfaces must cross the boundary normal to the boundary where $\partial E/\partial k_i = 0$. Finally for energies very close to the maximum energy present in the zone, the equal-energy surfaces are again spherical, this time being centered on the corner points of the basic cell. The symmetry of the equal-energy surfaces near the maximum energy points of the zone can be brought out by setting $\mathbf{k}' = \mathbf{k}_i \pm \pi/a$. For small $\mathbf{k}'$,

$$E(\mathbf{k}') = E_0 - E_0' + 6E' - E'a^2 \mid \mathbf{k}' \mid^2$$

Thus the symmetry is the same as for $\mathbf{k} = 0$, except that

$$m^* \Big]_{E_{\max}} = -\frac{\hbar^2}{2E'a^2} = -m^* \Big]_{E=0}$$

The effective mass corresponds to spherical equal-energy surfaces for the simple cubic lattice both near $\mathbf{k} = 0$ and near $\mathbf{k}_{\max}$ for the fundamental domain, but with opposite sign, being positive near $\mathbf{k} = 0$ and negative near $\mathbf{k}_{\max}$. A negative effective mass has been previously conceptually related to reflection of the electron with $\mathbf{k}$ corresponding to that of a zone face.

5.6 Summary

One of the major advances in the understanding of the properties of crystalline solids has been the development of techniques for the calculation of the allowed energy bands for electrons in crystals. The refinement and extension of these energy-band calculations is being actively pursued and will be the subject of interest for many years to come.

The appropriate wavefunction for an electron in a periodic potential is a Bloch function: a plane wave modulated by a periodic function with the periodicity of the crystal. The square of the amplitude of such a Bloch wavefunction is the same at points in the crystal separated by a lattice vector; the wavefunction itself differs at two such points only by the appropriate phase factor. Since the form of the Bloch function is not affected if $\mathbf{k}$ is increased by a reciprocal lattice vector, it is possible to describe all energy bands within a basic range of $\mathbf{k}$ values, called the basic Brillouin zone. The faces of the Brillouin zones occur for those values of $\mathbf{k}$ satisfying the Bragg reflection condition and hence corresponding to forbidden energy regions for the electron waves.

Two extreme cases were considered for the generation of energy bands in a crystal. In the first, a system of nearly free electrons was considered as affected by a small perturbing periodic potential. The effect of this periodic potential is to open up energy gaps in the allowed energy spectrum at the faces of the Brillouin zones. In the second case, a system of nearly isolated atoms was considered, and the effect on a specific atomic state was calculated as these atoms were brought close enough together to interact with one another. It was found that each discrete atomic energy level spreads into a band of allowed energies in the crystal, the width of the band depending on the extent of overlap of the atomic wavefunctions corresponding to the state in question between neighboring atoms, and the integrated difference between the periodic potential characterizing the crystal and the atomic potential characterizing the isolated atom.

The concept of effective mass of an electron was introduced as a helpful approximation. In many cases the motion of an electron subject both to external forces and the periodic potential of the crystal lattice can be treated as if the electron were free and subject only to the external forces, provided that all of the effects of the periodic potential are taken into account by replacing the actual electron mass m by an effective mass m^*. For valence electrons in the alkali metals, m^* differs only slight from m, and these electrons may be treated as almost free, but under other situations m^* can differ appreciably from m. When m^* is used to describe the effects of Bragg reflection of an electron wave with wave vector near the Brillouin zone boundary, even the sign must be changed and a negative effective mass considered. We considered explicitly the way in which effective masses can be calculated for valence electrons of alkali metals by the cellular method, and for s states of interacting atoms in the tight-binding approximation.

Properties of crystals because of energy bands and the kinds of energy bands and energy band calculations found for real crystals are treated in Chapter 6.

Problems

5.1 Show that the reciprocal lattice for a bcc crystal has fcc symmetry.

5.2 Construct the fourth Brillouin zone for a simple square lattice in both the extended- and reduced-zone representations.

5.3 Find the number of states per atom inside a sphere that just touches the faces of the first Brillouin zone of (a) the simple cubic lattice, (b) the body-centered cubic lattice, and (c) the face-centered cubic lattice.

5.4 Consider a one-dimensional crystal consisting of a series of periodically spaced potential wells, as in Fig. 5.16. Solve the Schroedinger equation inside and outside a well, and apply the boundary conditions that the wavefunction

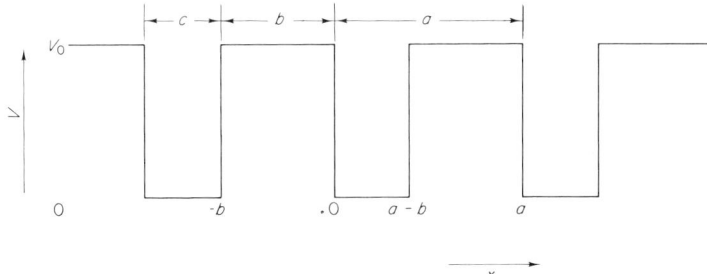

Fig. 5.16 Periodic series of potential wells as a model of a one-dimensional crystal.

must be continuous across the boundary at $x = 0$ and must satisfy the Bloch periodicity condition between $x = -b$ and $x = (a - b)$. Show that for $E < V_0$, the restrictions on allowed energy are given by

$$\cos ka = \frac{\alpha^2 - \beta^2}{2\alpha\beta} \sinh \alpha b \sin \beta(a - b) + \cosh \alpha b \cos \beta(a - b)$$

where $\beta = (2mE)^{1/2}/\hbar$, and $\alpha = [2m(V_0 - E)]^{1/2}/\hbar$.

5.5 In Problem 6.4, show that similar restrictions on the energy exist even when $E > V_0$.

5.6 Show that if in Fig. 5.16, b is allowed to decrease to zero while V_0 is increased without limit so as to keep $P \equiv [m(a - b)/\hbar^2]bV_0 =$ constant, the expression for the restriction on energy reduces to

$$\cos ka = \frac{P}{\beta a} \sin \beta a + \cos \beta a$$

This is the *Kronig–Penney approximation*. Show that the restrictions of this equation and their dependence on P leads to a series of allowed and forbidden energy bands, as shown in Fig. 5.17. What physical significance can be associated with the quantity P?

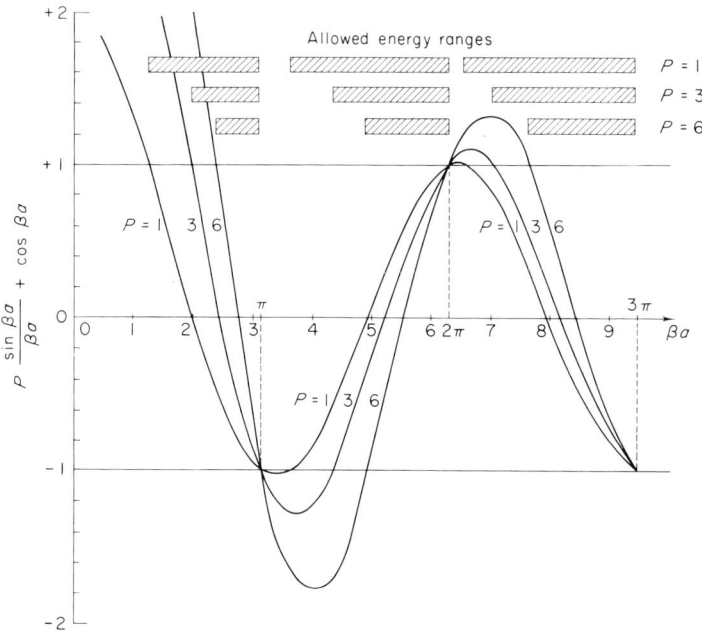

Fig. 5.17 Allowed and forbidden bands for the Kronig–Penney approximation to the periodic series of square-well potentials, for various values of the parameter P.

5.7 Calculate the width of the first allowed band and the first forbidden band for the Kronig–Penney model for $V_0 = 10$ eV, $c = 1$ Å, and $b = 3$ Å.

5.8 Plot E vs. k for the first allowed band of the Kronig–Penney model described in Problem 5.7.

5.9 Apply the tight-binding approximation of Section 5.5 for an s state to determine the form of the energy bands in a bcc and an fcc crystal.

5.10 The form for the energy bands given by the tight-binding approximation for s states in an fcc crystal is

$$E(\mathbf{k}) = E_0 - E'(\cos k_x a \cos k_y a + \cos k_y a \cos k_z a + \cos k_z a \cos k_x a)$$

(a) Show a plot of the energy bands starting at the L point of the zone, proceeding along the Λ axis to the Γ point, and then proceeding from the Γ point to the X point along the Δ axis. (b) Sketch the shape of the equal-energy surfaces near $\mathbf{k} = 0$ for $E' < 0$ and $E' > 0$.

5.11 Consider the general two-dimensional lattice shown in Fig. 5.18. (a) What is $\mathbf{R}_j$, the lattice vector of the real lattice? (b) What is $\mathbf{K}_m$, the position vector of the reciprocal lattice? (c) Calculate the form of the energy bands for this lattice in the tight-binding approximation using atomic s states.

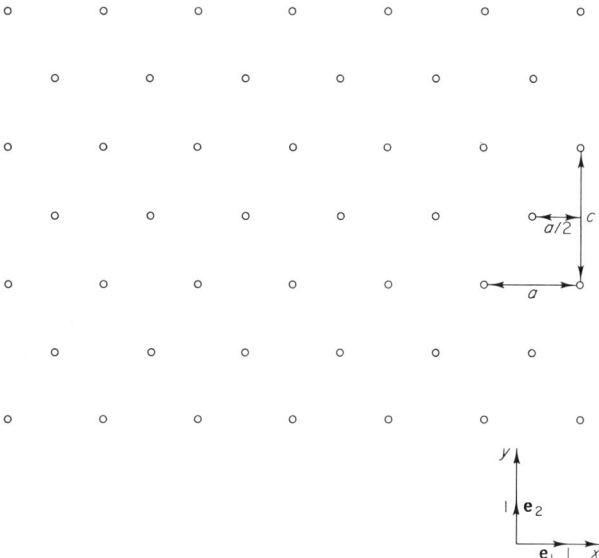

Fig. 5.18 Two-dimensional lattice for Problem 5.11.

5.12 An interesting kind of material involves linear chains of a few atoms of the same kind. As an extreme case, consider a linear "crystal" composed of just two atoms of the same kind separated by the distance a. Consider an s-like state of the isolated atom with an energy E_0 in the isolated atom. (a) What allowed energy spectrum is expected in the "crystal"? (b) Use the tight-binding approximation to calculate the allowed energy states for the "crystal." Write out all terms completely, remembering to include the Bloch periodicity condition and the periodic boundary condition. (c) Sketch the expected allowed energy states for linear chains of three atoms and of ten atoms.

Chapter 6

Properties of Energy Bands

A periodic potential gives rise to bands of allowed and forbidden energies for electrons in a solid. The exact form of the $E(\mathbf{k})$ curves depends in some detail on the actual three-dimensional crystal structure. Energy bands in real crystals seldom assume the simple forms given by the idealized calculations of the previous chapter. A variety of more sophisticated methods have been developed for the theoretical calculation of energy bands, and experimental results have revealed the nature of the energy-band structure in a number of different materials.

In order to use energy-band information in calculating the electronic properties of crystals, it is necessary to determine the variation of the density of states with energy within the bands. It is found that the density of states is highest near the middle of a band and decreases to zero at the band edges in a roughly analogous way at the lower and upper edges. Knowledge of the exact variation of $E(\mathbf{k})$ permits the calculation of the dependence of velocity and effective mass on $\mathbf{k}$.

Perhaps the most outstanding success of the general band model for solids has been its ability to give a theoretical reason for the difference between metals, semiconductors, and insulators. Closely tied to this success is the concept of the *hole*, a means of describing electronic properties in terms of missing electrons rather than in terms of electrons present. Carrier transport by electrons in nearly empty bands and by holes in nearly full bands can also be described by a formalism that permits them to be treated as if they were free carriers but with an effective mass different from the mass of a really free carrier.

6.1 Energy-Band Calculations

The energy-band calculation neglects the Coulomb force that exists between an electron and the residual hole when an electron is excited to the next higher allowed energy band in a semiconductor or insulator. This Coulomb force can lead to the existence of a series of bound states for the excited electron-and-hole pair; such a complex is called an *exciton* and is frequently an important step in the generation of free electron–hole pairs and in the recombination of free electron–hole pairs.

The very sophistication of modern energy-band calculations tends to increase their abstract nature and to remove them from the properties of atoms and chemical bonds to which they must also be related. Research on amorphous materials, in which the long-range periodic potential is completely removed, reveals that a basic short-range forbidden energy gap remains, associated physically with the breaking of an interatomic bond. Investigations of how to correlate the bond picture and the band picture are helpful and increasingly needed as materials of interest depart from the traditional crystalline materials.

6.1 Energy-Band Calculations

It is not our purpose to enter into a detailed discussion of the various methods and techniques that have been applied to the calculation of energy-band structures in real crystals. The number of books that treat this topic exclusively are increasing and examples are given in the Bibliography. We include here a summary of the general considerations involved and a few words about some particularly useful developments.

Two basic assumptions characterize most calculations of energy bands in real crystals. (1) A real crystal can be represented by a perfect periodic structure. (2) Each electron can be regarded as moving independently in a static potential field that in some way takes into account its average interaction with the rest of the crystal, i.e., the wavefunction for the total electronic system can be expressed in an approximate way in terms of one-electron wavefunctions, each of which depends on the coordinates of a single electron. If the many-particle Hamiltonian does not contain spin-dependent terms (such as those that arise from spin–orbit interaction), each one-electron wavefunction can be written as $\psi(x, y, z)S(\zeta)$. The set of one-electron wave equations is

$$\left[-\frac{\hbar^2}{2m}\nabla^2 + V_{\mathbf{k},l}(\mathbf{r})\right]\psi_{\mathbf{k},l}(\mathbf{r}) = E_l(\mathbf{k})\,\psi_{\mathbf{k},l}(\mathbf{r}) \qquad (6.1)$$

where as usual l is the band index, and the energy-band structure is described by the set of energy-band functions $E_l(\mathbf{k})$. One problem is the choice of the potential function $V_{\mathbf{k},l}(\mathbf{r})$. Simple one-electron potentials such as the Hartree and Hartree–Fock approaches have been used, as well as variations of these.

Many-electron approaches have also been used to include more of the correlation effects due to electron interactions.

In order to solve the wave equation, a number of approximate methods have been introduced. Most of these methods share the following common features. (a) A set of suitably chosen basis functions is selected (plane waves, atomic orbitals, etc.), and trial wavefunctions are expanded in terms of these basis functions. (b) The basis functions are held fixed and a variational calculation on the expansion coefficients is carried out. (c) Wavefunctions are required to have the Bloch form, boundary conditions are incorporated directly into the variational procedure, and the crystal potential is usually assumed to be spherically symmetric in the neighborhood of each nucleus and nearly constant elsewhere in the crystal.

The simplest wavefunctions to choose for the set of basis functions are plane waves. Unfortunately plane waves do not represent a practical choice because many hundred plane waves must be included in order to represent the sharp oscillations of the Bloch function in the neighborhood of the lattice ions. Improvements are achieved by using a much smaller number of plane waves to represent the valence electrons *and* a set of atomic-like ion-core wavefunctions to represent electrons in lower-energy states; this is the essence of OPW (orthogonalized plane waves) methods in which the plane waves are orthogonalized to the ion-core wavefunctions.

Variations of the free-electron perturbation, the tight-binding approach, and the cellular method, all discussed earlier in Chapter 5, are also possibilities.

PSEUDOPOTENTIALS

A method known as the *pseudopotential method* has proved to be of particular merit for a variety of problems. It is a way of treating the general problem so that it may be cast exactly into the form of a perturbation of a free-electron system.

A wavefunction for the actual system, ψ, is expressed as an expansion in both plane waves and core wavefunctions,

$$\psi = \sum_k a_k \psi_k + \sum_C a_C \psi_C \tag{6.2}$$

where ψ_k represents a plane wave, and ψ_C represents a core wavefunction.

6.1 Energy-Band Calculations

The expansion of Eq. (6.2) is overcomplete, since plane waves alone form a complete set, but is still legitimate. The part of the wavefunction involving plane waves is defined as a *pseudowavefunction* ϕ,

$$\phi \equiv \sum_k a_k \psi_k \tag{6.3}$$

When the wavefunction of Eq. (6.2) is substituted into the Schroedinger equation, we obtain

$$-\frac{\hbar^2}{2m}\nabla^2\phi + V\phi + \left(-\frac{\hbar^2}{2m}\nabla^2 + V\right)\sum_C a_C\psi_C = E\phi + E\sum_C a_C\psi_C \tag{6.4}$$

Now since

$$\left(-\frac{\hbar^2}{2m}\nabla^2 + V\right)\psi_C = E_C\psi_C \tag{6.5}$$

Eq. (6.4) becomes

$$-\frac{\hbar^2}{2m}\nabla^2\phi + V\phi + \sum_C (E_C - E)a_C\psi_C = E\phi \tag{6.6}$$

The coefficients a_C can be evaluated by multiplying through by a particular $\psi_{C'}^*$. The initial results are

$$-\frac{\hbar^2}{2m}\psi_{C'}^*\nabla^2\phi + \psi_{C'}^*V\phi + \psi_{C'}^*\sum_C (E_C - E)a_C\psi_C = E\psi_{C'}^*\phi \tag{6.7}$$

The first two terms can be evaluated by operating to the left since the Hamiltonian is a Hermitian operator, to give

$$\psi_{C'}^* \sum_C (E_C - E)a_C\psi_C = (E - E_{C'})\psi_{C'}^*\phi \tag{6.8}$$

Now if we integrate both sides over all coordinates, we retain a_C only for $C = C'$, and the result is

$$a_C = -\int \psi_C^*\phi \, d\mathbf{r} \tag{6.9}$$

If we substitute the value for a_C from Eq. (6.9) back into Eq. (6.6), we can write the Schroedinger equation as

$$-\frac{\hbar^2}{2m}\nabla^2\phi + \mathbf{W}\phi = E\phi \tag{6.10}$$

in terms of the *pseudopotential* operator **W**, where

$$\mathbf{W}\phi = V\phi + \sum_C (E - E_C)\psi_C \int \psi_C^* \phi \, d\mathbf{r} \qquad (6.11)$$

The pseudopotential equation given in Eq. (6.10) is both an exact equation and for small **W** becomes an almost-free-electron equation that can be treated with **W**ϕ as a perturbation. Schematic plots of the pseudopotential and the pseudowavefunction are given in Fig. 6.1. The pseudowavefunction is the true wavefunction ψ with the oscillations at the atom sites removed. That the pseudopotential term is small can be seen by recognizing that the energies being considered are much larger than the core energies E_C, so that the second term in Eq. (6.11) represents a large positive contribution to help cancel the large negative contribution of the potential V.

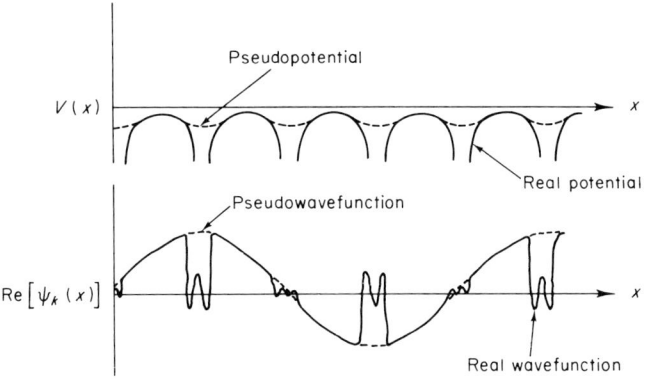

Fig. 6.1 Representation of the spatial variation of the pseudopotential and the pseudowavefunction, and their relation to the real potential and wavefunction.

Because the core wavefunctions are orthogonal to the true wavefunction, it turns out that any constant K_C can be used in place of $(E - E_C)$ in Eq. (6.11) and Eq. (6.10) remains an exact equation.[†] In actual practice, therefore, there is no correct pseudopotential; there are rather a set of good pseudopotentials that will yield reasonable answers when used in the perturbation calculation. Pseudopotential techniques have been used primarily in calculations on metal properties but show promise of extension to semiconductors and insulators as well. Once the pseudopotential is chosen, the need to know the band structure is much reduced; pseudopotential methods

---

[†] See, for example, W. A. Harrison, *Physics Today* **22**, No. 10, 23 (1969).

6.2 Density of States in Energy Bands

ought more properly to be considered as complementary to band-structure methods rather than as a method for determining energy bands themselves.

THE k · P APPROXIMATION

The Schroedinger equation written for Bloch waves can be arranged so that a term involving the scalar product of the wave vector and the momentum operator (hence k · P) can be treated as a perturbation to calculate the form of energy bands in the vicinity of extrema. Consider the wave equation for Bloch waves,

$$\left(-\frac{\hbar^2}{2m}\nabla^2 + V\right)u_k\, e^{i\mathbf{k}\cdot\mathbf{r}} = E(\mathbf{k})\, u_k\, e^{i\mathbf{k}\cdot\mathbf{r}} \tag{6.12}$$

From Eq. (5.19) and letting **P** represent the momentum operator $-i\hbar\nabla$, Eq. (6.12) becomes

$$\left(\frac{\mathbf{P}^2}{2m} + \frac{\hbar}{m}\mathbf{k}\cdot\mathbf{P} + \frac{\hbar^2 k^2}{2m} + V\right)u_k = E(\mathbf{k})\, u_k \tag{6.13}$$

Desired simplification is obtained by treating the second term as a perturbation. Near $\mathbf{k}=0$, the third term is still smaller and can be neglected.

Suppose, for example, that we are considering a crystal for which we know that the conduction and valence band extrema occur at $\mathbf{k}=0$. Then at $\mathbf{k}=0$, Eq. (6.13) becomes

$$\left(\frac{\mathbf{P}^2}{2m} + V\right)u_0 = E(0)\, u_0 \tag{6.14}$$

The solution of this equation involves a conduction band function u_c with a corresponding energy E_c, and a valence band function u_v with an energy E_v. The bandgap E_G is given by $(E_c - E_v)$ at $\mathbf{k}=0$. To treat the variation of $E(\mathbf{k})$ away from $\mathbf{k}=0$, express

$$u_k = a_c u_c + a_v u_v \tag{6.15}$$

and substitute into Eq. (6.13), proceeding via an evaluation of the coefficients a_c and a_v to a determination of the variation of $E(\mathbf{k})$ near $\mathbf{k}=0$. In Section 6.5 we apply the k · P approach to determining the variation of $E(\mathbf{k})$ near $\mathbf{k}=0$ for germanium.

6.2 Density of States in Energy Bands

For free electrons the density of states $N(E)$ is defined such that $N(E)\, dE$ is the number of orbital states with energies lying between E and $E + dE$

(see Section 4.2). The density of states in an energy band in a crystal can be defined in a similar manner. Since the number of **k** vectors with endpoints in a volume $d\mathbf{k}$ of **k** space is $(\mathscr{V}/8\pi^3)\,d\mathbf{k}$, as given in Eq. (5.85),

$$N(E)\,dE = \frac{\mathscr{V}}{8\pi^3}\int d\mathbf{k} \tag{6.16}$$

where the integration is over the volume of **k** space lying between the energy surfaces corresponding to E and $E+dE$.

The distance between the energy surfaces of E and $E+dE$ at the point **k** is

$$\frac{dE}{|\nabla_\mathbf{k}\,E(\mathbf{k})|} \tag{6.17}$$

where $\nabla_\mathbf{k}\,E(\mathbf{k})$ symbolizes the operation

$$\frac{\partial E}{\partial k_1}\mathbf{e}_1 + \frac{\partial E}{\partial k_2}\mathbf{e}_2 + \frac{\partial E}{\partial k_3}\mathbf{e}_3$$

i.e., the derivative of $E(\mathbf{k})$ in the direction normal to the energy surface through the point **k**. The density of states $N(E)$ is therefore given by the surface integral

$$N(E) = \frac{\mathscr{V}}{8\pi^3}\int \frac{dS}{|\nabla_\mathbf{k}\,E(\mathbf{k})|} \tag{6.18}$$

over the surface $E(\mathbf{k}) = E$. Accurate evaluation of $N(E)$ for a particular crystal requires a detailed knowledge of the contours of energy surfaces, something that is only occasionally available.

We do have available information on the energy contours for the simple cubic lattice in the tight-binding approximation, as given in Example 5.7. If the zero of energy is taken to be the lowest energy in the zone, E_0, the equal-energy surfaces near $\mathbf{k}=0$ are

$$E = \frac{\hbar^2 k^2}{2m^*} \tag{6.19}$$

Since this is identical with the energy dependence on k of a free electron with the simple substitution of the effective mass m^* for the electron mass m, we expect the density of states to be the same as for a free electron except for this substitution. From Eq. (6.18),

$$\begin{aligned}N(E) &= \frac{\mathscr{V}}{8\pi^3}\frac{4\pi k^2}{\hbar^2 k/m^*}\\ &= \frac{\mathscr{V}}{4\pi^2}\left(\frac{2m^*}{\hbar^2}\right)^{3/2}E^{1/2}\end{aligned} \tag{6.20}$$

giving the expected result.

6.2 Density of States in Energy Bands

If in the same approximation, we consider the equal-energy surfaces near the corner points of the reduced-zone cube, as given in Example 5.7, we have

$$E_{max} - E(\mathbf{k}) = \frac{\hbar^2 k'^2}{2m^*} \qquad (6.21)$$

where this m^* is the same as in Eq. (6.19). Thus the density of states near E_{max} is given by

$$N(E) = \frac{\mathscr{V}}{4\pi^2} \left(\frac{2m^*}{\hbar^2}\right)^{3/2} [E_{max} - E]^{1/2} \qquad (6.22)$$

The variation of the density of states with energy is therefore parabolic away from the band edge at both extremes of the energy range within a zone.

The general shapes of $N(E)$ vs. E for the free-electron approximation and for the tight-binding approximation are shown in Fig. 6.2. This figure also

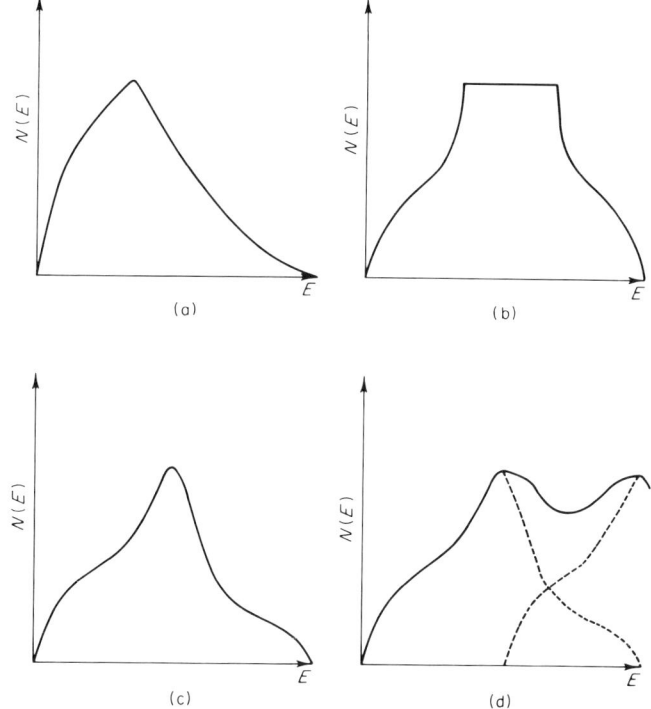

Fig. 6.2 Dependence of density of states on energy for several representative cases. (a) Free-electron approximation; (b) tight-binding approximation; (c) approximate $N(E)$ for a real crystal; (d) overlapping bands.

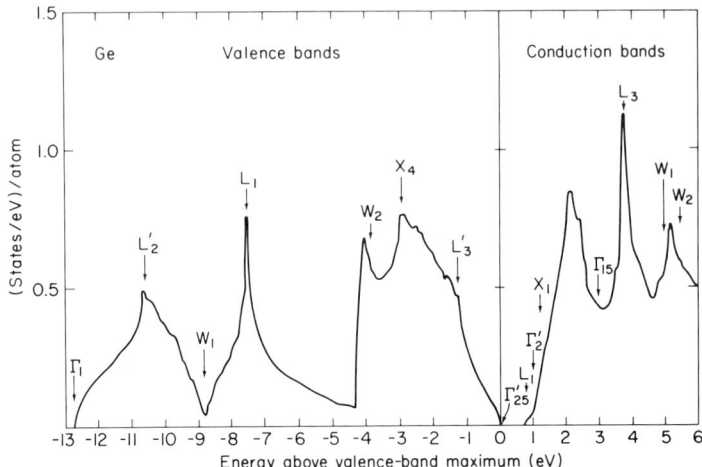

Fig. 6.3 Dependence of density of states on energy for the valence and conduction bands of Ge, i.e., the highest-lying filled bands (valence) and next-highest allowed empty bands (conduction). [After F. Herman, R. L. Kortum, C. D. Kuglin and J. L. Shay, in "II–VI Semiconducting Compounds, 1967 International Conference" (D. G. Thomas, ed.), Benjamin, New York, 1967, p. 503.]

shows the approximate shape of $N(E)$ vs. E for a real material, incorporating the possibility of different effective masses at the energy extremes of a zone, and showing a maximum $N(E)$ due to the bulging out of the equal-energy surfaces as they approach the zone boundary. The density of states for two overlapping bands, a possibility in the three-dimensional crystal, is represented by the fourth $N(E)$ vs. E plot in Fig. 6.2.

An approximation to the actual variation of density of states with energy is shown in Fig. 6.3 for the valence and conduction bands in germanium. The symbols represent points in the reduced zone for germanium as discussed in more detail in connection with Fig. 6.11. The curves of Fig. 6.3 correspond to three valence bands and to a series of overlapping conduction bands.

6.3 Electron Velocity and Effective Mass

Two important properties of electronic states in crystals, the corresponding velocity and effective mass, may be determined directly from the $E(\mathbf{k})$ curves describing the band structure. These curves are the dispersion (ω vs. k) curves for electron waves, corresponding to the dispersion curves for lattice waves discussed in Chapter 1.

6.3 Electron Velocity and Effective Mass

ONE-DIMENSIONAL TREATMENT

The velocity associated with a particular electron k state is the group velocity,

$$v_g = \frac{d\omega}{dk} = \frac{1}{\hbar}\frac{dE}{dk} \qquad (6.23)$$

If the electron is free, $E = \hbar^2 k^2/2m$, and $v_g = \hbar k/m$. Even when the $E(k)$ curves depart from that of a free electron, however, the velocity associated with a state with wavenumber k is given by Eq. (6.23).

In the parabolic $E(k)$ curves typical of the free electron, the mass m in a graphical sense determines the curvature of the variation. For a free electron the mass m is a constant and the $E(k)$ curves has the same curvature for all values of k. If in a real crystal with nonparabolic $E(k)$ curves, or with parabolic $E(k)$ curves with a different curvature than that given by m, we continue to describe the $E(k)$ as if we did have a free electron, we can be consistent by defining an effective mass m^* as determined by the variation of $E(k)$ vs. k. In previous discussions we have seen ways in which this effective mass can be calculated from different approximations. If in general $E(k) = \hbar^2 k^2/2m^*(k)$, indicating that m^* may be a function of k, then a general expression for m^* is

$$m^* = \frac{\hbar^2}{d^2E/dk^2} \qquad (6.24)$$

An alternative way of looking at the effective mass is to consider the effect of an *external* force F on an electron in a crystal. The work dE done in a time interval dt is given by

$$dE = Fv_g\, dt \qquad (6.25)$$

Since $dE = (dE/dk)\, dk = \hbar v_g\, dk$,

$$F = \hbar \frac{dk}{dt} \qquad (6.26)$$

This is a general relationship. For a free electron, for example, $p = \hbar k$, and Eq. (6.26) is simply Newton's law $F = dp/dt$. If we also write the force equation in analogy with the free-electron case as

$$F = m^* \frac{dv_g}{dt} \qquad (6.27)$$

it is evident that m^* in Eq. (6.27) must be given by Eq. (6.24).

Values of E, v_g and m^* are plotted for two specific simple $E(k)$ variations in the reduced zone in Fig. 6.4. The sign of the velocity is different for $+k$ and $-k$ regions of the $E(k)$ curve. In particular, in a completely filled band there are equal numbers of electrons moving with positive velocity and with negative velocity, thus making it impossible for a net transport of charge. In the wave picture, therefore, the absence of a net charge transport for a completely filled band does not result from a "parking lot" analogy, i.e., because of the absence of a place for an electron to move into, but because of a net cancellation of large numbers of electrons with opposite velocities.

Figure 6.4 also illustrates two unusual properties of the effective mass. First is the result that m^* goes to infinity at some point within the zone. In a three-dimensional picture this happens in only one crystal direction at a time, so that impossibility of changing velocity under these conditions does not prevail for all crystal directions simultaneously. It also marks the point where an increase in externally applied force F must begin to decelerate

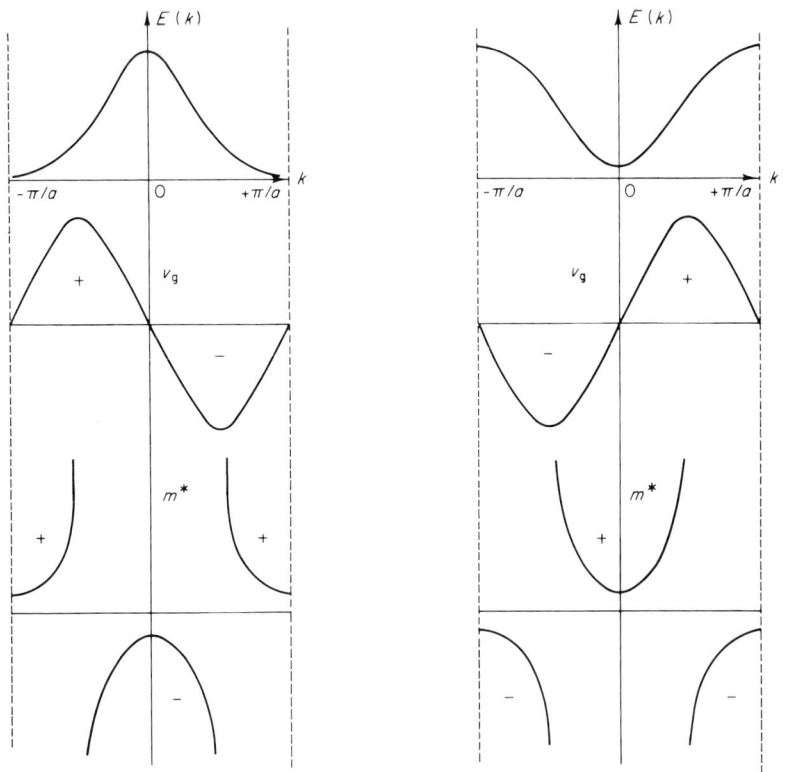

Fig. 6.4 Typical variations of group velocity v_g and effective mass m^* as a function of k.

6.3 Electron Velocity and Effective Mass

rather than accelerate the electron. The second peculiarity is that m^* has regions in which it is negative. A negative value of m^* means that the acceleration resulting from the application of an external force F is in a direction opposite to that of the applied force; this is the effect of electron reflection because of the periodic crystal lattice.

THREE-DIMENSIONAL TREATMENT

The group velocity is given in three dimensions by

$$v_\mathbf{k} = \frac{1}{\hbar} \nabla_\mathbf{k} E(\mathbf{k}) \tag{6.28}$$

with the $\nabla_\mathbf{k}$ as defined in Section 6.2.

Example 6.1 Calculate the dependence of the velocity of an electron on its wave vector in an energy band of a simple cubic lattice corresponding to an atomic s state derived by the tight-binding approximation.

From Example 5.7, we have the dependence of E on $\mathbf{k}$ for this system,

$$E(\mathbf{k}) = E_0 - E_0' - 2E'(\cos k_1 a + \cos k_2 a + \cos k_3 a)$$

From Eq. (6.28), we have

$$\hbar v_\mathbf{k} = (2E'a \sin k_1 a)\mathbf{e}_1 + (2E'a \sin k_2 a)\mathbf{e}_2 + (2E'a \sin k_3 a)\mathbf{e}_3$$

We might also inquire as to the actual value of $\mathbf{v}_\mathbf{k}$ at various points in the reduced zone, and construct a table like the following:

Point (k_1, k_2, k_3)	$\hbar v_\mathbf{k}$
$(0, 0, 0)$	0
$(\pi/a, \pi/a, \pi/a)$	0
$(\pi/2a, \pi/2a, \pi/2a)$	$2E'a(\mathbf{e}_1 + \mathbf{e}_2 + \mathbf{e}_3)$
$(0, \pi/2a, \pi/a)$	$2E'a\mathbf{e}_2$
$(\pi/2a, 0, \pi/2a)$	$2E'a(\mathbf{e}_1 + \mathbf{e}_2)$

The form of the effective mass in the three-dimensional case can be seen by calculating the acceleration, since the effective mass is the proportionality constant between the applied external force and the acceleration. The acceleration is given by

$$\frac{d\mathbf{v_k}}{dt} = \frac{1}{\hbar} \frac{d}{dt} \nabla_\mathbf{k} E(\mathbf{k})$$

$$= \frac{1}{\hbar} \nabla_\mathbf{k} \left[\frac{d\mathbf{k}}{dt} \cdot \nabla_\mathbf{k} E(\mathbf{k}) \right] \quad (6.29)$$

Substituting Eq. (6.26) for an external force **F**,

$$\frac{d\mathbf{v_k}}{dt} = \frac{1}{\hbar^2} \nabla_\mathbf{k} [\mathbf{F} \cdot \nabla_\mathbf{k} E(\mathbf{k})] \quad (6.30)$$

Since

$$\mathbf{F} \cdot \nabla_\mathbf{k} E(\mathbf{k}) = \frac{\partial E}{\partial k_1} F_1 + \frac{\partial E}{\partial k_2} F_2 + \frac{\partial E}{\partial k_3} F_3$$

each of the components of the acceleration has the form

$$a_i = \frac{dv_i}{dt} = \frac{1}{\hbar^2} \sum_j \frac{\partial^2 E}{\partial k_i \partial k_j} F_j \quad \text{with } i, j = 1, 2, 3 \quad (6.31)$$

If m_{ij}^* is the proportionality constant between the ith component of the acceleration and the jth component of the force **F**, it follows that

$$\frac{1}{m_{ij}^*} = \frac{1}{\hbar^2} \frac{\partial^2 E}{\partial k_i \partial k_j} \quad (6.32)$$

The effective mass is a tensor with the elements defined by this equation.

Example 6.2 Calculate the acceleration of an electron in the presence of an electric field $\mathscr{E}$ as a function of wave vector for a simple cubic lattice, assuming the $E(\mathbf{k})$ dependence given by the tight-binding approximation for an s state,

$$E(\mathbf{k}) = E_0 - E_0' - 2E'(\cos k_1 a + \cos k_2 a + \cos k_3 a)$$

According to Eq. (6.32) the elements of the tensor describing the proportionality between acceleration and electric field are given by

$$\left(\frac{1}{m^*}\right) = \begin{pmatrix} \dfrac{2E'a^2 \cos k_1 a}{\hbar^2} & 0 & 0 \\ 0 & \dfrac{2E'a^2 \cos k_2 a}{\hbar^2} & 0 \\ 0 & 0 & \dfrac{2E'a^2 \cos k_3 a}{\hbar^2} \end{pmatrix}$$

6.4 The Band Model and Electrical Properties

The acceleration is therefore

$$\mathbf{a} = (2E'a^2\mathscr{E}_1 \cos k_1 a/\hbar^2)\mathbf{e}_1 + (2E'a^2\mathscr{E}_2 \cos k_2 a/\hbar^2)\mathbf{e}_2$$
$$+ (2E'a^2\mathscr{E}_3 \cos k_3 a/\hbar^2)\mathbf{e}_3$$

for an electric field $\mathscr{E} = \mathscr{E}_1\mathbf{e}_1 + \mathscr{E}_2\mathbf{e}_2 + \mathscr{E}_3\mathbf{e}_3$. This is a simple example where there are no off-diagonal elements in the effective-mass tensor so that an electric field in the ith direction produces an acceleration only in the ith direction.

6.4 The Band Model and Electrical Properties

The greatest immediate success of the band model for electron energies in crystals was its ability to provide a simple and reasonably adequate explanation of the wide variety of electrical properties from metals to insulators.

For a simple Bravais lattice with cubic symmetry, a Brillouin zone accomodates two electrons per atom of the crystal. In a real crystal there are generally enough electrons per atom so that occupied states extend over several zones. In the simplest case, therefore, an even number of electrons per atom results in a series of completely filled energy bands (zones), and an odd number of electrons per atom results in the uppermost occupied band being half filled.

Since there can be no net transport of charge and no electrical conductivity in a completely filled band, those materials with a completely filled uppermost band are expected to be insulators. The only electrons able to contribute to electrical conductivity are those that are thermally excited from the uppermost filled band to the next-higher-lying empty allowed band; if the gap between allowed bands is much larger than kT, this number of thermally excited electrons is small and gives rise at most to semiconductor behavior at high temperatures and insulator behavior at low temperatures. Disregarding other complexities in their structure, Group IV elements such as C, Ge, and Si, and Group VI elements such as S, Se, and Te, all of which have even numbers of electrons per atom, are either insulators or semiconductors.

If the highest energy band containing electrons is only partially filled, metallic behavior is expected since empty allowed energy states are available immediately adjacent to occupied states. Since the only electrons that can contribute to electrical conductivity are the outermost or valence electrons, we may neglect inner-shell electrons (associated with filled bands) when

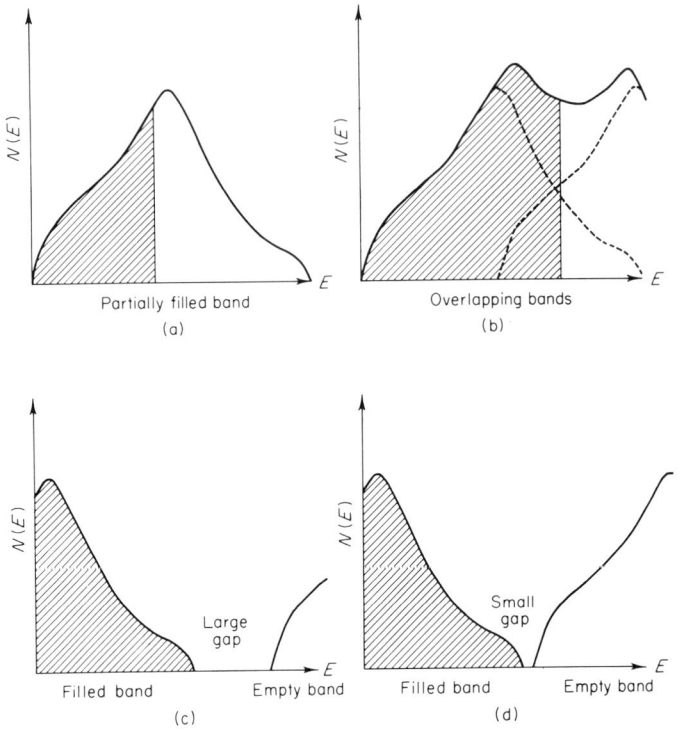

Fig. 6.5 Illustrations of the dependence of electrical properties on zone filling and spacing. (a) Metal with partially filled band; (b) metal with overlapping bands; (c) insulator; (d) intrinsic semiconductor.

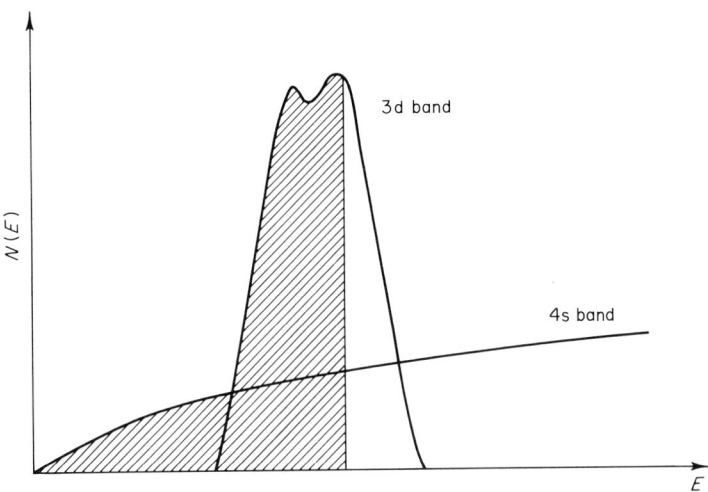

Fig. 6.6 Illustration of very high density of states available at the Fermi surface in Period 4 transition metals from Sc to Ni, because of an overlap between 3d and 4s bands.

6.4 The Band Model and Electrical Properties

discussing conductivity. Again disregarding complexities, Group I elements such as Na, K, Rb, and Group III elements such as Al, Ga, and In, have odd numbers of electrons per atom and are metals. Typical density of states versus energy curves are given in Fig. 6.5 for the various possibilities. Metallic properties can result in three-dimensional crystals of elements with even numbers of electrons per atom (such as the Group II elements) because of overlapping allowed bands.

A special case of some interest occurs for those metals with incomplete d shells such as scandium through nickel in the Periodic Table. The energy band derived from the 3d electrons overlaps the band derived from the 4s electrons in such a way that the Fermi level is located in a region with the very high density of states characteristic of the 3d states, which are derived from fivefold degenerate atomic states. Figure 6.6 illustrates this possibility schematically.

THE HOLE IN THE BAND MODEL

The electrical conductivity of a material with a partially filled conduction band can be conveniently described in terms of the current associated with the electrons in that band. The case sometimes arises, however, where a band is almost completely filled with electrons except for a few empty states near the top of the band. Since the current for a completely filled band is zero, the current due to a band with a single unoccupied state must be the negative of the current due to a band with a single occupied state. It is therefore possible to consider conductivity in a nearly full band in terms of the motion of equivalent particles with a positive charge; these equivalent particles are called *holes*. A hole, associated with an unoccupied electronic state, is not to be confused with a vacancy, an empty atomic or ionic state.

As illustrated in Fig. 6.7, electrons at the bottom of an almost empty band (a *conduction band*) have positive effective mass and negative charge. Electrons at the top of an almost full band (a *valence band*) have negative effective mass and negative charge. Holes at the top of an almost full (with electrons) valence band have positive effective mass and positive charge. In the presence of an electric field, electrons at the bottom of the conduction band and holes at the top of the valence band move in opposite directions, but electrons and holes both at the top of the valence band move in the same direction.

The motion of a hole in the valence band is shown in somewhat more detail in the one-dimensional model of Fig. 6.8. In order to fully represent

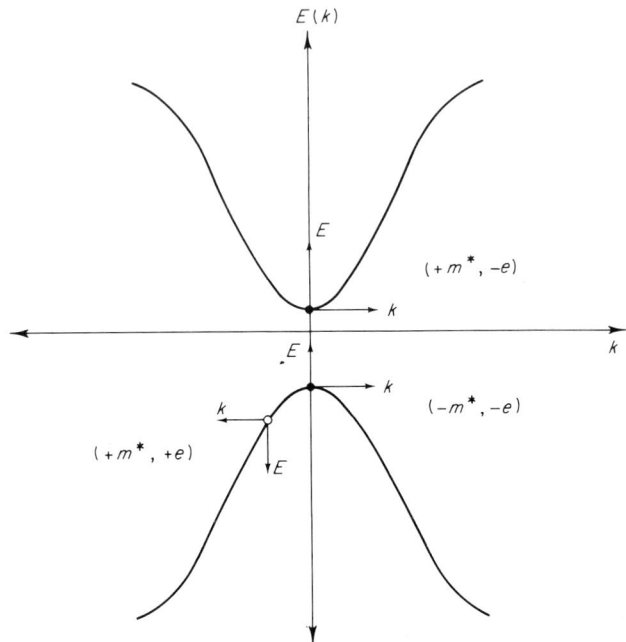

Fig. 6.7 Electrons at the bottom of the conduction band and top of the valence band, compared to holes at the top of the valence band. Arrows indicate positive directions of E and k for each of the species.

the situation, we need to consider simultaneously events as they occur in an electronic representation and in a hole representation. This is because the k for a hole state must be the negative of the k of the missing electronic state, and because the energy for a hole is the negative of the energy for an electron. At $t = 0$, we assume that all states are filled except one at the top of the band at $\mathbf{k} = 0$. An electric field $\mathscr{E}_x$ is applied in the $+x$ direction, so that the force on an electron is given by

$$F = \hbar \frac{dk_x}{dt} = -e\mathscr{E}_x \qquad (6.33)$$

Electrons are acclerated in the $-k_x$ direction. All electrons move together, with the result that the vacant electronic state also moves in the $-k_x$ direction. As the vacant electronic state moves in its $-k_x$ direction, the associated hole moves in its $+k_x$ direction. When the energy inversion is also considered, it is seen that the hole has moved from a region of zero velocity to a region of positive velocity; the hole has acquired velocity in the direc-

6.4 The Band Model and Electrical Properties

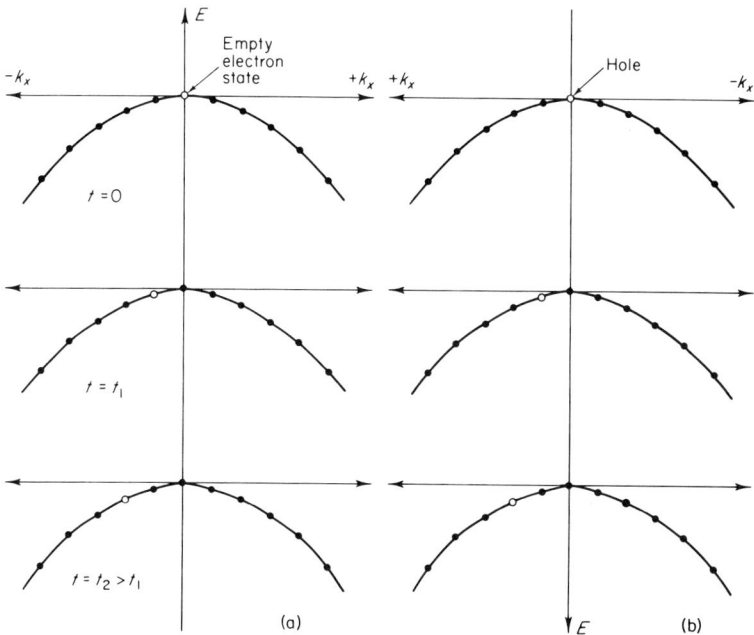

Fig. 6.8 The motion of an empty electron state and of the corresponding hole state in a nearly filled band in the presence of an electric field in the $+x$ direction. (a) Electron representation; (b) hole representation.

tion of the applied field. This is consistent with the assignment of positive charge and positive mass to the hole. The surrounding electrons also acquire velocity in the direction of the applied field, which is consistent with their negative charge and negative effective mass. Electrons in the conduction band under the same circumstances are also accelerated in the $-k_x$ direction, but for them this means acquiring velocity in the $-x$ direction, consistent with their negative charge and positive effective mass.

The occupancy of bands by holes can be described in a way similar to that used to describe the occupancy of bands by electrons. The probability that a state with energy E_n be occupied by an electron is given by

$$f(E_n) = \frac{1}{\exp[(E_n - E_F^e)/kT] + 1} \qquad (6.34)$$

The probability that a state with energy E_n *not* be occupied by an electron, i.e., that it be occupied by a hole, is

$$1 - f(E_n) = \frac{1}{1 + \exp[-(E_n - E_F^e)/kT]} = \frac{1}{\exp\{[-E_n - (-E_F^e)]/kT\} + 1} \quad (6.35)$$

or in terms of hole energies,

$$f(E_p) = 1 - f(E_n) = \frac{1}{\exp[(E_p - E_F^h)/kT] + 1} \quad (6.36)$$

E_n and E_F^e are conveniently measured from the bottom of a band upwards; E_p and E_F^e are conveniently measured from the top of a band downwards. These relationships are illustrated in Fig. 6.9.

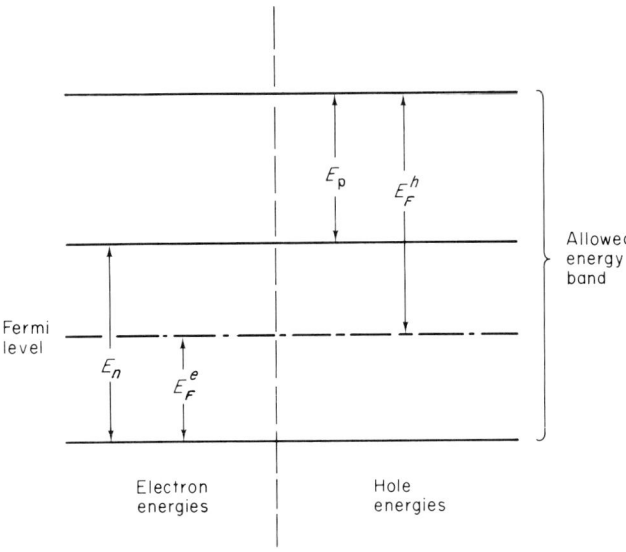

Fig. 6.9 Illustration of energies involved in Eqs. (6.34) and (6.36).

6.5 Energy Bands in Real Crystals

Simple energy bands are characterized by extrema at $\mathbf{k} = 0$, spherically symmetric equal-energy surfaces, scalar effective mass not a function of $\mathbf{k}$, and nondegenerate extrema in the conduction and valence bands. Real crystals may be characterized by these same features in specific instances, but they usually show a variety of features that do not follow these simple structures. In real crystals it is not uncommon to find extrema at values of $\mathbf{k}$

6.5 Energy Bands in Real Crystals

other than $\mathbf{k} = 0$, nonspherically symmetric equal-energy surfaces, tensor effective mass which is a function of $\mathbf{k}$, and degenerate extrema, i.e., more than one band with the same energy for the same point or range in $\mathbf{k}$ space. The properties of bands in real crystals can be conveniently summarized by examining the possible situations for extrema at $\mathbf{k} = 0$, then considering extrema for $\mathbf{k} \neq 0$, and finally illustrating these possibilities by examining the actual band structure in several groups of materials.

Extrema at $\mathbf{k} = 0$

The general expression for a conduction band with minimum at $\mathbf{k} = 0$ is

$$E = E_0 + Ak_1^2 + Bk_2^2 + Ck_3^2$$
$$= E_0 + \frac{\hbar^2}{2}\left(\frac{k_1^2}{m_1^*} + \frac{k_2^2}{m_2^*} + \frac{k_3^2}{m_3^*}\right) \quad (6.37)$$

where as usual $\mathbf{k} = k_1\mathbf{e}_1 + k_2\mathbf{e}_2 + k_3\mathbf{e}_3$. In this general expression there is a different value of effective mass [i.e., a different curvature of the $E(\mathbf{k})$ vs. $\mathbf{k}$ dependence] in each of the three orthogonal directions. Similarly the general expression for a valence band with maximum at $\mathbf{k} = 0$ is

$$E = E_0 - \frac{\hbar^2}{2}\left(\frac{k_1^2}{m_1^*} + \frac{k_2^2}{m_2^*} + \frac{k_3^2}{m_3^*}\right) \quad (6.38)$$

Of course the effective masses in Eq. (6.37) refer to electrons, and the effective masses in Eq. (6.38) to holes.

If the equal-energy surfaces are spherically symmetric, the conduction band, for example, becomes

$$E = E_0 + \frac{\hbar^2}{2m^*}k^2 \quad (6.39)$$

since in this case $m_1^* = m_2^* = m_3^* = m^*$.

If, instead of spherical symmetry, the equal-energy surfaces have cylindrical symmetry, the equal-energy surfaces are ellipsoidal (cigar-like) in shape, and may be written as

$$E = E_0 + \frac{\hbar^2}{2}\left(\frac{k_1^2}{m_1^*} + \frac{k_2^2 + k_3^2}{m_2^*}\right) \quad (6.40)$$

since $m_2^* = m_3^*$ for this particular case of symmetry about the x axis.

Extrema at $\mathbf{k} \neq 0$

The number and nature of extrema not at $\mathbf{k} = 0$ depends on the crystal symmetry and the actual location of the extrema in $\mathbf{k}$ space. For a cubic

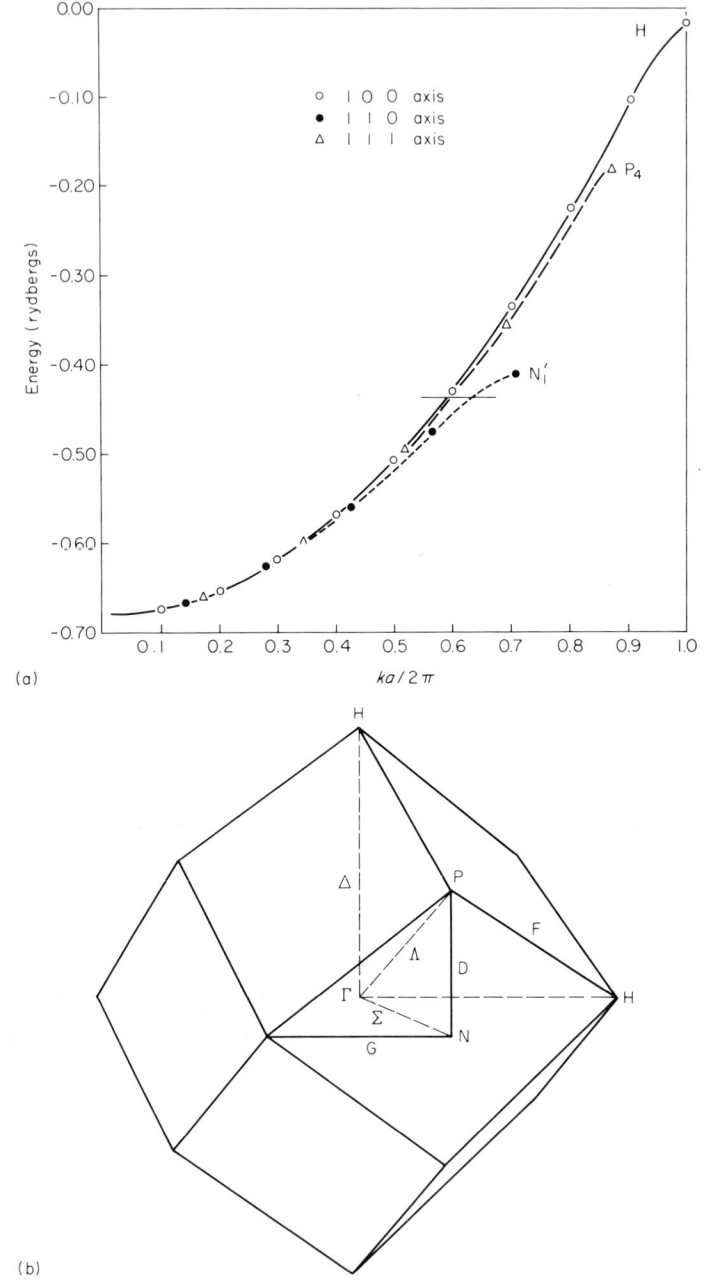

Fig. 6.10 (a) Energy bands for valence electrons in Li metal for several different crystallographic directions. (b) Basic Brillouin zone for a bcc crystal. (After J. Callaway, "Energy Band Theory," Academic Press, New York, 1964, p. 149.)

6.5 Energy Bands in Real Crystals

crystal where the extrema occur at the zone edge along 100 cube axes, there are three equivalent minima at $(\pi/a, 0, 0)$, $(0, \pi/a, 0)$, and $(0, 0, \pi/a)$. For cylindrical symmetry about the corresponding axis, the bands are described by expressions like the following for the x axis,

$$E = \frac{\hbar^2}{2} \left[\frac{(\pi/a - k_1^2)}{m_1^*} + \frac{k_2^2 + k_3^2}{m_2^*} \right] \quad (6.41)$$

If the extrema fall along the 100 axes, but not at the zone edge, there are six equivalent minima at $(\pm k_0, 0, 0)$, $(0, \pm k_0, 0)$, and $(0, 0, \pm k_0)$ if the extremum occurs at $|k| = k_0$.

If the extrema fall along the 111 directions (cube diagonals), there are four equivalent minima if the extrema occur at the zone edge, or eight equivalent minima if the extrema occur within the zone.

Energy Bands in Lithium

As an example of energy bands in a metal in which the electrons are almost describable in terms of a free-electron model, consider the metal lithium. Figure 6.10a gives the calculated energy bands for the valence electrons in lithium for three different directions in the reduced zone; lithium is a bcc crystal and the reduced zone for the bcc structure is given in Fig. 6.10b. The behavior is almost isotropic and departs little from the free-electron picture of parabolic energy bands and spherical equal-energy surfaces.

Energy Bands in Germanium and Silicon

Energy bands for Ge and Si are given in Fig. 6.11 for different directions in the reduced zone. These are fcc crystals and the reduced zone is given in Fig. 5.8. The energy bands shown in Fig. 6.11 are calculated with the omission of an energy term which affects the final quantitative form of the bands: the spin–orbit interaction, resulting from the interaction between the electronic orbital magnetic moment and spin magnetic moment. Figure 6.12 indicates the changes near the edges of the forbidden gap that arise when this spin–orbit interaction is included in the calculation; the energies cited are for $0°K$.

Ge and Si are quite similar materials, Si having an electronic configuration of ... $3s^2\, 3p^2$ and Ge of ... $4s^2\, 4p^2$. In order to understand the interpretation of the somewhat complicated curves of Fig. 6.11, consider in more detail the case of Ge.

Ge crystallizes in the diamond cubic structure and exhibits a band

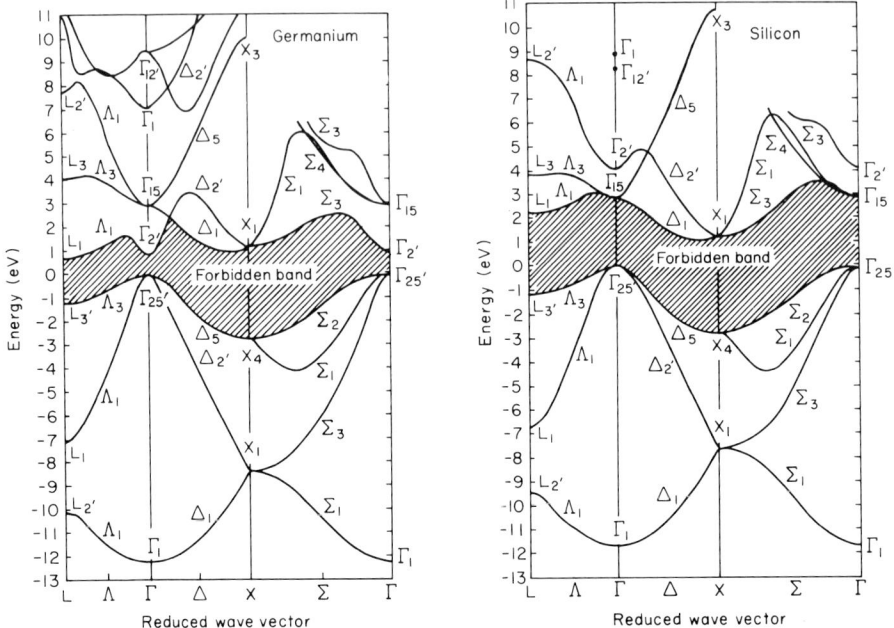

Fig. 6.11 Energy bands $E(\mathbf{k})$ vs. $\mathbf{k}$ in different crystal directions for Ge and Si. Spin–orbit splitting energies are neglected. [After F. Herman, R. L. Kortum and C. D. Kuglin, *Intern. J. Quant. Chem.* **1S**, 533 (1967).]

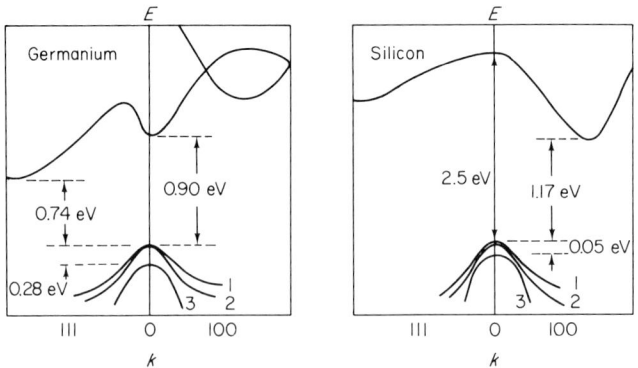

Fig. 6.12 Energy bands $E(\mathbf{k})$ vs. $\mathbf{k}$ near the conduction and valence band extrema including spin–orbit splitting energies for Ge and Si. Energies given are for $0°K$.

6.5 Energy Bands in Real Crystals

structure the general features of which are found in all other diamond cubic and zincblende cubic materials. The variation of $E(\mathbf{k})$ with $\mathbf{k}$ is plotted in Figs. 6.11 and 6.12 in the 100 direction (along the Δ axes) and in the 111 direction (along the Λ axes). The conduction band has several different kinds of minima, and the valence band has a maximum at $\mathbf{k} = 0$, i.e., at Γ. The energy gap, i.e., the minimum separation between conduction and valence bands, is given by the difference in energy between the conduction band minimum at the L points of the 111 zone face and the valence band maximum at the Γ point. Because the extrema of conduction and valence bands lie at different values of $\mathbf{k}$, this kind of bandgap is called an *indirect* bandgap, i.e., a change in the value of $\mathbf{k}$ is required for an electron to make a transition from the top of the valence band to the bottom of the conduction band. A second conduction-band minimum occurs at the Γ point. The energy difference between the conduction- and valence-band extrema at Γ is called the *direct bandgap*. In Ge the direct bandgap is 0.16 eV larger than the indirect bandgap. Measurement of the bandgap by optical absorption techniques is able to detect the presence of both direct and indirect bandgaps; measurement of the bandgap by the temperature dependence of electrical conductivity (i.e., for thermal excitation of electrons from valence to conduction band) reveals only the indirect bangdap.

A third conduction-band minimum occurs in Ge along the Δ axes. All the members of the cubic family similar to Ge exhibit similar energy-band structures, but for some materials the conduction-band minimum at Γ or along Δ is lower than the conduction-band minimum at L. The energy-band diagrams for Si in Figs. 6.11 and 6.12 show that the minimum along the Δ axes are the lowest conduction-band minima in Si.

The conduction band in Ge at L has four equivalent minima. The conduction band in Si along Δ has six equivalent minima occurring about 85 per cent of the way to the zone face. In terms of a set of axes in $\mathbf{k}$ space such that the components of $\mathbf{k}$ are along the principal axes of the equal-energy ellipsoids of revolution, the energy near the band edge in either Ge or Si is

$$E = \frac{\hbar^2}{2}\left(\frac{2k_t^2}{m_t^*} + \frac{k_\ell^2}{m_\ell^*}\right) \tag{6.42}$$

Here E is measured from the bottom of the energy minimum. Subscripts t and ℓ stand for transverse and longitudinal, and are defined with respect to the principal axes of the equal-energy ellipsoid. In units of the free-electron mass, the following values of effective mass are found for Ge and Si.

	m_t^*	m_ℓ^*
Ge	0.0815	1.59
Si	0.191	0.92

These equal-energy ellipsoids are elongated markedly in the longitudinal direction.

The conduction-band minimum in E at Γ has

$$E = \frac{\hbar^2}{2m_0^*} k^2 \qquad (6.43)$$

with $m_0^* = 0.036m$. In the following section we show how this can be derived from a $\mathbf{k} \cdot \mathbf{P}$ calculation at $\mathbf{k} = 0$ for Ge.

The valence electrons in Ge are the electrons that take part in forming the covalent bonds that hold the crystal together. Each Ge atom is surrounded by four other Ge atoms arranged on the vertices of a tetrahedron. The states of the electrons in the covalent bonds are described by orbitals derived from one s and three p atomic orbitals combined to give four equivalent electron distributions oriented in the tetrahedral symmetry. Since there are two atoms per unit cell, the valence band contains eight electronic states per unit cell, including the degeneracy due to spin. Two electrons occupy the s-like states, which lie so low in energy that they do not appear in Fig. 6.12. The other six electrons occupy the p-like states which are degenerate at Γ except for the contribution of the spin–orbit splitting interaction which lowers one of these bands by 0.28 eV at Γ.

Close to the valence-band maximum, the form of the energy bands for both Ge and Si can be expressed as

$$E(\mathbf{k}) = -\frac{\hbar^2}{2m} \{Ak^2 \pm [B^2k^4 + C^2(k_1^2k_2^2 + k_2^2k_3^2 + k_3^2k_1^2)]^{1/2}\} \qquad (6.44)$$

where E is again measured from the band edge. The constants A, B, and C have the following values for Ge and Si:

	A	B	C
Ge	13.1	8.3	12.5
Si	4.0	1.1	4.1

6.5 Energy Bands in Real Crystals

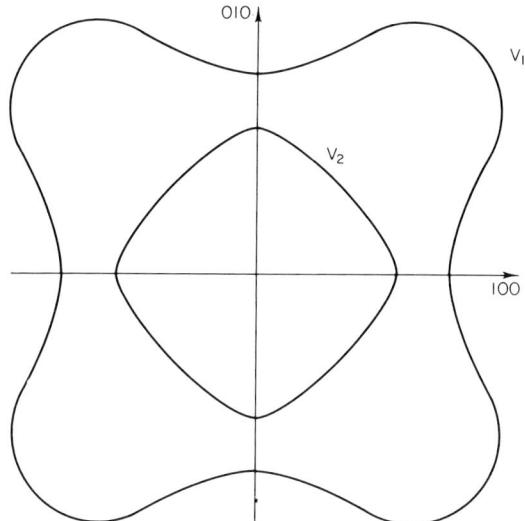

Fig. 6.13 Exaggerated equal-energy contours for the V_1 and V_2 valence bands in Ge and Si.

The $\pm$ signs in Eq. (6.44) refer respectively to the V_2 and V_1 valence bands. Typical equal-energy contours for these two bands in the (001) plane are shown in Fig. 6.13. These surfaces show cubic symmetry in three dimensions. Effective mass is different in different directions, resulting from the C^2 term in Eq. (6.44). The variation of $E(\mathbf{k})$ with $\mathbf{k}$ for the V_1 band, illustrated by the reentrant nature of the equal-energy surfaces in Fig. 6.13, makes it possible to have negative-mass holes at relatively low energies within the valence band. Sometimes the "warped" energy surfaces of Fig. 6.13 are approximated by spherical surfaces with effective masses equal to the average of the directionally dependent effective masses,

$$E(\mathbf{k}) = -\frac{\hbar^2}{2m}\left[A \pm \left(B^2 + \frac{C^2}{6}\right)^{1/2}\right]k^2 \quad (6.45)$$

Here the term in brackets is the approximation to the average effective mass. Values, in units of the free mass, are as follows:

	$\overline{m^*_{V_1}}$	$\overline{m^*_{V_2}}$
Ge	0.30	0.044
Si	0.50	0.17

In contrast, the V_3 valence band has approximately spherical equal-energy surfaces with an effective mass of $0.066m$ in Ge and $0.25m$ in Si. Because of the magnitude of the averaged effective masses in the V_1 and V_2 valence bands, the V_1 band is usually called the "heavy hole" band, and the V_2 band is called the "light hole" band. This nomenclature leads to the semantic curiosity of "heavy holes with negative mass!"

APPLICATION OF THE $\mathbf{k} \cdot \mathbf{P}$ APPROXIMATION TO GERMANIUM AT Γ

As an illustration of one technique discussed in Section 6.1, consider here the specific determination of the $E(\mathbf{k})$ vs. $\mathbf{k}$ dependence near Γ according to the $\mathbf{k} \cdot \mathbf{P}$ approximation in Ge.

At Γ, $\mathbf{k} = 0$, and we have the same equation as Eq. (6.14). Now, however, there are four eigenfunctions to consider. For the conduction band, $u_c(\mathbf{r})$ has the symmetry of an atomic s state; for the valence band there are three functions $u_v(\mathbf{r})$ having the symmetry of atomic p states, i.e., with angular dependences of x/r, y/r, and z/r as given in Eqs. (3.63) through (3.65). For the present calculation, we neglect the spin–orbit interaction. The energy gap at $\mathbf{k} = 0$ (the direct energy gap) is given by $E_{G0} = E_c - E_v$, where E_c is the energy corresponding to $u_c(\mathbf{r})$ in Eq. (6.14) and E_v is the energy corresponding to $u_v(\mathbf{r})$.

For $\mathbf{k} \neq 0$, write

$$u_\mathbf{k}(\mathbf{r}) = \sum_i^4 a_{ki} u_i(\mathbf{r}) \qquad (6.46)$$

where the sum is over the four eigenfunctions of Eq. (6.14). If we substitute Eq. (6.46) with the corresponding eigenvalues $E_i = E_c$, E_v into Eq. (6.13), we obtain

$$\sum_i^4 \left[E_i + \frac{\hbar}{m} \mathbf{k} \cdot \mathbf{P} - E(\mathbf{k}) \right] a_{ki} u_i(\mathbf{r}) = 0 \qquad (6.47)$$

if the $\hbar^2 k^2/2m$ term is neglected. If we multiply Eq. (6.47) by the complex conjugate of each of the functions $u_i(\mathbf{r})$ and integrate over all space, we obtain four simultaneous equations in the coefficients a_{ki}. For these four equations to have a consistent solution, the determinant of the quantities multiplying the a_{ki} in each equation must be zero. From this condition we obtain the variation of $E(\mathbf{k})$ with $\mathbf{k}$ near $\mathbf{k} = 0$.

As an example of this calculation, consider the effects of multiplying Eq. (6.47) by $u_c^*(\mathbf{r})$ and integrating over all space. The first term is that

6.5 Energy Bands in Real Crystals

involving a_{kc} multiplied by

$$\int u_c^*(\mathbf{r}) E_c u_c(\mathbf{r}) \, d\mathbf{r} + \frac{\hbar}{m} \left[k_x \int u_c^* \mathbf{P}_x u_c \, d\mathbf{r} + k_y \int u_c^* \mathbf{P}_y u_c \, d\mathbf{r} \right.$$

$$\left. + k_z \int u_c^* \mathbf{P}_z u_c \, d\mathbf{r} \right] - \int u_c^* E(\mathbf{k}) u_c \, d\mathbf{r}$$

$$= E_c - E(\mathbf{k}) - \frac{i\hbar^2}{m} \left(k_x \int u_c^* \frac{\partial u_c}{\partial x} \, d\mathbf{r} + k_y \int u_c^* \frac{\partial u_c}{\partial y} \, d\mathbf{r} \right.$$

$$\left. + k_z \int u_c^* \frac{\partial u_c}{\partial z} \, d\mathbf{r} \right) = E_c - E(\mathbf{k}) \qquad (6.48)$$

because of the spherical symmetry of $u_c(\mathbf{r})$. The second term resulting from multiplying Eq. (6.47) by $u_c^*(\mathbf{r})$ and integrating over all space gives the multiplier of a_{kx}, the coefficient of the valence band function $u_v(\mathbf{r})$ with angular dependence x/r, which we write simply as u_x. The result is

$$\int u_c^* E_v u_x \, d\mathbf{r} + \frac{\hbar}{m} \left[k_x \int u_c^* \mathbf{P}_x u_x \, d\mathbf{r} + k_y \int u_c^* \mathbf{P}_y u_x \, d\mathbf{r} + k_z \int u_c^* \mathbf{P}_z u_x \right]$$

$$- \int u_c^* E(\mathbf{k}) u_x \, d\mathbf{r}$$

$$= \frac{\hbar}{m} k_x \int u_c^* \mathbf{P}_x u_x \, d\mathbf{r}$$

$$\equiv k_x P \qquad (6.49)$$

because of the orthogonality of the u_c and u_v functions, and because of the dependence of u_x on x only. The quantity P defined in Eq. (6.49) is a *momentum matrix element*. The third and fourth terms resulting from multiplying Eq. (6.47) by $u_c^*(\mathbf{r})$ and integrating over all space, giving the multipliers of a_{ky} and a_{kz}, are evaluated in a similar way, and are respectively $k_y P$ and $k_z P$. The quantity P is the same for any nonzero combination of the components of $\mathbf{P}$ and the functions $u_i(\mathbf{r})$.

The multipliers of a_{kc}, a_{kx}, a_{ky}, and a_{kz} in the other three equations obtained by multiplying Eq. (6.47) by u_x^*, u_y^*, and u_z^* in turn, and integrating over all space, can be obtained by procedures similar to those just illustrated. The determinant of the multipliers obtained in this way is

$$\begin{vmatrix} E_c - E(\mathbf{k}) & k_x P & k_y P & k_z P \\ k_x P & E_v - E(\mathbf{k}) & 0 & 0 \\ k_y P & 0 & E_v - E(\mathbf{k}) & 0 \\ k_z P & 0 & 0 & E_v - E(\mathbf{k}) \end{vmatrix} \qquad (6.50)$$

Setting this determinant equal to zero and expanding via the elements of the top row, we obtain

$$[E_c - E(\mathbf{k})][E_v - E(\mathbf{k})]^3 - k_x^2 P^2 [E_v - E(\mathbf{k})]^2 - k_y^2 P^2 [E_v - E(\mathbf{k})]^2$$
$$- k_z^2 P^2 [E_v - E(\mathbf{k})]^2 = 0 \qquad (6.51)$$

or

$$[E_v - E(\mathbf{k})]^2 \{[E_c - E(\mathbf{k})][E_v - E(\mathbf{k})] - k^2 P^2\} = 0 \qquad (6.52)$$

The four roots of $E(\mathbf{k})$ are

$$E(\mathbf{k}) = \begin{cases} E_v \\ E_v \\ \dfrac{E_c + E_v}{2} + \left[\dfrac{(E_c - E_v)^2}{4} + k^2 P^2\right]^{1/2} \\ \dfrac{E_c + E_v}{2} - \left[\dfrac{(E_c - E_v)^2}{4} + k^2 P^2\right]^{1/2} \end{cases} \qquad (6.53)$$

For small values of $\mathbf{k}$, the square-root terms can be approximated to give

$$E(\mathbf{k}) = E_v, \quad E_v, \quad E_v - \frac{k^2 P^2}{E_{G0}}, \quad E_c + \frac{k^2 P^2}{E_{G0}} \qquad (6.54)$$

These four solutions of $E(\mathbf{k})$ correspond in turn to the two heavy-hole bands, the light-hole band, and the conduction band. The calculation carried out here does not give $E(\mathbf{k})$ as a function of $\mathbf{k}$ for the two heavy-hole bands; the $E(\mathbf{k})$ dependence for these bands depends on the properties of conduction bands lying at higher energies than the one included here. If these bands are included in the calculation, the $\mathbf{k} \cdot \mathbf{P}$ method can be used to determine $E(\mathbf{k})$ for the heavy-hole bands as well.

In the approximation calculated above, without spin–orbit interaction effects, it is seen that the effective mass for the conduction band minimum at Γ and for the light-hole band maximum at Γ is the same, each being also proportional to E_{G0}^{-1}. These bands are parabolic near Γ, but away from Γ they become nonparabolic, as shown by Eqs. (6.53). Values of the momentum matrix P have been determined empirically for many materials by using experimentally determined values of effective mass and bandgap; it is not strongly dependent on the particular material, and has values fairly constant within a given class of materials, e.g., Group IV elements, III–V compounds, II–VI compounds, and so on.

6.5 Energy Bands in Real Crystals

Energy Bands in Other Materials

Energy-band structures have also been rather extensively investigated for other types of simple compounds and for elements. In most cases theoretical calculations of the band structure have been guided and informed by experimental data, particularly from optical absorption or photoemission, techniques in which the absorption process is interpreted to indicate the energy and density of initial and final states.

The III–V compound GaAs and the II–VI compound ZnSe are isoelectronic with the element Ge; they all have the same crystal structure and the same total number of valence electrons. Similarities in band structure are expected, as indicated by the energy-band curves given in Fig. 6.14 for GaAs and ZnSe.

All III–V compounds except Al compounds and GaP, and all II–VI compounds of Zn or Cd with S, Se, or Te have conduction-band minima at $\mathbf{k} = 0$. The detailed energy-band structure of the III–V compound InSb, showing the details near the band edges and the effects of spin–orbit splitting, is given in Fig. 6.15. The $E(\mathbf{k})$ vs. $\mathbf{k}$ dependence is nonparabolic, with m^* increasing with electron energy,

$$E(\mathbf{k}) = \frac{\hbar^2}{2m_0^*} k^2 - \alpha \hbar^4 k^4 + \cdots \qquad (6.55)$$

TABLE 6.1 Bandgap, Effective Mass, and Spin–Orbit Splitting in III–V Compounds

	E_G^0 (eV)	m_0^*/m	Spin–orbit splitting energy (eV)
InSb	0.23	0.0155	0.9
InAs	0.43	0.023	0.43
GaSb	0.81	0.047	0.8
InP	1.41	0.077	0.24
GaAs	1.53	0.07	0.34
AlSb	1.60	—	0.75
GaP	2.4	—	—

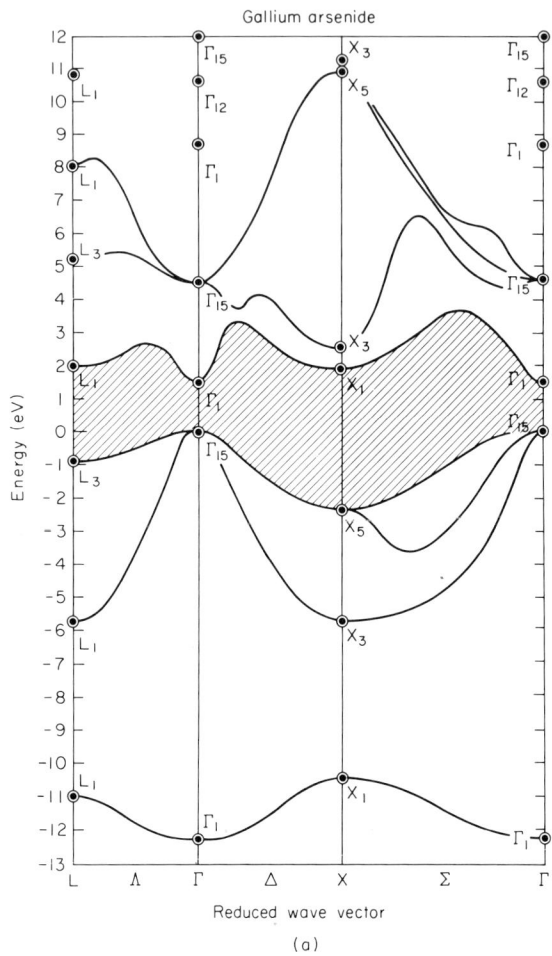

Fig. 6.14 Energy bands $E(\mathbf{k})$ vs. $\mathbf{k}$ in different crystal directions with spin–orbit splitting energies neglected. (a) GaAs [after F. Herman, W. E. Spicer, *Phys. Rev.* **174** 906 (1968)]; (b) (*Opposite*) ZnSe [after D. J. Stukel, R. N. Euwema, T. C. Collins, F. Herman, and R. L. Kortum, *Phys. Rev.* **179**, 740 (1969)].

In the valence-band structure, the twofold spin degeneracy of the V_1 band is lifted, as compared with Ge or Si, essentially because adjacent atoms are different chemical entities in the compounds. There is a similar very small lifting of the degeneracy of the V_2 band, not shown in Fig. 6.15. The eight equivalent maxima of V_1 for InSb lie at about 3 per cent of the way from Γ to the zone edge, and are not more than 0.01 eV higher in energy than

6.5 *Energy Bands in Real Crystals* 197

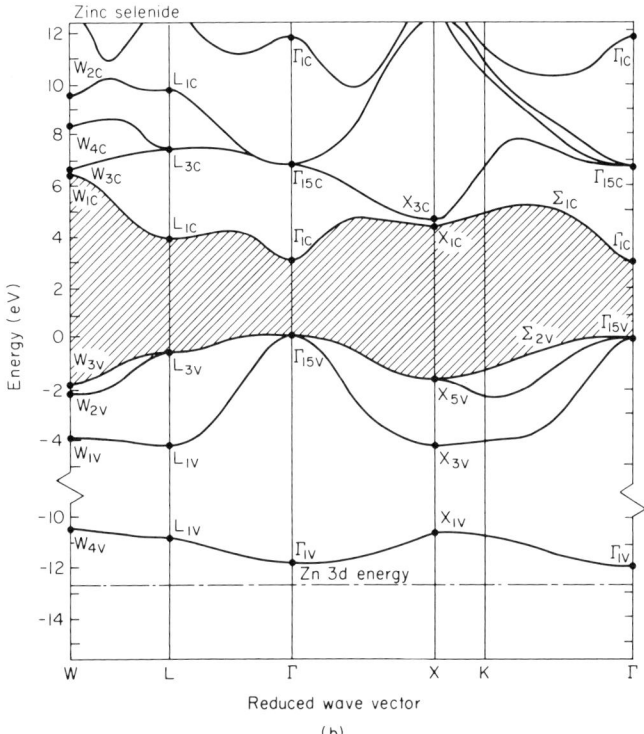

(b)

at Γ. The third valence band V_3 is depressed by the spin–orbit splitting. Values of bandgap at $0°K$, effective mass of electrons at Γ, and magnitude of the spin–orbit splitting are given in Table 6.1 for III–V compounds. For comparison, the average effective mass of holes in InSb, for example, is $0.2m$ for V_1 and $0.015m$ for V_2.

Fig. 6.16 shows the energy-band structure near the band edge for CdS and CdSe, two II–VI compounds with a hexagonal (wurtzite) crystal structure. The extrema of the bands occur at Γ. The degeneracy of the valence bands is completely removed by a combination of hexagonal crystal field splitting, most important contribution to the offset of the V_3 band, and of spin–orbit splitting, producing the V_1 and V_2 band separation. The effective mass of electrons near Γ in both CdS and CdSe is near $0.2m$ and isotropic.

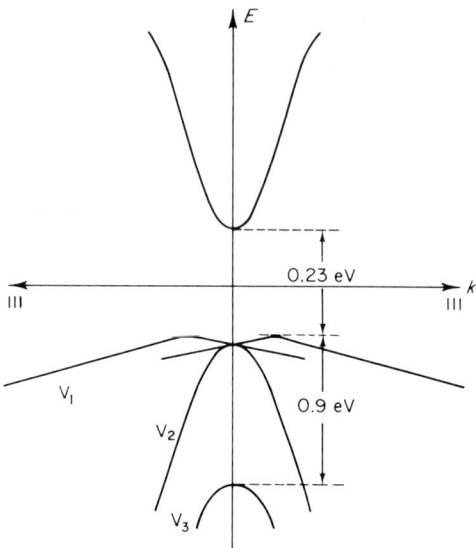

Fig. 6.15 Energy bands $E(\mathbf{k})$ vs. $\mathbf{k}$ near the conduction- and valence-band extrema including spin–orbit splitting energies for InSb.

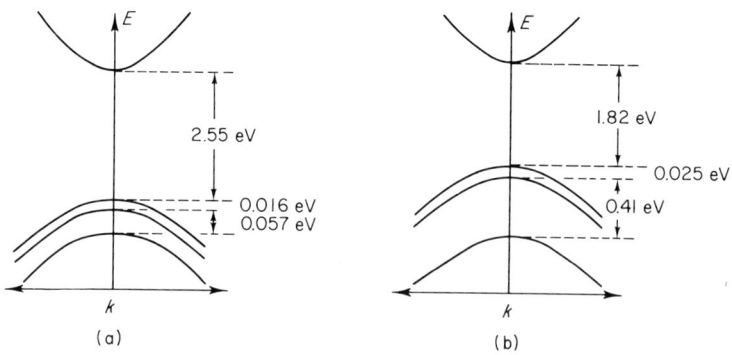

Fig. 6.16 Energy bands $E(\mathbf{k})$ vs. $\mathbf{k}$ near the conduction- and valence-band extrema including hexagonal crystal field splitting and spin–orbit splitting energies in (a) CdS and (b) CdSe.

6.6 Excitons and Polarons

The energy bands discussed so far have been those available to an *external* electron added to the system. There are additional levels, however, available to the electrons and holes of the crystal itself.

Excitons arise because of the possibility of binding between a free electron and a free hole by Coulomb attraction, the bound pair moving together through the crystal, thus transporting energy but no net charge. Such a pair is called an *exciton*. It represents a quantum of the excited crystal.

The binding energy of an exciton in which the electron and hole are weakly bound is quite similar to that for a hydrogen atom, a negative charge bound to a positive charge. There are therefore a number of hydrogenic energy levels available to the exciton with energies given by

$$E_{\text{ex},n} = \frac{(M_r/m)E_H}{\varepsilon^2 n^2} \qquad (6.56)$$

where M_r is the reduced mass for the exciton,

$$\frac{1}{M_r} = \frac{1}{m_e^*} + \frac{1}{m_h^*} \qquad (6.57)$$

E_H is the ionization energy of the hydrogen atom (13.5 eV), and ε is the dielectric constant of the particular material. The concept of dielectric constant is valid for these relatively weakly bound excitons, for which the effective "orbit" embraces many crystal atoms. The energy required to form an exciton is $E_G - E_{\text{ex},1}$, where E_G is the bandgap. As an illustrative example, with $m_e^* = m_h^* = m$, and $\varepsilon = 10$,

$$E_{\text{ex},n} = \frac{0.067}{n^2} \text{ eV} \qquad (6.58)$$

Thus such excitons will not normally be stable except at low temperatures, but will be thermally dissociated into a free electron and a free hole after formation.

The motion of the center of gravity of an exciton through the crystal can be represented by a wavefunction like that of a free electron,

$$\Psi_{\text{ex}}(\mathbf{r}) = A\, e^{i\mathbf{K}_{\text{ex}} \cdot \mathbf{r}} \qquad (6.59)$$

where $\mathbf{r}$ is the position vector of the center of gravity of the exciton, and $\mathbf{K}_{\text{ex}}$ is a wave vector for the exciton. For small values of $\mathbf{K}_{\text{ex}}$, the kinetic energy of an exciton may be written as

$$E_{\text{ex}} = \frac{\hbar^2 K^2}{2(m_e^* + m_h^*)} \tag{6.60}$$

Since $\mathbf{K}_{\text{ex}}$ is influenced in a manner similar to $\mathbf{k}$ for an electron by the periodicity of the crystal lattice, there are in general exciton bands rather than simply discrete exciton levels. When excitation of an exciton involves only a photon, however, the momentum acquired from the photon is sufficiently small that the exciton may be regarded as having an energy equal to one of the $E_{\text{ex},n}$ of Eq. (6.56). Further discussion of the optical excitation of excitons is given in Chapter 10.

There is also a type of excited state of the crystal, also called an exciton, that is much more tightly bound in appropriate crystals. In molecular crystals, for example, the binding within a molecule may be greater than the binding between molecules. Excited states of a particular molecule often exist with little change even when the molecule is part of a molecular crystal. It is even possible to observe exciton absorption to molecular energy levels lying above the bottom of the conduction band formed by the interaction between the molecules of the crystal. Ionic crystals also show this type of tightly bound exciton.

Polarons result from the interaction of either electron or hole with lattice phonons. If the motion of a free carrier is sufficiently slow, the surrounding lattice atoms become polarized by the charge of the carrier and relax into a new configuration through the emission or absorption of phonons. A polaron consists of the carrier and this associated polarization of the lattice atoms. The motion of the polaron can be treated in an effective-mass approximation, in which clearly the polaron effective mass is greater than the corresponding free-carrier effective mass. The motion of the polaron cannot be considered as the continuous motion of a free carrier, but must be treated as a kind of hopping process from one polarization site to the next.

6.7 Bands and Bonds

A question of continued interest is what is the relation between these energy bands, $E(\mathbf{k})$ versus $\mathbf{k}$, and the chemical bonds and order of the crystal itself, i.e., the configuration in coordinate space? It is found, for example, that an energy gap between uppermost filled and the next allowed states exists even in an amorphous material in which the periodic potential on which the existence of energy bands is based is absent. Physically it seems

6.7 Bands and Bonds

appropriate to associate the properties of this energy gap with the short-range order in the material, with the energy to remove an electron from a localized bond, and so on, while the detailed higher-energy band structure is much more determined by the long-range order of the material and tends to disappear as this long-range periodic order disappears. Interest in a variety of materials with spatial potential variation that does not conform to the ideal periodic potential of the previous sections leads to renewed awareness of the desirability of correlating bands and bonds. In this section we briefly summarize a few approaches that have been taken.

The binding forces between atoms that hold a crystal together are essentially electrostatic in nature, having their origin in the electronic and nuclear charges. It is common to distinguish between three main types of binding: (1) *ionic*, in which the atoms making up the crystal actually transfer electrons to form positively charged cations and negatively charged anions with relatively small electronic density between atoms; (2) *covalent*, in which bonding electrons are shared between atoms and the electronic density between atoms is high; and (3) *metallic*, in which the valence electrons are essentially free and the positively charged ion cores can be visualized as being embedded in a "sea" of negative charge, as discussed in Chapter 4. For most insulators and semiconductors, no clear-cut definition of binding in terms of ionic and covalent is possible; it is evident that in some materials ionic bonding predominates, in others covalent bonding predominates, and in others an ill-defined mixture of both is operative.

ELECTRONEGATIVITY

One convenient index to the type of binding to be expected in a particular material is provided by the concept of electronegativity, as formulated by Pauling. Electronegativity is a measure of the ability of an atom to attract electrons to itself. The electronegativity difference between two atoms forming a compound is proposed as a measure of the type of binding; the larger the electronegativity difference, the greater the ionic contribution to the binding. Suitable values for electronegativity have been investigated by many workers in recent years; a useful summary is shown in Fig. 6.17. The electronegativity values themselves can be well correlated with such properties as the work function of metals; empirical data fit well a relationship

$$X = 0.44\phi - 0.15 \tag{6.61}$$

where X is the electronegativity in Pauling's units, and ϕ is the work function in eV.

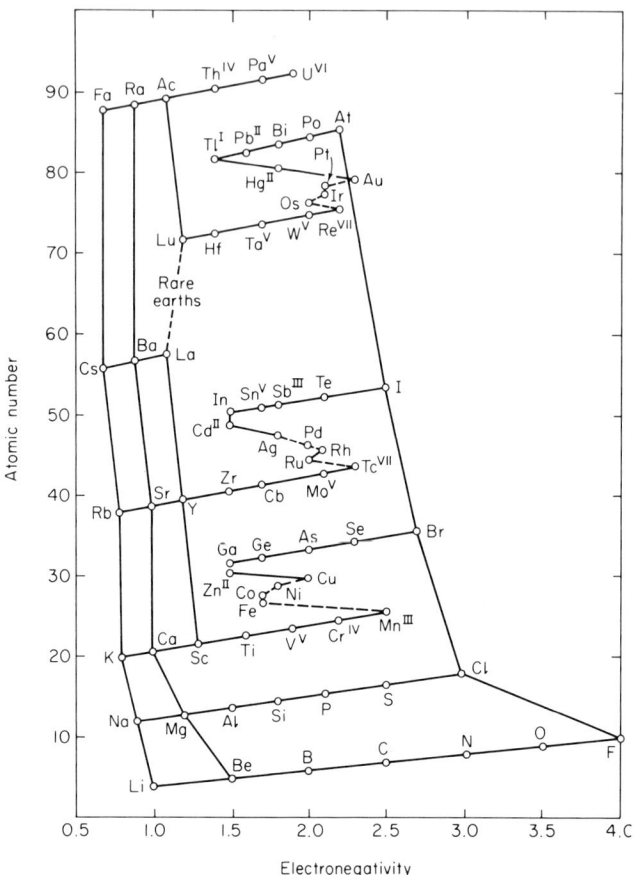

Fig. 6.17 Electronegativities of the elements. [After W. Gordy and W. J. O. Thomas, *J. Chem. Phys.* **24**, 439 (1956).]

From Fig. 6.17 it is evident that electronegativity differences are greatest for compounds in which the component atoms come from widely separated columns in the Periodic Table. It is also evident that for atoms in a given column, electronegativity tends to decrease with increasing atomic number. This empirical observation can be translated into a simple linear relationship between bandgap and position of the components of a compound in the Periodic Table on the assumption that increased electronegativity difference means increased ionic contribution to the bonding, which in turn means increased bandgap. A simple formulation is

$$E_G = C \frac{N_X - N_M}{A_X + A_M} \text{ eV} \tag{6.62}$$

6.7 Bands and Bonds

where E_G is the bandgap, C is a constant set empirically equal to 43, N_X is the number of valence electrons of the anion, N_M is the number of valence electrons of the cation, A_M is the atomic number of the cation, and A_X is the atomic number of the anion. This formulation is good for a number of compounds as long as elements with either very small or very large atomic number are avoided. A typical plot is given in Fig. 6.18.

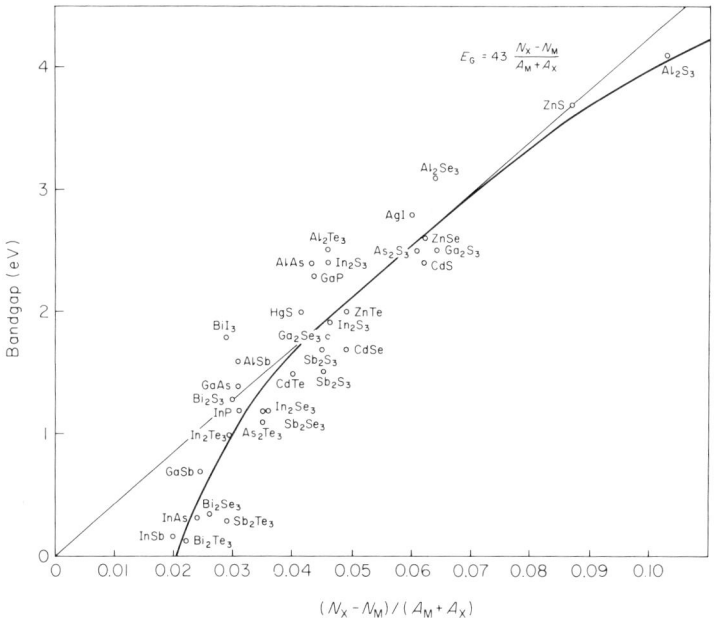

Fig. 6.18 Variation of bandgap for a number of compounds compared to Eq. (6.62).

BOND LENGTH

Goodman[†] assumed that the covalent bond energy is inversely proportional to the bond length. By plotting the bandgap for purely covalent C (diamond), Si, Ge, and Sn against reciprocal bond length, a relationship is obtained between covalent bond energy and bond length that can in turn be applied to other materials in which the binding is not purely co-

---

[†] C. H. L. Goodman, *J. Electronics* **1**, 115 (1955).

valent. For any other material, the difference between the experimental bandgap and the covalent bandgap corresponding to its bond length gives the ionic contribution to the bandgap.

Phillips[†] has developed a new scale of ionicity, using both the bond-length dependence of the covalent contribution and optical determination of the electronic dielectric constant to determine the ionic contribution. The following table compares ionicity for several compounds as predicted by Pauling's, Goodman's, and Phillips' approaches.

Compound	Ionicity (percent)		
	Pauling	Goodman	Phillips
GaAs	9	43	30
CdTe	11	94	70
AgI	13	97	77

It appears that Pauling's definition of ionicity seriously underestimates the ionic contribution to bonding, and that Goodman's definition somewhat overestimates the ionic contribution.

BOND ENERGY

Manca[‡] has proposed a simple linear empirical relationship between the energy gap and the single bond energy,

$$E_G = a(E_s - b) \quad (6.63)$$

where E_s is the single bond energy, and a and b are characteristic constants for II–VI, III–V, and IV–IV compounds. The values of single-bond energy are obtained as the sum of a covalent and ionic contribution. The covalent contribution is taken as the geometric mean of the covalent bond energies of the two atoms, and the ionic contribution is taken as the square of the difference in electronegativities. Figure 6.19 shows the nature of this correlation.

[†] J. C. Phillips, *Phys. Rev. Letters* **20**, 550 (1968); **22**, 645 (1969).
[‡] P. Manca, *J. Phys. Chem. Solids* **20**, 268 (1961).

6.7 Bands and Bonds

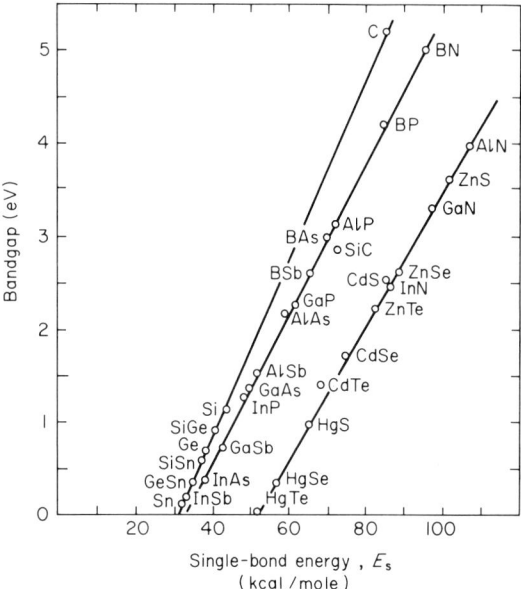

Fig. 6.19 Illustration of the relationship between energy bandgap and single-bond energy for various compounds of Group IV, III–V, and II–VI classes. [After P. Manca, *J. Phys. Chem. Solids* **20**, 268 (1961).]

ELECTRONIC CONFIGURATION

Mooser and Pearson[†] have attempted to define the electronic configuration conditions present in binding that lead to a semiconductor rather than a metal. They argue that semiconductivity is the result of predominantly covalent bonds; such bonds correspond to completely filled s and p orbitals in the valence shells of all elemental semiconductors, or to completely filled s and p orbitals in either one of the two bonded atoms in a compound. Empty orbitals lead to metallic properties only when atoms with empty orbitals are bonded together. It is observed that elemental semiconductors obey the $(8 - N)$ rule, i.e., each atom has $(8 - N)$ nearest neighbors where N is the ordinal number of the Periodic-Table column, in agreement with the above statements. A more general relationship for the criterion for semiconducting behavior in compounds is that

$$\frac{n_e}{n_a} + b = 8 \quad (6.64)$$

[†] E. Mooser and W. B. Pearson, *J. Electronics* **1**, 629 (1956).

TABLE 6.2 ILLUSTRATION THAT $(n_e/n_a) + b = 8$ FOR TYPICAL SEMICONDUCTORS

Compound	n_e	n_a	b
Ge	4	1	4
Se	6	1	2
InSb	8	1	0
CdSb	7	1	1
SiC	8	2	4
In_2Te_3	24	3	0

where n_e is the number of valence electrons per molecule, n_a is the number of Group IV to Group VII atoms per molecule, and b is the average number of bonds formed by one of these atoms with other atoms of Group IV to Group VII. This relationship is a direct result of the requirement that at least one of two bonded atoms have completed s and p orbitals, and the observation that completed s and p orbitals are possible only for atoms in Group IV to Group VII. Illustrations of the fulfillment of this rule are given in Table 6.2.

6.8 Summary

Refinement and extension of energy-band calculations to a variety of materials is certain to be an on-going enterprise in the future. Further investigations are likely also to provide closer correlations between the band (extended or group) properties of crystalline and noncrystalline solids and the nature of the chemical bond between localized specific atomic pairs.

Among the various more general approaches to energy-band calculations, specific attention has been given to the pseudopotential method, in which the problem is reduced to one with the form of a free-electron system disturbed by a relatively small perturbing potential, and to the $\mathbf{k} \cdot \mathbf{P}$ approximation, particularly useful for determining the variation of $E(\mathbf{k})$ with $\mathbf{k}$ near an extremum.

Knowledge of the shape of the $E(\mathbf{k})$ vs. $\mathbf{k}$ curves provides also knowledge of the effective velocity and the effective mass for an electron in a given $\mathbf{k}$ state. The velocity is the group velocity for the wave system. The effective mass is a useful construct that permits lumping of all crystal-potential forces on the electron into one defined parameter, so that only external forces need be considered in the equations of motion; the effective mass is inversely proportional to the local curvature of the $E(\mathbf{k})$ vs. $\mathbf{k}$ curves.

Simple descriptions of ideal crystals are usually depicted in terms of band extrema at $\mathbf{k} = 0$, spherically symmetric equal-energy surfaces, scalar effective mass not a function of $\mathbf{k}$, and nondegenerate bands, but examples drawn from real crystals illustrate departures from all these conditions.

Charge transport can be described either in terms of electron-occupied states or in terms of electron-unoccupied (hole) states in an allowed band. In an almost empty band, the former description is simpler, while in an almost full band, the latter description in terms of holes is simpler. The description in terms of holes replaces a description in terms of electron states with negative charge and negative effective mass, with a description in terms of hole states with positive charge and positive effective mass. Other "particles" encountered in describing electron and hole transport are the exciton, a bound electron–hole pair, and a polaron, an electron (or a hole) and an associated lattice polarization.

In most of the remainder of this book, we take the band picture for granted. At various points we point out particular phenomena or devices (such as p–n junction, transistor, tunnel diode, Gunn diode, and solid-state light emitter and laser) which depend in some unique way on the band picture of solids. Successes based on this approach indicate that future improvements in theory must include the band properties of solids as one definite domain of approximation.

Problems

6.1 For a free electron $m^* v_g = \hbar k$, where $m^* = m$ is the effective mass for the free electron, and v_g is the group velocity of waves with wavenumber k. (a) Under what conditions does $m^* v_g = \hbar k$ also in a crystal? (b) What is the value of $m^* v_g$ at k_0 for electrons if the conduction band minimum occurs at k_0 (e.g., in the 100 direction of a cubic crystal)? (c) Write $E(k)$ and $v_g(k)$ for a parabolic conduction band minimum occurring at k_0, as in (b). (d) Sketch the equal-energy surfaces for values of k near k_0 for the situation described in (b) and (c).

6.2 A hypothetical energy band can be fit approximately by the relation $E(k) = E_0(1 - e^{-a^2 k^2})$, where a is the lattice constant. Calculate: (a) The effective mass at $k = 0$. (b) The value of k for maximum electron velocity. (c) The effective mass at the zone boundary.

6.3 Assume a material for which the conduction band can be described by

$$E(k) = \frac{\hbar^2 k^2}{2m_0^*} - C\hbar^4 k^4 + \cdots$$

where m_0^* is the effective mass at $k = 0$, and C is a small constant. Plot $E(k)$, $v_g(k)$ and $m^*(k)$ vs. k. If this relationship between $E(k)$ and k extends all the way to the zone edge, what must the value of C be? (v_g must equal zero at zone edge.)

6.4 Discuss why it is "normal" for the effective mass of carriers in a band to vary inversely as the width of the band. Calculate and compare the effective mass near $k = 0$ in terms of the bandwidth for the tight-binding approximation and for the free-electron approximation discussed in Chapter 5.

6.5 Calculate the components of the effective-mass tensor for the V_1 valence band in silicon.

6.6 Mechanical acceleration **a** of a conductor gives rise to a current caused by the inertial force $-m\mathbf{a}$ with a current density $\mathbf{j} = (m/e)\sigma\mathbf{a}$, where σ is the conductivity of the conductor. What mass enters this effect? Can this experiment be used to distinguish between a conductor in which only electrons contribute to the conductivity and a conductor in which only holes contribute to the conductivity?

6.7 Consider the energy bands predicted by the tight-binding approximation for s states for a bcc crystal,

$$E(\mathbf{k}) = E_0 - E' \cos k_x a \cos k_y a \cos k_z a$$

(a) Describe the extrema found in the 100 and 111 directions. (b) At $k = 0$, what is the ratio between the effective mass in the 100 direction and the effective mass in the 111 direction?

6.8 The energy band structure of a hypothetical cubic crystal has the characteristics listed in the next paragraph. Draw an $E(\mathbf{k})$ vs. **k** diagram for this crystal illustrating all of the below features.

The conduction band has four nonequivalent minima. The lowest of these (C_1) has eight equivalent minima, the next lowest (C_2) has one minimum, the next lowest (C_3) has six equivalent minima, and the highest-lying (C_4) has four equivalent minima. The valence-band structure consists of one band (V_1) with maximum at $\mathbf{k} = 0$ with spherical symmetry, and a second band (V_2) with its maximum also at $\mathbf{k} = 0$ with spherical symmetry, located below V_1 by the spin–orbit splitting energy. In this particular crystal, the spin–orbit splitting energy is equal to one-half the thermally determined bandgap, which in turn is equal to one-half of the direct optically determined bandgap. The effective masses of the carriers in the various bands are ordered as follows:

$$m^*_{C_2} < m^*_{V_1} = m^*_{V_2} < m^*_{C_1} = m^*_{C_3} = m^*_{C_4}$$

6.9 Using the expression for $E(\mathbf{k})$ for the tight-binding approximation for s states for a bcc crystal given in Problem 6.7: (a) Calculate the general expression for the magnitude of the velocity $|\mathbf{v}(\mathbf{k})|$. (b) What is the value of the maximum velocity and at what points of the zone does it occur? (c) Calculate the acceleration in the y direction as the result of an electric field applied in the x direction.

6.10 Given the fact that the conduction and valence band structure at $\mathbf{k} = 0$ is quite similar for Ge and the III–V compounds, calculate the ratio of the average momentum matrix element at $\mathbf{k} = 0$ for the III–V compounds to that for Ge.

Estimate the conduction band m^* at $\mathbf{k} = 0$ for ZnS with a direct bandgap of 3.7 eV.

6.11 In a semiconductor, the Fermi level frequently lies in the forbidden energy gap between the uppermost filled band and the next-higher empty allowed band. At a temperature T it is possible to write the *total density n* of electrons in the conduction band as

$$n = N_c e^{-(E_c - E_F)/kT}$$

where E_c is the conduction band edge energy and E_F is the Fermi level energy, both measured from the top of the valence band, and N_c is not a function of energy. Consider $(E_c - E_F) \gg kT$.

(a) Show that this is indeed the case by calculating n assuming a scalar effective mass m^* for the states near the bottom of the conduction band, and thus derive the value for N_c.

(b) What would N_c be if equal-energy surfaces for small k in the conduction band were characterized by effective masses m_1^*, m_2^*, and m_3^* rather than a single scalar effective mass m^*? (Hint: volume of ellipsoid with axes a, b, c is $4\pi abc/3$.)

6.12 It is found that two materials have about the same bandgap if

$$\Delta G_M + \Delta P_M + \Delta P_X - \Delta G_X = 0$$

where ΔG_M is the difference between the effective groups of the cations in the two compounds according to the Periodic Table, ΔP_M is the difference between the periods of the cations, ΔP_X is the difference between the periods of the anions, and ΔG_X is the difference between the groups of the anions. Explain the reason for this approximate rule. Use the rule to separate the II–VI and III–V compounds (ZnS, ZnSe, ZnTe, CdS, CdSe, CdTe, AlP, AlAs, AlSb, GaP, GaAs, GaSb, InP, InAs, InSb) into groups with the same bandgap and order these groups according to decreasing bandgap. Check the results with experimentally reported values of bandgap for these materials.

Chapter 7

Carrier Transport

The application of an electric or magnetic field or of a thermal gradient to a crystal results in a variety of carrier transport phenomena. These phenomena are associated with the motion of electrons and holes in conduction or valence bands. In this chapter we consider some of the general properties of transport in crystals, with particular application to electrical and thermal conductivity. In Chapter 10 we consider additional processes associated with the presence of a magnetic field.

The description of the motion of an electron (or hole) in terms of a *wave packet* permits an equation of motion that has the same form as the classical treatment of the electron as a particle, provided that the expectation values of acceleration and force for the packet are used. This possibility permits a simplification in the whole area of carrier transport. A wave-packet constructed of Bloch waves has a velocity given by the group velocity associated with the waves used to construct the packet.

The governing equation for the description of transport is the *Boltzmann equation*. This equation describes the total rate of change of the occupancy of allowed states as a result of external fields, carrier diffusion, and carrier scattering. Many different types of scattering process are possible in a crystal and these are discussed in more detail in Chapter 8. If the change in energy of a carrier upon scattering is less than kT, it becomes appropriate to treat the solution of the Boltzmann equation in the *relaxation-time approximation*. For some scattering processes, however, the change in energy of the carrier upon scattering is greater than kT and a more general solution of the Boltzmann equation must be sought.

In those cases where it is applicable, the relaxation time for scattering is a useful parameter in describing transport. The relaxation time is defined as the average time that a carrier moves freely between collisions or scattering events. If the specific scattering mechanism is known, then detailed calculations may be made to calculate the relaxation time from fairly basic considerations. For most scattering processes, the relaxation time is a function of the energy of the carrier. In a real situation there is a distribution of such energies and hence an average relaxation time over energy must be calculated to compare with experimental values, the appropriate averaging process being determined by the Boltzmann equation. The particular dependence of relaxation time on carrier energy, which is characteristic of a given scattering process, is manifest also as a characteristic temperature dependence of the *mobility*, defined as the velocity of the carrier in the direction of the electric field per unit field.

General expressions for electrical conductivity, thermal conductivity, and the thermoelectric effect are derivable which are applicable to transport processes in both metals and semiconductors.

7.1 Wave Packets

According to wave mechanics, a free particle moving along the x axis is described by the Schroedinger equation,

$$-\frac{\hbar^2}{2m}\frac{\partial^2 \Psi}{\partial x^2} = i\hbar \frac{\partial \Psi}{\partial t} \tag{7.1}$$

with the solution in the form of a plane wave,

$$\Psi_k(x, t) = e^{i(kx-\omega t)} \tag{7.2}$$

Since for a free electron k may take on all values, the general solution of Eq. (7.1) is

$$\Psi(x, t) = \int_{-\infty}^{+\infty} A(k)\, e^{i(kx-\omega t)}\, dk \tag{7.3}$$

where $A(k)$ is an arbitary function of k as long as the integral of Eq. (7.3) does not diverge, and the function $\Psi(x, t)$ is normalizable.

The general solution given in Eq. (7.3), known as a *wave packet*, provides a way to draw a correlation between the *particle* with a position x and a momentum mv, and the wave formulation in terms of a wavelength $\lambda = 2\pi/k$ and a frequency $\omega = E/\hbar$. By restricting the values of $A(k)$ to be large only

7.1 Wave Packets

over a limited range of values of k, it becomes possible to represent the particle as a superposition of plane waves; the location of the maximum of the wave packet can be related to the particle position x, and the velocity of the wave packet can be related to the particle velocity v. Thus the wave packet becomes a kind of extended particle with properties that are in accord with the Schroedinger wave picture but which can be treated in terms of classical laws of motion, i.e., the *mean* values of acceleration and force for a wave packet can be shown to satisfy the classical equation of motion for a particle.

One of the most useful forms for $A(k)$ is that of a Gaussian distribution,

$$A(k) = C\, e^{-a^2(k-K)^2} \tag{7.4}$$

In this form $A(k)$ has its maximum value at $k = K$, and falls off to $1/e$ of its maximum value for $(k - K) = \pm 1/a$, as indicated in Fig. 7.1a. The corresponding wave packet is

$$\Psi(x, t) = C \int_{-\infty}^{+\infty} \exp\left\{-a^2(k - K)^2 + i\left[kx - \frac{\hbar^2 k^2}{2m} t\right]\right\} dk \tag{7.5}$$

Let $\Delta_{kK} = (k - K)$. Then the integrand of Eq. (7.5) becomes

$$\exp\left\{-a^2 \Delta_{kK}^2 + ix(K + \Delta_{kK}) - \frac{i\hbar t}{2m}(K + \Delta_{kK})^2\right\} \tag{7.6}$$

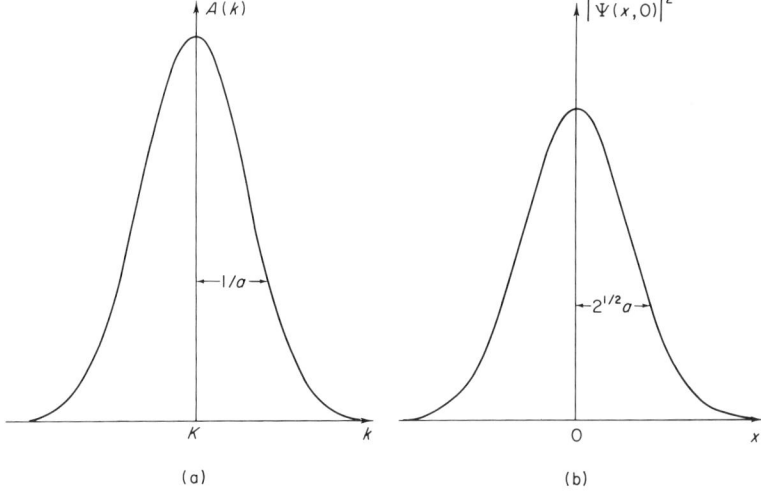

Fig. 7.1 A comparison of the square of the amplitude of the wavefunction for a wave packet at $t = 0$, $|\Psi(x, 0)|^2$, and a Gaussian form of the coefficient modulating factor, $A(k)$: (a) is Eq. (7.4); (b) is Eq. (7.14).

which we can write as

$$\exp\{-\alpha \Delta_{kK}^2 - 2\beta \Delta_{kK} - \gamma\} \tag{7.7}$$

if we define

$$\alpha \equiv a^2 + \frac{i\hbar t}{2m} \tag{7.8}$$

$$\beta \equiv \frac{i}{2}\left(-x + \frac{\hbar tK}{2m}\right) \tag{7.9}$$

$$\gamma \equiv iK\left(-x + \frac{\hbar tK}{2m}\right) \tag{7.10}$$

With these substitutions, Eq. (7.5) becomes

$$\Psi(x, t) = C \int_{-\infty}^{+\infty} \exp(-\alpha \Delta_{kK}^2 - 2\beta \Delta_{kK} - \gamma)\, d\Delta_{kK}$$

$$= C \exp\left[\frac{\beta^2}{\alpha} - \gamma\right] \int_{-\infty}^{+\infty} \exp\left[-\alpha\left(\Delta_{kK} + \frac{\beta}{\alpha}\right)^2\right] d\Delta_{kK} \tag{7.11}$$

The integrand can be written as $e^{-\alpha u^2}$ with $u \equiv [\Delta_{kK} + (\beta/\alpha)]$, giving for the result

$$\Psi(x, t) = C\left(\frac{\pi}{\alpha}\right)^{1/2} \exp\left(\frac{\beta^2}{\alpha} - \gamma\right) \tag{7.12}$$

or, in terms of the original variables,

$$\Psi(x, t) = \frac{C\pi^{1/2}}{(a^2 + i\hbar t/2m)^{1/2}} \exp\left[-\frac{(x - \hbar tK/m)^2}{4[a^2 + (i\hbar t/2m)]} + iK\left(x - \frac{\hbar tK}{2m}\right)\right] \tag{7.13}$$

In order to interpret $\Psi(x, t)$ as representing the motion of a particle, we must investigate the value of the probability $|\Psi(x, 0)|^2$ at $t = 0$, and compare it with the value of the probability $|\Psi(x, t)|^2$ at some later time. From Eq. (7.13) it follows that

$$|\Psi(x, 0)|^2 = \frac{|C|^2 \pi}{a^2} e^{-x^2/2a^2} \tag{7.14}$$

and that

$$|\Psi(x, t)|^2 = \frac{|C|^2 \pi}{[a^4 + (\hbar^2 t^2/4m^2)]^{1/2}} \exp\left[-\frac{a^2(x - \hbar tK/m)^2}{2[a^4 + (\hbar^2 t^2/4m^2)]}\right] \tag{7.15}$$

7.1 Wave Packets

The wavefunction $\Psi(x, 0)$ represents a wave characterized by the wave vector K, and with amplitude modulated by $e^{-x^2/4a^2}$. Figure 7.1b shows a plot of Eq. (7.14) for $|\Psi(x,0)|^2$ for comparison with $A(k)$. $|\Psi(x,0)|^2$ gives the probability of finding the particle in a range dx at x at $t = 0$. The constant C must be chosen so that the total area under the $|\Psi(x,0)|^2$ curve is unity. The probability falls to $1/e$ of its maximum value for $x = \pm(2a^2)^{1/2}$.

At $t = 0$, the probability is quite high that the particle will lie within a distance $(2a^2)^{1/2}$ of the origin. If we take the uncertainty in the location of the particle, Δx, to be $(2a^2)^{1/2}$ at $t = 0$, therefore, then we can check the Heisenberg uncertainty relationship if we obtain a related value for Δp_x. From Eq. (7.4) we note that $A(k)$ falls to $1/e$ of its maximum value for $\Delta_{kK} = 1/a$. The quantity Δ_{kK} is really Δk, and since $p_x = \hbar k$, $\Delta p_x = \hbar/a$. At $t = 0$, then, we have

$$\Delta p_x \cdot \Delta x = 2^{1/2}\hbar \tag{7.16}$$

The fact that $\Delta p_x \cdot \Delta x$ for the Gaussian form of the wave packet is larger than the minimum value required by the Heisenberg relationship shows that a somewhat sharper distribution than the Gaussian can be used to define the wave packet and still maintain consistency with the uncertainty requirements.

Another form of the Heisenberg uncertainty relation can be derived by considering the uncertainty in time at which a particle, described by a wave packet, passes a given point on the x axis,

$$\Delta t = \frac{\Delta x}{v_g} = \frac{\hbar \Delta x}{\partial E/\partial k} \tag{7.17}$$

For small Δk, we may write

$$\Delta E = \frac{\partial E}{\partial k} \Delta k$$

and therefore

$$\Delta t \cdot \Delta E = \hbar \Delta x \cdot \Delta k = \Delta x \cdot \Delta p_x \tag{7.18}$$

Thus the indeterminacy relationship between the complementary quantities, time and energy, is directly connected with the relationship between the complementary quantities, position and momentum.

The behavior of the wave packet as a function of time for $t > 0$ is described by Eq. (7.15). The maximum value of $|\Psi(x,t)|^2$ occurs for

$x = \hbar t K/m$ at time t. This means that the maximum of the wave packet travels in the x direction with a velocity

$$v_g = \frac{\hbar}{m} K = \left(\frac{\partial \omega}{\partial k}\right)_{k=K} \tag{7.19}$$

since $\omega = \hbar k^2/2m$. Thus the maximum of the wave packet moves with the average velocity of the particle. Inspection of Eq. (7.15) also shows that the maximum value of $|\Psi(x, t)|^2$ decreases and that the $1/e$ width of $|\Psi(x, t)|^2$ increases with time. Thus the passage of time sees the wave packet moving in the x direction, but with decreasing maximum amplitude and with increasing spread away from the maximum, as indicated in Fig. 7.2. Of course the area under the various curves of Fig. 7.2 must remain constant, but information on the probable location of the particle is lost with increasing time. For very large values of t (i.e., $t \gg 2ma^2/\hbar$), the maximum value is given by

$$|\Psi(x, t)|^2_{\max} = \frac{2m |C|^2 \pi}{\hbar t} \tag{7.20}$$

and the $1/e$ width is given by

$$\text{Width} = \frac{\hbar}{2^{1/2} ma} t \tag{7.21}$$

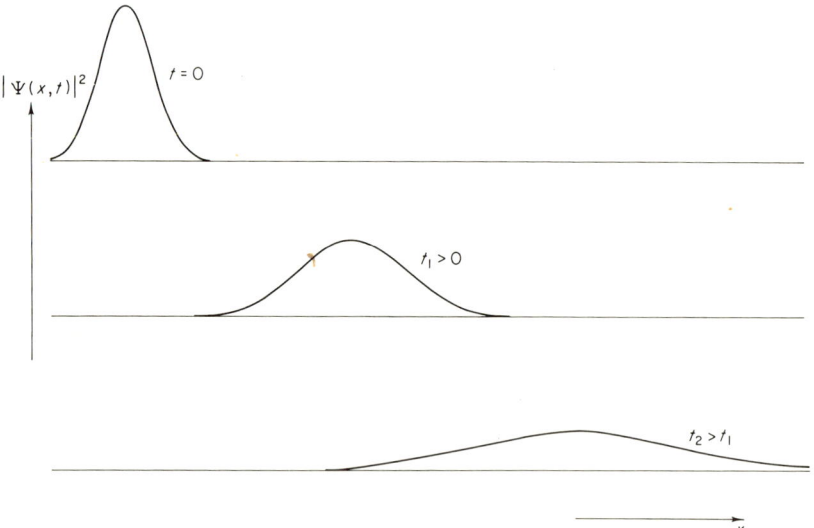

Fig. 7.2 Behavior of the square of the amplitude of the wavefunction for a wave packet $|\Psi(x, t)|^2$ with the passage of time, according to Eq. (7.15).

7.2 Description of Particle Motion Using Wave Packets

BLOCH-FUNCTION WAVE PACKETS

If we construct a wave packet from Bloch waves, we can show directly the relationship between velocity and the gradient of $E(\mathbf{k})$.

A wave packet constructed from time-dependent Bloch waves has the form

$$\Psi(\mathbf{r}, t) = \int A(\mathbf{k}') u_{\mathbf{k}'}(\mathbf{r}) \exp\left[i\left(\mathbf{k}' \cdot \mathbf{r} - \frac{E(\mathbf{k}') t}{\hbar}\right)\right] d\mathbf{k}' \quad (7.22)$$

The wave packet is constructed from Bloch waves corresponding to wave vectors $\mathbf{k}'$ in the vicinity of $\mathbf{k}$. The function $A(\mathbf{k}')$ is chosen to have a sharp maximum magnitude at $\mathbf{k}' = \mathbf{k}$, and to fall off rapidly for $\mathbf{k}'$ different from $\mathbf{k}$.

Because of this property of the $A(\mathbf{k}')$ function, we can expand $E(\mathbf{k}')$ about $\mathbf{k}$ in a Taylor series and retain only the first two terms,

$$E(\mathbf{k}') = E(\mathbf{k}) + (\mathbf{k}' - \mathbf{k}) \cdot \nabla_{\mathbf{k}} E(\mathbf{k}) \quad (7.23)$$

Using this expansion, $\Psi(\mathbf{r}, t)$ given in Eq. (7.22) becomes

$$\Psi(\mathbf{r}, t) = \exp\left[i\left(\mathbf{k} \cdot \mathbf{r} - \frac{E(\mathbf{k}) t}{\hbar}\right)\right] \int A(\mathbf{k}') u_{\mathbf{k}'}(\mathbf{r})$$
$$\times \exp\left\{i\left[(\mathbf{k}' - \mathbf{k}) \cdot \left(\mathbf{r} - \frac{t \nabla_{\mathbf{k}} E}{\hbar}\right)\right]\right\} d\mathbf{k}' \quad (7.24)$$

If we assume that $u_{\mathbf{k}}(\mathbf{r})$ does not change appreciably over the effective range of the integral, we may set $u_{\mathbf{k}'}(\mathbf{r}) = u_{\mathbf{k}}(\mathbf{r})$ and remove it from the integrand of Eq. (7.24). Then the result is

$$\Psi(\mathbf{r}, t) = \Psi_{\mathbf{k}}(\mathbf{r}, t) \int A(\mathbf{k}') \exp\left\{i\left[(\mathbf{k}' - \mathbf{k}) \cdot \left(\mathbf{r} - \frac{t \nabla_{\mathbf{k}} E}{\hbar}\right)\right]\right\} d\mathbf{k}' \quad (7.25)$$

which corresponds to a wave packet moving with velocity $\mathbf{v} = (1/\hbar) \nabla_{\mathbf{k}} E$.

7.2 Description of Particle Motion Using Wave Packets

In this section we desire to demonstrate the following basic result. If a particle is represented by a wave packet, its motion under the action of an external force satisfies the classical equation of motion, i.e., $\mathbf{F} = m\mathbf{a}$, provided that the expectation values of acceleration and force defined in terms of the wave packet are used.

Restating this in more quantitative terms, we desire to show that

$$F = -\frac{\partial V}{\partial x} = ma = m\frac{d^2 x}{dt^2}$$

is still valid for the particle in the wave-packet treatment if we replace x by its expectation value $\langle x \rangle$, and if we replace $-\partial V/\partial x$ by its expectation value $\langle -\partial V/\partial x \rangle$. Our approach is to calculate $d^2\langle x \rangle/dt^2$, and show that it is indeed equal to $\langle -\partial V/\partial x \rangle/m$.

The expectation value of x is given by

$$\langle x \rangle = \int_{-\infty}^{+\infty} \Psi^*(x, t)\, x\Psi(x, t)\, dx \tag{7.26}$$

where

$$\Psi(x, t) = \int A(k)\, \psi_k(x)\, e^{-iEt/\hbar}\, dk \tag{7.27}$$

is a suitable wave-packet general solution of

$$-\frac{\hbar^2}{2m}\frac{\partial^2 \Psi}{\partial x^2} + V(x)\,\Psi = i\hbar\frac{\partial \Psi}{\partial t} \tag{7.28}$$

The velocity is given by

$$v = \frac{d\langle x \rangle}{dt} = \int_{-\infty}^{+\infty} \left[x\Psi^* \frac{\partial \Psi}{\partial t} + x\Psi \frac{\partial \Psi^*}{\partial t} \right] dx \tag{7.29}$$

Since x is an operator in Eq. (7.29), its time derivative does not enter. The time derivatives of Eq. (7.29) can be transformed into x derivatives by using the Schroedinger equation. We multiply Eq. (7.28), written for the complex conjugate wavefunction Ψ^*, by $x\Psi$, and subtract from Eq. (7.28) multiplied by $x\Psi^*$. Integrate the difference over all x, and obtain

$$-\frac{\hbar^2}{2m}\int_{-\infty}^{+\infty} \left[x\Psi^* \frac{\partial^2 \Psi}{\partial x^2} - x\Psi \frac{\partial^2 \Psi^*}{\partial x^2} \right] dx$$

$$= i\hbar \int_{-\infty}^{+\infty} \left[x\Psi^* \frac{\partial \Psi}{\partial t} + x\Psi \frac{\partial \Psi^*}{\partial t} \right] dx$$

$$= i\hbar \frac{d\langle x \rangle}{dt} \tag{7.30}$$

Integrating by parts,

$$\int_{-\infty}^{+\infty} x\Psi^* \frac{\partial^2 \Psi}{\partial x^2}\, dx = x\Psi^* \frac{\partial \Psi}{\partial x}\bigg|_{-\infty}^{+\infty} - \int_{-\infty}^{+\infty} \frac{\partial \Psi}{\partial x}\left(x\frac{\partial \Psi^*}{\partial x} + \Psi^* \right) dx \tag{7.31}$$

7.2 Description of Particle Motion Using Wave Packets

with a similar expression for $\int_{-\infty}^{+\infty} x\Psi(\partial^2\Psi^*/\partial x^2)\, dx$. The first term on the right of Eq. (7.31) is equal to zero since the wavefunction goes to zero at infinity. The evaluation of Eq. (7.30) therefore leads to

$$i\hbar \frac{d\langle x \rangle}{dt} = \frac{\hbar^2}{2m} \int_{-\infty}^{+\infty} \left(\Psi^* \frac{\partial \Psi}{\partial x} - \Psi \frac{\partial \Psi^*}{\partial x} \right) dx \quad (7.32)$$

Now

$$\int_{-\infty}^{+\infty} \Psi^* \frac{\partial \Psi}{\partial x}\, dx = \Psi^*\Psi \Big|_{-\infty}^{\infty} - \int_{-\infty}^{+\infty} \Psi \frac{\partial \Psi^*}{\partial x}\, dx \quad (7.33)$$

so that Eq. (7.32) becomes

$$i\hbar \frac{d\langle x \rangle}{dt} = \frac{\hbar^2}{m} \int_{-\infty}^{+\infty} \Psi^* \frac{\partial \Psi}{\partial x}\, dx \quad (7.34)$$

Now we can continue with our calculation of

$$m \frac{d^2\langle x \rangle}{dt^2} = -i\hbar \int_{-\infty}^{+\infty} \left(\frac{\partial \Psi^*}{\partial t} \frac{\partial \Psi}{\partial x} + \Psi^* \frac{\partial^2 \Psi}{\partial x\, \partial t} \right) dx \quad (7.35)$$

The form of Eq. (7.35) can be made more symmetric by realizing that

$$\int_{-\infty}^{+\infty} \Psi^* \frac{\partial^2 \Psi}{\partial x\, \partial t}\, dx = \Psi^* \frac{\partial \Psi}{\partial t} \Big|_{-\infty}^{+\infty} - \int_{-\infty}^{+\infty} \frac{\partial \Psi}{\partial t} \frac{\partial \Psi^*}{\partial x}\, dx \quad (7.36)$$

so that

$$m \frac{d^2\langle x \rangle}{dt^2} = -i\hbar \int_{-\infty}^{+\infty} \left(\frac{\partial \Psi^*}{\partial t} \frac{\partial \Psi}{\partial x} - \frac{\partial \Psi}{\partial t} \frac{\partial \Psi^*}{\partial x} \right) dx \quad (7.37)$$

In view of our initial goal, we need to introduce the potential energy $V(x)$ explicitly into the calculation. We can do this by substituting for the time derivatives of Ψ and Ψ^* from Eq. (7.28) and its complex conjugate. We obtain

$$m \frac{d^2\langle x \rangle}{dt^2} = -\frac{\hbar^2}{2m} \int_{-\infty}^{+\infty} \left(\frac{\partial^2 \Psi^*}{\partial x^2} \frac{\partial \Psi}{\partial x} + \frac{\partial^2 \Psi}{\partial x^2} \frac{\partial \Psi^*}{\partial x} \right) dx$$

$$+ \int_{-\infty}^{+\infty} V(x) \left(\Psi^* \frac{\partial \Psi}{\partial x} + \Psi \frac{\partial \Psi^*}{\partial x} \right) dx \quad (7.38)$$

The first integral in Eq. (7.38) is zero since

$$\int_{-\infty}^{+\infty} \frac{\partial \Psi}{\partial x} \frac{\partial^2 \Psi^*}{\partial x^2}\, dx = \frac{\partial \Psi}{\partial x} \frac{\partial \Psi^*}{\partial x} \Big|_{-\infty}^{+\infty} - \int_{-\infty}^{+\infty} \frac{\partial \Psi^*}{\partial x} \frac{\partial^2 \Psi}{\partial x^2}\, dx \quad (7.39)$$

The second integral in Eq. (7.38) can be simplified since

$$\int_{-\infty}^{+\infty} V(x) \Psi^* \frac{\partial \Psi}{\partial x} dx = V(x) \Psi^* \Psi \Big|_{-\infty}^{+\infty}$$
$$- \int_{-\infty}^{+\infty} \Psi \left[V(x) \frac{\partial \Psi^*}{\partial x} + \Psi^* \frac{\partial V(x)}{\partial x} \right] dx \qquad (7.40)$$

Finally we obtain

$$m \frac{d^2 \langle x \rangle}{dt^2} = - \int_{-\infty}^{+\infty} \Psi^* \frac{\partial V(x)}{\partial x} \Psi \, dx$$
$$= \left\langle - \frac{\partial V(x)}{\partial x} \right\rangle \qquad (7.41)$$

Although this calculation has been carried out for the one-dimensional case, generalization of the result to three dimensions leads to the same conclusions. As long as the equation of motion is discussed in terms of the expectation values of acceleration and force, therefore, the wave-mechanical equation utilizing wave packets leads to the same results as classical physics for the equivalent particles. This result affords us the possibility of extensive simplification in the discussion of the motion of electrons in crystals. Instead of it being necessary for us constantly to revert to the solution of the entire Schroedinger equation, we can in many cases continue to treat the motion of electrons in crystals in a quasi-classical particle framework, remembering the wave-mechanical justification for this is the wave-packet context.

7.3 The Boltzmann Equation

What the Schroedinger equation is for describing energy levels in crystals, the Boltzmann equation is for describing carrier transport in crystals. The Boltzmann equation expresses the total time rate of change of the distribution function $f(\mathbf{k}, \mathbf{r}, t)$, which describes the occupancy of allowed energy states involved in transport processes:

$$\frac{df}{dt} = \frac{\partial f}{\partial t}\bigg]_{\text{external field}} + \frac{\partial f}{\partial t}\bigg]_{\text{diffusion}} + \frac{\partial f}{\partial t} + \frac{\partial f}{\partial t}\bigg]_{\text{scattering}} \qquad (7.42)$$

In steady state $df/dt = \partial f/\partial t = 0$. For an external force $\mathbf{F}$, we can write

$$\frac{\partial f}{\partial t}\bigg]_{\text{external field}} = - \frac{\partial \mathbf{k}}{\partial t} \nabla_{\mathbf{k}} f$$
$$= - \frac{1}{\hbar} \mathbf{F} \cdot \nabla_{\mathbf{k}} f \qquad (7.43)$$

7.3 The Boltzmann Equation

in view of Eq. (6.26). Similarly we can write

$$\left.\frac{\partial f}{\partial t}\right]_{\text{diffusion}} = -\frac{d\mathbf{r}}{dt} \nabla_\mathbf{r} f$$
$$= -\mathbf{v} \cdot \nabla_\mathbf{r} f \qquad (7.44)$$

The steady-state equation is therefore

$$\frac{1}{\hbar} \mathbf{F} \cdot \nabla_\mathbf{k} f + \mathbf{v} \cdot \nabla_\mathbf{r} f = \left.\frac{\partial f}{\partial t}\right]_{\text{scattering}} \qquad (7.45)$$

The form of $(\partial f/\partial t)_{\text{scattering}}$ depends on the nature of the specific scattering process.

Consider the general situation where $P_{\mathbf{k}\mathbf{k}'}$ is the probability per unit time for scattering from $\mathbf{k}$ to $\mathbf{k}'$. Then the formal relation holds,

$$\left.\frac{\partial f}{\partial t}\right]_{\text{scattering}} = \int \{P_{\mathbf{k}'\mathbf{k}} f(\mathbf{k}')[1 - f(\mathbf{k})] - P_{\mathbf{k}\mathbf{k}'} f(\mathbf{k})[1 - f(\mathbf{k}')]\} d\mathbf{k}' \qquad (7.46)$$

At equilibrium the distribution function is given by its equilibrium value, which is a function only of the energy E,

$$f(\mathbf{k}, \mathbf{r}, t)_{\text{equil.}} = f_0(E) \qquad (7.47)$$

where $f_0(E)$ is given by Eq. (4.84). Since $(\partial f/\partial t)_{\text{scattering}} = 0$ at equilibrium,

$$P_{\mathbf{k}'\mathbf{k}} = P_{\mathbf{k}\mathbf{k}'} \frac{f_0(E)[1 - f_0(E')]}{f_0(E')[1 - f_0(E)]} \qquad (7.48)$$

as long as $P_{\mathbf{k}\mathbf{k}'}$ is not a function of $\mathbf{F}$, which we are assuming to be the case.

It is customary to express the departure of the distribution function from equilibrium as

$$f(\mathbf{k}, \mathbf{r}) = f_0(E) + f'(\mathbf{k}, \mathbf{r}) \qquad (7.49)$$

and to write in addition

$$f'(\mathbf{k}, \mathbf{r}) = -\phi(\mathbf{k}, \mathbf{r}) \frac{\partial f_0}{\partial E} \qquad (7.50)$$

If Eqs. (7.48) through (7.50) are substituted into Eq. (7.46), it becomes

$$\left.\frac{\partial f}{\partial t}\right]_{\text{scattering}} = \frac{1}{kT} \int P_{\mathbf{k}\mathbf{k}'} f_0(E)[1 - f_0(E')][\phi(\mathbf{k}') - \phi(\mathbf{k})] d\mathbf{k}' \qquad (7.51)$$

Finally, by setting **F** in Eq. (7.45) explicitly equal to the Lorentz force for a carrier with charge e,

$$\mathbf{F} = e\left(\mathscr{E} + \frac{1}{c}\mathbf{v} \times \mathbf{B}\right) = e\left(\mathscr{E} + \frac{1}{c\hbar}\nabla_{\mathbf{k}} E(\mathbf{k}) \times \mathbf{B}\right) \quad (7.52)$$

and using $\mathbf{v} = (1/\hbar)\nabla_{\mathbf{k}} E(\mathbf{k})$ in the second term of Eq. (7.45), we convert Eq. (7.45) with Eq. (7.51) into the Bloch equation,

$$e\left(\mathscr{E} + \frac{1}{c\hbar}\nabla_{\mathbf{k}} E(\mathbf{k}) \times \mathbf{B}\right) \cdot \frac{1}{\hbar}\nabla_{\mathbf{k}} f(\mathbf{k}, \mathbf{r}) + \frac{1}{\hbar}\nabla_{\mathbf{k}} E(\mathbf{k}) \cdot \nabla_{\mathbf{r}} f(\mathbf{k}, \mathbf{r})$$

$$= \frac{1}{kT}\int P_{\mathbf{kk}'} f_0(E)[1 - f_0(E')][\phi(\mathbf{k}') - \phi(\mathbf{k})]\, d\mathbf{k}' \quad (7.53)$$

7.4 Solution of the Boltzmann Equation

A particularly simple solution of the Boltzmann equation becomes possible in those particular cases where the effects of scattering can be described in terms of a relaxation time τ.

$$\left.\frac{\partial f}{\partial t}\right|_{\text{scattering}} = -\frac{f(\mathbf{k},\mathbf{r}) - f_0(E)}{\tau} = -\frac{f'(\mathbf{k},\mathbf{r})}{\tau} \quad (7.54)$$

We consider the solution of the Boltzmann equation in the relaxation time approximation in the following section. Here we consider briefly the criterion for using the relaxation-time approximation justifiably.

A formal relaxation time $\tau(\mathbf{k})$ can always be defined,

$$\tau(\mathbf{k}) \equiv \phi(\mathbf{k})\frac{(\partial f_0/\partial E)}{(\partial f/\partial t)_{\text{scattering}}} \quad (7.55)$$

following Eq. (7.54). Using Eq. (7.51) this becomes

$$\frac{1}{\tau(\mathbf{k})} = \frac{1}{\phi(\mathbf{k})}\frac{1}{(\partial f_0/\partial E)}\frac{1}{kT}\int P_{\mathbf{kk}'} f_0(E)[1 - f_0(E')][\phi(\mathbf{k}') - \phi(\mathbf{k})]\, d\mathbf{k}'$$

$$= \int P_{\mathbf{kk}'}\left[\frac{1 - f_0(E')}{1 - f_0(E)}\right]\left[1 - \frac{\phi(\mathbf{k}')}{\phi(\mathbf{k})}\right] d\mathbf{k}' \quad (7.56)$$

since $(\partial f_0/\partial E) = -f_0(E)[1 - f_0(E)]/kT$.

The relaxation time $\tau(\mathbf{k})$ is a meaningful quantity provided that it is independent of the strength and type of the perturbation causing f to depart from f_0. If the relaxation time is a function of the type of perturbation (for example, electric field, thermal gradient), then it is not a physically useful quantity.

Equation (7.56) shows that two conditions are imposed on the particular process in order for the relaxation-time approximation to be justified. (1) $f_0(E') = f_0(E)$. This means that $E(\mathbf{k}') = E(\mathbf{k})$, or that the scattering process is *elastic*, i.e., involves no energy change, or at least in practical considerations an energy change much less than the average carrier energy of kT. (2) $[\phi(\mathbf{k}')/\phi(\mathbf{k})]$ must be independent of the type of perturbation. This condition can be checked for particular types of perturbation, as is demonstrated in the following section.

In those cases where the relaxation-time approximation is not justified, e.g., for inelastic scattering, another method must be used for the solution of the Boltzmann equation. A variational calculation is used based on the principle that, for a particular perturbation, the steady-state distribution achieved is such that, were the perturbation removed suddenly, the return to equilibrium would be the most rapid for a particular relaxation mechanism. This corresponds physically to maximizing the time rate of change of entropy production associated with the scattering processes. We will not pursue the details of this process further here, but will cite some of its results where applicable in later sections.

7.5 Relaxation-Time Solution of the Boltzmann Equation

In the relaxation-time approximation, the Bloch equation of Eq. (7.53) can be written

$$e\left[\mathscr{E} + \frac{1}{c}\mathbf{v}\times\mathbf{B}\right]\cdot\frac{1}{\hbar}\nabla_{\mathbf{k}}f + \mathbf{v}\cdot\nabla_{\mathbf{r}}f = -\frac{f'}{\tau}$$

$$= \frac{\phi}{\tau}\frac{\partial f_0}{\partial E} \quad (7.57)$$

We have the following relationships:

$$\nabla_{\mathbf{r}}f \approx \nabla_{\mathbf{r}}f_0 = \frac{\partial f_0}{\partial T}\nabla_{\mathbf{r}}T \quad (7.58)$$

$$\frac{\partial f_0}{\partial T} = -\frac{E - E_{\mathrm{F}}}{T}\frac{\partial f_0}{\partial E} - \frac{\partial E_{\mathrm{F}}}{\partial T}\frac{\partial f_0}{\partial E} \quad (7.59)$$

so that

$$\nabla f \approx -\left[\frac{E - E_{\mathrm{F}}}{T}\nabla T + \nabla E_{\mathrm{F}}\right]\frac{\partial f_0}{\partial E} \quad (7.60)$$

where, in keeping with ordinary practice, we have set $\nabla_r \equiv \nabla$. In the processes of this section, we are discussing steady-state conditions, and we may therefore drop the term in Eq. (7.60) involving ∇E_F. In a later discussion of thermal conductivity, the term involving ∇E_F must be included. We also have that

$$\nabla_k f = \frac{\partial f_0}{\partial E} \nabla_k E + \nabla_k f' \qquad (7.61)$$

where we neglect the second term on the right as long as the first term does not make an identically zero contribution. With the substitution of the results of Eqs. (7.58) through (7.61) into Eq. (7.57), we obtain

$$e\mathscr{E} \cdot \mathbf{v} \frac{\partial f_0}{\partial E} + \frac{e}{c}[\mathbf{v} \times \mathbf{B}] \cdot \mathbf{v} \frac{\partial f_0}{\partial E} + \frac{e}{c\hbar}[\mathbf{v} \times \mathbf{B}] \cdot \nabla_k f'$$
$$- (E - E_F)\frac{\partial f_0}{\partial E}\mathbf{v} \cdot \nabla \ln T = \frac{\phi}{\tau}\frac{\partial f_0}{\partial E} \qquad (7.62)$$

The second term in this equation is identically zero, and hence we retain the term involving f'. Collecting terms,

$$[e\mathscr{E} - (E - E_F)\nabla \ln T] \cdot \mathbf{v}\frac{\partial f_0}{\partial E} + \frac{e}{c\hbar}[\mathbf{v} \times \mathbf{B}] \cdot \nabla_k f' = \frac{\phi}{\tau}\frac{\partial f_0}{\partial E} \qquad (7.63)$$

In view of Eq. (7.50),

$$\frac{e}{c\hbar}[\mathbf{v} \times \mathbf{B}] \cdot \nabla_k f' = \frac{e}{c\hbar}\mathbf{B} \cdot [\nabla_k f' \times \mathbf{v}]$$
$$= -\frac{e}{c\hbar^2}\mathbf{B} \cdot [\nabla_k E \times \nabla_k f']$$
$$= \frac{e}{c\hbar^2}\mathbf{B} \cdot [\nabla_k E \times \nabla_k \phi]\frac{\partial f_0}{\partial E} \qquad (7.64)$$

Substitution of Eq. (7.64) into Eq. (7.63) permits the solution for ϕ,

$$\phi = \tau[e\mathscr{E} - (E - E_F)\nabla \ln T] \cdot \mathbf{v} + \frac{e\tau}{c\hbar^2}\mathbf{B} \cdot [\nabla_k E \times \nabla_k \phi] \qquad (7.65)$$

This is the basic equation for the treatment of transport problems involving steady state and a meaningful relaxation time.

7.5 Relaxation-Time Solution of Boltzmann Equation

ELECTRIC FIELD ONLY

For an electric field only, $\mathbf{B} = 0$ and $\nabla T = 0$. From Eq. (7.65),

$$\phi = \tau e \mathscr{E} \cdot \mathbf{v} \tag{7.66}$$

and

$$f = f_0 - \tau e \mathscr{E} \cdot \mathbf{v} \frac{\partial f_0}{\partial E} \tag{7.67}$$

If the electric field is applied in the x direction, i.e., $\mathscr{E} = \mathscr{E}_x$,

$$\begin{aligned} f &= f_0 - \frac{e\tau \mathscr{E}_x}{\hbar} \frac{\partial E}{\partial k_x} \frac{\partial f_0}{\partial E} \\ &= f_0 - \frac{e\tau \mathscr{E}_x}{\hbar} \frac{\partial f_0}{\partial k_x} \end{aligned} \tag{7.68}$$

and the total distribution function is expressible as

$$f(k_x, k_y, k_z) = f\left[k_x - \frac{e\mathscr{E}_x \tau}{\hbar}, k_y, k_z\right] \tag{7.69}$$

The displacement of the Fermi sphere in $\mathbf{k}$ space and of the distribution function as it depends on k_x, with respect to the equilibrium positions, are shown in Fig. 7.3. All points on these surfaces undergo the same displacement, i.e., no change in shape occurs in the simple case of an electric field only.

With reference to Eq. (7.56), the suitability of the relaxation-time approximation to this case can be tested by seeing whether $\phi(\mathbf{k}')/\phi(\mathbf{k})$ is independent of the type of the perturbation. From Eq. (7.66),

$$\frac{\phi(\mathbf{k}')}{\phi(\mathbf{k})} = \frac{\tau(\mathbf{k}') e\mathscr{E}_x v_x(\mathbf{k}')}{\tau(\mathbf{k}) e\mathscr{E}_x v_x(\mathbf{k})} = \frac{\tau(\mathbf{k}') v_x(\mathbf{k}')}{\tau(\mathbf{k}) v_x(\mathbf{k})} \tag{7.70}$$

We conclude that scattering processes for an electric field only can be described by a relaxation time provided that the scattering processes are elastic.

THERMAL GRADIENT ONLY

For a thermal gradient only, $\mathbf{B} = 0$ and $\mathscr{E} = 0$. From Eq. (7.65),

$$\phi = \tau[-(E - E_\text{F})\nabla \ln T] \cdot \mathbf{v} \tag{7.71}$$

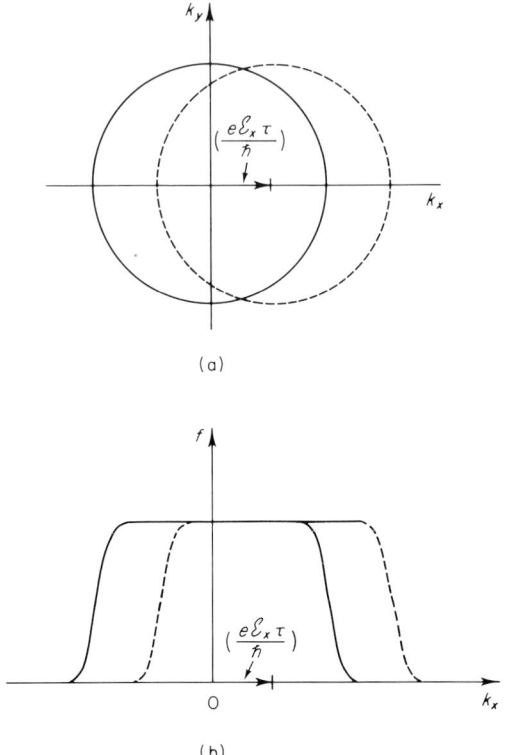

Fig. 7.3 (a) Displacement of the Fermi sphere in **k** space as the result of the application of an electric field only, in the x direction. (b) Displacement of the distribution function as the result of the application of an electric field only, in the x direction.

For a thermal gradient in the x direction,

$$f' = -\phi \frac{\partial f_0}{\partial E} = \tau(E - E_F) \frac{1}{T} \frac{\partial T}{\partial x} \frac{1}{\hbar} \frac{\partial f_0}{\partial E} \frac{\partial E}{\partial k_x} \quad (7.72)$$

Consider the simple case of spherical equal-energy surfaces such that

$$(E - E_F) = \frac{\hbar^2 k_0}{m^*}(k - k_0) \quad (7.73)$$

which follows from $E_F = \hbar^2 k_0^2/2m^*$, $|E - E_F| \ll E_F$, and $E = E_F \pm (k - k_0)(\partial E_F/\partial k)$. Then the expression for f' in Eq. (7.72) becomes

$$f' = \frac{\tau \hbar k_0}{m^* T}(k - k_0) \frac{\partial T}{\partial x} \frac{\partial f_0}{\partial k_x} \quad (7.74)$$

7.5 Relaxation-Time Solution of Boltzmann Equation

The total distribution function is expressible as

$$f(k_x, k_y, k_z) = f\left[k_x + \frac{\tau \hbar k_0}{m^* T}(k - k_0)\frac{\partial T}{\partial x}, k_y, k_z\right] \quad (7.75)$$

This is a more complicated variation from the equilibrium distribution than found for the case of an electric field only. Displacements of the Fermi surface and of the distribution function are shown in Fig. 7.4. For the case of a thermal gradient, the displacement depends on the energy of the surface being considered with respect to the Fermi energy E_F. For $E = E_F$,

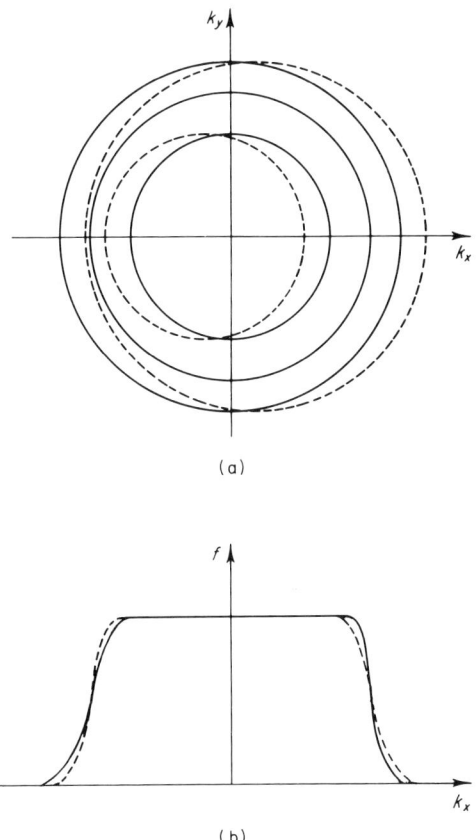

Fig. 7.4 (a) Displacement of the Fermi sphere in **k** space as the result of the application of a thermal gradient only, in the x direction. The magnitude and direction of the displacement depends on the energy of the surface with respect to the Fermi energy. (b) Displacement of the distribution function as the result of the application of a thermal gradient only, in the x direction.

there is no shift, while shifts in opposite directions are found for $E > E_F$ and $E < E_F$. Comparison of Figs. 7.3 and 7.4 show how strongly the departure from equilibrium can depend on the nature of the specific perturbation causing this departure.

Again with Eq. (7.56) in mind, we can check the suitability of a relaxation-time approximation to the case of a thermal gradient only.

$$\frac{\phi(\mathbf{k}')}{\phi(\mathbf{k})} = \frac{\tau(\mathbf{k}')[E(\mathbf{k}') - E_F]\,\nabla \ln T \cdot \mathbf{v}(\mathbf{k}')}{\tau(\mathbf{k})[E(\mathbf{k}) - E_F]\,\nabla \ln T \cdot \mathbf{v}(\mathbf{k})} = \frac{\tau(\mathbf{k}')\,v_x(\mathbf{k}')[E(\mathbf{k}') - E_F]}{\tau(\mathbf{k})\,v_x(\mathbf{k})[E(\mathbf{k}) - E_F]} \quad (7.76)$$

This ratio is indeed independent of the thermal gradient. Comparison of Eq. (7.76) for the case of a thermal gradient only with Eq. (7.70) for the case of an electric field only shows that the *same* relaxation time can be used for both processes, however, only in the case of elastic scattering, i.e., only if $E(\mathbf{k}') = E(\mathbf{k})$.

GENERAL SOLUTION

A general solution of the Boltzmann equation in the relaxation-time approximation can be obtained in fairly simple form if spherical equal-energy surfaces are assumed, i.e., $E = \hbar^2 k^2/2m^*$, and $\mathbf{v} = \hbar \mathbf{k}/m^*$. Then Eq. (7.65) becomes

$$\phi = \tau[e\mathscr{E} - (E - E_F)\,\nabla \ln T] \cdot \frac{\hbar \mathbf{k}}{m^*} + \frac{e\tau}{c\hbar^2}\mathbf{B} \cdot \left[\frac{\hbar^2 \mathbf{k}}{m^*} \times \nabla_\mathbf{k} \phi \right] \quad (7.77)$$

The solution of this equation can be expressed as

$$\phi = \mathbf{k} \cdot \boldsymbol{\theta}(E, \mathscr{E}, \mathbf{B}, \nabla T) \quad (7.78)$$

where $\boldsymbol{\theta}$ is not a function of $\mathbf{k}$. The equation for $\boldsymbol{\theta}$ is

$$\boldsymbol{\theta} = \tau \frac{\hbar}{m^*}[e\mathscr{E} - (E - E_F)\,\nabla \ln T] - \frac{e\tau}{cm^*}[\mathbf{B}\times\boldsymbol{\theta}] \quad (7.79)$$

which has the solution

$$\boldsymbol{\theta} = \frac{\dfrac{\hbar\tau}{m^*}\left[[e\mathscr{E} - (E - E_F)\,\nabla \ln T] - \dfrac{e\tau}{m^*c}\mathbf{B}\times[e\mathscr{E} - (E - E_F)\,\nabla \ln T] + \left(\dfrac{e\tau}{m^*c}\right)^2 \mathbf{B}\{\mathbf{B}\cdot[e\mathscr{E} - (E - E_F)\,\nabla \ln T]\}\right]}{1 + \left(\dfrac{e\tau B}{m^*c}\right)^2} \quad (7.80)$$

7.6 Electrical Conductivity

We consider this equation in more detail in Chapter 10 in connection with galvanomagnetic effects. We note here only the simplification that results if **B** is in the same direction as $\mathscr{E}$ and ∇T. Multiplication of Eq. (7.80) by **k** in this case (the second term in the numerator is zero) shows that ϕ reduces to the same form as that obtained from Eq. (7.65) with **B** = 0. It is possible to conclude directly without further calculation, therefore, that *longitudinal* galvanomagnetic and thermomagnetic effects are not observed in the case of spherical equal-energy surfaces and sufficiently small **B** that the relaxation-time approximation holds.

7.6 Electrical Conductivity in the Relaxation-Time Approximation

For the case of electrical conductivity we are concerned with the effect of applying an electric field only; and hence the distribution function is given by Eq. (7.67). This relationship written explicitly for electrons with charge of $-e$ and for an electric field in the x direction is

$$f = f_0 + \tau e \mathscr{E}_x v_x \frac{\partial f_0}{\partial E} \tag{7.81}$$

The electric current density (current per unit area) j_x is given by the product of the electron charge and electron velocity, summed over all the electrons contributing,

$$j_x = -e \int f(\mathbf{k}') N(\mathbf{k}') v_x \, d\mathbf{k}'$$

$$= -\frac{e}{4\pi^3} \int f(\mathbf{k}') v_x \, d\mathbf{k}' \tag{7.82}$$

since $(1/4\pi^3)$ is the density per unit volume of allowed values of $\mathbf{k}'$, including a factor of 2 for electron spin. The integration of Eq. (7.82) is to be taken over all partially filled bands. Inserting f from Eq. (7.81) gives

$$j_x = -\frac{e^2 \mathscr{E}_x}{4\pi^3} \int \tau v_x^2 \frac{\partial f_0}{\partial E} \, d\mathbf{k}' \tag{7.83}$$

since the integration over the equilibrium distribution f_0 is zero. Sometimes j_x is written simply as

$$j_x = e^2 \mathscr{E}_x K_1 \tag{7.84}$$

in terms of the transport integral K_n, defined by

$$K_n \equiv -\frac{1}{4\pi^3} \int \tau v_x^2 E^{n-1} \frac{\partial f_0}{\partial E} \, d\mathbf{k}' \tag{7.85}$$

In most cases of electrical conductivity, only one partially occupied band is involved. In such a case, if the lower edge of the band is at $E = 0$ and the upper edge of the band is at $E = E_m$, the band contains n electrons per unit volume, given by

$$n = \frac{1}{4\pi^3} \int_0^{E_m} f_0 \, d\mathbf{k}' \tag{7.86}$$

In terms of n, Eq. (7.83) can be rewritten as

$$j_x = -e^2 \mathscr{E}_x n \frac{\int_0^{E_m} \tau v_x^2 (\partial f_0/\partial E) \, d\mathbf{k}'}{\int_0^{E_m} f_0 \, d\mathbf{k}'} \tag{7.87}$$

This expression may be rewritten using $(\partial f_0/\partial E) = -f_0(1 - f_0)/kT$:

$$j_x = \frac{e^2 \mathscr{E}_x n}{kT} \frac{\int_0^{E_m} \tau v_x^2 f_0 (1 - f_0) \, d\mathbf{k}'}{\int_0^{E_m} f_0 \, d\mathbf{k}'} \tag{7.88}$$

Further simplification is possible if the equal-energy surfaces involved may be assumed to be spherical, so that

$$E = \frac{m^* |\mathbf{v}|^2}{2} = \frac{\hbar^2 |\mathbf{k}'|^2}{2m^*}$$

$$d\mathbf{k}' \propto k'^2 \, dk' \propto E^{1/2} \, dE$$

$$v_x^2 \approx \frac{|\mathbf{v}|^2}{3} = \frac{2E}{3m^*}$$

The final result for the current density for spherical equal-energy surfaces is

$$j_x = \frac{2e^2 \mathscr{E}_x n}{3m^* kT} \frac{\int_0^{E_m} \tau(E) E^{3/2} f_0 (1 - f_0) \, dE}{\int_0^{E_m} E^{1/2} f_0 \, dE}$$

$$= \frac{e^2 \mathscr{E}_x n}{m^*} \frac{\int_0^{E_m} \tau(E) E^{3/2} f_0 (1 - f_0) \, dE}{\int_0^{E_m} E^{3/2} f_0 (1 - f_0) \, dE} \tag{7.89}$$

since

$$\int_0^{E_m} E^{3/2} f_0 (1 - f_0) \, dE = -kT \int_0^{E_m} E^{3/2} \frac{\partial f_0}{\partial E} \, dE = \frac{3kT}{2} \int_0^{E_m} E^{1/2} f_0 \, dE$$

A relatively simple physical interpretation of Eq. (7.89) can be made. It can be rewritten as

$$j_x = \frac{e^2 \mathscr{E}_x n}{m^*} \langle \tau(E) \rangle \tag{7.90}$$

7.7 Semiconductors and Metals

where the average value of the relaxation time $\langle \tau(E) \rangle$ is defined by the expression of Eq. (7.89). Physically the relaxation time is defined as the mean free time of the carrier between scattering events. Thus if an electron makes a scattering collision at $t = 0$ in the presence of an electric field $\mathscr{E}_x$ and loses all previous velocity in the direction of $\mathscr{E}_x$ (this velocity is called the *drift velocity* v_x and is expressible as $v_x = -\mu \mathscr{E}_x$, where μ is the electron *mobility*), its velocity at some later time t is

$$v_x = -\frac{e\mathscr{E}_x t}{m_e^*} \tag{7.91}$$

where m_e^* is the effective mass of the electron in the crystal. The average value of the velocity over time is

$$\bar{v}_x = -\frac{e\mathscr{E}_x}{m_e^*} \tau_e \tag{7.92}$$

where τ_e is the relaxation time for the electron. In view of the definition of mobility,

$$\mu_e = \frac{e}{m_e^*} \tau_e \tag{7.93}$$

The current density is given by

$$j_x = -ne\bar{v}_x = \frac{e^2 \mathscr{E}_x n}{m_e^*} \tau_e \tag{7.94}$$

which is identical to Eq. (7.90). In any real crystal we must consider also the average of the relaxation time τ_e over the distribution of electron energies. When this average is taken into account, τ_e in Eq. (7.94) is replaced by $\langle \tau(E) \rangle$, according to the procedure set forth in Eq. (7.89).

7.7 Electrical Conductivity in Semiconductors and Metals

Many different kinds of scattering processes are possible in semiconductors and metals, some of which allow a description in terms of a relaxation-time approximation and some of which do not. In this section we consider only two of the most generally encountered mechanisms: (1) scattering by acoustic phonons, and (2) scattering by charged imperfections. Other types of scattering are discussed in Chapter 8. The nature of these two scattering processes is such that a relaxation-time approximation in semiconductors is appropriate for both processes over the entire temperature range of interest. In metals, however, scattering by acoustic phonons becomes inelastic at low temperatures and cannot be adequately described by the relaxation-time approximation.

SEMICONDUCTORS

In semiconductors the carriers contributing to electrical conductivity have **k** values near $\mathbf{k}_{\text{ext}}$ for the conduction or valence band extrema. Initial and final states have **k** values not greatly different for scattering by acoustic phonons, and it is found in general that scattering by acoustic phonons in nondegenerate semiconductors (i.e., with Fermi level in the forbidden gap between valence and conduction bands) is elastic. Scattering by charged imperfections involves no change in carrier energy, and hence is always elastic. Thus, for the two scattering processes under consideration here, the relaxation-time approximation is appropriate for semiconductors.

For a nondegenerate semiconductor, $f_0(1 - f_0) \approx f_0$, and Eq. (7.89) simplifies to

$$j_x = \frac{e^2 \mathscr{E}_x n}{m^*} \frac{\int_0^{E_m} \tau(E) \, E^{3/2} f_0 \, dE}{\int_0^{E_m} E^{3/2} f_0 \, dE} \tag{7.95}$$

If $\tau(E)$ can be expressed in the form of a simple power, e.g., $\tau(E) = AE^{-s}$, where A is a constant, an expression for $\langle \tau(E) \rangle$ can be calculated as a function of s. For the nondegenerate semiconductor, the Fermi function f_0 reduces to the Boltzmann distribution, and the integrals of Eq. (7.95) can be evaluated directly to give

$$\langle \tau(E) \rangle = \frac{A}{(kT)^s} \frac{\Gamma(\tfrac{5}{2} - s)}{\Gamma(\tfrac{5}{2})} \tag{7.96}$$

As we show in Chapter 8, $s = \tfrac{1}{2}$ for acoustic phonon scattering, and $s = -\tfrac{3}{2}$ for charged imperfection scattering. When the relaxation time $\tau(E)$ is written in the form $\tau(E) = AE^{-s}$, it follows that

$$\langle \tau(E) \rangle_{\text{ac.ph.}} = \frac{4A}{3(\pi kT)^{1/2}} \tag{7.97}$$

$$\langle \tau(E) \rangle_{\text{ch.imp.}} = \frac{8A'(kT)^{3/2}}{\pi^{1/2}} \tag{7.98}$$

An alternative way of some physical interest to look at the calculation of $\langle \tau(E) \rangle$ for a nondegenerate semiconductor can be obtained by returning to Eq. (7.88) rewritten as

$$j_x = \frac{e^2 \mathscr{E}_x n}{3kT} \langle \tau v^2 \rangle \tag{7.99}$$

7.7 Semiconductors and Metals

Now we may associate $m^*\langle v^2\rangle/2 = 3kT/2$, where $\langle v^2\rangle$ is the root-mean-square velocity of a Maxwellian distribution, and we conclude that

$$\langle\tau\rangle = \frac{\langle\tau v^2\rangle}{\langle v^2\rangle} \qquad (7.100)$$

The average relaxation time $\langle\tau\rangle$ is a weighted average of the relaxation times over the particles of the distribution, each particle being given a weight v^2, proportional to the energy E.

If more than one type of scattering process is contributing appreciably to the actual mobility of a charge carrier, we may expect as a first approximation that the rates of scattering add linearly, i.e., that the relaxation times or mobilities add reciprocally

$$\left(\frac{1}{\mu}\right)_{\text{total}} \approx \left(\frac{1}{\mu}\right)_{\text{ac.ph.}} + \left(\frac{1}{\mu}\right)_{\text{ch.imp.}} \qquad (7.101)$$

This type of linear combination is not strictly correct, because of the different energy dependence of the relaxation time for the two processes. A more exact calculation taking proper averages results in the type of correction shown in Fig. 7.5. At the point of the largest correction, which is when 40% of the resistance is due to charged imperfection scattering and 60% due to acoustic phonon scattering, the total mobility calculated according to Eq. (7.101) is about 30% too large.

Whether or not the electrical conductivity is a simple scalar or a more complex tensor quantity depends on the nature of the energy bands of the

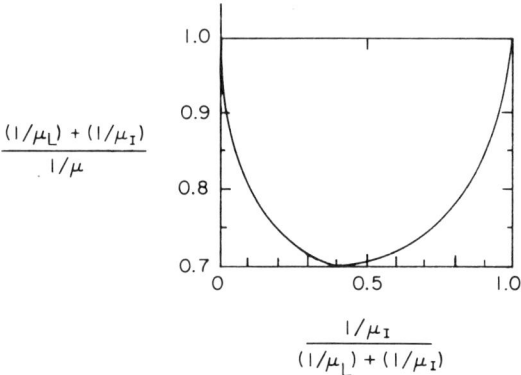

Fig. 7.5 Correction to be applied to the total mobility, as calculated by the simple sum of Eq. (7.101), in the presence of both acoustic lattice scattering (L) and charged-impurity scattering (I), when a proper average over energy is used. [After V. A. Johnson and K. Lark-Horovitz, *Phys. Rev.* **82**, 977 (1951).]

crystal. In the calculations following Eq. (7.88) we have been assuming spherical equal-energy surfaces with a scalar effective mass, $E(\mathbf{k}) = \hbar^2 k^2/2m^*$. As an example of a more complicated situation, consider an n-type semiconductor with a single conduction band minimum at $\mathbf{k} = 0$ with non-spherical equal-energy surfaces:

$$E(\mathbf{k}) = \frac{\hbar^2}{2}\left(\frac{k_1^2}{m_1^*} + \frac{k_2^2}{m_2^*} + \frac{k_3^2}{m_3^*}\right)$$

Electrical conductivity is conveniently described if the axes of the equal-energy surfaces are chosen as the coordinate axes for the description of the components of the electric field. An electric field applied in the $+x$ direction produces a current flow in the x direction,

$$j_x = -en\bar{v}_{xe} = en\frac{e}{m_1^*}\langle\tau_e\rangle\mathscr{E}_x = en\mu_{e1}\mathscr{E}_x \tag{7.102}$$

A similar result is obtained for electric fields applied in the y or z directions. Under these conditions, therefore, the conductivity takes the form of a diagonalized tensor,

$$\sigma = \begin{pmatrix} ne\mu_{e1} & 0 & 0 \\ 0 & ne\mu_{e2} & 0 \\ 0 & 0 & ne\mu_{e3} \end{pmatrix} \tag{7.103}$$

with

$$j_i = \sum_j \sigma_{ij}\mathscr{E}_j \tag{7.104}$$

Such a crystal exhibits anisotropic conductivity, and except for electric fields in the direction of the equal-energy surface axes, the current flow is not in the same direction as the applied electric field.

Another example is a semiconductor with several equivalent minima in the conduction band, like that found for Ge or Si. Equal-energy surfaces for a band structure like that of Si are represented by the ellipsoids of Fig. 7.6. In this case the sum of the electrons in all the minima must be made. If, for example, the applied electric field is in the x direction,

$$\begin{aligned} j_x &= \tfrac{1}{6}(2ne\mu_{e1} + 2ne\mu_{e2} + 2ne\mu_{e3})\mathscr{E}_x \\ &= \tfrac{1}{3}ne(\mu_{e1} + \mu_{e2} + \mu_{e3})\mathscr{E}_x \end{aligned} \tag{7.105}$$

Each ellipsoid corresponds to one sixth of the total number of electrons, and there are two ellipsoids of each type to be included in the sum. For an electric field in the x direction, $j_y = j_z = 0$. For a general orientation

7.7 Semiconductors and Metals

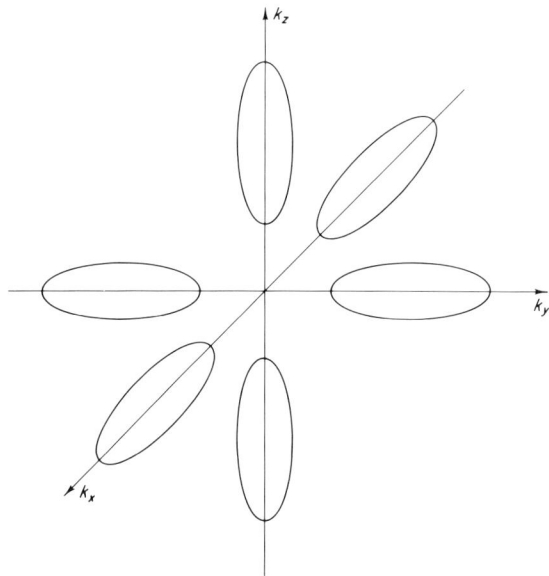

Fig. 7.6 Ellipsoidal equal-energy surfaces found for electrons in silicon.

of $\mathscr{E}$, the conductivity for this case remains a scalar, and it is common practice to define a *conductivity mobility*,

$$\mu_{e,\text{con}} \equiv \tfrac{1}{3}(\mu_{e1} + \mu_{e2} + \mu_{e3}) \tag{7.106}$$

so that it is possible to write

$$\sigma_e = ne\mu_{e,\text{con}} \tag{7.107}$$

It is also common to define a *conductivity effective mass*,

$$\frac{1}{m^*_{e,\text{con}}} \equiv \frac{1}{3}\left(\frac{1}{m^*_{e1}} + \frac{1}{m^*_{e2}} + \frac{1}{m^*_{e3}}\right) \tag{7.108}$$

so that it is also possible to write

$$\mu_{e,\text{con}} = \frac{e}{m^*_{e,\text{con}}} \langle \tau_e \rangle \tag{7.109}$$

In general, of course, the conductivity effective mass is not the same as the effective mass that enters the density of states, this latter *density-of-states effective mass* being given by

$$m^* = (m_1^* m_2^* m_3^*)^{1/3} \tag{7.110}$$

as derived in Problem 6.11b.

In a crystal with cubic symmetry, the equal-energy surfaces shown in Fig. 7.6 are spheroids, so that the two transverse effective masses are equal and may be set equal to a transverse mass m_t^*, the longitudinal mass being represented by m_ℓ^*, as discussed previously in Eq. (6.40).

Example 7.1 If a maximum uncertainty in energy of the order of kT is to be allowed in order for the suitability of the band picture to be justified, what is the minimum mobility that is consistent with a band interpretation?

From the Heisenberg indeterminacy principle, we have

$$\Delta E \cdot \Delta t \geq \frac{\hbar}{2}$$

or

$$\Delta t \geq \frac{\hbar}{2kT}$$

Now we may take

$$\Delta t \approx \langle \tau \rangle$$

so that the smallest value of mobility that would be consistent is

$$\mu = \frac{e\hbar}{2m_e^* kT}$$

in view of Eq. (7.93). If we take $m_e^* = m$, this minimum value of mobility estimated in this way is about 20 cm^2/V-sec at room temperature.

Example 7.2 In a cubic semiconductor crystal with spheroidal equal-energy surfaces, a mobility of 1000 cm^2/V-sec is measured for electrons, for which the average scattering relaxation time is 10^{-13} sec. If the transverse effective mass is $0.5m$, what is the longitudinal effective mass?

If the measured mobility is determined from conductivity, then $\mu_{\text{con}} = 1000$ cm^2/V-sec. From Eq. (7.109)

$$m_{\text{con}}^* = \frac{e\langle\tau_e\rangle}{\mu_{\text{con}}} = \frac{4.8 \times 10^{-10} \times 10^{-13}}{3 \times 10^2 \times 10^3}$$

$$= 1.6 \times 10^{-28} \text{ g}$$

$$= 0.18m$$

7.7 Semiconductors and Metals

Now from Eq. (7.108),

$$\frac{1}{m^*_{\text{con}}} = \frac{1}{0.18m} = \frac{1}{3}\left(\frac{1}{m_\ell^*} + \frac{2}{0.5m}\right)$$

Solution gives

$$m_\ell^* = 0.079m$$

These equal-energy surfaces are pancake shaped.

METALS

In metals there are free electrons at the Fermi surface, and scattering by acoustic phonons is not necessarily elastic. The wave vector for the most energetic phonons is of the same order of magnitude as the wave vector at the Fermi surface, but the energy of the most energetic phonons is much less than the Fermi energy. If the energy of the most energetic phonons is written as $k\Theta_\text{D}$, in terms of the Debye temperature Θ_D, then large-angle elastic scattering by acoustic phonons is expected at higher temperatures such that $kT > k\Theta_\text{D}$, and small-angle inelastic scattering by acoustic phonons is expected at lower temperatures such that $kT < k\Theta_\text{D}$. This means that the relaxation-time approximation for electrical conductivity is appropriate for metals at temperatures above Θ_D, and again at sufficiently low temperatures that charged imperfection scattering dominates over acoustic phonon scattering. But it is not appropriate over the intermediate temperature range, nor of course at very low temperatures where *superconductivity* may become a possibility.

Where the relaxation-time approximation is applicable, Eq. (7.89) can be evaluated for a metal (or degenerate semiconductor), for which the Fermi level lies in the midst of an allowed energy band. The quantity $\partial f_0/\partial E$ is appreciable only over a narrow range of energies within about kT of the Fermi level. The integrals of Eq. (7.89) can be approximated to first order by

$$\int_0^{E_m} \tau(E)\, E^{3/2} \left(\frac{\partial f_0}{\partial E}\right) dE \approx -\tau(E_\text{F})\, E_\text{F}^{3/2} \tag{7.111}$$

considering $(\partial f_0/\partial E)$ to be essentially a delta function at $E = E_\text{F}$, and

$$\int_0^{E_m} E^{3/2} \left(\frac{\partial f_0}{\partial E}\right) dE = -E_\text{F}^{3/2} \tag{7.112}$$

so that

$$j_x = \frac{e^2 \mathscr{E}_x n}{m^*}\, \tau(E_\text{F}) \tag{7.113}$$

Although all the electrons take part in the conductivity process, as shown by the appearance of n in the expression for the current density, the mobility with which they contribute to the conductivity is that appropriate for an electron with energy equal to the Fermi energy. If it turns out in a specific case that $\tau(E)$, $N(E)$, or $\nabla_{\mathbf{k}} E(\mathbf{k})$ is a rapidly varying function of E at E_F, then the inclusion of second-order terms in the integration of Eq. (7.89) may be desirable [see Eq. (7.131)].

In the temperature range in which the relaxation-time approximation is not valid, the electrical conductivity can be calculated from a variational solution of the Boltzmann equation, as referred to earlier. Such a calculation, the details of which we will not enter into here, shows that for $T \ll \Theta_D$,

$$\sigma(T)_{\text{lo}} \propto \left(\frac{\Theta_D}{T}\right)^5 \qquad (7.114)$$

This can be compared to the high-temperature range in which the relaxation-time approximation is valid in which

$$\sigma(T)_{\text{hi}} \propto \left(\frac{\Theta_D}{T}\right) \qquad (7.115)$$

since for acoustic phonon scattering in a metal or degenerate semiconductor, the relaxation time is inversely proportional to the average lattice energy, i.e., inversely proportional to kT, as we see in more detail in Chapter 8.

The total resistivity of a metal can be expressed, following Eq. (7.101) as

$$\varrho_{\text{total}} = \varrho_{\text{lattice}} + \varrho_{\text{impurities}} \qquad (7.116)$$

This expression is sometimes known as *Matthiessen's rule*. At low temperatures, scattering by lattice acoustic phonons becomes negligible compared to scattering by charged impurities. Purity of a metal can therefore be specified by quoting the ratio of the measured resistance at room temperature to that at liquid-helium temperature,

$$p = \frac{R_{300°\text{K}}}{R_{4°\text{K}}} \qquad (7.117)$$

For 99.999% pure Cu, for example, $p = 10^3$.

Considerable interest in metals is directed toward the effects of alloying on electrical conductivity. *Linde's rule* provides an empirical guide to the effects of alloying,

$$\varrho_{\text{I}} = a + bZ^2 \qquad (7.118)$$

where ϱ_{I} is the "residual resistivity" (the resistivity due to charged imperfections only) per atomic percent of solute, and Z is the valence difference between solute and solvent atoms. Linde's rule is suitable only for small concentrations of solute. The effects of alloying on the electrical conductivity may be attributed to a variety of effects, including, in addition to an increase in impurity scattering directly, changes in band structure, Fermi energy, density of states, effective mass, and phonon spectrum. For concentrated alloys, the empirical *Nordheim rule* provides a reasonable description, indicating that the residual resistivity is proportional to $c(1-c)$ for a concentration c of solute which is not $\ll 1$, provided that no phase transitions occur that would change the band structure, and provided that transition elements are not involved as solute.

Electrical resistivity of metals is also increased by plastic deformation of the metal. Small effects are caused by dislocations introduced by the plastic deformation; larger effects are caused by vacancy and interstitial atoms produced (see Chapter 9).

At very low temperatures metals may become *superconducting*. The superconducting state is a new state of matter in which the normal scattering mechanisms that contribute to electrical resistivity are made ineffective. The superconducting state is the result of a mutual attraction between pairs of electrons, coupled by a phonon-interaction process involving the local polarization of the metal (similar to that described for a polaron in Section 6.6), which becomes greater than the mutual repulsion between the pair of electrons due to their charge. When such pairs of electrons become stable (one with $+\mathbf{k}$, and the other with $-\mathbf{k}$), a *new state* of the system is produced that is separated by a small energy gap from the normal free electrons. The magnitude of this superconducting energy gap is sufficient, compared to the energy of available phonons, to prevent the normal scattering processes that result in electrical resistivity. The highest temperature at which a known metallic alloy is superconducting is about $20°K$ to date, and this appears to be close to an upper limit for the phonon-interaction mechanism. Efforts are being made to see if materials can be made in which an excitonic interaction between electrons might replace a phonon interaction, and thus permit superconductivity to exist at much higher temperatures.

7.8 Thermal Conductivity Due to Electrons

Thermal conductivity in solids is due both to lattice vibrations, i.e., phonons, and to electrons. In insulators and semiconductors, in which the "free" carrier density is small, thermal conductivity is usually dominated by phonons. In metals or degenerate semiconductors, electrons can make an appreciable contribution to the thermal conductivity. In this section we calculate from the Boltzmann equation the electronic thermal conductivity.

For the calculation of thermal conductivity, we must begin with Eq. (7.60) for ∇f, with the term involving ∇E_F included. With this addition, a combination of Eqs. (7.66) and (7.71), allowing for the simultaneous presence of an electric field and a thermal gradient, gives

$$\phi = \left\{ \tau e \mathscr{E} - \tau \left[(E - E_F) + T \frac{\partial E_F}{\partial T} \right] \nabla \ln T \right\} \cdot \mathbf{v} \qquad (7.119)$$

$$j_x = \frac{e}{4\pi^3} \left\{ \int -\tau e \mathscr{E}_x \frac{\partial f_0}{\partial E} v_x^2 \, d\mathbf{k}' \right.$$

$$\left. + \int \tau \left(E - E_F + T \frac{\partial E_F}{\partial T} \right) \nabla \ln T \, v_x^2 \frac{\partial f_0}{\partial E} \, d\mathbf{k}' \right\} \qquad (7.120)$$

replacing Eq. (7.83) for this case of a temperature gradient in the x direction, an associated electric field in the x direction, and in the relaxation-time approximation. In terms of the transport integrals K_n defined in Eq. (7.85),

$$j_x = K_1 \left[e^2 \mathscr{E}_x - eT \nabla \left(\frac{E_F}{T} \right) \right] - K_2 \left(\frac{e}{T} \right) \nabla T \qquad (7.121)$$

since $\nabla(E_F/T) = -(E_F/T^2) \nabla T + (1/T) \nabla E_F$.

The heat current can similarly be written as

$$Q_x = \int f(\mathbf{k}') N(\mathbf{k}') E v_x \, d\mathbf{k}'$$

$$= K_2 \left[e \mathscr{E}_x - T \nabla \left(\frac{E_F}{T} \right) \right] - K_3 \left(\frac{1}{T} \right) \nabla T \qquad (7.122)$$

The measurement of Q_x usually involves the requirement that $j_x = 0$, i.e.,

$$\mathscr{E}_x = \frac{1}{e} \left[\nabla E_F + \left(\frac{K_2}{K_1 T} - \frac{E_F}{T} \right) \nabla T \right] \qquad (7.123)$$

which gives, for the heat current,

$$Q_x = \frac{K_2^2 - K_1 K_3}{K_1 T} \nabla T \qquad (7.124)$$

The thermal conductivity $\varkappa$ is defined by

$$Q_x = -\varkappa T \qquad (7.125)$$

7.8 Thermal Conductivity Due to Electrons

so that

$$\varkappa = \frac{K_1 K_3 - K_2^2}{K_1 T} \tag{7.126}$$

The transport integrals K_n for spherical equal energy surfaces become

$$K_n = -\frac{4}{3m^*} \int \tau(E) \, N(E) \, E^n \, \frac{\partial f_0}{\partial E} \, dE \tag{7.127}$$

Under the first-order approximation for metals used to derive Eq. (7.113), it follows therefore that the thermal conductivity $\varkappa$ given in Eq. (7.126) is identically zero. Higher-order terms in the integration must be retained, therefore, to evaluate the thermal conductivity.

Such a second-order approximation can be accomplished by considering the integral

$$\int_0^\infty f(E) \, \frac{d F(E)}{dE} \, dE \tag{7.128}$$

where $F(E)$ is a function of E such that $F(0) = 0$. The procedure is to integrate Eq. (7.128) by parts, expand $F(E)$ in a Taylor series about E_F, place this expansion in the result of the integration by parts, and obtain

$$\int_0^\infty f(E) \, \frac{d F(E)}{dE} \, dE = F(E_F) + \frac{\pi^2}{6} (kT)^2 \, \frac{d^2 F(E)}{dE^2} \bigg|_{E_F} \tag{7.129}$$

If we set

$$F(E) \equiv \int_0^E N_v^t(E) \, dE$$

so that Eq. (7.128) is the expression for the total electron density, we can derive an expression for the temperature dependence of the Fermi energy,

$$E_F(T) = E_F(0) \left[1 - \frac{\pi^2}{12} \left(\frac{kT}{E_F(0)} \right)^2 \right] \tag{7.130}$$

If we set

$$F(E) \equiv \int_0^E E \, N_v^t(E) \, dE$$

we have the means of evaluating Eq. (4.88) in Section 4.6 for the electronic contribution to the heat capacity.

Applying the expansion of Eq. (7.129) to the electrical conductivity for a metal gives, in place of Eq. (7.113),

$$\sigma = \sigma(E_F) + \frac{(\pi kT)^2}{6} \, \frac{\partial^2 \sigma(E)}{\partial E^2} \bigg|_{E_F} \tag{7.131}$$

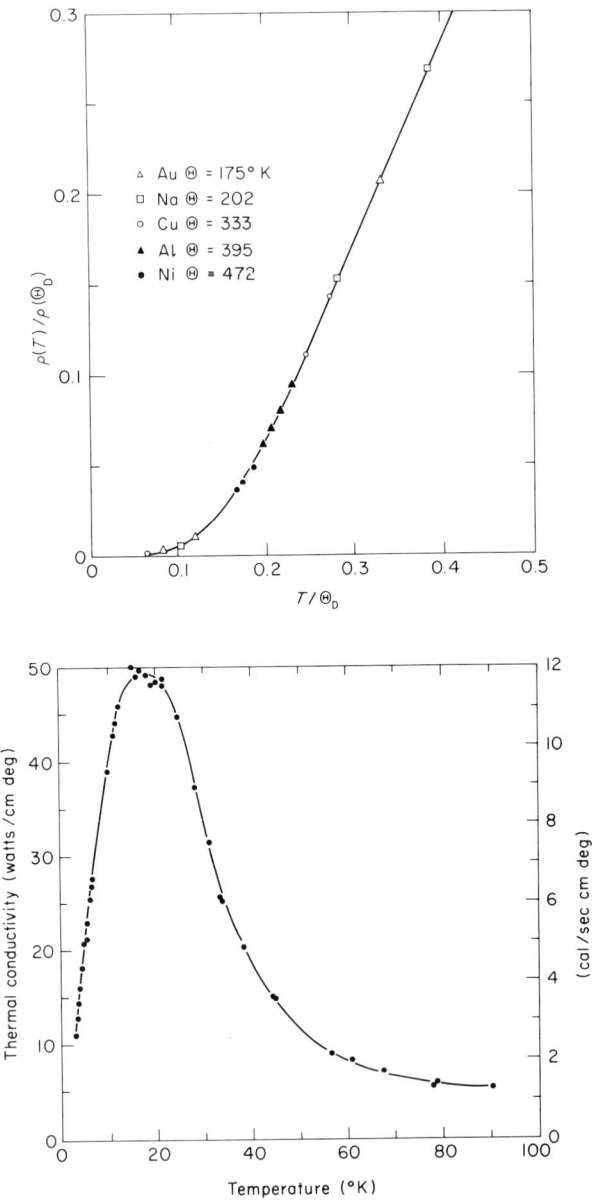

Fig. 7.7 Temperature dependence of electrical resistivity and thermal conductivity for copper (Debye temperature = 333°K) and other metals with different Debye temperature. The electrical resistivity follows the semiempirical Grüneisen relation, which

7.8 Thermal Conductivity Due to Electrons

Applying this expansion to the evaluation of the thermal conductivity gives

$$\varkappa = \frac{\pi^2 k^2 T}{3e^2} \sigma \tag{7.132}$$

If we write

$$\sigma = \frac{ne^2}{m^*} \tau_\sigma \tag{7.133}$$

and

$$\varkappa_e = \frac{n\pi^2 k^2 T}{3m^*} \tau_\varkappa \tag{7.134}$$

where the subscript e is added to $\varkappa$ to emphasize that only the electronic contribution is being considered, then the following correlation between electrical and thermal conductivity at high and low temperatures can be drawn.

At high temperatures the relaxation-time approximation holds and $\tau_\sigma = \tau_\varkappa$. The electrical conductivity σ is proportional to $1/T$ (density of scattering phonons is proportional to T) and the thermal conductivity is independent of T. The *Wiedemann–Franz ratio* is a meaningful constant quantity and is given by

$$\frac{\varkappa_e}{\sigma T} = \frac{\pi^2 k^2}{3e^2} = 0.245 \text{ erg-sec-ohm/degree}^2 \tag{7.135}$$

At low temperatures $\tau_\sigma \propto T^{-5}$ as stated in Eq. (7.114), and $\tau_\varkappa \propto T^{-3}$ (since in this range the total energy varies as T^4, the average energy per phonon $\propto T$, making the number of phonons $\propto T^3$). Thus the electrical conductivity $\propto T^{-5}$ and the thermal conductivity $\propto T^{-2}$. Over this temperature range, therefore, the Wiedemann–Franz ratio increases as T^2. Figure 7.7 shows the temperature dependence of electrical resistivity and thermal conductivity for copper.

For a semiconductor with spherical equal-energy surfaces and a relaxation time $\tau \propto E^{-s}$, the thermal conductivity can be calculated directly from

---

predicts $\varrho \propto TG(\Theta_D/T)$, with

$$G(x) = x^{-4} \int_0^x \frac{y^2\, dy}{(e^y - 1)(1 - e^{-y})}$$

which predicts $\varrho \propto T$ for $T > \Theta_D$, and $\varrho \propto T^5$ for $T < \Theta_D$. [Resistivity plot after J. Bardeen, *J. Appl. Phys.* **11**, 88 (1940); thermal conductivity after R. Berman and D. K. C. MacDonald, *Proc. Roy. Soc. (London)* **A209**, 368 (1951); **A211**, 122 (1952).]

Eq. (7.126). The result is

$$\varkappa_e = \frac{k^2 T}{e^2} (\tfrac{5}{2} - s)\sigma \qquad (7.136)$$

The Wiedemann–Franz ratio is

$$\frac{\varkappa_e}{\sigma T} = \frac{k^2}{e^2} (\tfrac{5}{2} - s) \qquad (7.137)$$

7.9 Thermoelectric Effect

If a temperature gradient is imposed on a material under conditions in which no current is drawn, a potential difference is detectable between the ends of the material. Physically the effect can be seen as the result of increased energy of electrons at the hot end of the material causing a diffusion of electrons toward the cold end; the charge imbalance caused by this diffusion sets up an electric field sufficient to make the net current flow zero. The *thermoelectric power* α_e is defined as the ratio between the electric field and the temperature gradient. From Eq. (7.123), setting $\nabla E_F = 0$ in the steady state,

$$\alpha_e = \frac{K_2 - E_F K_1}{e K_1 T} \qquad (7.138)$$

For metals, the electronic thermoelectric power is given by

$$\alpha_e = \frac{\pi^2 k^2 T}{3e} \left. \frac{\partial \ln \sigma(E)}{\partial E} \right]_{E_F} \qquad (7.139)$$

For an ideal metal this leads to

$$\alpha_e = \frac{\pi^2 k^2 T}{e E_F} \quad \text{for} \quad T > \Theta_D \qquad (7.140)$$

and

$$\alpha_e = \frac{\pi^2 k^2 T}{3 e E_F} \quad \text{for} \quad T < \Theta_D \qquad (7.141)$$

For spherical equal-energy surfaces in a semiconductor and $\tau \propto E^{-s}$, Eq. (7.138) leads directly to

$$\alpha_e = -\frac{k}{e} \left[(\tfrac{5}{2} - s) + \frac{(E_c - E_F)}{kT} \right] \qquad (7.142)$$

7.9 Thermoelectric Effect

where E_c is the energy of the bottom of the band in which the electrons are moving, and $E_c > E_F$. Comparison of Eqs. (7.142) and (7.140) shows that the thermoelectric power in a nondegenerate semiconductor is of the order of (E_F/kT) larger than that of a metal; for a metal at 300°K the thermoelectric power is of the order of 1 μV/°K, whereas for a typical semiconductor it is some 10^2–10^3 times larger.

If optical-mode scattering of carriers dominates, the relaxation-time approximation used here is no longer applicable. The thermoelectric power, however, can still be written in the form of Eq. (7.142), provided that the quantity $(\frac{5}{2} - s)$ is replaced by the quantity A, which is a function of temperature and is given in Fig. 7.8. The range of values for A over the normal range of temperatures does not differ widely from the value of 2 for acoustic-mode scattering.

At low temperatures another contribution to the measured thermoelectric power may arise in both metals and semiconductors in addition to that caused by the diffusion of free electrons. This additional contribution arises when the implicit assumption above, that the lattice vibrations are at equilibrium even with a temperature gradient on the material, is violated. When the interaction between phonons and electrons dominates over that between phonons and other phonons, then phonons moving in a thermal gradient tend to drag electrons along with them. This additional charge motion is called *phonon drag*, and the measured thermoelectric power is the sum of

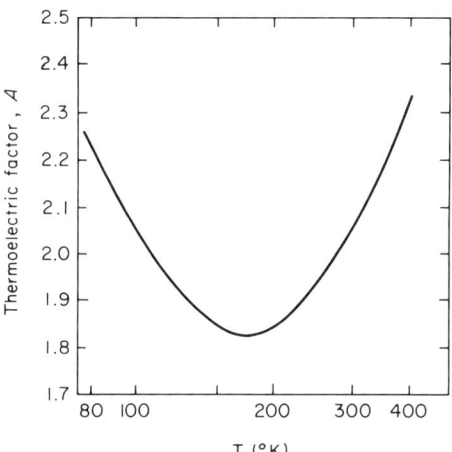

Fig. 7.8 The thermoelectric factor A in the expression $\alpha = \pm(k/e)[A + (E_c - E_F)/kT]$ for optical-mode scattering in a nondegenerate semiconductor. [After S. S. Devlin, *in* "Physics and Chemistry of II–VI Compounds," (M. Aven and J. Prener, eds.), Wiley, New York, 1967, p. 561.]

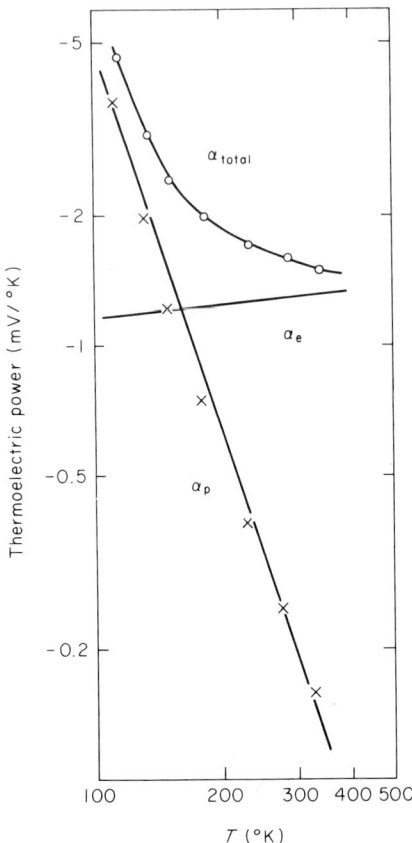

Fig. 7.9 Temperature dependence of the total measured thermoelectric power, the calculated electronic contribution to the thermoelectric power, and the phonon-drag contribution obtained by subtraction, for 10^2 Ω-cm n-type silicon. [After J. G. Harper, H. E. Matthews and R. H. Bube, *J. Appl. Phys.* **41**, 765 (1970).]

the diffusion contribution and the phonon-drag contribution. Figure 7.9 shows the temperature dependence of electron-diffusion and phonon-drag contributions to the total measured thermoelectric power for n-type silicon.

7.10 Summary

The utility of wave packets to describe the transport of particles in a wave mechanical framework rests on the fact that classical laws of motion can be applied to the wave packet as a pseudoparticle with a position given

7.10 Summary

by the expectation value of x for the packet and acted on by a force given by the expectation value of the external force. A wave packet constructed of Bloch waves corresponding to wave vectors $\mathbf{k}'$ in the vicinity of $\mathbf{k}$ has a velocity given by the group velocity $\mathbf{v} = \hbar^{-1} \nabla_{\mathbf{k}} E(\mathbf{k})$.

Steady-state transport properties of metals and semiconductors, such as electrical and thermal conductivity, are derived in terms of a departure from and a return to the equilibrium distribution of state occupancy. The Boltzmann equation is the key to this analysis, and the form of the relaxation of the perturbation from equilibrium back to the equilibrium state through a scattering process determines the type of solution. In those cases where scattering involves a change in energy of the carrier much less than its thermal energy, the relaxation can be described in terms of a simple relaxation-time approximation. In those cases where larger changes in energy are involved in scattering, a more general solution of the Boltzmann equation via a variational approach is required.

In this chapter we have been concerned only with scattering by acoustic-mode lattice vibrations or by charged impurities. In the following chapter we consider the actual calculation of the scattering probability for these two processes, as well as for others. Here we have seen that in semi-conductors, both of these processes may be adequately represented by the relaxation-time approximation. In metals, on the other hand, acoustic-mode scattering becomes inelastic at low temperatures and can no longer be adequately described in the relaxation-time approximation. This means that in metals the properties of electrical and thermal conductivity can be described in terms of a single relaxation time that is proportional to T^{-1} at high temperatures, but that at low temperatures rather more complicated behavior is encountered in which the electrical conductivity varies as T^{-5} while thermal conductivity varies as T^{-2}. The constancy of the Wiedemann-Franz ratio (the ratio of thermal conductivity due to electrons to the product of electrical conductivity and temperature) holds only in the higher temperature range.

If a temperature gradient is imposed on a material, an electric field results. The ratio between the electric field and the thermal gradient is called the thermoelectric power. Contributions to the thermoelectric power arise both from the diffusion of electrons in the temperature gradient and, at low temperatures, from the interaction between phonons moving in the temperature gradient and free electrons. In general the thermoelectric power is several orders of magnitude larger in semiconductors than in metals. This results from the fact that only a small fraction (about kT/E_F) of the free electrons in a metal are affected by the presence of a thermal gradient, a condition that is also shown by the effective relaxation time for a metal being to first order just the relaxation time corresponding to electrons at the Fermi energy.

Problems

7.1 If the maximum value of a Gaussian wave packet is 10^2 times the width of the packet at $t = 0$, how far will the wave packet travel before the maximum value is equal to the width for a free electron?

7.2 The mobilities and effective masses of carriers in GaAs at room temperature are typically as follows:

	Mobility, cm^2/V-sec	Effective-mass ratio
Electrons	7000	0.07
Holes	300	0.5

Calculate (a) the relaxation time for electrons and for holes, (b) the mean free path for electrons and holes, (c) the drift path length in an electric field of 100 V/cm in one relaxation time, (d) the electrical conductivity for a sample with 10^{14} cm^{-3} free electrons and 2.3×10^{15} cm^{-3} free holes.

7.3 If scattering relaxation times for electrons and holes in a semiconductor were about equal, would you in general expect electrons or holes to have the higher mobility? Why? (How do the widths of conduction and valence bands usually compare?)

7.4 In a crystal with scalar effective mass of $0.2m$ for electrons, the mobility is measured at 30°K, where it is dominated by impurity scattering, to be 20 cm^2/V-sec, and at 600°K, where it is dominated by lattice scattering, to be 70 cm^2/V-sec. What is the relaxation time for scattering at 200°K?

7.5 In a crystal in which scattering by both lattice vibrations and by charged imperfections is present, the following measurements were made of mobility versus T:

Mobility, cm^2/V-sec	T, °K
500	100
125	400

What is the measured mobility expected for 50°K?

7.6 Suppose that the relaxation time for a particular scattering process is directly proportional to the electron velocity. Calculate the temperature dependence of the mobility for a nondegenerate semiconductor.

7.7 The measured thermal conductivity of germanium at 300°K is 0.63 W/°K, in a range where acoustic phonon scattering dominates.
(a) What fraction of this thermal conductivity is caused by free electrons if the free-electron density is 10^{17} cm^{-3} and the electron mobility is 4×10^3 cm^2/V-sec?
(b) What must the concentration of free electrons in germanium at 300°K be in order for the electronic contribution to the thermal conductivity to be equal to the lattice contribution?
(c) The thermoelectric power corresponding to (a) is measured to be 0.46 mV/°K. How far below the conduction-band edge does the Fermi level lie?

Chapter 8

Scattering Processes

In the previous chapter we considered some of the characteristics of carrier transport in the presence of scattering by acoustic-mode vibrations or by charged imperfections. In this chapter we consider the way in which the scattering cross section for these two processes can be calculated from the nature of the processes themselves.

The scattering effects of lattice waves can be pictured in terms of local changes in the potential experienced by free carriers, describable in terms of a *deformation potential* characteristic of the material. A more complete treatment can be given in the same perspective through a first-order time-dependent perturbation calculation in which deviations in the local potential because of lattice waves are treated as perturbations to the potential of the perfectly periodic lattice.

Scattering of carriers by charged imperfections is the result of a Coulomb interaction, which may be shielded by the presence of other free carriers. Determination of the scattering cross section can be carried out by a geometrical analysis of the scattering process in analogy with the Rutherford scattering law, leading to the Conwell–Weisskopf expression. A quantum mechanical treatment in terms of a screened scattering potential leads to the Brooks–Herring relationship.

Many other possibilities exist for scattering beside those due to acoustic-mode lattice waves and charged imperfections. These include scattering by optical-mode lattice waves, neutral imperfections, local inhomogeneities and strain-induced piezoelectric fields, to cite a few examples.

As the magnitude of the electric field is increased in a semiconductor, a

8.1 Simple Model of Wave Reflection

condition is eventually reached where the mobility of the carriers becomes dependent on the electric field. The form of this variation can be determined for specific scattering processes.

8.1 Scattering by Acoustic-Mode Lattice Waves: Simple Model of Wave Reflection

Consider that the principal effect of a lattice wave is to distort the local potential of the lattice. A simple way to estimate carrier scattering effects is to calculate the reflection of carrier waves because of these variations in local potential. We consider specifically longitudinal acoustic-mode waves for the purposes of this model.

The simple energy bands that are drawn for crystals are based upon a spacing between atoms corresponding to a fixed equilibrium situation. The presence of a longitudinal lattice wave alters the spacings between atoms, and therefore produces local alterations in the effective band gap. The crystal is composed of alternating regions of compression, where the atoms have a smaller-than-equilibrium spacing, and of expansion, where the atoms have a larger-than-equilibrium spacing. We may picture the dependence of the energy bands on lattice spacing in terms of Fig. 8.1, and the resultant effect on the potential that an electron experiences in Fig. 8.2. Consider such a potential pattern to be frozen in time and the effects of such a potential distribution on a traveling electron wave. Because of the large dif-

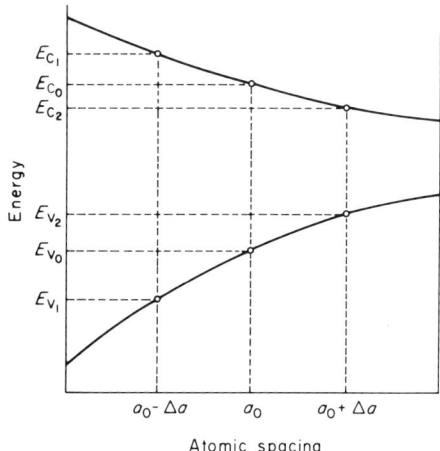

Fig. 8.1 Schematic representation of the variation of the conduction-band edge and the valence-band edge as a function of lattice spacing.

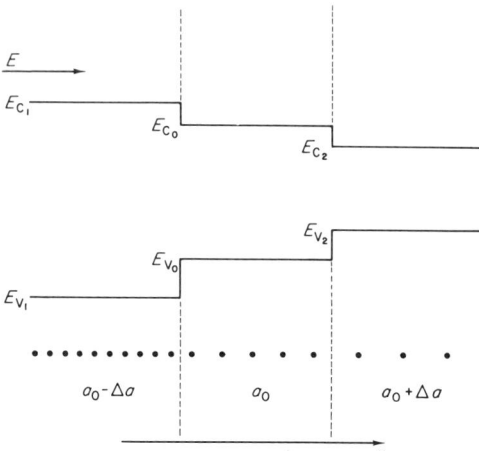

Fig. 8.2 Potential distribution experienced by an electron with energy E, and the corresponding potential distribution in the conduction and valence bands, if the possible changes in lattice spacing as indicated in Fig. 8.1 are assumed abrupt between regions of different lattice spacing and constant within regions of different lattice spacing.

ferences in the velocities of electron and lattice waves, this is not such a gross assumption as might appear on first examination.

We know from Problem 3.3 that there is a reflection of the electron wave associated with these potential discontinuities. An electron with energy E' for example, experiences a varying potential energy depending on its position in the crystal. An electron wave in one region A where the kinetic energy is $E_A = \hbar^2 k_A^2/2m^*$, upon passing into a second region B where the kinetic energy is $E_B = \hbar^2 k_B^2/2m^*$, is characterized by a reflection coefficient

$$|R|^2 = \left(\frac{k_A - k_B}{k_A + k_B}\right)^2 \tag{8.1}$$

If the difference between k_A and k_B is small enough, we may approximate by writing $(k_A - k_B) \approx dk$, and $(k_A + k_B) \approx 2k$, giving

$$|R|^2 = \frac{dk^2}{4k^2} = \frac{m^{*2}}{4\hbar^4 k^4} dE^2 = \frac{m^{*2}}{4p^4} dE^2 \tag{8.2}$$

since for a small change in the energy,

$$dE = \frac{\hbar^2 k}{m^*} dk \tag{8.3}$$

8.1 Simple Model of Wave Reflection

Equation (8.2) provides a relationship between the fraction of the electron current reflected and the difference in energy associated with the potential discontinuity.

In order to generalize this result to include the whole crystal, we extend our picture by considering the crystal as a whole to be divided into small individual cubes, each of which may be considered to expand or contract independently of the others. We wish to relate the energy required to bring about the distortion of each of these cubes to the energy available thermally to lattice waves, and hence to relate the reflection coefficient to the temperature. To accomplish this, define a characteristic energy change for small changes in lattice constant, E^*, by the expression

$$dE \equiv E^* \frac{dV}{V_0} \tag{8.4}$$

where dV is the change in volume accompanying the distortion, and V_0 is the undistorted volume. E^* is related to the slope of the energy versus lattice constant curves of Fig. 8.1, at the equilibrium lattice spacing a_0. The stored elastic energy in an individual cube can be written as

$$-\frac{1}{2} \mathscr{P} \, dV = \frac{1}{2} \frac{1}{\gamma} \frac{dV}{V_0} dV \tag{8.5}$$

where the one-half indicates an average energy, $\mathscr{P}$ is a differential pressure, and γ is the compressibility. This average elastic energy must be equal to the average potential energy in lattice vibrations, which is $kT/2$ for low-energy phonon waves. Combining Eqs. (8.4) and (8.5) gives

$$dE^2 = E^{*2} \frac{kT\gamma}{V_0} \tag{8.6}$$

Therefore, from Eq. (8.2),

$$|R|^2 = \frac{m^{*2}}{4p^4} \frac{E^{*2}kT\gamma}{V_0} \tag{8.7}$$

To complete the relationship between the mean free path and the reflection coefficient, consider that these hypothetical individual cubes have a side that is a multiple η of the wavelength λ of the electron wave. Since the probability for an electron to be scattered in a distance dx is dx/l, the probability for an electron to be scattered in a distance $\eta\lambda$ is $\eta\lambda/l$. This expression is to be equated to $2|R|^2$, since this is the probability of an electron being scattered in traversing the elementary cube, one reflection occurring at the front and one at the rear of the cube. The mean free path is

$$l = \frac{\eta\lambda}{2\,|\,R\,|^2} = \frac{2(\eta\lambda)^4 p^4}{m^{*2}E^{*2}kT\gamma} = \frac{2h^4\eta^4}{m^{*2}E^{*2}kT\gamma} \tag{8.8}$$

The mean free path is inversely proportional to the temperature, and does not depend on the electron energy. A reasonable value for η is $\frac{1}{4}$, which corresponds to constructive interference between electron waves reflected from the front and rear faces of the individual cubes, since the reflection condition at the two faces introduces a phase difference of π and a round-trip through the cube adds another phase difference of π, making a total of 2π in phase difference altogether. If $\eta \ll 1$, effects would be expected to average out, and if $\eta \gg 1$, the electron wave could adjust without much reflection. A choice $\eta = \frac{1}{4}$ gives a value of l from Eq. (8.8) close to that obtained from a quantum mechanical calculation.

Taking $l = A/T$ from Eq. (8.8), we may calculate the average value of the relaxation time for scattering by longitudinal acoustic waves, and hence the corresponding temperature dependence of the mobility in a nondegenerate semiconductor. Since

$$\tau = \frac{l}{v} = \frac{Am^{*1/2}}{2^{1/2}T}E^{-1/2} \tag{8.9}$$

we obtain from Eq. (7.95),

$$\langle\tau\rangle_{\text{LA lattice scattering}} = \frac{4Am^{*1/2}}{3(2\pi kT)^{1/2}T} \tag{8.10}$$

We conclude therefore that for scattering by LA phonons in a semiconductor,

$$\mu_\text{L} \propto T^{-3/2} \tag{8.11}$$

For a metal, $\langle\tau\rangle \approx \tau(E_\text{F})$ and both $\langle\tau\rangle$ and $\mu_\text{L} \propto 1/T$.

The proper averaged effective mass to use in place of a scalar effective mass in the case of nonspherical equal-energy surfaces in a semiconductor deserves brief mention at this point. From Eqs. (8.8) and (8.10) we see that

$$\langle\tau\rangle \propto m^{*-3/2}T^{-3/2} \tag{8.12}$$

The presence of the $m^{*-3/2}$ term is connected with the density of states to which the electron may be scattered. The scalar $m^{*-3/2}$ is therefore to be replaced by $(m_1^* m_2^* m_3^*)^{-1/2}$ for nonspherical equal-energy surfaces. Since the mobility for lattice scattering

$$\mu_\text{L} = \frac{e}{m^*}\langle\tau\rangle \tag{8.13}$$

8.1 Simple Model of Wave Reflection

it follows that

$$\mu_L \propto m^{*-5/2} T^{-3/2} \qquad (8.14)$$

Since a similar relationship can be derived for electrons or holes, a simple correlation is suggested between the mobility ratio for lattice scattering and the effective mass ratio,

$$\frac{\mu_e}{\mu_h} \approx \left(\frac{m_h^*}{m_e^*}\right)^{5/2} \qquad (8.15)$$

assuming that the constants entering Eq. (8.14) for electrons and holes effectively cancel. For a crystal with ellipsoidal equal-energy surfaces, Eq. (8.15) must be applied to an averaged effective mass. The effective mass entering the relationship via the equation between mobility and relaxation time is the conductivity effective mass, previously defined in Eq. (7.108). Thus the appropriate averaged effective mass to be used in Eq. (8.15) for ellipsoidal equal-energy surfaces is

$$\frac{1}{m^{*5/2}} = \frac{1}{3(m_1^* m_2^* m_3^*)^{1/2}} \left(\frac{1}{m_1^*} + \frac{1}{m_2^*} + \frac{1}{m_3^*}\right) \qquad (8.16)$$

The variation of mobility with T^{-1} for a metal has already been illustrated in Fig. 7.7 for elastic scattering in the range of $T > \Theta_D$. The ideal variation of mobility with $T^{-3/2}$ predicted for a semiconductor is somewhat more difficult to find in real materials because of the presence of a variety of other scattering processes. Semiconductors with predominantly covalent bonding seem the most likely candidates for such ideal behavior, and measurements indicating a $T^{-3/2}$ temperature variation of mobility have been reported for electrons in such materials as diamond and InSb. The temperature dependence of the electron and hole mobility in Ge and Si is shown in Fig. 8.3. The mobility varies approximately as T^{-n} in the high-temperature range (the bend-over at low temperatures is due to charged-impurity scattering) with $n = 1.66$ for electrons and 2.33 for holes in Ge, 2.5 for electrons and 2.7 for holes in Si. As we saw in Chapter 6, these covalent elemental semiconductors have a fairly complex energy-band structure with several equivalent conduction-band extrema and multiple valence bands. Scattering processes involving not only low-energy acoustic phonons (elastic scattering) but higher-energy phonons (inelastic scattering), as well as scattering between energy bands in the valence-band structure, are believed necessary to describe the experimentally observed variations of mobility in Ge and Si.

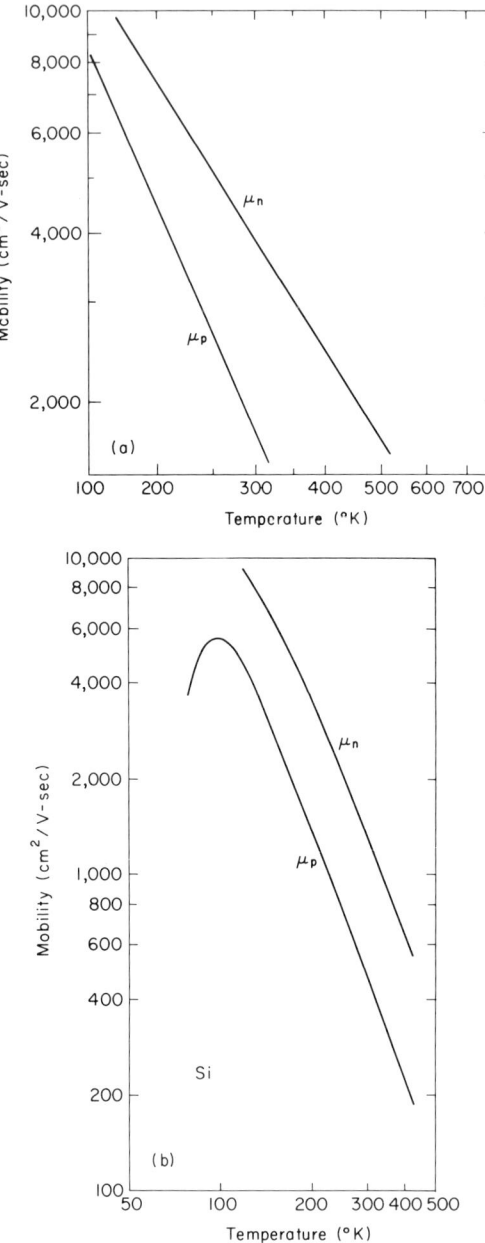

Fig. 8.3 Temperature dependence of the mobility in Ge (*top*) [after M. B. Prince, *Phys. Rev.* **92**, 681 (1953)] and Si (*bottom*) [after G. W. Ludwig and R. L. Watters, *Phys. Rev.* **101**, 1699 (1956)] for both electrons and holes.

8.2 Perturbation Calculation

Example 8.1 Effective mass ratios in Ge are as follows:

Conduction band: Longitudinal mass $= 1.3m$
Transverse mass $= 0.082m$

Valence band: $0.34m$

From the knowledge of these effective masses only, estimate the ratio of electron to hole mobility and compare with known values.

Using Eqs. (8.15) and (8.16),

$$\frac{1}{m_e^{*5/2}} = \frac{1}{3(0.082)(1.3)^{1/2}} \left(\frac{1}{1.3} + \frac{2}{0.082}\right)$$

$$= 90$$

$$m_h^{*5/2} = (0.34)^{5/2} = 6.8 \times 10^{-2}$$

Therefore

$$\frac{\mu_e}{\mu_h} = (6.8 \times 10^{-2})(90) = 6.1$$

Experimental values of mobilities for Ge give a ratio approximately

$$\frac{\mu_e}{\mu_h} = \frac{3900}{1900} = 2.1$$

Thus the estimated value of the mobility ratio from Eq. (8.15) is about three times larger than the experimental value. If the origin of Eq. (8.15) is examined a little more closely from Eqs. (8.8) and (8.10), it is seen that one term entering these equations for electrons and holes may well be different, namely the E^* term. If this is retained, Eq. (8.15) becomes

$$\frac{\mu_e}{\mu_h} = \left(\frac{m_h^*}{m_e^*}\right)^{5/2} \left(\frac{E_h^*}{E_e^*}\right)^2$$

If this idealized formulation were to be applied to Ge, therefore, our above calculation indicates that the factor $(E_h^*/E_e^*)^2 \approx \frac{1}{3}$.

8.2 Scattering by Acoustic-Mode Lattice Waves: Perturbation Calculation

In order to consider the effects of perturbation of the potential caused by the lattice wave, we need to know a little about the general calculation of a transition probability from state **k** to state **k**' under the effects of a perturbation. We summarize briefly first-order time-dependent perturbation theory for a periodic perturbation, therefore, and then apply this to the problem of scattering by acoustic lattice waves.

FIRST-ORDER TIME-DEPENDENT PERTURBATION THEORY

Consider a system for which the stationary states are known as the result of the solution of the time-independent problem,

$$\mathbf{H}_0 \psi_n = E_n \psi_n \tag{8.17}$$

In the presence of a perturbation $\mathbf{H}'$ which may cause a transition from the state with energy E_n to the state with energy E_m, the time-dependent Schroedinger equation is

$$(\mathbf{H}_0 + \mathbf{H}') = i\hbar \frac{\partial \Psi}{\partial t} \tag{8.18}$$

In order to calculate the effect of the perturbation $\mathbf{H}'$, we express the general solution of the perturbed time-dependent Schroedinger equation as a linear combination of the eigenfunctions of the unperturbed system,

$$\Psi = \sum_n A_n(t) \psi_n e^{-iE_n t/\hbar} \tag{8.19}$$

where the $A_n(t)$ are functions of time.

If we substitute Eq. (8.19) into Eq. (8.18), we obtain

$$\sum_n A_n (\mathbf{H}_0 \psi_n + \mathbf{H}' \psi_n) e^{-iE_n t/\hbar} = \sum_n \left(A_n E_n \psi_n + i\hbar \frac{dA_n}{dt} \psi_n \right) e^{-iE_n t/\hbar} \tag{8.20}$$

Use of Eq. (8.17) permits simplification to

$$\sum_n A_n \mathbf{H}' \psi_n e^{-iE_n t/\hbar} = i\hbar \sum_n \frac{dA_n}{dt} \psi_n e^{-iE_n t/\hbar} \tag{8.21}$$

In order to solve for dA_n/dt, multiply both sides of Eq. (8.21) by $(\psi_m e^{-iE_m t/\hbar})^*$ and integrate over the coordinates

$$\sum_n A_n \int \psi_m^* \mathbf{H}' \psi_n e^{i(E_m - E_n)t/\hbar} \, d\mathbf{r} = i\hbar \sum_n \frac{dA_n}{dt} \int \psi_m^* \psi_n e^{i(E_m - E_n)t/\hbar} \, d\mathbf{r} \tag{8.22}$$

The exponential terms on both sides of the equation can be brought outside the integral, since the integration is over the spatial coordinates only. The integral on the right side of Eq. (8.22) is δ_{mn} because of the orthonormality

8.2 Perturbation Calculation

of the ψ_n functions, and the expression for the time rate of change of the A coefficients is

$$\frac{dA_m}{dt} = \frac{1}{i\hbar} \sum_n A_n H_{mn} e^{i(E_m - E_n)t/\hbar} \qquad (8.23)$$

where H_{mn} is the *matrix element* for the transition,

$$H_{mn} \equiv \int \psi_m^* \mathbf{H}' \psi_n \, d\mathbf{r} \qquad (8.24)$$

The probability of a transition from a state with energy E_n to a state with energy E_m depends on the magnitude of the matrix element linking the initial and final states of the transition.

Suppose that the system is in the stationary state with energy E_0 at $t = 0$. This means that at $t = 0$,

$$\begin{aligned} A_n(0) &= 1 \quad \text{if} \quad E_n = E_0 \\ &= 0 \quad \text{if} \quad E_n \neq E_0 \end{aligned} \qquad (8.25)$$

At $t = 0$, Eq. (8.23) is

$$\begin{aligned} \frac{dA_m}{dt} &= \frac{1}{i\hbar} A_0 H_{m0} e^{i(E_m - E_0)t/\hbar} \\ &= \frac{1}{i\hbar} H_{m0} e^{i(E_m - E_0)t/\hbar} \end{aligned} \qquad (8.26)$$

since $A_0 = 1$ for $E_n = E_0$.

In first-order time-dependent perturbation theory, we make the assumption that the coefficient A_0 *remains* approximately equal to 1 for times not too long. For times greater than zero, we write

$$\begin{aligned} A_0 &= 1 + A_0' \\ A_m &= A_m' \quad (m \neq 0) \end{aligned} \qquad (8.27)$$

where the primed coefficients represent small terms. Substitution of Eq. (8.27) into the upper form of Eq. (8.26) yields two terms; the second of these contains $A_0' H_{m0}$, the product of two small terms, which can be neglected in a first-order calculation. Integration of Eq. (8.26) gives the desired result for A_m,

$$A_m(t) = \frac{1}{i\hbar} \int_0^t H_{m0} e^{i(E_m - E_0)t/\hbar} \qquad (8.28)$$

for $m \neq 0$. The quantity $|A_m(t)|^2$ is the probability of the transition between the state with energy E_0 and the state with energy E_m in the time t.

A periodic perturbation can be written as

$$\mathbf{H}'(\mathbf{r}, t) = \mathbf{H}_0'(\mathbf{r}) \cos \omega t = \frac{\mathbf{H}_0'(\mathbf{r})}{2} (e^{i\omega t} + e^{-i\omega t}) \quad (8.29)$$

Substitution of such a periodic perturbation into Eq. (8.28) gives

$$A_m(t) = \frac{1}{i\hbar} \int_0^t H_{m0} [e^{i(E_m - E_0 + \hbar\omega)t/\hbar} + e^{i(E_m - E_0 - \hbar\omega)t/\hbar}] \, dt$$

$$= H_{m0} \left[\frac{1 - e^{i(E_m - E_0 + \hbar\omega)t/\hbar}}{(E_m - E_0 + \hbar\omega)} + \frac{1 - e^{i(E_m - E_0 - \hbar\omega)t/\hbar}}{(E_m - E_0 - \hbar\omega)} \right] \quad (8.30)$$

where the matrix element is given by

$$H_{m0} = \frac{1}{2} \int \psi_m^* \, \mathbf{H}_0'(\mathbf{r}) \, \psi_0 \, d\mathbf{r} \quad (8.31)$$

The coefficient $A_m(t)$ given in Eq. (8.30) has a large value only if

$$E_m \approx E_0 \pm \hbar\omega \quad (8.32)$$

This is also an expression of the conservation of energy; in this case the energy of the final state can differ from that of the initial state by the energy of the quantum $\hbar\omega$ associated with the perturbation. If $E_m > E_0$, the principal contribution to $A_m(t)$ in Eq. (8.30) comes from the second term; in making the transition from E_0 to E_m, a quantum of energy $\hbar\omega$ is *absorbed* from the perturbation. If $E_m < E_0$, the principal contribution to $A_m(t)$ comes from the first term; in making the transition a quantum of energy $\hbar\omega$ is *emitted* to the perturbation.

For $E_m > E_0$, for example, the transition probability is

$$|A_m(t)|^2 = \frac{4 |H_{m0}|^2 \sin^2[(E_m - E_0 - \hbar\omega)t/2\hbar]}{(E_m - E_0 - \hbar\omega)^2} \quad (8.33)$$

CALCULATION OF LATTICE SCATTERING

In applying first-order time-dependent perturbation theory to the calculation of the scattering cross section for acoustic-mode vibrations, the following schedule is followed.

1. Formulate a perturbation potential in terms of the displacement of atoms as the result of the presence of lattice waves.

8.2 Perturbation Calculation

2. Divide the crystal into Wigner–Seitz cells.
3. Take Bloch functions for the initial and final states entering the matrix element. Integrate over a single cell and sum over the cells that make up the crystal.
4. Recognize that the matrix element vanishes unless appropriate wave vector conservation is present in the scattering process.
5. Recognize that conservation of energy is also required in order to obtain a large transition probability.
6. Calculate the explicit form of the matrix element.
7. Calculate the scattering probability from the matrix element.
8. Relate the scattering probability to the relaxation time for scattering.

1. Formulate the perturbation potential. Consider an acoustic lattice wave with frequency $\bar{\nu} = \bar{\omega}/2\pi$. Figure 8.4 pictures the displacement $\mathbf{X}_j$ of the atom at the site in the crystal with lattice vector $\mathbf{R}_j$. This displacement may be expressed as

$$\mathbf{X}_j = \frac{1}{2} \mathbf{A}\, e^{i(\mathbf{K}\cdot\mathbf{R}_j - \bar{\omega}t)} + \frac{1}{2} \mathbf{A}^* \, e^{-i(\mathbf{K}_j\cdot\mathbf{R} + \bar{\omega}t)} \quad (8.34)$$

where the amplitude of the wave is $|\mathbf{A}|$, the first term represents a wave moving in the $+\mathbf{K}$ direction, and the second term represents a wave moving in the $-\mathbf{K}$ direction. What kind of perturbation potential is generated by this lattice wave? Conceptually we might expect this perturbation potential to be obtainable by multiplying the actual displacement of the atom by the

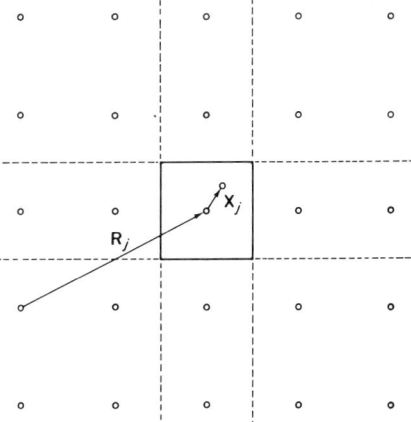

Fig. 8.4 In the *j*th Wigner–Seitz cell, the *j*th atom with lattice vector $\mathbf{R}_j$ undergoes a displacement $\mathbf{X}_j$ because of an acoustic lattice wave.

change in the periodic crystal potential at $\mathbf{R}_j$ per unit atomic displacement. This choice is confirmed by the following considerations.

2. *Divide the crystal into Wigner–Seitz cells.* Consider the Wigner–Seitz cell (Section 5.2) centered on the atom displaced by $\mathbf{X}_j$, keeping all other atoms at their equilibrium position, as indicated in Fig. 8.4. Then the potential in the *j*th cell is

$$V(\mathbf{r}) = V_0(\mathbf{r} - \mathbf{R}_j - \mathbf{X}_j)$$
$$= V_0(\mathbf{r} - \mathbf{R}_j) + U_j(\mathbf{r}) \quad (8.35)$$

where $V_0(\mathbf{r} - \mathbf{R}_j)$ is the periodic crystal potential of the unperturbed crystal at $\mathbf{R}_j$, and $U_j(\mathbf{r})$ is the perturbation potential. $U_j(\mathbf{r}) = 0$ outside of the *j*th cell by definition. Combining the two forms of Eq. (8.35), we obtain

$$U_j(\mathbf{r}) = V_0(\mathbf{r} - \mathbf{R}_j - \mathbf{X}_j) - V_0(\mathbf{r} - \mathbf{R}_j)$$
$$\approx \mathbf{X}_j \cdot \nabla V_0(\mathbf{r} - \mathbf{R}_j) \quad (8.36)$$

for small $\mathbf{X}_j$. Equation (8.36) holds for each cell in the crystal. Combining Eq. (8.36) with Eq. (8.34) gives for the desired perturbation potential

$$U_j(\mathbf{r}) = \frac{1}{2} \mathbf{A} \cdot \nabla V_0(\mathbf{r} - \mathbf{R}_j) \, e^{i(\mathbf{K}\cdot\mathbf{R}_j - \bar{\omega}t)} + \frac{1}{2} \mathbf{A}^* \cdot \nabla V_0(\mathbf{r} - \mathbf{R}_j) \, e^{-i(\mathbf{K}\cdot\mathbf{R}_j + \bar{\omega}t)}$$
(8.37)

This $U_j(\mathbf{r})$ is a periodic perturbation potential with frequency $\bar{\omega}/2\pi$. The matrix element H_{m0} corresponding to the first term in Eq. (8.37) is therefore given by Eq. (8.31),

$$H_{m0} = \frac{1}{2} \int \psi_m^* \, \mathbf{H}_0'(\mathbf{r}) \, \psi_0 \, d\mathbf{r}$$
$$= \frac{1}{2} \sum_j e^{i\mathbf{K}\cdot\mathbf{R}_j} \int_{j\text{th cell}} \psi_m^* \, \mathbf{A} \cdot \nabla V_0(\mathbf{r} - \mathbf{R}_j) \, \psi_0 \, d\mathbf{r} \quad (8.38)$$

3. *Choose Bloch functions.* For the initial and final states we assume Bloch wave functions,

$$\psi_{\mathbf{k}_0} = N^{-1/2} \, u_{\mathbf{k}_0}(\mathbf{r}) \, e^{i\mathbf{k}_0 \cdot \mathbf{r}} \quad (8.39)$$
$$\psi_{\mathbf{k}_m} = N^{-1/2} \, u_{\mathbf{k}_m}(\mathbf{r}) \, e^{i\mathbf{k}_m \cdot \mathbf{r}} \quad (8.40)$$

where N is the number of unit cells per crystal, included so that $u_{\mathbf{k}_0}(\mathbf{r})$ and $u_{\mathbf{k}_m}(\mathbf{r})$ are normalized for a unit cell. Substitution of these wave-

8.2 Perturbation Calculation

functions into Eq. (8.38) gives, for the matrix element,

$$H_{m0} = \frac{1}{2N} \sum_j e^{i\mathbf{K}\cdot\mathbf{R}_j}$$
$$\times \int_{j\text{th cell}} e^{i(\mathbf{k}_0-\mathbf{k}_m)\cdot\mathbf{r}} u^*_{\mathbf{k}_m}(\mathbf{r}) u_{\mathbf{k}_0}(\mathbf{r}) \mathbf{A}\cdot\nabla V_0(\mathbf{r}-\mathbf{R}_j)\, d\mathbf{r} \quad (8.41)$$

4. *Recognize conditions for* $H_{m0} \neq 0$. In order to recognize what is required so that H_{m0} is not zero, make a transformation of coordinates from $\mathbf{r}$ to $(\mathbf{r}+\mathbf{R}_j)$. This transformation gives

$$H_{m0} = \frac{1}{2N} \sum_j e^{i(\mathbf{k}_0+\mathbf{K}-\mathbf{k}_m)\cdot\mathbf{R}_j}$$
$$\times \int_{j\text{th cell}} e^{i(\mathbf{k}_0-\mathbf{k}_m)\cdot\mathbf{r}} u^*_{\mathbf{k}_m}(\mathbf{r}) u_{\mathbf{k}_0}(\mathbf{r}) \mathbf{A}\cdot\nabla V_0(\mathbf{r})\, d\mathbf{r} \quad (8.42)$$

since the functions $u^*_{\mathbf{k}_m}(\mathbf{r})$ and $u_{\mathbf{k}_0}(\mathbf{r})$ are unchanged because of their periodicity.

The sum in Eq. (8.42) is of the same form as that met previously in treating one- and three-dimensional band gaps in terms of first-order perturbation theory (e.g., see Eq. (5.66) ff.). All of the summation takes place over phase factors outside of the integral over a unit cell. The value of the sum is given by

$$\sum_j e^{i(\mathbf{k}_0+\mathbf{K}-\mathbf{k}_m)\cdot\mathbf{R}_j} = 0 \quad \text{if} \quad (\mathbf{k}_0+\mathbf{K}-\mathbf{k}_m)\cdot\mathbf{R}_j \neq 0$$
$$= N \quad \text{if} \quad (\mathbf{k}_0+\mathbf{K}-\mathbf{k}_m)\cdot\mathbf{R}_j = 0 \quad (8.43)$$

Equation (8.43) for the nonzero value of the sum is satisfied if

$$\mathbf{k}_m - \mathbf{k}_0 = \mathbf{K} \quad (8.44)$$

or if

$$(\mathbf{k}_0 + \mathbf{K} - \mathbf{k}_m) = \mathbf{b}_j \quad (8.45)$$

where $\mathbf{b}_j$ is a vector of the reciprocal lattice, since $\mathbf{b}_j \cdot \mathbf{R}_j = 2n\pi$. The relationship of Eq. (8.44) represents what may be called *normal* scattering processes. In semiconductors, where only small values of $\mathbf{k}_0$ and $\mathbf{k}_m$ are involved, only normal scattering processes are present. The relationship of Eq. (8.45) represents what have been called *Umklapp* scattering processes; such a process can be considered as the result of absorption or emission of a phonon with simultaneous Bragg reflection of the carrier. Normal and Umklapp processes are indicated schematically in Fig. 8.5. In metals

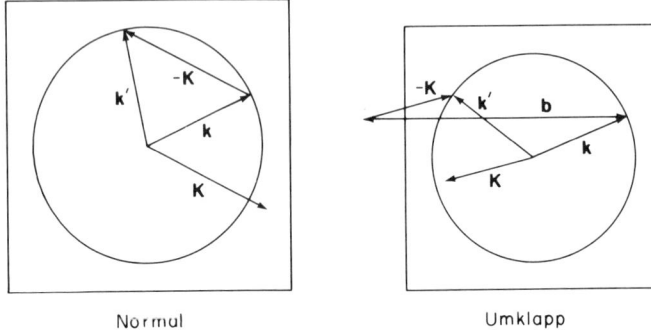

Fig. 8.5 Normal and Umklapp scattering effects on electron wave vector, shown in the reduced-zone representation for a simple cubic crystal. In the normal scattering process (shown as an elastic process), the electron wave vector is changed from **k** to **k'** by emission of a phonon with wave vector **K**. In the Umklapp scattering process, the electron wave vector is changed from **k** to **k'** by simultaneous emission of a phonon with wave vector **K** and translation by a reciprocal lattice vector **b**. Similar processes with phonon absorption can also occur.

both normal and Umklapp scattering processes must be considered. In our previous discussion of thermoelectric power and the phenomenon of phonon drag (see Section 7.9), we did not include the possibility of both types of scattering process; indeed it can be shown that Umklapp processes can lead to a phonon drag of opposite sign to that of normal processes.

5. *Recognize conservation of energy.* From Eq. (8.33) we saw earlier that a large value for the transition probability requires conservation of energy in the scattering process,

$$E(\mathbf{k}_m) - E(\mathbf{k}_0) = \hbar\bar{\omega} \qquad (8.46)$$

If instead of the transition probability given in Eq. (8.33) for transitions between two discrete energy states, we instead consider the integrated transition probability of a transition from a given initial state to one of a group of final possible states, with density $N(E_m)\,dE_m$ for states with energy between E_m and $E_m + dE_m$, we find that the transition probability increases linearly with time if Eq. (8.46) is satisfied. The probability that an electron initially in the state with energy E_0 and wave vector $\mathbf{k}_0$ will make a transition to one of the group of available final states in the time t is given by

$$P = \int |A_m(t)|^2 \, N(E_m) \, dE_m \, dS \qquad (8.47)$$

8.2 Perturbation Calculation

where $dE_m \, dS$ is the volume element in **k** space corresponding to $k_m^2 \, dk_m \sin\theta \, d\theta \, d\phi$. Substituting from Eq. (8.33) and recognizing that the major contribution requires the condition of Eq. (8.46), we obtain

$$dP \approx 4 \mid H_{m0} \mid^2 N(E_0) \int \frac{\sin^2(E_m - E_0 - \hbar\bar{\omega})t/2\hbar}{(E_m - E_0 - \hbar\bar{\omega})^2} d(E_m - E_0 - \hbar\bar{\omega})$$

$$= \frac{2\pi}{\hbar} \mid H_{m0} \mid^2 N(E_0) \, t \qquad (8.48)$$

per unit area of the surface in **k** space. Equation (8.48) assumes a small change in energy upon scattering and the permissibility of taking the limits of the integral from $-\infty$ to $+\infty$ because of the narrow range of energies making a significant contribution.

6. *Explicit calculation of the matrix element.* In view of the previous discussion, Eq. (8.42) for H_{m0} becomes

$$H_{m0} = \frac{1}{2} \int_{\text{jth cell}} e^{i(\mathbf{k}_0 - \mathbf{k}_m)\cdot\mathbf{r}} u_{\mathbf{k}_m}^*(\mathbf{r}) u_{\mathbf{k}_0}(\mathbf{r}) \, \mathbf{A} \cdot \boldsymbol{\nabla} V_0(\mathbf{r}) \, d\mathbf{r} \qquad (8.49)$$

This equation represents the matrix element derived for the first term of the perturbation potential of Eq. (8.37), corresponding to phonon absorption. A similar matrix element can be derived from the second term of Eq. (8.37) corresponding to phonon emission. The corresponding conditions to Eqs. (8.44) and (8.46) for phonon emission are

$$\mathbf{k}_m - \mathbf{k}_0 = -\mathbf{K} \qquad (8.50)$$

$$E(\mathbf{k}_m) - E(\mathbf{k}_0) = -\hbar\bar{\omega} \qquad (8.51)$$

The evaluation of H_{m0} from Eq. (8.49) still represents a formidable problem, especially since the quantities $u_{\mathbf{k}_m}(\mathbf{r})$, $u_{\mathbf{k}_0}(\mathbf{r})$ and $V_0(\mathbf{r})$ are not generally known quantities, even in approximate form. One approximate method of overcoming these difficulties is to replace the Wigner–Seitz cell by an equivalent sphere, using an approximation similar to that described in Section 5.2. The radius of the equivalent sphere r_s is chosen according to Eq. (5.14) and the pertinent boundary condition is given by Eq. (5.13).

A second step to remove the difficulties in calculating H_{m0} is to relate $V_0(\mathbf{r})$ to the wavefunctions of the initial and final states through the Schroedinger equation relationships, and hence to remove the explicit dependence of H_{m0} on the unknown $V_0(\mathbf{r})$. Consider first the Schroedinger equation,

$$\frac{\hbar^2}{2m} \nabla^2 \psi_{\mathbf{k}_0} - (V_0 - E_{\mathbf{k}_0})\psi_{\mathbf{k}_0} = 0 \qquad (8.52)$$

Multiply Eq. (8.52) by $\psi_{\mathbf{k}_m}^*$ and then operate on all terms with ∇.

$$\frac{\hbar^2}{2m}\nabla\psi_{\mathbf{k}_m}^*\nabla^2\psi_{\mathbf{k}_0} + \frac{\hbar^2}{2m}\psi_{\mathbf{k}_m}^*\nabla^2\nabla\psi_{\mathbf{k}_0} - V_0\psi_{\mathbf{k}_m}^*\nabla\psi_{\mathbf{k}_0} - V_0\nabla\psi_{\mathbf{k}_m}^*\psi_{\mathbf{k}_0}$$
$$- \psi_{\mathbf{k}_m}^*\psi_{\mathbf{k}_0}\nabla V_0 + E_{\mathbf{k}_0}\psi_{\mathbf{k}_0}\nabla\psi_{\mathbf{k}_m}^* + E_{\mathbf{k}_0}\psi_{\mathbf{k}_m}^*\nabla\psi_{\mathbf{k}_0} = 0 \qquad (8.53)$$

Also simply multiply Eq. (8.52) by $\nabla q_{\mathbf{k}_m}^*$:

$$E_{\mathbf{k}_0}\psi_{\mathbf{k}_0}\nabla\psi_{\mathbf{k}_m}^* = V_0\psi_{\mathbf{k}_0}\nabla\psi_{\mathbf{k}_m}^* - \frac{\hbar^2}{2m}\nabla\psi_{\mathbf{k}_m}^*\nabla^2\psi_{\mathbf{k}_0} \qquad (8.54)$$

Similarly if we multiply the Schroedinger equation like Eq. (8.52) but written for $\psi_{\mathbf{k}_m}^*$ with eigenvalue $E_{\mathbf{k}_m}$, by $\nabla\psi_{\mathbf{k}_0}$, we obtain

$$E_{\mathbf{k}_m}\psi_{\mathbf{k}_m}^*\nabla\psi_{\mathbf{k}_0} = V_0\psi_{\mathbf{k}_m}^*\nabla\psi_{\mathbf{k}_0} - \frac{\hbar^2}{2m}\nabla\psi_{\mathbf{k}_0}\nabla^2\psi_{\mathbf{k}_m}^* \qquad (8.55)$$

If we combine Eqs. (8.53), (8.54), and (8.55), and if we assume that the energy difference between the initial and final states is sufficiently small that we may write $E_{\mathbf{k}_0} \approx E_{\mathbf{k}_m} \approx E_{\mathbf{k}}$, the following expression for V_0 results:

$$\nabla V_0 = \frac{\hbar^2}{2m\psi_{\mathbf{k}_0}\psi_{\mathbf{k}_m}^*}(\psi_{\mathbf{k}_m}^*\nabla^2\nabla\psi_{\mathbf{k}_0} - \nabla\psi_{\mathbf{k}_0}\nabla^2\psi_{\mathbf{k}_m}^*) \qquad (8.56)$$

This expression for ∇V_0 can now be used in the calculation of H_{m0} according to Eq. (8.49).

For the appropriate Wigner–Seitz function, $\psi_{\mathbf{k}}(\mathbf{r}) = u(\mathbf{r})\,e^{i\mathbf{k}\cdot\mathbf{r}}$, the value of ∇V_0 given by Eq. (8.56) would be exactly zero if the $u(\mathbf{r})$ functions were really constant over the whole cell, and if $\mathbf{k}_0 = \mathbf{k}_m$ as would be the case for spherical equal-energy surfaces in the equivalent-sphere approximation to the Wigner–Seitz cell. The major contribution to ∇V_0 and hence to H_{m0} arises, therefore, from the variable contribution of $u(\mathbf{r})$.

Substituting Eq. (8.56) into Eq. (8.49) gives

$$H_{m0} = \frac{\hbar^2}{4m}\int_{\text{cell}} \mathbf{A}\cdot(\psi_{\mathbf{k}_m}^*\nabla^2\nabla\psi_{\mathbf{k}_0} - \nabla\psi_{\mathbf{k}_0}\nabla^2\psi_{\mathbf{k}_m}^*)\,d\mathbf{r} \qquad (8.57)$$

in view of Eqs. (8.39) and (8.40). This integral may be converted from an integral over the volume of the cell to an integral over the surface of the cell by Green's theorem:

$$H_{m0} = \frac{\hbar^2}{4m}\int_{\text{cell surface}}\left(\psi_{\mathbf{k}_m}^*\frac{\partial}{\partial r}\mathbf{A}\cdot\nabla\psi_{\mathbf{k}_0} - \mathbf{A}\cdot\nabla\psi_{\mathbf{k}_0}\frac{\partial}{\partial r}\psi_{\mathbf{k}_m}^*\right)dS \qquad (8.58)$$

8.2 Perturbation Calculation

where the operator $\partial/\partial r$ indicates the derivative in the direction normal to the surface, i.e., the direction of the radius **r** in the spherical approximation, the integration being over the surface of the equivalent sphere with radius r_s.

As a result of the periodic boundary condition imposed on the Wigner–Seitz cell, the first derivative $\partial u/\partial r = 0$ at the surface. The only appreciable contribution to the integral of Eq. (8.58) comes from those terms involving the second derivative $\partial^2 u/\partial r^2$; there is only one such term in the integrand of Eq. (8.58). We then have for H_{m0}

$$H_{m0} = \frac{\hbar^2}{4m} \int_{\text{cell surface}} u(r_s)\, e^{-i\mathbf{k}_m \cdot \mathbf{r}}\, \frac{\partial^2 u(r_s)}{\partial r^2}\, e^{i\mathbf{k}_0 \cdot \mathbf{r}}\, \mathbf{A} \cdot \left(\frac{\mathbf{r}_X}{r_s}\right) dS \quad (8.59)$$

where $(\mathbf{r}_X/r_s)$ is a unit vector in the displacement direction given by $\mathbf{X}_j$, and where we have retained only that single term involving the second derivative.

According to the approximation of the Wigner–Seitz approach, if E_0 is the energy for $\mathbf{k} = 0$,

$$\nabla^2 u(\mathbf{r}) + \frac{2m}{\hbar^2} [E_0 - V_0(\mathbf{r})]\, u(\mathbf{r}) = 0 \quad (8.60)$$

or if E_0 is taken equal to zero, and for $r = r_s$,

$$\frac{\partial^2 u(r_s)}{\partial r^2} = \frac{2m}{\hbar^2} V_0(r_s)\, u(r_s) \quad (8.61)$$

since

$$\nabla^2 u(\mathbf{r}) = \frac{1}{r^2} \frac{\partial}{\partial r}\left[r^2 \frac{\partial}{\partial r} u(\mathbf{r})\right]$$

and the term involving the first derivative vanishes at $r = r_s$. Substitution of Eq. (8.61) into Eq. (8.59) gives for H_{m0}

$$H_{m0} = \frac{V_0(r_s)\, u(r_s)^2}{2r_s} \int_{\text{cell surface}} (\mathbf{A} \cdot \mathbf{r}_X)\, e^{i(\mathbf{k}_0 - \mathbf{k}_m) \cdot \mathbf{r}}\, dS \quad (8.62)$$

Finally we may take $u(r_s) = (N/\mathscr{V})^{1/2}$ so that $\psi_{\mathbf{k}_0}$ and $\psi_{\mathbf{k}_m}$ are normalized, $\mathscr{V}$ being the volume of the crystal:

$$H_{m0} = \frac{V_0(r_s)\, N}{2r_s \mathscr{V}} \int_{\text{cell surface}} (\mathbf{A} \cdot \mathbf{r}_X)\, e^{i(\mathbf{k}_0 - \mathbf{k}_m) \cdot \mathbf{r}}\, dS \quad (8.63)$$

To evaluate this integral it is convenient to transform to a polar coordinate system with the direction of $(\mathbf{k}_0 - \mathbf{k}_m) = \mathbf{K}$ (for normal scattering).

The angle between **r** and **K** is taken to be θ, and the angle between **A** and **K** is γ. Then

$$dS = 2\pi r_s^2 \int_0^\pi \sin\theta \, d\theta$$

The component along the polar axis is

$$H_{m0} = \frac{V_0(r_s) \, N 2\pi r_s^2}{2 r_s \mathscr{V}} \int_0^\pi A \cos\gamma \, r_s \cos\theta \, e^{i|\mathbf{k}_0 - \mathbf{k}_m| r_s \cos\theta} \sin\theta \, d\theta$$

$$= \frac{\pi V_0 N r_s^2 A}{\mathscr{V}} \int_0^\pi \cos\gamma \cos\theta \, e^{i|\mathbf{k}_0 - \mathbf{k}_m| r_s \cos\theta} \sin\theta \, d\theta \quad (8.64)$$

For a transverse acoustic wave, $\cos\gamma = 0$, whereas for a longitudinal acoustic wave, $\cos\gamma = 1$. We conclude that, for the case of normal scattering and spherical equal-energy surfaces being considered here, there is no scattering by transverse acoustic modes but only by longitudinal acoustic modes. For Umklapp scattering, $(\mathbf{k}_0 - \mathbf{k}_m)$ need not have the same direction as **K**, and scattering by both longitudinal and transverse acoustic modes is possible.

The integral of Eq. (8.64) can be evaluated easily for a nondegenerate semiconductor where the quantity $|\mathbf{k}_0 - \mathbf{k}_m| r_s$ is small. Then the exponential in the integrand can be expanded and the integral becomes approximately

$$\int_0^\pi \cos\theta \, [1 + i |\mathbf{k}_0 - \mathbf{k}_m| r_s \cos\theta] \sin\theta \, d\theta = \frac{2i |\mathbf{k}_0 - \mathbf{k}_m| r_s}{3} \quad (8.65)$$

Using Eq. (5.14) to eliminate r_s,

$$H_{m0} = \frac{1}{2} i \, V_0(r_s) \, KA \quad (8.66)$$

The matrix element for scattering by longitudinal acoustic waves in a nondegenerate semiconductor with spherical equal-energy surfaces depends on three basic quantities: the crystal potential evaluated at the surface of the Wigner–Seitz cell, the wave vector of the vibrational wave, and the amplitude of the vibrational wave.

For the case of a metal, even normal scattering requires a more complete evaluation of Eq. (8.64) since $|\mathbf{k}_0 - \mathbf{k}_m| r_s$ is in general no longer small. The integral of Eq. (8.64) can be calculated exactly with the result

$$H_{m0} = \frac{2\pi \, V_0(r_s) \, NA [\sin K r_s - K r_s \cos K r_s]}{i \mathscr{V} K^2} \quad (8.67)$$

for the condition $\mathbf{k}_0 - \mathbf{k}_m = \mathbf{K}$ for normal scattering.

8.2 Perturbation Calculation

7. *Calculate the scattering probability.* Because of its simpler mathematical form, we develop the rest of this discussion in terms of the matrix element for scattering associated with longitudinal acoustic modes in a nondegenerate semiconductor, given in Eq. (8.66).

If the change in energy upon scattering is small, the probability of a scattering transition can be calculated from Eq. (8.48):

$$dP = \frac{4\pi}{\hbar} N(E) \mid H_{m0} \mid^2 dS$$

where dP is the probability of a transition *per unit time* to a final state such that $\mathbf{k}_m$ ends in a surface element dS. The additional factor of 2, as compared with Eq. (8.48), is included to take account of the possibility of both phonon absorption and emission in the scattering process.

To be specific about the surface element, suppose that dS corresponds to a solid angle $\sin\theta\, d\theta\, d\phi$. Then the density of states may be introduced explicitly by recognizing that

$$N(E)\, dE\, dS = \frac{\mathscr{V}}{8\pi^3} k_m^2\, dk_m \sin\theta\, d\theta\, d\phi$$

$$= \frac{\mathscr{V} k_m}{8\pi^3} \frac{m^*}{\hbar^2} \sin\theta\, d\theta\, d\phi\, dE \qquad (8.68)$$

since $dk_m = (m^*/\hbar^2 k_m)\, dE$. Then, since $k_m \approx k_0$,

$$dP = \frac{k_0 m^* \mathscr{V}}{2\pi^2 \hbar^3} \mid H_{m0} \mid^2 \sin\theta\, d\theta\, d\phi \qquad (8.69)$$

Inserting H_{m0} from Eq. (8.66) gives

$$dP = \frac{\mathscr{V} k_0 m^* K^2 V_0^2 A^2}{8\pi^2 \hbar^3} \sin\theta\, d\theta\, d\phi \qquad (8.70)$$

The amplitude of the vibrations A can be related to the temperature of the crystal. The mean kinetic energy of a vibrational mode is equal to $kT/2$ when kT is much larger than the phonon energy characteristic of that mode. The mean kinetic energy of a single atom is $M\langle(d\mathbf{X}_j/dt)^2\rangle_{\text{av}}/2$, and

$$\left\langle \left(\frac{d\mathbf{X}_j}{dt}\right)^2 \right\rangle_{\text{av}} = \frac{1}{2} A^2 \bar\omega^2 \qquad (8.71)$$

from Eq. (8.34). The amplitude of the vibration is therefore

$$A^2 = \frac{2kT}{N'M\bar\omega^2} \qquad (8.72)$$

where N' is the number of atoms. The transition probability of Eq. (8.70) becomes

$$dP = \frac{\mathscr{V} k_0(kT)m^* V_0^2}{4\pi^2 \hbar^3 \bar{v}^2 N' M} \sin \theta \, d\theta \, d\phi \tag{8.73}$$

where we have used $\bar{\omega} = K\bar{v}$, so that $\bar{v}$ is the velocity of sound in the material.

The expression of Eq. (8.73) expresses the *total scattering probability* for all the scattering centers (atoms) in the crystal.

8. *Relating scattering probability to relaxation time.* A discussion o- scattering is often facilitated by the introduction of the concept of a *scatterf ing cross section*. Such a cross section has the dimensions of area and describes in a geometrical way how effective the scattering center is. The probability of scattering per second is defined as

$$\text{Probability of scattering} = \left(\frac{N'}{\mathscr{V}}\right) S^* v \tag{8.74}$$

where $(N'/\mathscr{V})$ is the number of atoms per unit volume, hence also the number of scattering centers for lattice scattering per unit volume; S^* is an appropriate cross section per scattering center; and v is the velocity of a charge carrier. The product (S^*v) represents the volume swept out in relative motion between a scattering center with cross section S^* and a charge carrier moving with velocity v with respect to it. The assumption is that scattering occurs whenever a carrier falls within this volume.

In order to describe the angular dependence of the cross section, define $S(\theta, \phi)$ as the cross section for a collision that takes an electron from motion in the direction described by $\theta = 0$ to motion in the direction (θ, ϕ) in the solid angle $\sin \theta \, d\theta \, d\phi$. A total cross section S_t can then be defined as

$$S_t = \iint S(\theta, \phi) \sin \theta \, d\theta \, d\phi \tag{8.75}$$

The question arises whether such a total cross section is the relevant quantity in describing scattering effects related to electrical conductivity.

Consider a charge carrier moving with velocity v_{z0} in the z direction ($\theta = 0$), which is scattered in the direction (θ, ϕ). The effect on the transport properties depends on how the velocity of the carrier in the z direction is affected by the scattering process. The change of velocity of the carrier in the z direction is given by

$$dv_z = v_{z0}(1 - \cos \theta) \tag{8.76}$$

8.2 Perturbation Calculation

If $S(\theta, \phi)$ is a function only of θ, the average change in v_z per scattering collision is

$$\langle dv_z \rangle_{av} = v_{z0} \frac{\int_0^\pi (1 - \cos \theta) S(\theta) \sin \theta \, d\theta}{\int_0^\pi S(\theta) \sin \theta \, d\theta} \tag{8.77}$$

If we now choose to define another cross section S_c by the relation

$$S_c \equiv 2\pi \int_0^\pi (1 - \cos \theta) S(\theta) \sin \theta \, d\theta \tag{8.78}$$

the average change in v_z per scattering collision is given by

$$(dv_z)_{av} = v_{z0} \left(\frac{S_c}{S_t} \right) \tag{8.79}$$

In other words, (S_t/S_c) scattering collisions are required to completely destroy the original component of electron velocity in the z direction. Thus the probability of a scattering collision in the time dt is given by $(N'/\mathscr{V})S_t v \, dt$, whereas the probability of a velocity-destroying scattering collision in the time dt is given by $(N'/\mathscr{V})S_c v \, dt$. The cross section S_c is the significant cross section for problems of electrical transport.

In terms of the total scattering probability of Eq. (8.73), the cross section $S_a(\theta, \phi)$ for scattering into the direction (θ, ϕ) per unit solid angle *per atom* of the crystal is defined after Eq. (8.74) as

$$\frac{N'}{\mathscr{V}} [S_a(\theta, \phi) \sin \theta \, d\theta \, d\phi] v = dP \tag{8.80}$$

The resulting value for $S_a(\theta, \phi)$ is

$$S_a = \frac{\Omega(kT)m^{*2}V_0^2}{4\pi^2 \hbar^4 \varrho \bar{v}^2} \tag{8.81}$$

in terms of the following quantites:

$$\text{density } \varrho = \frac{N'M}{\mathscr{V}}$$

$$\text{unit cell volume } \Omega = \frac{\mathscr{V}}{N'}$$

$$\text{crystal momentum } \hbar k_0 = m^* v$$

We conclude that S_a is not a function of θ, i.e., the elastic lattice scattering described here is isotropic.

Example 8.2 Demonstrate that S_a in Eq. (8.81) does indeed have the units of a cross section.

A dimensional analysis gives

$$S_a = \frac{cm^3 \, erg \, g^2 \, erg^2}{erg^4 \, sec^4 \, g \, cm^{-3} \, cm^2 \, sec^{-1}} = \frac{cm^4 \, g}{erg \, sec^2} = \frac{cm^4 \, g}{g \, cm^2 \, sec^{-2} \, sec^2}$$

$$= cm^2$$

The relaxation time appropriate for scattering processes affecting transport properties is

$$\frac{1}{\tau} = \frac{N'}{\mathscr{V}} S_c v$$

$$= \frac{2\pi}{\Omega} \int_0^\pi v S_a (1 - \cos\theta) \sin\theta \, d\theta$$

$$= \frac{4\pi v}{\Omega} S_a$$

$$= \frac{(kT) m^{*2} V_0^2 v}{\pi \hbar^4 \rho \bar{v}^2} \tag{8.82}$$

This gives finally, for the mean free path,

$$l = \tau v = \frac{\pi \hbar^4 \rho \bar{v}^2}{m^{*2} V_0^2 (kT)} \tag{8.83}$$

This result may be compared with Eq. (8.8) where the mean free path was derived from a semiqualitative application of the deformation-potential approach. The problem can also be treated quantum mechanically in detail using the deformation potential with results quite similar to those above.

Example 8.3 The room-temperature mobility for electrons in Ge is about 3900 cm²/V-sec. If this mobility is limited by scattering by longitudinal acoustic lattice waves, determine the value of V_0 applicable to Ge.

The procedure is as follows: from the mobility calculate the relaxation time for scattering; from the relaxation time calculate the atomic scattering cross section; from the cross section calculate the value of V_0.

From Eq. (7.107) and the data of Example 8.1, determine that the conductivity effective mass for Ge is $0.12m$.

From the basic relationship between mobility and relaxation time,

8.3 Charged-Imperfection Scattering

such as that given in Eq. (7.108), calculate the relaxation time,

$$3.9 \times 10^3 = \frac{1}{300} \frac{4.8 \times 10^{-10}}{(0.12)(9 \times 10^{-28})} \tau$$

giving $\tau = 2.6 \times 10^{-13}$ sec.

The relaxation time and the cross section are related by Eq. (8.82), which contains Ω. Since the density of Ge is 5.35 g/cm³ and the atomic weight is 72.6 g, we conclude that $\Omega = 2.3 \times 10^{-23}$ cm³,

$$\Omega = \frac{72.6}{(5.35)(6 \times 10^{23})} \text{ cm}^3$$

From Eq. (8.82) with $v = 2.8 \times 10^7$ cm/sec for Ge at room temperature, we obtain

$$S_a = 2.5 \times 10^{-19} \text{ cm}^2$$

Finally we can solve Eq. (8.81) for V_0, using $\bar{v} = 3 \times 10^5$ cm/sec, and we obtain

$$V_0 = 14 \text{ eV}$$

Values of V_0 generally lie in the range of several electron volts.

8.3 Charged-Imperfection Scattering

If imperfections are present in a crystal (see Chapter 9 for more detail), such as impurity ions, missing ions of the crystal, or displaced ions of the crystal, as well as more complex combinations of these, local regions of excess charge may be set up. A neutral imperfection at low temperatures, for example, may become positively charged at temperatures sufficiently high to excite an electron from the imperfection into the conduction band of the crystal. Because of the positive charge on the imperfection, there will be a Coulombic interaction between this imperfection and charge carriers drifting through the crystal under the action of an applied electric field. This Coulombic interaction causes scattering of the charge carriers.

Approximations to the scattering cross section of a charged imperfection can be made on a very elementary level, which differ surprisingly little in qualitative details from much more sophisticated calculations. Each approach depends to a considerable extent on the assumption made as to where the effects of the Coulomb field should be effectively terminated in the calculation. This is equivalent to asking the question, "How far must a carrier pass from a charged center before we consider it, for the sake of the calculation, not to be scattered?"

CUT-OFF COULOMB POTENTIAL WHERE COULOMB ENERGY EQUALS THERMAL ENERGY

The most elementary approach makes the ad hoc assumption that an electron is scattered if the Coulomb interaction energy is as large as the thermal energy of the electron. If the charge on the scattering center is Ze, this criterion results in

$$\frac{Ze^2}{\varepsilon r_0} = kT \tag{8.84}$$

If we define the scattering cross section $S_0 \equiv \pi r_0^2$,

$$S_0 = \frac{\pi Z^2 e^4}{\varepsilon^2 (kT)^2} = \frac{10^{-10} Z^2}{\varepsilon^2} \left(\frac{300}{T}\right)^2 \text{ cm}^2 \tag{8.85}$$

For a dielectric constant $\varepsilon = 10$, and for $Z = 1$, $S_0 = 10^{-12}$ cm² at 300°K.

CUT-OFF POTENTIAL AT HALF THE DISTANCE BETWEEN IMPERFECTIONS

The next degree of approximation involves a consideration of the actual scattering mechanism: a change in the direction of carrier motion because of the Coulomb interaction. This calculation can be made in a classical formulation by treating the scattering of electrons and holes by charged imperfections, using the wave-packet construct for the free carriers.

The approach we discuss here is known as the *Conwell–Weisskopf* calculation. The basic assumption is that the Coulomb field is cut off at half the distance between charged imperfections. If there are N^+ charged imperfections per unit volume distributed in a crystal with cubic symmetry, the average separation between imperfections is

$$a_\text{I} = (N^+)^{-1/3} \tag{8.86}$$

This assumption is equivalent to proceeding as if a charge carrier sees only one charged imperfection at a time, the effect of other charged imperfections being sufficiently screened as to be negligible. Therefore each passaga of a carrier through a unit cell of the imperfection lattice corresponds teo single scattering event,

$$\tau = \frac{a_\text{I}}{v} \tag{8.87}$$

8.3 Charged-Imperfection Scattering

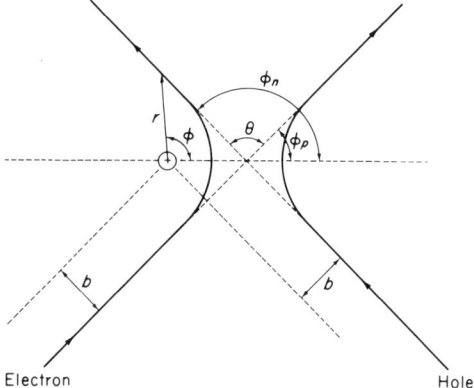

Fig. 8.6 Typical scattering trajectories for an electron and a hole in interaction with a positively charged imperfection.

and it may be assumed that the angle of deflection θ (see Fig. 8.6) is the same for the single scattering event as it would be if the carrier were traveling on an ideal infinite hyperbolic path. This last assumption overestimates the scattering and hence compensates for the underestimation of scattering resulting from the assumption of scattering by only one imperfection at a time.

Typical trajectories for electrons and holes in the presence of a positively charged imperfection are shown in Fig. 8.6. The equation of an orbit in a central force field where the forces varies inversely as r^2 is known from classical mechanics, and the appropriate equation for an electrostatic field is given in terms of the radius r and the angle ϕ by

$$\frac{1}{r} = \pm \frac{e^2 m}{\varepsilon L^2} + A \cos \phi \tag{8.88}$$

where the plus sign holds for attractive forces (electrons in the present case of a positive imperfection), and the minus sign holds for repulsive forces. The constant A is a collection of several terms independent of r or θ, and L is the angular momentum,

$$L = mr^2 \frac{d\phi}{dt} = \mathbf{r} \times \mathbf{p} = xp_y - yp_x = mbv \tag{8.89}$$

Since momentum is conserved, the asymptotic value of $L = mbv$ is preserved throughout the whole trajectory.

The problem is to evaluate the angle θ as a function of the impact parameter b so that we can calculate the probability of scattering through an angle θ in terms of the relative separation of carrier path and scattering center before scattering. This is exactly the calculation carried out by Rutherford to describe the scattering of alpha particles by atomic nuclei. The result is well known and gives, for the scattering process,

$$S_a(\theta) = \left(\frac{e^2}{2\varepsilon mv^2}\right)^2 \operatorname{cosec}^4(\theta/2) \tag{8.90}$$

A brief derivation of this result may be given as follows. Inspection of Fig. 8.6 shows that $\theta = \phi_n - \phi_p$ and that $\phi_n = \pi - \phi_p$, so that

$$\theta = 2\phi_n - \pi = \pi - 2\phi_p \tag{8.91}$$

If r goes to infinity, Eq. (8.88) becomes for electrons

$$0 = \frac{e^2 m}{\varepsilon L^2} + A \cos \phi_n \tag{8.92}$$

Subtracting Eq. (8.92) from Eq. (8.88) gives

$$\frac{1}{r} = A(\cos \phi - \cos \phi_n)$$
$$= -2A \sin\left(\frac{\phi + \phi_n}{2}\right) \sin\left(\frac{\phi - \phi_n}{2}\right) \tag{8.93}$$

For all but small r, $\phi \approx \phi_n$, and Eq. (8.93) can be approximated by

$$\frac{1}{r} = -2A \sin \phi_n \left(\frac{\phi - \phi_n}{2}\right) = -(\phi - \phi_n) A \sin \phi_n \tag{8.94}$$

Also for large values of r, where $(\phi_n - \phi)$ is a small angle,

$$(\phi_n - \phi)r = b \tag{8.95}$$

We can solve now for b in terms of A and ϕ_n from Eqs. (8.94) and (8.95), and then eliminate A by using Eq. (8.92), to obtain

$$\cot \phi_n = -\frac{e^2}{\varepsilon mv^2 b} \tag{8.96}$$

Recalling the relationship between θ and ϕ_n given in Eq. (8.91) permits us to write

$$\tan \frac{\theta}{2} = \tan\left(\phi_n - \frac{\pi}{2}\right) = -\cot \phi_n = \frac{e^2}{\varepsilon mv^2 b} \tag{8.97}$$

8.3 Charged-Imperfection Scattering

This is the desired expression relating the angle of scattering θ and the impact parameter b. Exactly the same expression is obtained if the calculation is carried out for holes instead of electrons.

In order for an electron to be deflected through an angle lying between θ and $\theta + d\theta$ into the solid angle $2\pi \sin \theta \, d\theta$, it must have an impact parameter lying between b and $b + db$, where db can be calculated from Eq. (8.97) to be

$$db = \frac{e^2}{\varepsilon m v^2} d\cot\left(\frac{\theta}{2}\right) = -\frac{e^2}{2\varepsilon m v^2} \csc^2\left(\frac{\theta}{2}\right) d\theta \qquad (8.98)$$

The cross section for scattering into the angle θ per unit solid angle, $S_a(\theta)$, is given by

$$S_a(\theta) = \frac{2\pi b \, db}{2\pi \sin \theta \, d\theta} \qquad (8.99)$$

As illustrated in Fig. 8.7, the numerator of Eq. (8.99) gives the number of electrons passing through an annular ring of radius b and thickness db, if one electron passes through 1 cm² per second of the plane in which the ring lies. Therefore also the number of electrons deviated through an angle lying between θ and $\theta + d\theta$ is given by the numerator. These electrons are uniformly distributed over a zone of a unit sphere with area given by the denominator of Eq. (8.99). Substitution of Eqs. (8.97) and (8.98) into Eq. (8.99) leads directly to the Rutherford scattering relationship given in Eq. (8.90).

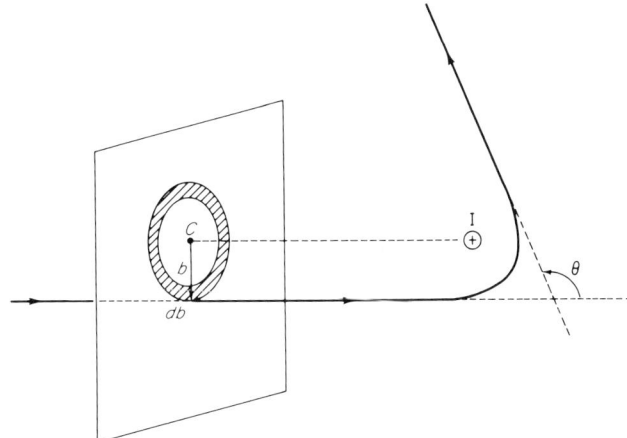

Fig. 8.7 An electron passing through an annular ring of radius b and thickness db of a plane perpendicular to the approaching electron's velocity (line IC is perpendicular to the plane) is scattered through an angle between θ and $\theta + d\theta$.

In contrast to normal scattering by acoustic lattice vibrations where S_a is not a function of θ, for impurity scattering $S_a(\theta)$ is a function of θ. For small deflections θ, $S_a(\theta)$ becomes very large. As θ approaches π, $S_a(\theta)$ approaches $r^{*2}/16$, where r^* is the radius at which the potential energy due to interaction equals the kinetic energy of the electron,

$$\frac{e^2}{\varepsilon r^*} = \frac{1}{2} mv^2 \tag{8.100}$$

giving

$$r^* = \frac{2e^2}{\varepsilon mv^2} \tag{8.101}$$

The scattering cross section relevant for discussions of transport is

$$S_c = 2\pi \int_{\theta_m}^{\pi} (1 - \cos\theta) S_a(\theta) \sin\theta \, d\theta \tag{8.102}$$

It is through the lower limit θ_m that the cut-off distance enters the calculation explicitly; the deflection θ_m corresponds to $b = a_I/2$. Such a cut-off is necessary because $S_a(\theta) \to \infty$ as $\theta \to 0$. It is assumed that if an electron passes an imperfection with $b > a_I/2$, it is not scattered by that particular imperfection. Substituting $b = a_I/2$ into Eq. (8.97) gives

$$\tan\frac{\theta_m}{2} = \frac{2e^2}{\varepsilon mv^2 a_I} \tag{8.103}$$

The integration of Eq. (8.102) for S_c proceeds as follows:

$$S_c = 2\pi \left(\frac{e^2}{2\varepsilon mv^2}\right)^2 \int_{\theta_m}^{\pi} (1 - \cos\theta) \operatorname{cosec}^4\left(\frac{\theta}{2}\right) \sin\theta \, d\theta \tag{8.104}$$

Since $\sin^2(\theta/2) = (1 - \cos\theta)/2$, $\operatorname{cosec}^4(\theta/2) = 4/(1 - \cos\theta)^2$, and defining

$$R \equiv \left(\frac{e^2}{2\varepsilon mv^2}\right)^2 \tag{8.105}$$

we obtain

$$S_c = 8\pi R^2 \int_{\theta_m}^{\pi} \frac{\sin\theta}{1 - \cos\theta} \, d\theta$$

$$= 8\pi R^2 \ln(1 - \cos\theta)\Big|_{\theta_m}^{\pi}$$

$$= -8\pi R^2 \ln\left(\frac{1 - \cos\theta_m}{2}\right)$$

$$= -8\pi R^2 \ln\left[\sin^2\left(\frac{\theta_m}{2}\right)\right] \tag{8.106}$$

8.3 Charged-Imperfection Scattering

If we set $r_m = a_I/2$, the cut-off distance, Eq. (8.103) gives

$$\cot^2\left(\frac{\theta_m}{2}\right) = \frac{\cos^2(\theta_m/2)}{\sin^2(\theta_m/2)} = \left(\frac{r_m}{2R}\right)^2 \tag{8.107}$$

Adding unity to both sides of Eq. (8.179), we obtain

$$\frac{1}{\sin^2(\theta_m/2)} = 1 + \left(\frac{r_m}{2R}\right)^2 \tag{8.108}$$

which in turn converts Eq. (8.106) to

$$S_c = 8\pi R^2 \ln\left[1 + \left(\frac{r_m}{2R}\right)^2\right] = \frac{\pi e^4}{2\varepsilon^2 E^2} \ln\left(1 + \frac{\varepsilon^2 E^2}{e^4 (N^+)^{2/3}}\right) \tag{8.109}$$

if we set $E = mv^2/2$.

The relaxation time may be calculated from the cross section,

$$\frac{1}{\tau} = N^+ v S_c \tag{8.110}$$

giving

$$\frac{1}{\tau} = \frac{\pi e^4 N^+}{2^{1/2} \varepsilon^2 m^{1/2} E^{3/2}} \ln\left(1 + \frac{\varepsilon^2 E^2}{e^4 (N^+)^{2/3}}\right) \tag{8.111}$$

using $v = (2E/m)^{1/2}$. We have arrived at an expression giving the energy dependence of the relaxation time for charged imperfection scattering, namely that

$$\tau(E) \propto E^{3/2} \tag{8.112}$$

The average value of $\tau \approx A' E^{3/2}$ is given by Eq. (7.97). Since the proportionality coefficient A' is essentially a constant, varying inversely only as the logarithm of the temperature, the characteristic temperature dependence for imperfection scattering is

$$\langle \tau \rangle \propto \mu_I \propto T^{+3/2} \tag{8.113}$$

The mobility limited by charged-imperfection scattering increases with temperature, as the increased energy of the carriers makes scattering less effective, whereas the mobility limited by lattice scattering decreases with temperature, as the increased energy of the lattice vibrations makes scattering more effective.

The expression for the average relaxation time for charged-imperfection scattering is frequently written as follows:

$$\langle \tau \rangle = \frac{2^{7/2} \varepsilon^2 (kT)^{3/2} m^{1/2}}{\pi^{3/2} e^4 N^+} \frac{1}{\ln\left[1 + \left(\dfrac{3\varepsilon kT}{e^2 (N^+)^{1/3}}\right)^2\right]} \qquad (8.114)$$

taking the logarithmic term of Eq. (8.111) outside the integral in the calculation of $\langle \tau \rangle$, and replacing E by $3kT$, its value at the maximum of the remainder of the integrand. The Conwell–Weisskopf formulae for mobility and scattering cross section can be obtained from Eq. (8.114) as $\mu_1 = (e/m^*)\langle \tau \rangle$, and $S_{CW} = (v\langle \tau \rangle N^+)^{-1}$. The Conwell–Weisskopf cross section can be written in terms of the simple cross section S_0 of Eq. (8.85) as follows:

$$S_{CW} = S_0 \left\{ \frac{\pi^{1/2}}{16} \ln\left[1 + \left(\frac{3\varepsilon kT}{e^2 (N^+)^{1/3}}\right)^2\right] \right\} \qquad (8.115)$$

A plot of the factor in curly brackets as a function of T for several values of N^+ is given in Fig. 8.8, assuming $\varepsilon = 10$. The correction factor calculated from the Conwell–Weisskopf approach is generally of order unity.

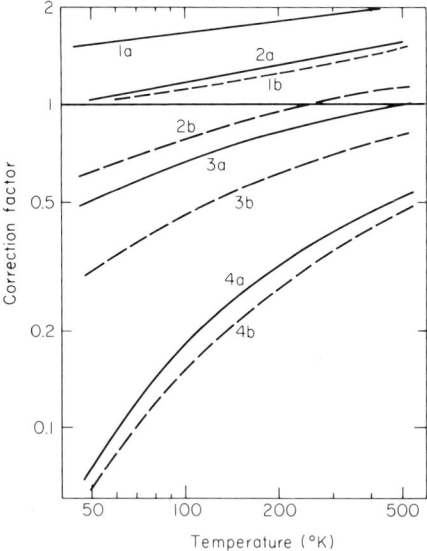

Fig. 8.8 Magnitude of the correction factor in the curly brackets of Eqs. (8.115) and (8.118). Curves marked "a" are for the Brooks–Herring model; curves marked "b" are for the Conwell–Weisskopf model. Different curves are for different ionized imperfection densities (Conwell–Weisskopf) or for different free-electron densities (Brooks–Herring) of (1) 10^{12} cm^{-3}, (2) 10^{14} cm^{-3}, (3) 10^{16} cm^{-3}, and (4) 10^{18} cm^{-3}.

8.3 Charged-Imperfection Scattering

CUT-OFF POTENTIAL ASSOCIATED WITH SCREENING DISTANCE

Finally we consider the results of a third type of calculation of charged-imperfection scattering by Brooks and Herring. In this quantum mechanical approach, the cut-off of the Coulomb potential is associated with a screening distance, the free electrons being assumed to provide screening against the charge of the imperfections. The potential function is chosen to be of the form

$$V(r) = \frac{e}{\varepsilon r} e^{-r/d} \qquad (8.116)$$

where d is the Debye length,

$$d = \left(\frac{\varepsilon kT}{4\pi e^2 n} \right)^{1/2} \qquad (8.117)$$

Thus the scattering cross section, which depended on the density of charged imperfections N^+ in the Conwell–Weisskopf calculation because of the cut-off of the Coulomb potential at $\frac{1}{2}(N^+)^{-1/3}$, depends in the Brooks–Herring calculation on the free carrier density n because of the cut-off of the Coulomb potential related to the Debye length.

We do not take the space here to develop this quantum mechanical calculation in detail, but cite the result for comparison with the previous approximations. The Brooks–Herring cross section for a semiconductor is given by

$$S_{\text{BH}} = S_0 \left\{ \frac{\pi^{1/2}}{16} \left[\ln(1+b) - \frac{b}{1+b} \right] \right\} \qquad (8.118)$$

where

$$b = \frac{6\varepsilon m^*(kT)^2}{\pi \hbar^2 e^2 n} \qquad (8.119)$$

A plot of the factor in curly brackets in Eq. (8.118), a correction factor for the simple S_0 cross section of Eq. (8.85), is given in Fig. 8.8, assuming $\varepsilon = 10$ and $m^* = m$. As in the case of the Conwell–Weisskopf calculation, the correction factor that must be applied to S_0 to make it equal to the Brooks–Herring cross section is of the order of unity over a wide range of temperatures and free-carrier densities.

Example 8.4 For a particular semiconductor crystal at 300°K, the relaxation time for scattering by lattice waves is 10^{-12} sec, and the relaxation time for scattering by charged imperfections is 10^{-13} sec. At what temperature is the maximum mobility found? Effective mass of the charge carriers is $0.5m$.

The total mobility is given approximately by Eq. (7.100), which we can write as

$$\frac{1}{\mu} = \frac{1}{AT^{-3/2}} + \frac{1}{BT^{+3/2}}$$

where the denominator of the first term on the right represents the lattice-limited mobility and the denominator of the second term on the right represents the imperfection-limited mobility. By differentiation, we can show that the maximum value of μ given by the above equation occurs for

$$T^3 = \frac{A}{B}$$

The magnitudes of A and B can be obtained from the given information about relaxation times.

At 300°K,

$$\mu_L = \frac{1}{300} \frac{e}{m^*} \tau = \frac{4.8 \times 10^{-10} \times 10^{-12}}{3 \times 10^2 \times 0.5 \times 9 \times 10^{-28}}$$

$$= 3.53 \times 10^3 \text{ cm}^2/\text{V-sec}$$

$$\mu_I = 3.53 \times 10^2 \text{ cm}^2/\text{V-sec}$$

Therefore at temperature T,

$$\mu_L = 3.53 \times 10^3 \left(\frac{T}{300}\right)^{-3/2} = 1.8 \times 10^7 \, T^{-3/2}$$

$$\mu_I = 3.53 \times 10^2 \left(\frac{T}{300}\right)^{+3/2} = 6.8 \times 10^{-2} \, T^{+3/2}$$

Since $A = 1.8 \times 10^7$ and $B = 6.8 \times 10^{-2}$, the temperature for maximum mobility is 640°K. The dependence of mobility on temperature for this problem is illustrated in Fig. 8.9.

Example 8.5 Obtain a better approximation using the simplest cut-off criterion, i.e., the Coulomb interaction energy $= kT$, by using the screened potential of Eq. (8.116) instead of an unscreened potential.

The cross section is defined by the radius r given by

$$\frac{e^2}{\varepsilon r} e^{-r/d} = kT$$

or

$$r = d \ln\left(\frac{e^2}{\varepsilon r k T}\right)$$

8.3 Charged-Imperfection Scattering

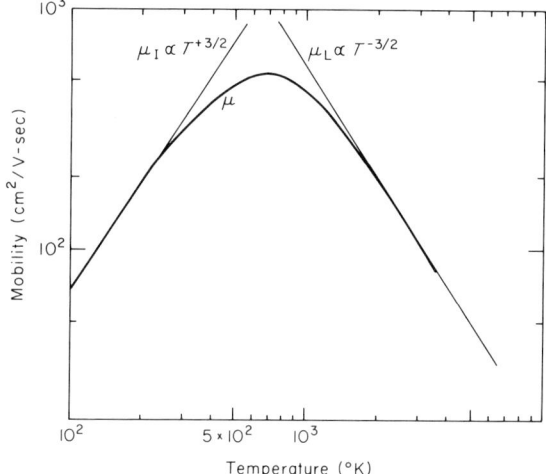

Fig. 8.9 Temperature dependence of mobility according to the data of Example 8.4.

As an approximation to the r in the log term, expand the exponential of the defining equation and retain the first term,

$$\frac{e^2}{\varepsilon r'}\left(1 - \frac{r'}{d}\right) = kT$$

which gives, for r',

$$r' = \frac{e^2}{\varepsilon[kT + (e^2/\varepsilon d)]}$$

Replace the r in the logarithmic term for r above, by this value of r' and obtain

$$r = d\ln\left(1 + \frac{e^2}{\varepsilon d kT}\right) = d\ln\left(1 + \frac{r_\infty}{d}\right)$$

where $r_\infty = e^2/(\varepsilon kT)$ is the value of the scattering radius for $d = \infty$, i.e., for no screening, the value corresponding to S_0; $r_\infty = 4.5 \times 10^{-7}$ cm for $\varepsilon = 10$, at 300°K.

Now from Eq. (8.117),

$$d = \left(\frac{420}{n^{1/2}}\right) \text{ cm}$$

It follows that for $r_\infty \ll d$,

$$r = d\left(\frac{r_\infty}{d}\right) = r_\infty$$

whereas for $r_\infty \gg d$,

$$r = d \ln\left(\frac{r_\infty}{d}\right)$$

For example, if $n = 10^{14}$ cm^{-3}, a relatively low density, $d = 4.2 \times 10^{-5}$ cm, which is much larger than r_∞; it follows that for $n = 10^{14}$ cm^{-3} the screening effects are totally negligible. If $n = 10^{18}$ cm^{-3}, on the other hand, $d = 4.2 \times 10^{-7}$ cm, and $r = 0.74d = 3.1 \times 10^{-7}$ cm, reduced from the value of r_∞ by the effects of screening. This simple correction includes some of the correction introduced by the more complete Brooks–Herring approach.

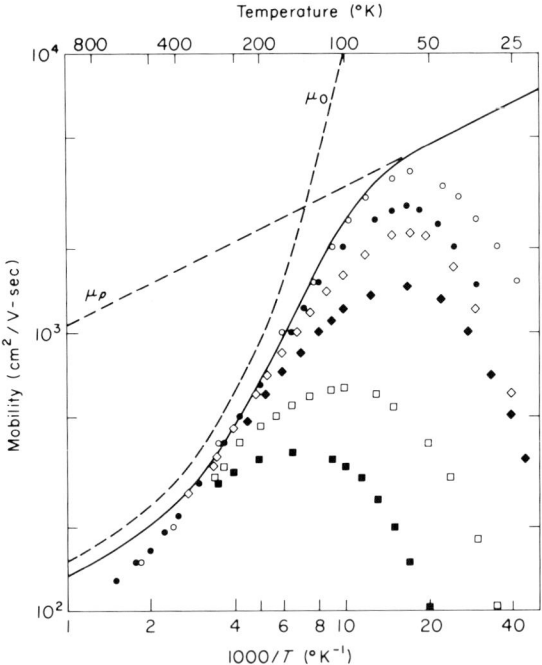

Fig. 8.10 Temperature dependence of the mobility for CdS crystals with different densities of Ga impurity. In order of increasing low-temperature mobility magnitude, the Ga densities are 6×10^{17} cm^{-3}, 2×10^{17} cm^{-3}, and 2×10^{16} cm^{-3}. The three largest low-temperature mobility curves are for crystals without any intentionally added impurity; the impurity scattering observed is associated with unintentional charged impurities in the "pure" CdS. (After W. W. Piper and R. E. Halsted, "Proc. Intern. Conf. Semiconductor Physics, Prague, 1960," Czechoslovak Academy of Sciences, Prague, 1961, pp. 1046–1048.) In the high-temperature range scattering for all crystals can be described by a combination of optical-mode lattice scattering and piezoelectric scattering.

8.4 Other Scattering Mechanisms

Figure 8.10 shows the effect on the low-temperature mobility of CdS of a varying concentration of charged Ga impurities.

8.4 Other Scattering Mechanisms

Although space does not allow the development of detailed treatments for a variety of other possible scattering mechanisms, it is important to realize that a number of other possibilities exist and that in real crystals these other possibilities may dominate.

SCATTERING BY OPTICAL-MODE LATTICE WAVES[†]

In crystals with an ionic contribution to the bonding, a strong electric dipole is set up in the crystal as the result of optical vibration modes, and optical-mode scattering can become important. Unlike the acoustic-mode phonons which are at higher temperatures comparable in energy to or less than kT, the optical-mode phonons have an energy $\hbar\omega_0$ that is in general larger than kT (producing inelastic scattering). The temperature dependence of the mobility under optical-mode scattering is an approximately exponential one, because of the temperature dependence of the number of such optical phonons available for scattering,

$$\bar{n} = \frac{1}{e^{\hbar\omega_0/kT} - 1} \approx e^{-\hbar\omega_0/kT} \tag{8.120}$$

The strength of the interaction with the optical-mode vibrations in a crystal is proportional to $(1/\varepsilon_\infty - 1/\varepsilon_0)$, a measure of the ionic polarizability, where ε_∞ is the high-frequency dielectric constant and ε_0 is the static dielectric constant.

NEUTRAL IMPERFECTIONS[‡]

At very low temperatures, neutral impurities may outnumber ionized impurities. The problem of scattering from such neutral impurities may be treated as a variation of the problem of the scattering of electrons by neutral hydrogen atoms. Two processes are significant: (1) direct elastic scattering, and (2) exchange scattering. Calculation yields a relaxation time that is independent of electron energy as long as the electron energy is less

---

[†] D. Howarth and E. Sondheimer, *Proc. Roy. Soc. London* **219A**, 53 (1953).
[‡] C. Erginsoy, *Phys. Rev.* **79**, 1013 (1950); N. Sclar, *Phys. Rev.* **104**, 1559 (1956).

than one fourth of the ionization energy of the imperfection. At 100°K, the scattering cross section for a neutral imperfection is about one fiftieth of that for a charged imperfection.

PIEZOELECTRIC SCATTERING[†]

Crystals that are piezoelectric show an electric polarization associated with the acoustic modes of lattice vibrations. Such polarization can lead to a periodic electric perturbation potential, and hence to electron scattering. The mobility varies as $T^{-1/2}$ and the effects of piezoelectric scattering may be sufficiently large to be important in determining the mobility in piezoelectric crystals. Experimental data on the temperature dependence of mobility for CdS crystals in Fig. 8.10 show that contributions from optical mode scattering and piezoelectric scattering dominate at higher temperatures.

DISLOCATIONS[‡]

Since dislocations cause a deformation of the crystal lattice, their presence affects the mobility, and a calculation of their effect may be made using the concept of a deformation potential. For normal dislocation densities, $N \leq 10^8$ cm^{-2}, such scattering is quite small. The density of dislocations necessary to give scattering of this type at temperature T equal to the scattering by acoustic vibrations is $N = 6 \times 10^1 T^{5/2}$ cm^{-2}.

Another mechanism by which dislocations affect mobility is by presenting a line of charge in the crystal. If sites along a dislocation act like acceptors in an n-type crystal, a line of negative charge surrounded by a positive space-charge region results. Such a dislocation is effectively a cylinder of radius R that repels electrons, the radius R being the extent of the space-charge region. Such scattering should be anisotropic, and calculation of the effect must average over various directions. Once again the magnitude of the scattering is small unless $N > 10^8$ cm^{-2}, at all but the lowest temperatures.

INTERVALLEY SCATTERING[§]

In a semiconductor with a number of equivalent minima in the conduction band, for example, an electron with a wave vector near to that cor-

[†] H. J. G. Meijer and D. Polder, *Physica* **19**, 255 (1953); W. A. Harrison, *Phys. Rev.* **101**, 903 (1956); A. R. Hutson, *Phys. Rev. Letters* **4**, 505 (1960).
[‡] D. L. Dexter and F. Seitz, *Phys. Rev.* **86**, 964 (1952); W. T. Read, *Phil. Mag.* **46**, 111 (1955).
[§] C. Herring, *Bell Syst. Tech. J.* **34**, 237 (1955).

8.4 Other Scattering Mechanisms

responding to the value of **k** at one minimum may change its wave vector to another near to that corresponding to the value of **k** at another minimum. Since the momentum change involved is $\hbar(\mathbf{k}_1 - \mathbf{k}_2)$, such scattering usually involves the participation of energetic phonons and hence is inelastic and strongly temperature dependent.

INHOMOGENEITY EFFECTS[†]

It is not unusual for real crystals to be characterized by certain deviations from homogeneity that can have a large effect on the mobility. Such inhomogeneities may result from (a) anisotropic impurity segregation, (b) impurity complex formation, (c) formation of clusters in solid-solution crystals, or (d) composition fluctuations during growth. Three basic types of such inhomogeneity may be proposed: (1) reversal in electrical type, (2) change in resistivity, and (3) change in band gap. The neighborhood of an inhomogeneity is usually characterized by a space-charge region extending over a much larger volume than the specific inhomogeneity itself.

If the assumption is made that charge carriers are simply excluded from a space-charge region in the material, and that the scattering can be approximated by a hard-sphere collision process with these space-charge regions, the mobility is

$$\mu = \frac{e}{NS(2m^*kT)^{1/2}} \qquad (8.121)$$

since $1/\tau$ for the process is $NS(2kT/m^*)^{1/2}$, where N is the density of space-charge regions, each with an effective scattering cross section of S. Since the size of a space-charge region depends on the density of free carriers, the effective cross section S is appreciably larger in high-resistivity than in low-resistivity material. This is consistent with the observation that scattering apparently due to inhomogeneities is much more prominent in high-resistivity material.

An interesting feature of such inhomogeneous imperfection scattering is that the *apparent* scattering cross section *per unit imperfection* can be much larger than a Coulomb cross section as calculated in Eq. (8.85). Consider for example a p-type region containing 10^{15} acceptors cm^{-3} in an n-type material. Even if the space-charge region is considered to be no larger than the p-type regions themselves, the *apparent* cross section *per acceptor* is $(\pi/4)$(distance between acceptors)^2, taking the radius as one-half the distance between acceptors. For 10^{15} acceptors, this gives a cross

---

[†] L. R. Weisberg, *J. Appl. Phys.* **33**, 1817 (1962).

section of 8×10^{-11} cm², or almost two orders of magnitude larger than a Coulomb atomic scattering cross section.

Example 8.6 Consider a metal precipitate in a crystal. If N metal atoms form the precipitate, how large may N be before the apparent cross section per atom is no larger than a Coulomb atomic scattering cross section?

Assume that the principle "size" of the precipitate as far as scattering is concerned results from the space-charge layer surrounding the precipitate. The space-charge layer has a radius r given by

$$r = \left(\frac{\varepsilon\phi}{2\pi e n}\right)^{1/2}$$

[see Eq. (12.36)], where ϕ is the potential drop at the barrier formed between the metal and the semiconductor crystal (depending on the relative work functions and the nature of the imperfections formed at the metal–semiconductor interface), and n is the density of free carriers. A reasonable value for the radius of the space-charge region in a material with $\varepsilon = 10$ and $n = 10^{15}$ cm^{-3} is 10^{-4} cm, giving a cross section S of about 3×10^{-8} cm². If each precipitate is made up of N atoms, the effective cross section per metal atom will be larger than the Coulomb cross section for an isolated metal atom as long as $S/N > 10^{-12}$ cm², i.e., as long as less than 10^4 metal atoms are required to form the precipitate.

8.5 High-Electric-Field Effects in Semiconductors[†]

Previous discussions of electrical conductivity have been based on the proportionality between current density and electric-field intensity,

$$\mathbf{j} = \sigma \mathscr{E} = ne\mu\mathscr{E}$$

If it is experimentally found that $\mathbf{j}$ is not proportional to $\mathscr{E}$, this disproportionality must arise either because n becomes a function of $\mathscr{E}$ or because μ becomes a function of $\mathscr{E}$. It is possible for n to become a function of $\mathscr{E}$ if (a) the electric field is high enough to increase the energy of free carriers by an amount sufficient for them to ionize other imperfections or crystal

---

[†] Our discussion here is representative of a simple possibility. For a comprehensive treatment, see E. Conwell, "High Field Transport in Semiconductors," Academic Press, New York, 1967.

8.5 High-Electric-Field Effects in Semiconductors

atoms themselves upon impact (*impact ionization*), (b) the electric field is high enough to permit ionization of imperfections by quantum mechanical tunneling to the nearest band (*field ionization*), or (c) the electric field is high enough to inject free carriers from the contacts into the crystal (*electrical injection*). These processes all lead to a variation of **j** *more* rapidly with increasing $\mathscr{E}$ than that predicted by the linear proportionality found at low fields. In the present section we confine the discussion to the opposite effect, a variation of **j** *less* rapidly than proportional to $\mathscr{E}$ with increasing $\mathscr{E}$, because of a field dependence of the mobility. The reason for this choice is that it is this field dependence of mobility that results from the lattice scattering of electrons as the attempt is made to increase their energy by the application of an electric field.

Under the effect of an applied electric field, the energy of an electron tends to increase except for the effects of scattering. In working out the specific effects in the following calculation, we limit ourselves specifically to the properties of scattering by acoustic lattice waves. Such scattering involves either the absorption or the emission of phonons. In order to calculate the increase in energy of an electron because of an electric field in the presence of scattering, the average energy resulting from both absorption and emission must be determined. If there is a net gain in energy, then the electrons increase in energy.

The energy of an electron is often described in terms of an equivalent *electron "temperature."* For an electron in an insulator or nondegenerate semiconductor, an increase in energy on the average by an amount kT represents a large change in the mean energy. An effective "temperature" T_e for such energetic electrons may be defined assuming a Maxwellian distribution, by setting the mean velocity

$$v = \left(\frac{2kT_e}{m^*}\right)^{1/2} \tag{8.122}$$

The mobility is independent of $\mathscr{E}$ only if $T_e \approx T$, i.e., only if the electrons are not "hot." If there is a net gain of energy when both the effects of electric field and lattice scattering are included, the electrons "heat up," T_e becomes larger than T, and **j** is not proportional to $\mathscr{E}$.

The calculation is carried out in the following steps.

1. Calculate the rate of energy gain from the electric field.
2. Calculate the increase in electron energy in a phonon-absorption scattering process.

3. Calculate the decrease in electron energy in a phonon-emission scattering process.
4. Calculate the average change in electron energy upon scattering.
5. Calculate the net rate of energy change for all electrons, assuming a Maxwellian distribution.
6. Set up the steady-state relation in which the rate of energy gain from the electric field is equated to the rate of energy loss from scattering.
7. Express the mobility as a function of electric field.
8. Consider other likely interactions at high electron energies.

1. Rate of Energy Gain from the Electric Field. The rate of energy gain by an electron with a mobility μ in the presence of an electric field is given by

$$\frac{dE}{dt} = e\mathscr{E} \cdot \mathbf{v} = e\mu\mathscr{E} \cdot \mathscr{E} = e\mu \mid \mathscr{E} \mid^2 \quad (8.123)$$

2. Increase in Electron Energy in Phonon-Absorption Process. Consider the scattering of an electron with initial wave vector $\mathbf{k}$ and final wave vector $\mathbf{k}'$ by a lattice wave with wave vector $\mathbf{K}$. Conservation of $\mathbf{k}$ gives

$$\mathbf{k} - \mathbf{k}' = -\mathbf{K} \quad (8.124)$$

and conservation of energy gives

$$\hbar\omega' - \hbar\omega = \hbar\bar{\omega} \quad (8.125)$$

For small values of $\mathbf{k}$ and $\mathbf{k}'$, $\bar{\omega} = K\bar{v}$, where $\bar{v}$ is the velocity of sound in the material. Also for the electron waves, $\omega = kv$, and $\omega' = k'v$. Eq. (8.125) becomes

$$k' = k + K\left(\frac{\bar{v}}{v}\right) \quad (8.126)$$

We may express K in terms of k and k' by applying the cosine law to the triangle of Fig. 8.11

$$K^2 = k^2 + k'^2 - 2kk' \cos \theta'$$
$$= k^2 + \left[k + K\left(\frac{\bar{v}}{v}\right)\right]^2 - 2k\left[k + K\left(\frac{\bar{v}}{v}\right)\right]\cos \theta'$$
$$\approx 2k\left[k + K\left(\frac{\bar{v}}{v}\right)\right](1 - \cos \theta') \quad (8.127)$$

if we neglect $K^2(\bar{v}/v)^2$ since $\bar{v} \ll v$. Replacing $(1 - \cos \theta')$ by $2\sin^2(\theta'/2)$, and solving Eq. (8.127) for K gives

$$K = \frac{2k\sin(\theta'/2)}{1 - (\bar{v}/v)\sin(\theta'/2)} \quad (8.128)$$

8.5 High-Electric-Field Effects in Semiconductors

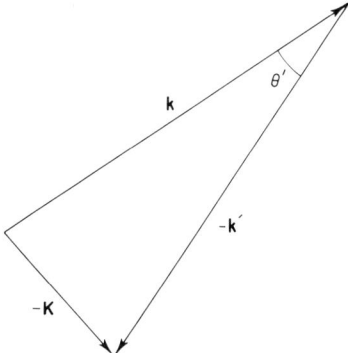

Fig. 8.11 Scattering of electron with initial wave vector **k** into final wave vector **k'** = **k** + **K** by absorption of a phonon from lattice waves, through an angle θ'.

after approximating $[1 + (K/k)(\bar{v}/v)]^{1/2} \approx [1 + (K/2k)(\bar{v}/v)]$. The gain in energy by an electron in a scattering event in which a phonon is absorbed is therefore given by

$$\Delta E = \hbar\bar{\omega} = \hbar K\bar{v} = \frac{2\hbar k \bar{v} \sin(\theta'/2)}{[1 - (\bar{v}/v)\sin(\theta'/2)]}$$

$$\approx 2m^* v \bar{v} \sin\left(\frac{\theta'}{2}\right)\left[1 + \frac{\bar{v}}{v}\sin\left(\frac{\theta'}{2}\right)\right] \quad (8.129)$$

3. Decrease in Electron Energy in Phonon-Emission Process. In an analogous way, the loss in energy of an electron in a scattering event involving phonon emission is

$$\Delta E \approx 2m^* v \bar{v} \sin\left(\frac{\theta}{2}\right)\left[1 - \left(\frac{\bar{v}}{v}\right)\sin\left(\frac{\theta}{2}\right)\right] \quad (8.130)$$

Note that since the change in energy upon scattering must be the same for both absorption and emission of a phonon, namely $\hbar\bar{\omega}$, the angle of scattering must be different, i.e., $\theta' \neq \theta$.

4. Average Change in Electron Energy upon Scattering. Since Eqs. (8.129) and (8.130) give the change in energy of an electron in a scattering process for either phonon absorption or phonon emission respectively, the average change in electron energy depends on the relative probability of a scattering event with phonon absorption as compared to a scattering event with phonon emission. We need then to introduce a probability for absorption of a phonon in a scattering event, P_a, and a probability for emission of a phonon in a scattering event, P_e. It is convenient if P_a and P_e are formulated so that $(P_a + P_e) = 1$, i.e., in any scattering event a phonon is either absorbed or emitted.

Now P_a is proportional to the average number of phonons of frequency $\bar{\omega}$,

$$P_a \propto \bar{n} = \frac{1}{e^{\hbar\bar{\omega}/kT} - 1} \qquad (8.131)$$

and P_e is proportional to $(\bar{n} + 1)$,

$$P_e \propto 1 + \frac{1}{e^{\hbar\bar{\omega}/kT} - 1} \qquad (8.132)$$

(That P_e is proportional to $(\bar{n} + 1)$ can be considered physically to represent the probability of a stimulated emission (proportional to $\bar{n}$) plus a spontaneous emission. Quantum mechanically it derives from the calculation of matrix elements for transition probabilities involving states with oscillator-like properties.[†])

We may assure that $P_a + P_e = 1$ by setting $A(P_a + P_e) = 1$, and solving to obtain

$$A = \frac{e^{\hbar\bar{\omega}/kT} - 1}{e^{\hbar\bar{\omega}/kT} + 1} \qquad (8.133)$$

Substituting this value for A back into $A(P_a + P_e) = 1$ gives

$$P_a = \frac{1}{e^{\hbar\bar{\omega}/kT} + 1} \approx \frac{1}{2 + \hbar\bar{\omega}/kT} \approx \frac{1 - \hbar\bar{\omega}/2kT}{2} \qquad (8.134)$$

$$P_e = \frac{e^{\hbar\bar{\omega}/kT}}{e^{\hbar\bar{\omega}/kT} + 1} \approx \frac{1}{2 - \hbar\bar{\omega}/kT} \approx \frac{1 + \hbar\bar{\omega}/2kT}{2} \qquad (8.135)$$

under the assumption that $\hbar\bar{\omega} \ll kT$ for the longitudinal acoustic vibrations under consideration.

The contribution to the loss of energy due to emission of a phonon upon scattering through an angle θ into the solid angle $d\Omega = 2\pi \sin \theta \, d\theta$ is $\hbar\bar{\omega}P_e \, d\Omega/4\pi$, as weighted by the probability of an emission process occurring. The contribution to the gain of energy due to absorption of a phonon upon scattering through an angle θ' into the solid angle $d\Omega' = 2\pi \sin \theta' \, d\theta'$ is $\hbar\bar{\omega}P_a \, d\Omega'/4\pi$, as weighted by the probability of an absorption process occurring. In order to combine these two quantities it is necessary to correlate θ and θ'. This may be accomplished by replacing the energy gain upon absorption, $\hbar\bar{\omega} = \Delta E$, by $\Delta E \sin(\theta/2)/\sin(\theta'/2)$, as obtained approx-

---

[†] See for example, R. A. Smith, "Wave Mechanics of Crystalline Solids," Wiley, New York, 1961, pp. 429, 430.

8.5 High-Electric-Field Effects in Semiconductors

imately from a combination of Eqs. (8.129) and (8.130), and by replacing the solid angle $d\Omega'$ by $d\Omega \sin\theta \, d\theta/(\sin\theta' \, d\theta')$. Therefore the contribution to the mean change in energy per scattering event may be written as

$$\langle \delta E(\theta) \rangle \, d\Omega = \frac{d\Omega}{8\pi} \left[-\varDelta E \left(1 + \frac{\varDelta E}{2kT}\right) \right.$$
$$\left. + \varDelta E \left(1 - \frac{\varDelta E}{2kT}\right) \frac{\sin(\theta/2) \sin\theta \, d\theta}{\sin(\theta'/2) \sin\theta' \, d\theta'} \right] \quad (8.136)$$

By a series of approximations the ratio involving θ and θ' becomes

$$\frac{\sin(\theta/2) \sin\theta \, d\theta}{\sin(\theta'/2) \sin\theta' \, d\theta'} \approx \left[1 + 8\left(\frac{\bar{v}}{v}\right) \sin\left(\frac{\theta}{2}\right)\right] \quad (8.137)$$

This approximation is obtained in the following way. Divide Eq. (8.129) by Eq. (8.130) to obtain

$$1 = \frac{\sin(\theta'/2)}{\sin(\theta/2)} \frac{1 + (\bar{v}/v) \sin(\theta'/2)}{1 - (\bar{v}/v) \sin(\theta/2)} \quad (8.138)$$

As an approximation, set $\theta' = \theta$ in the small correction term and obtain

$$\sin\left(\frac{\theta'}{2}\right) \approx \sin\left(\frac{\theta}{2}\right) \frac{1 - (\bar{v}/v) \sin(\theta/2)}{1 + (\bar{v}/v) \sin(\theta/2)}$$
$$\approx \sin\left(\frac{\theta}{2}\right) \left[1 - \left(\frac{\bar{v}}{v}\right) \sin\left(\frac{\theta}{2}\right)\right]^2$$
$$\approx \sin\left(\frac{\theta}{2}\right) \left[1 - 2\left(\frac{\bar{v}}{v}\right) \sin\left(\frac{\theta}{2}\right)\right] \quad (8.139)$$

depending on the small magnitude of $(\bar{v}/v)$ for each successive approximation. The differential of Eq. (8.139) is

$$\cos\left(\frac{\theta'}{2}\right) d\theta' = \cos\left(\frac{\theta}{2}\right) \left[1 - 4\left(\frac{\bar{v}}{v}\right) \sin\left(\frac{\theta}{2}\right)\right] d\theta \quad (8.140)$$

Since $\sin\theta = 2 \sin(\theta/2) \cos(\theta/2)$, the cosine terms in Eq. (8.140) can be replaced to obtain

$$\frac{\sin\theta' \, d\theta'}{\sin(\theta'/2)} = \frac{\sin\theta \, d\theta}{\sin(\theta/2)} \left[1 - 4\left(\frac{\bar{v}}{v}\right) \sin\left(\frac{\theta}{2}\right)\right] \quad (8.141)$$

which simplifies to

$$\sin\left(\frac{\theta}{2}\right) \sin\theta' \, d\theta' = \sin\left(\frac{\theta'}{2}\right) \sin\theta \, d\theta \left[1 - 4\left(\frac{\bar{v}}{v}\right) \sin\left(\frac{\theta}{2}\right)\right] \quad (8.142)$$

Now we may substitute from Eq. (8.139) for $\sin(\theta/2)$ on the left of Eq. (8.142) and for $\sin(\theta'/2)$ on the right of Eq. (8.142) to obtain

$$\frac{\sin(\theta'/2) \sin\theta' \, d\theta'}{1 - 2(\bar{v}/v)\sin(\theta/2)}$$
$$= \sin\left(\frac{\theta}{2}\right) \sin\theta \, d\theta \left[1 - 4\left(\frac{\bar{v}}{v}\right)\sin\left(\frac{\theta}{2}\right)\right]\left[1 - 2\left(\frac{\bar{v}}{v}\right)\sin\left(\frac{\theta}{2}\right)\right]$$
(8.143)

which reduces, with neglect of terms in $(\bar{v}/v)^2$, to the relation of Eq. (8.137).

When the approximation of Eq. (8.137) is inserted into Eq. (8.136) for the contribution to the mean change in energy per scattering event, we obtain

$$\langle \delta E(\theta) \rangle_{\text{av}} \, d\Omega = \frac{d\Omega}{8\pi} \left\{ \Delta E\left[8\left(\frac{\bar{v}}{v}\right)\sin\left(\frac{\theta}{2}\right)\right] - \frac{(\Delta E)^2}{kT}\left[1 + 4\left(\frac{\bar{v}}{v}\right)\sin\left(\frac{\theta}{2}\right)\right]\right\}$$
(8.144)

A specific value of $\Delta E \approx 2m^*\bar{v}v\sin(\theta/2)$ can be obtained approximately from either Eq. (8.129) or Eq. (8.130), which when substituted into Eq. (8.144) with neglect of the term in $\bar{v}^3$ gives

$$\langle \delta E(\theta) \rangle_{\text{av}} \, d\Omega = \frac{d\Omega}{4\pi}\left[8m^*\bar{v}^2\left(1 - \frac{m^*v^2}{4kT}\right)\sin^2\left(\frac{\theta}{2}\right)\right] \quad (8.145)$$

The total average energy change upon scattering is obtained by integrating over θ,

$$\langle \delta E \rangle_{\text{av}} = 2\pi \int_0^\pi \langle \delta E(\theta) \rangle_{\text{av}} \sin\theta \, d\theta$$
$$= 4m^*\bar{v}^2\left(1 - \frac{m^*v^2}{4kT}\right)\int_0^\pi \sin^2\left(\frac{\theta}{2}\right)\sin\theta \, d\theta$$
$$= 4m^*\bar{v}^2\left(1 - \frac{m^*v^2}{4kT}\right)$$
$$= 4m^*\bar{v}^2\left(1 - \frac{E}{2kT}\right) \quad (8.146)$$

since the value of the integral is unity. If $E = m^*v^2/2$, the kinetic energy of an electron before scattering, is less than $2kT$, the electron increases its energy on the average upon scattering, whereas if its energy is greater than $2kT$, it loses energy on the average upon scattering.

8.5 High-Electric-Field Effects in Semiconductors

5. Net Rate of Energy Change for All Electrons. We have now determined the change in energy per scattering process on the average as a function of the energy of the electron. In order to calculate an average effect for all the electrons, we assume a Maxwellian distribution of electron velocities corresponding to an electron temperature T_e,

$$N(v)\, dv = Av^2\, e^{-E/kT_e}\, dv \tag{8.147}$$

The mean rate at which an electron gains energy is $v \langle \delta E \rangle_{av}/l$, where l is the mean free path, the ratio v/l giving the number of scattering events per second. The average rate of energy change for all the electrons involves an average over the Maxwellian distribution,

$$\frac{dE}{dt} = \frac{\int_0^\infty (v \langle \delta E \rangle_{av}/l)\, N(v)\, dv}{\int_0^\infty N(v)\, dv}$$

$$= \frac{4m^*\bar{v}^2}{l}\, \frac{\int_0^\infty v^3 [1 - (m^*v^2/4kT)]\, e^{-m^*v^2/2kT_e}\, dv}{\int_0^\infty v^2\, e^{-m^*v^2/2kT_e}\, dv} \tag{8.148}$$

The evaluation of Eq. (8.148) involves three integrals with integrands $v^n\, e^{-av^2}$ with $n = 2$, 3, and 5, and $a = (m^*/2kT_e)$. Results are

$$\int_0^\infty v^3\, e^{-av^2}\, dv = \frac{(2kT_e/m^*)^2}{2} \tag{8.149}$$

$$\int_0^\infty v^5\, e^{-av^2}\, dv = (2kT_e/m^*)^3 \tag{8.150}$$

$$\int_0^\infty v^2\, e^{-av^2}\, dv = \frac{\pi^{1/2}(2kT_e/m^*)^{3/2}}{2} \tag{8.151}$$

Putting these results into Eq. (8.148) gives, for the mean rate of energy change due to scattering for all the electrons,

$$\frac{dE}{dt} = \frac{8\bar{v}^2(2\pi m^*kT_e)^{1/2}}{\pi lT} (T - T_e) \tag{8.152}$$

Therefore if T_e does not differ from T, there is no change in the energy on the average, whereas if $T_e > T$, there is a net energy loss.

6. Steady-State Conditions. In the steady state, the net rate of energy gain from the electric field given in Eq. (8.123) must be equal to the net rate of energy loss due to scattering according to Eq. (8.152),

$$e\mu \mathscr{E}^2 = -\frac{8\bar{v}^2(2\pi m^*kT_e)^{1/2}}{\pi lT} (T - T_e) \tag{8.153}$$

7. *Mobility as a Function of Electric Field.* The final goal of our consideration is an expression for the mobility as a function of the electric field in the steady state. It is convenient to use the effective electron temperature T_e as an intermediate parameter. The relation between the low-field mobility and the temperature is given by Eq. (8.10), expressible as

$$\mu_0 = \frac{4el}{3(2\pi m^* kT)^{1/2}} \tag{8.154}$$

where the subscript 0 is used to designate the low-field, field-independent mobility. If the mobility corresponding to an effective electron temperature T_e is written in the same way,

$$\mu = \frac{4el}{3(2\pi m^* kT_e)^{1/2}}$$

$$= \mu_0 \left(\frac{T}{T_e}\right)^{1/2} \tag{8.155}$$

Since it is known that $\mu_0 \propto T^{-3/2}$ when the dependence of $l \propto T^{-1}$ is included, it follows that $\mu \propto T^{-1} T_e^{-1/2}$.

Equations (8.154) and (8.155) can be used to relate T_e and $\mathscr{E}$ by substitution on both sides of Eq. (8.153) to obtain

$$e\mu_0 \left(\frac{T}{T_e}\right)^{1/2} \mathscr{E}^2 = -\frac{32\bar{v}^2 e}{3\pi\mu_0} \left(\frac{T_e}{T}\right)^{1/2} \frac{T - T_e}{T} \tag{8.156}$$

or dividing through by $[32\bar{v}^2 e(T/T_e)^{1/2}/(3\pi\mu_0)]$,

$$\frac{3\pi}{32} \left(\frac{\mu_0 \mathscr{E}}{\bar{v}}\right)^2 = \left(\frac{T_e}{T}\right)^2 - \frac{T_e}{T} \tag{8.157}$$

Here we have the condition for T_e to exceed T; it is that

$$\mu_0 \mathscr{E} > \bar{v} \tag{8.158}$$

This means that the effective temperature of the electrons starts to increase when the drift velocity (calculated with the low-field mobility) of the electrons becomes comparable to the velocity of sound in the crystal.

Eq. (8.157) can be solved for (T_e/T) to obtain

$$\frac{T_e}{T} = \frac{1}{2} \left\{ 1 + \left[1 + \frac{3\pi}{8} \left(\frac{\mu_0 \mathscr{E}}{\bar{v}}\right)^2 \right]^{1/2} \right\} \tag{8.159}$$

8.5 High-Electric-Field Effects in Semiconductors

In the relatively low-field regime, $\mu_0 \mathscr{E} \ll \bar{v}$, and

$$\frac{T_e}{T} = 1 + \frac{3\pi}{32}\left(\frac{\mu_0 \mathscr{E}}{\bar{v}}\right)^2 \tag{8.160}$$

giving, for the mobility from Eq. (8.155),

$$\mu = \mu_0\left[1 - \frac{3\pi}{64}\left(\frac{\mu_0 \mathscr{E}}{\bar{v}}\right)^2\right] \tag{8.161}$$

In this low-field range, the difference $(\mu - \mu_0)$ varies as the square of the electric field.

In the high-field region, where $\mu_0 \mathscr{E} \gg \bar{v}$, Eq. (8.159) becomes

$$\frac{T_e}{T} = \left(\frac{3\pi}{32}\right)^{1/2} \frac{\mu_0 \mathscr{E}}{\bar{v}} \tag{8.162}$$

corresponding to the mobility

$$\mu = \left(\frac{32}{3\pi}\right)^{1/4}\left(\frac{\mu_0 \bar{v}}{\mathscr{E}}\right)^{1/2} \tag{8.163}$$

The mobility in the high-field range varies inversely as the square root of the electric field strength. Since the drift velocity is equal to the product of mobility and field strength, the drift velocity increases as $\mathscr{E}^{1/2}$ instead of being proportional to $\mathscr{E}$ as in the Ohm's-law low-field range.

Of course it must be remembered that this result applies only to the case of scattering by acoustic lattice waves. For such scattering, an increase in energy of electrons because of the electric field results in increased scattering, since the probability of scattering by acoustic lattice waves increases with electron energy. If charged-impurity scattering were active, for example, an increase in electron energy with electric field would result in an *increased* mobility, since for charged-impurity scattering the probability of scattering decreases with increasing electron energy.

8. Other Processes at High Fields. As the magnitude of the electric field $\mathscr{E}$ is increased further, other effects come into play to remove energy from the "hot" electrons. One such process is the excitation of optical modes of lattice vibration with frequency ω_0 when $m^* v_d^2/2 = \hbar \omega_0$, where $v_d = \mu \mathscr{E}$ is the drift velocity. An electron reaching such an energy loses almost all of its energy in exciting an optical-mode phonon, and this leads to a saturation effect in the dependence of drift mobility on $\mathscr{E}$. If the field strength is increased still further, some process increasing the magnitude of n, as mentioned at the beginning of this section, becomes dominant, resulting

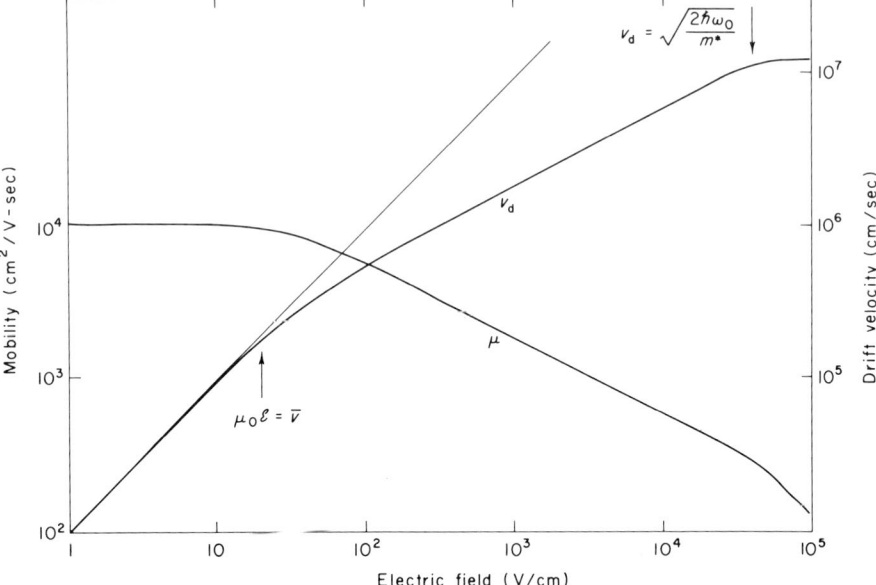

Fig. 8.12 Typical variation of mobility and drift velocity for electrons under the condition that scattering is dominated at low fields by LA phonons, and at high fields by optical-mode phonons. Calculated for the following values of the parameters: $\bar{v} = 2 \times 10^5$ cm/sec, $\mu_0 = 10^4$ cm^2/V-sec, $\hbar\omega_0 = 0.04$ eV, $m^* = m$.

in the experimental observation of a rapid increase in current. Unless the current is limited externally, destruction of the crystal is likely. A representative dependence of mobility and drift velocity on electric field is shown in Fig. 8.12 for the prebreakdown region.

Example 8.7 A certain semiconductor crystal has a room-temperature electron mobility of 3000 cm^2/V-sec for low electric fields. The velocity of sound in the crystal is 3×10^5 cm/sec, and the Reststrahlen absorption band is at 35 μm. (a) At what field strength is the mobility 10 per cent less than its low-field value? (b) At what field strength is the mobility one-half of its low-field value? (c) At what field strength does the current density saturate?

(a) Since only a 10-per-cent decrease is involved, it is possible that the low-field regime of Eq. (8.161) is applicable. If the value of $\mathscr{E}$ is calculated from this equation, the result is $\mathscr{E} = 82$ V/cm for $\mu = 0.9\mu_0$. If the value of $\mu_0\mathscr{E}$ is calculated, it is found to be 2.5×10^5 cm/sec, which does not fulfill the criterion of being either much less than or much greater than $\bar{v}$. In order to obtain an accurate value it is necessary to obtain the general expression resulting from the combination of Eqs. (8.155) and (8.159),

$$\left(\frac{\mu_0}{\mu}\right)^2 = \frac{1}{2}\left\{1 + \left[1 + \frac{3\pi}{8}\left(\frac{\mu_0 \mathscr{E}}{\bar{v}}\right)^2\right]^{1/2}\right\}$$

If we solve for $\mathscr{E}$ from this equation, we find that for $\mu = 0.9\mu_0$, $\mathscr{E} = 92$ V/cm.

(b) In view of the experience in (a) it is likely that the value of $\mathscr{E}$ necessary for $\mu = 0.5\mu_0$ lies in the high-field regime covered by Eq. (8.163). If this equation is solved for $\mathscr{E}$, we obtain $\mathscr{E} = 730$ V/cm. The corresponding drift velocity $(\mu_0 \mathscr{E}) = 2.2 \times 10^6$ cm/sec, which does satisfy the requirement that $\mu_0 \mathscr{E} \gg \bar{v}$ for the applicability of Eq. (8.163).

(c) A Reststrahlen absorption band at 35 μm corresponds to an optical phonon energy of 0.035 eV. If for our hypothetical crystal, $m^* = 0.07m$, for example, $v_d = 4.2 \times 10^7$ cm/sec. This means that the current density saturates when

$$\left(\frac{32}{3\pi}\right)^{1/4}(\mu_0 \bar{v})\mathscr{E}^{1/2} = 4.2 \times 10^7$$

or for a value $\mathscr{E} = 10^6$ V/cm. This is the same order of magnitude as the electric field for dielectric breakdown, and it is likely that some other process will occur before the transfer of energy from "hot" electrons to optical phonons.

Effects of high electric fields on the mobility of electrons in n-type Ge are shown in Fig. 8.13.

8.6 Summary

In this rather lengthy chapter we have selected a few possible problems out of the many that might be treated in the area of scattering processes, and have developed the solution of these particular problems in some detail. It is our hope that the opportunity to see the complete solution of particular problems, rather than simply the beginning and ending steps in the calculation, is of sufficient help to the student to compensate for the lack of treatment in other areas. A general solution to the problem of scattering processes would have to include general degeneracy (capable of describing cases from that of the nondegenerate semiconductor, through the degenerate

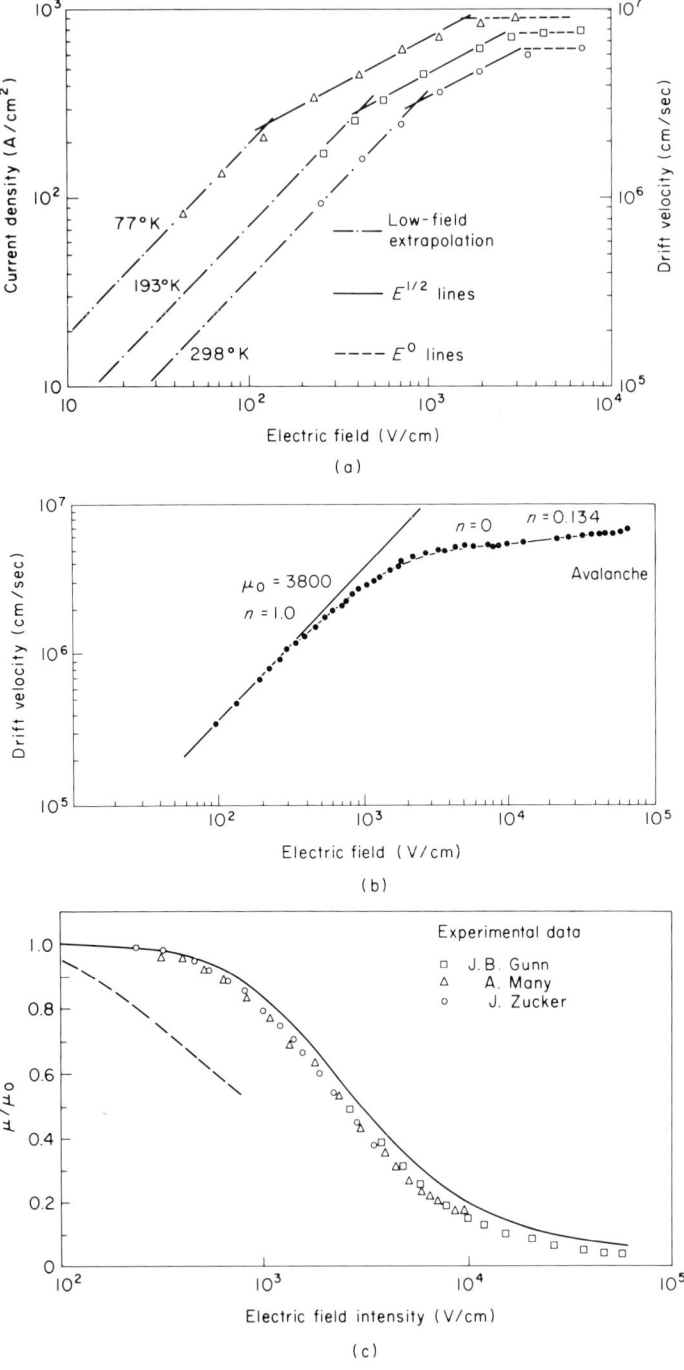

8.6 Summary

semiconductor, to the metal), and appropriate treatments for elastic scattering with a meaningful relaxation-time approximation and for inelastic scattering which cannot be described by the relaxation-time approach.

Two different approaches were applied to the problem of electron scattering by acoustic lattice waves: (1) reflection of electrons due to a deformation of the lattice potential by lattice waves, and (2) a quantum mechanical perturbation treatment of the effect of lattice waves on the periodic lattice potential. The result in each case is that the mean free path of an electron, as determined by elastic acoustic-wave scattering, is inversely proportional to the temperature. This result arises physically because the energy of the acoustic phonons is small compared to kT at normal temperatures, and therefore the average energy carried by lattice waves is proportional to kT. The scattering cross section for acoustic lattice scattering of this type is independent of the angle of scattering. Other considerations necessary for a more complete treatment would include inelastic scattering in metals at temperatures much less than the Debye temperature, as discussed in Chapter 7, and Umklapp scattering.

The scattering of carriers by charged imperfections arises from the Coulomb interaction between the charged electron or hole and the charged imperfection. Treatment of this problem requires some assumption about the maximum distance from a charged imperfection that the Coulomb field is still large enough to be effective in scattering, for the scattering cross section approaches infinity for very small scattering angles. Assumptions discussed include (1) when the Coulomb energy is equal to kT, (2) half the distance between charged imperfections (Conwell–Weisskopf), and (3) a Debye length from the charged imperfection characterized by a screened Coulomb potential (Brooks–Herring). The first assumption leads directly to a scattering cross section of the order of 10^{-12} cm² at room temperature, increasing as T^{-2} with decreasing temperature. The second assumption leads through the Rutherford scattering formula applied to this problem to a similar conclusion except for a correction term depending on the density of charged imperfections and the temperature, which is generally of order

---

Fig. 8.13 High-electric-field effects on mobility in n-type Ge. (a) Current density versus electric field at three temperatures. [After E. J. Ryder, *Phys. Rev.* **90**, 766 (1953).] (b) Drift velocity versus electric field. [After J. B. Gunn, *J. Electron.* **2**, 87 (1956).] (c) Ratio of high-field to low-field mobility versus electric field. Experimental data are shown by data points, theoretically calculated curves by a dashed line (for acoustic-mode scattering only) and by a solid line (for acoustic- and optical-mode scattering both present). [After E. M. Conwell, *Phys. Chem. Solids* **8**, 234 (1959).]

302   8 Scattering Processes

unity. The third assumption leads through a quantum mechanical calculation again to a similar conclusion, except for a correction term depending on the temperature and the density of free electrons, which is also generally of order unity. Refinements to these calculations are necessary to include the effects of degeneracy.

Other scattering mechanisms may also play a significant role in determining the mobility over certain ranges of temperature. These include scattering by optical-mode lattice vibrations, neutral imperfections, piezoelectric effects, dislocations, and inhomogeneities.

When the electric field becomes large enough, the mobility of carriers in semiconductors varies with the magnitude of the electric field. The nature of the variation depends on the variation of scattering probability with increasing electron energy. In the case of scattering by acoustic-mode lattice vibrations, the mobility of carriers begins to decrease as the electron energy is increased, and the high-field mobility varies inversely with the square root of the electric field. For still higher electric fields and higher carrier energy, the "hot" carriers may interact with optical-mode lattice vibrations to provide a limiting drift velocity and an effective saturation of the current density.

Problems

8.1 The evaluation of the matrix element for phonon scattering in insulators or nondegenerate semiconductors involves the assumption that $|\mathbf{k}_0 - \mathbf{k}_m| r_s$ is small so that the expansion of an exponential term in the matrix element expression can be terminated after the second term of the expansion. Carry the calculation out for the next nonzero term of the expansion, and determine the correction to the scattering probability due to this additional term.

8.2 An electron moving in the z direction ($\theta = 0$) is scattered into the direction given by θ. If the probability of this scattering due to a hypothetical mechanism is inversely proportional to $\sin \theta$, calculate the average number of scattering collisions required to destroy completely the electron's original drift velocity in the z direction.

8.3 Given the compounds shown in the tabulation, if acoustic-wave lattice scattering dominates, which compound will show a deviation from Ohm's law with increasing electric field first? Which compound will show saturation of current density with increasing electric field first?

Problems

Compound	Low-field mobility (cm²/V-sec)	Reststrahlen wavelength (μm)
A	300	35
B	100	50
C	500	40
D	1000	25

8.4 Consider a material in which scattering is due to N^+ singly charged impurities. Photoexcitation of free carriers is used to fill these charged impurities with electrons, just making them neutral. Sketch the expected behavior of mobility as a function of photoexcitation intensity. How could the scattering cross section of the centers be obtained from such an experiment? (This problem can be done quantitatively on the basis of the discussion of imperfection level occupation given in Chapter 9.)

8.5 Within the range of small electron and lattice vibration wave vectors' what is the error made by estimating the change in energy of an electron upon scattering from a longitudinal acoustic lattice wave through an angle θ by assuming that the magnitude of the electron wave vector is unaffected by the scattering event?

8.6 Given the acoustical and optical vibrational branches shown in Fig.

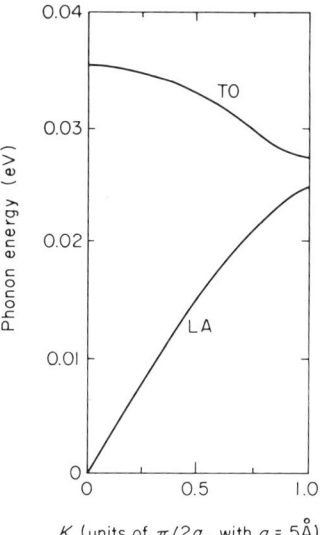

Fig. 8.14 Phonon dispersion curves. See Problem 8.6.

8.14, calculate (a) the critical electric field for high-field effects on the mobility due to interaction with LA phonons, and (b) the critical electric field for saturation of the drift velocity due to interaction with long-wavelength optical phonons. Assume a low-field electron mobility of 10^4 cm^2/V-sec and an effective mass of $0.1m$.

8.7 Outline the procedure by which the mobility for high electric fields could be calculated in the presence of charged-impurity scattering only.

Chapter 9

Localized Energy Levels

The existence of crystalline imperfections has been indirectly introduced through the discussion of scattering of charge carriers in the previous chapter. Such imperfections give rise to localized energy states, an electron associated with the imperfection being able to have an energy that is forbidden in the perfect crystal. In general it is found that local deviations from the periodic potential of the perfect crystal give rise to localized energy states, characterized by wavefunctions that decay exponentially in amplitude with distance from the imperfection site.

Many of the electronic properties of real crystals, semiconductors in particular, depend in some way on the presence of imperfections, which may act as the source of free carriers or as local regions where recombination of free carriers is favored. Many of the most important, most difficult to control, and most technologically significant properties of electronically active solids can be traced to the effects of imperfections, often present in concentrations of only parts per million or less.

One of the most evident imperfections in a crystal is the surface, where a whole plane of atoms has only half of its usual neighbors. Localized *surface states* are expected both because of the interruption of the periodicity of the lattice at the surface, and because of specific chemical, atomic, or molecular properties of the surface atoms.

The periodicity of the perfect crystal potential can also be interrupted in the volume of the crystal by the existence of any of a family of substitutional, interstitial, or complex imperfections. The term *imperfection* is used to describe any departure from the structure of the perfect crystal. Imperfections may be crystal *defects*, such as vacancies, interstitial crystal atoms, misplaced crystal atoms, or dislocations, or they may be *impurity atoms* present in the crystal either substitutionally or interstitially. This terminology is not always universally accepted, and some authors speak of intrinsic defects (simply defects in our terminology) and extrinsic defects (impurities).

The description of imperfections and their charge states in a crystal is achieved through the use of a particular set of symbols. This symbolism has for many years been widely variant from author to author, but in recent years there is a tendency to conform more or less to a single system. The key to this system is that the normal atom in the lattice is taken as the reference state; deviations from this charge state toward more positive are represented by a superscript dot, and deviations toward more negative are represented by a superscript slash.

Shallow imperfection levels in crystals can be treated as if the imperfection constituted a hydrogenic system, as adjusted for the dielectric constant of the material and the effective mass of the charge carriers in the nearby band. Theoretical treatments of deep imperfection levels are still the subject of active research.

The electronic occupancy of imperfection levels is described by a Fermi level defined for the semiconductor or insulator in thermal equilibrium. This Fermi level provides a convenient concept for the description of electrical conductivity as related to the giving up or taking away of electrons by crystal imperfections.

9.1 Energy Levels in an Imperfect Crystal

Whenever the periodicity of the crystal lattice is disturbed, new energy levels become possible in a localized region of the crystal near the disturbance. When considered with respect to the energies of the bands, these localized levels frequently lie in what is normally the forbidden gap between the valence and conduction bands. In terms of wave mechanics, the presence of a perturbation potential at some particular imperfection site in the crystal allows a new solution of the Schroedinger equation which has a finite value over a localized region in the crystal, and then decays exponentially with distance away from the imperfection site. Figure 9.1 illustrates this possibility. Important imperfections may be chemical impurities, vacancies, interstitials, dislocations, or more complex imperfections formed by the coalescence of two or more of the simple imperfections.

An example of the possibility of such imperfection states can be drawn

9.1 Energy Levels in an Imperfect Crystal

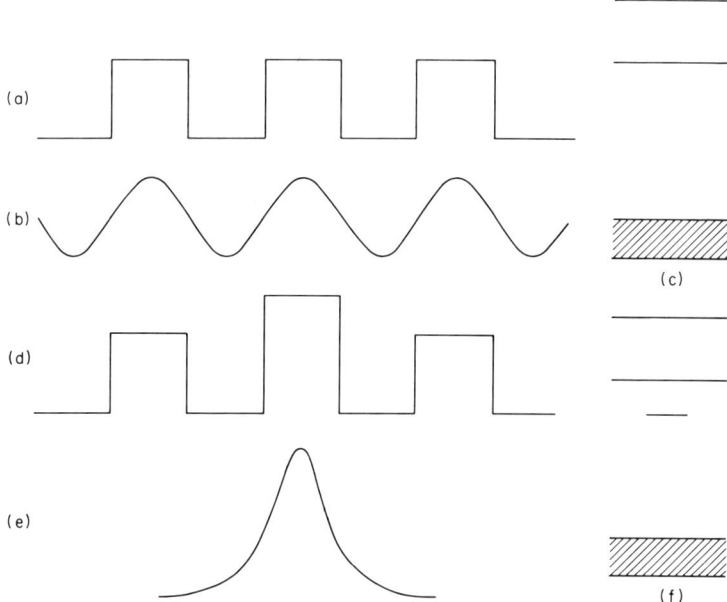

Fig. 9.1 Schematic representation of (a) periodic potential in a perfect crystal, (b) wavefunction for electrons in a perfect crystal (neglecting the local effects of the $u_k(\mathbf{r})$ function; see Fig. 6.3), (c) energy bands for a perfect crystal, (d) disturbance in periodic potential due to an imperfection, (e) localized wavefunction at the imperfection with exponential decay with distance, (f) localized energy level in the forbidden gap associated with the imperfection.

first from the surface of the crystal, which represents a major departure from the periodicity of the perfect lattice. A possible potential distribution suitable for calculations in the form of the Kronig–Penney model (Problem 5.6) is given in Fig. 9.2. The conditional relation for allowed energies on the basis of the Kronig–Penney model for a one-dimensional periodic potential without consideration of the surface is

$$\cos ka = \frac{P}{\beta a} \sin \beta a + \cos \beta a \tag{9.1}$$

where $P = mV_0 b(a-b)/\hbar^2$ and $\beta = (2mE/\hbar^2)^{1/2}$, as defined in Problem 5.6. The wavefunction solutions are of the form $\psi \propto e^{ikx}$, and it is necessary that k be real in order for ψ to be well-behaved. Because of this limitation, the right-hand side of Eq. (9.1) is restricted to values lying between -1 and $+1$. This limitation that k must be real no longer applies for $x < 0$ if the effect of the surface is considered.

For $x < 0$, the solution is

$$\psi = Ce^{\gamma x} \tag{9.2}$$

where $\gamma = [2m(W - E)/\hbar^2]^{1/2}$. These solutions must be joined to the Kronig–Penney solutions for the periodic potential in the crystal volume.[†] For real values of k the solutions may be joined at $x = 0$ without difficulty

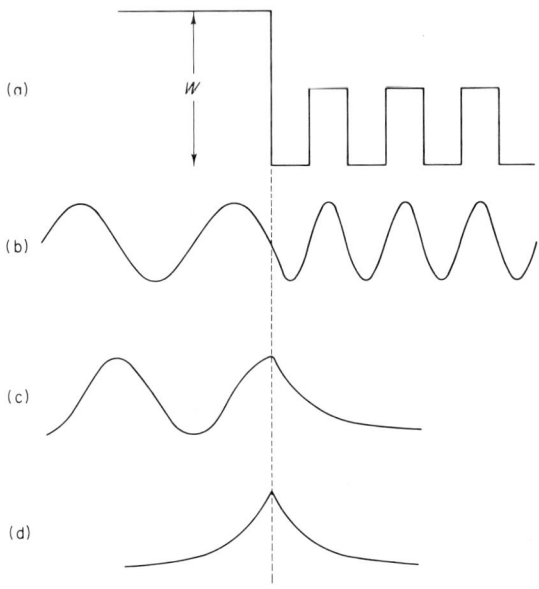

Fig. 9.2 Schematic representation of (a) periodic potential terminating at a surface, (b) wavefunction for $E > W$, real k, (c) wavefunction for $E > W$, imaginary k, (d) wavefunction for $E < W$, imaginary k.

and the band states within the crystal are unaffected. When $E > W$, solutions corresponding to imaginary k for $x < 0$ are exponentially damped for $x > 0$. When $E < W$, a new state becomes allowed in the forbidden gap (one for each surface in the one-dimensional model) which is exponentially damped for both $x < 0$ and $x > 0$; it corresponds to an electron localized at the surface in a *surface state*. Such surface states, although usually of a more complicated origin than the simple argument presented here, can be detected experimentally in most materials.

A similar effect is associated with any perturbation of the periodic po-

---

[†] See, for example, F. Seitz, "The Modern Theory of Solids," McGraw-Hill, New York, 1940, p. 32.

9.1 Energy Levels in an Imperfect Crystal

tential within the volume of a crystal. Since the condition for the growth of real crystals is frequently under a thermodynamic environment favoring the existence of imperfections, such departures from the periodic potential of the pure and perfect crystal are the rule rather than the exception. Concentrations of such imperfections usually lie in the range of 10^{15}–10^{19} cm^{-3}; the result of the most extensive materials purification program attempted to date has produced a lower limit of about 10^{11} cm^{-3} imperfections in Ge crystals.

The presence of a localized energy level due to a vacancy in the volume of a crystal is indicated schematically in Fig. 9.3. The energy separation between valence band and conduction band represents the energy required to break a bond in the crystal; this energy is expected to be less in the immediate vicinity of the vacancy, which represents a missing positively charged ion core. The existence of this localized region where less energy is required to free an electron from a bond is represented on an energy-band diagram as a localized level lying a distance below the conduction-band edge corresponding to the reduced energy to break a bond.

These examples are concerned with localized levels that come into being because of the perturbation of the perfect periodic potential of the crystal; in a sense, they are simply displaced levels of the crystal. Localized levels can also be directly associated with the impurity itself and correspond to atomic levels of the impurity as altered by the dielectric constant and interaction of the crystal; such levels are extra levels, not present at all in the pure and perfect crystal. Examples are the energy levels introduced by impurity elements with incomplete d or f shells.

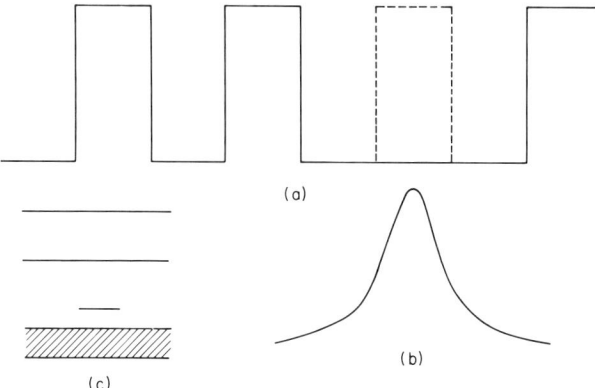

Fig. 9.3 Schematic representation of (a) periodic potential disturbed by a vacancy (b) localized wavefunction at the vacancy, (c) localized energy level in the forbidden gap associated with the vacancy.

9.2 Imperfection Terminology

Imperfections in crystals represent a field of research that has been approached from several different discipline-oriented perspectives, each with its own peculiar historically developed terminology. There are at least three types of designation commonly applied to imperfections, involving categorization according to (1) physical identity, (2) structural location, and (3) electronic behavior.

Categorization by *physical identity* involves a statement as to whether the imperfection results from an impurity atom (i.e., an atom not a member of the pure host crystal), a host crystal atom (i.e., one of the atoms of the pure crystal), or a complex involving both impurities and host crystal atoms (e.g., an impurity–vacancy pair). It is important to know also if possible the nature of the impurity atom (what element it is, what charge state it has), or of the host crystal atom (whether a host crystal atom is missing or displaced, and the effective charge in its environment).

Categorization by *structural location* involves a statement as to whether the imperfection can be considered as a point defect located either substitutionally or interstitially in the crystal, or whether the imperfection is a complex with certain symmetry properties.

Categorization by *electronic behavior* is the least explicit of all, since a given imperfection as defined by its identity and structural location may contribute a wide variety of electronic behavior. Typical terms in common usage are donor, acceptor, trap, recombination center, absorption center, activator, coactivator, sensitizing center, poison or killer center, scattering center, and so on. Not only may a single type of imperfection contribute a variety of electronic behavior, but the type of the behavior for such an imperfection may vary widely depending on the operating conditions as described by such parameters as applied electric field, applied magnetic field, incident light intensity, temperature, wavelength of incident light, and so on.

9.3 Description of Imperfection Incorporation

In this section we consider in a conceptual way the major processes that occur when imperfections are introduced deliberately into a crystal. Consider, for example, a II–VI compound such as CdS, where the cation has two outer-shell electrons and the anion has six outer-shell electrons. In describing the role of imperfections in such a crystal, and recognizing that

9.3 Description of Imperfection Incorporation

the binding is neither purely ionic nor purely covalent, it is nevertheless possible to think either in terms of an ionic or a covalent model with equivalent models for the qualitative features of imperfection incorporation in many cases. We illustrate this by first choosing an ionic binding model, and later considering the equivalent covalent binding model.

On an ionic binding model, the Cd in CdS transfers its two electrons to the S, resulting in a $Cd^{+2}S^{-2}$ crystal. We designate the Cd^{+2} ion in the crystal as the reference state for the cation sublattice, and write $Cd^\times$; similarly we designate the S^{-2} ion in the crystal as the reference state for the anion sublattice, and write $S^\times$. The host crystal CdS is then represented as $Cd^\times S^\times$. Suppose now that we introduce an impurity atom like Cl with seven outer-shell electrons substitutionally in place of S. Since Cl requires only one electron to complete its outer shell, there is one electron left over that will be weakly bound in the vicinity of the Cl imperfection at sufficiently low temperatures. This system of Cl on a S site plus a bound electron may be represented as $Cl^\times$ in the crystal, since the total charge in the vicinity of the replaced S is the same as before the impurity substitution. Now at a sufficiently high temperature, the bound electron is thermally freed to the conduction band, leaving a positively charged Cl center (with respect to the reference $S^\times$ state). The ionization process is described by

$$Cl^\times \rightarrow Cl^\cdot + e' \qquad (9.3)$$

The positive charge on the Cl site is represented by the superscript dot, and the negative charge of the electron is represented by the superscript slash. To emphasize that Cl has replaced S substitutionally, a subscript noting the sublattice is sometimes used, e.g.,

$$Cl_S^\times \rightarrow Cl_S^\cdot + e' \qquad (9.4)$$

A certain amount of energy must be supplied to cause the process shown in Eq. (9.4), and this energy may be found in the crystal vibrations due to the temperature of the crystal. On an energy-band diagram, the imperfection center is represented by a short line, located an energy separation ΔE below the bottom of the conduction band, where ΔE is the energy required to free the electron from the $Cl_S^\times$ center. The short line is used to emphasize the localized nature of the allowed energy state, and the location of the energy level for the electron in the forbidden gap shows how much energy is required to raise the electron into the conduction band.

The concept of an imperfection *center* has been introduced in the above discussion to imply that the actual physical imperfection is the center of a perturbation that reaches out in its effects beyond the atomic dimensions of the imperfection itself. Thus the symbol $Cl_S^\times$ must not be taken to mean a Cl^{-2} ion, but rather simply a Cl^- ion *plus* an electron bound in a relatively large "orbit" of hydrogenic type embracing many of the neighboring atoms of the crystals. (The bandgap of CdS is 2.4 eV, whereas the binding energy of an electron to a $Cl_S^\times$ center is only about 0.03 eV.) In the following section, we consider the quantum mechanical calculation of the energy levels of such weakly bound systems.

The center represented by $Cl_S^\times$ is called a neutral center. Since the Cl impurity is capable of giving rise to free electrons at sufficiently high temperatures, it is called a *donor* impurity in CdS. The $Cl_S^\cdot$ center is called an *ionized donor* since the electron bound to the $Cl_S^\times$ center has been freed, thus ionizing the $Cl_S^\times$ center.

A second possibility exists for the effect of incorporating Cl impurity substitutionally for S in CdS. This possibility is that a Cd vacancy may be formed for every two Cl substituted for S, in this way maintaining charge neutrality in the crystal

$$2S^\times + Cd^\times + 2Cl \rightarrow 2Cl_S^\cdot + V_{Cd}'' + 2S + Cd \tag{9.5}$$

where V_{Cd} represents a cadmium vacancy. In this case no free electrons are involved, the extra electrons available after substitution of 2Cl for 2S having been utilized in the formation of the Cd vacancy,

$$Cd^\times + 2e' \rightarrow V_{Cd}'' + Cd \tag{9.6}$$

The quantities in Eqs. (9.5) and (9.6) without superscripts represent displaced atoms removed from the crystal, hence no longer part of the charge balance in the crystal. Inside the crystal, the following charge neutrality relations define the processes of Eqs. (9.4) and (9.5) respectively:

$$Cl_S^\cdot = e' \tag{9.7}$$

$$2Cl_S^\cdot = V_{Cd}'' \tag{9.8}$$

The Cd vacancy has accepted two electrons from the Cl impurity. It is therefore doubly negative with respect to the reference state. Because it is capable of accepting electrons, it is called an *acceptor* imperfection. An acceptor imperfection if incorporated alone into CdS would accept electrons from the valence band, thus producing free holes and ionized acceptors,

$$V_{Cd}^\times \rightarrow V_{Cd}'' + 2h^\cdot \tag{9.9}$$

9.3 Description of Imperfection Incorporation

governed by the neutrality condition,

$$V_{Cd}'' = 2h^{\bullet} \tag{9.10}$$

In the example of Eq. (9.5) the acceptors are formed at the same time as a number of donors are incorporated, just sufficient to supply all the electrons the acceptor imperfections are capable of accepting. When the electrons accepted come from a donor rather than from the valence band, the acceptor is said to be *compensated*. Therefore, in Eq. (9.5) the $Cl_S^{\bullet}$ are *compensated donors* (the corresponding electron of the $Cl_S^{\times}$ center is not free but has been transferred to an acceptor), and the V_{Cd}'' *are compensated acceptors* (the corresponding holes of the $V_{Cd}^{\times}$ are not free but have been replaced by electrons from donors).

Equation (9.5) contains examples of both impurity imperfections and defect imperfections in a common interaction process. Because of the positive charge on the $Cl_S^{\bullet}$ and the negative charge on the V_{Cd}'', there is an attractive Coulomb force acting to associate these defects. If pairing between donors and acceptors takes place, a $(V_{Cd}Cl_S)'$ complex may be formed that would be expected to have an energy level at about the same location as the level of a V_{Cd}' center.

Whether the substitution of Cl for S gives rise to Eq. (9.4) or (9.5) depends on the environment present during the incorporation. An environment that prevents Cd vacancies from forming, e.g., a high Cd pressure, favors the process of Eq. (9.4). Likewise an environment that encourages Cd vacancy formation, e.g., a high S pressure, favors the process of Eq. (9.5). Another factor in the relative probability of Eq. (9.5) is the magnitude of the energy separation between $Cl_S^{\bullet}$ and V_{Cd}'' energy levels in the forbidden gap, as compared to the energy required to form a Cd vacancy. If the energy which can be recovered by transferring an electron from the $Cl_S^{\bullet}$ level to the V_{Cd}'' level is larger than the energy required for Cd vacancy formation, it is energetically favorable for the crystal to proceed by the process of Eq. (9.5) in providing *self-compensation* by forming Cd vacancies to compensate the incorporated Cl donors. These considerations are consistent with the common observation that the probability of self-compensation becomes high in large-bandgap materials.

A defect such as V_{Cd} is multivalent, and may exist in the crystal in several different ionization states. The defect V_{Cd}'' corresponds to a vacancy with two captured electrons, V_{Cd}' to a vacancy with one captured electron, and $V_{Cd}^{\times}$ to a vacancy with no captured electrons. Sometimes this terminology is turned around and is expressed in terms of holes rather than electrons. Then V_{Cd}'' corresponds to no captured holes, V_{Cd}' to one captured hole, and V_{Cd}^x to two captured holes. Thinking in terms of the lattice in which V_{Cd} is surrounded by four tetrahedrally arranged S^{-2} ions, it is perhaps more appropriate to speak in terms of the number of captured holes rather than captured electrons. These possibilities are illustrated in Fig. 9.4.

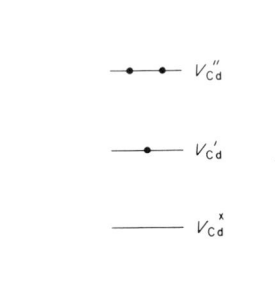

Fig. 9.4 Various energy levels associated with the multivalent cadmium vacancy imperfection in CdS. The uppermost level's distance below the conduction band represents the energy required to free one of the two electrons at a V''_{Cd} center, to form a V'_{Cd} center. The distance of the next lower level below the conduction band represents the energy required to remove the one electron at a V'_{Cd} center, to form a V_{Cd}^x center. The lowest level's distance below the conduction band represents the energy required to free another electron from the V_{Cd}^x center (i.e., remove an electron from one of the four neighboring S ions) to form a $V_{Cd}^{\cdot}$ center.

The type of substitutions described above for replacement of S by Cl occurs also if we replace Cd by a trivalent cation such as In. The neutrality conditions governing these substitutions are

$$In_{Cd}^{\cdot} = e' \qquad (9.11)$$

and

$$2In_{Cd}^{\cdot} = V''_{Cd} \qquad (9.12)$$

Analogous processes result upon the substitutional incorporation of a trivalent anion such as As in place of S, or of a monovalent cation such as Cu in place of Cd. These processes are like those described above, except than they involve holes rather than electrons, and S vacancies rather than Cd vacancies. The corresponding neutrality conditions are

$$As_S' = h^{\cdot} \qquad (9.13)$$

$$2As_S' = V_S^{\cdot\cdot} \qquad (9.14)$$

and

$$Cu'_{Cd} = h^{\cdot} \qquad (9.15)$$

$$2Cu'_{Cd} = V_S^{\cdot\cdot} \qquad (9.16)$$

9.3 Description of Imperfection Incorporation

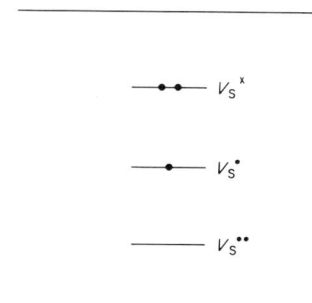

Fig. 9.5 Various levels associated with the multivalent sulfur vacancy in CdS. The interpretation of the levels is the same as that given for the V_{Cd} center in Fig. 9.6.

As and Cu are acceptor impurities in CdS, but if they are incorporated in an environment that encourages the formation of sulfur vacancies, their substitution may result in the formation of compensated donor sulfur vacancies and compensated acceptor impurities. There are three possible ionization states for the multivalent sulfur vacancies, $V_S^{\cdot\cdot}$ without any electrons, $V_S^{\cdot}$ with one electron, and $V_S^{\times}$ with two electrons. These are illustrated in Fig. 9.5.

All of the preceding discussion has been on the basis of an ionic model of the crystal. On a covalent model, many of the same types of results are obtained. In the covalent picture, as indicated in Fig. 9.6, Cd and S each

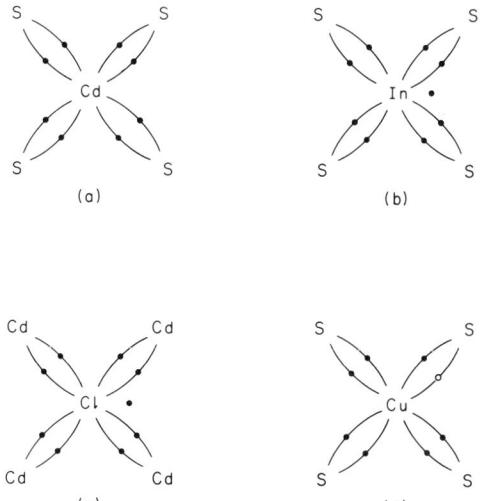

Fig. 9.6 Representations of covalent binding for (a) a perfect crystal of CdS, (b) an In donor in CdS, (c) a Cl donor in CdS, (d) a Cu acceptor in CdS.

make four bonds because of the tetrahedral arrangement of the crystal structure in the sphalerite and wurtzite lattices. Each bond can be considered as effectively composed of $\frac{1}{2}$ electron from the Cd and $1\frac{1}{2}$ electrons from the S. If Cd is replaced by In with three electrons available for bonding, for example, only two are required, and one is left over to become a free electron at sufficiently high temperatures. Similarly if S is replaced by Cl with seven electrons, one electron is again in excess. If Cd is replaced by Cu with one electron, one more electron is required for bonding and this electron is taken from a neighboring S, freeing a hole in the process. Thus the donor and acceptor behavior of impurities with differing valencies from the host crystal atoms is not affected by the choice of ionic or covalent binding model.

Example 9.1 Give the neutrality conditions that govern the possible ways of incorporating Cd and Te impurities in AgI.

In the ionic model, AgI becomes a Ag^+I^- crystal. Cd impurity substituted for Ag has one excess electron and therefore acts as a donor or as a driving force for the generation of Ag vacancies. Te impurity substituted for I has a deficiency of one electron and therefore acts as an acceptor or as a driving force for the generation of I vacancies.

The relevant neutrality conditions are as follows.

$$Cd_{Ag}^{\cdot} = e' \qquad Te_I{'} = h^{\cdot}$$
$$Cd_{Ag}^{\cdot} = V_{Ag}' \qquad Te_I{'} = V_I^{\cdot}$$

9.4 Description of Electronic Behavior

As an illustration of the description of the electronic behavior associated with imperfections in crystals, consider just the simplest type of defects to be expected in a CdS crystal with Cl impurity: $Cl_S^{\cdot}$ and V_{Cd}'. The electronic behavior of the $Cl_S^{\cdot}$ can be described as that of

(a) an ionized donor, since it represents an impurity that has given up an electron either to the conduction band or to acceptors;

(b) a compensated donor, since at least part of the $Cl_S^{\cdot}$ donors are compensated by the V_{Cd}' acceptors;

(c) an optical absorption center, since excitation of an electron from the valence band to the center is possible by absorption of a photon (for Cl_S^x, absorption exciting an electron from the level to the conduction band would be possible);

9.4 Description of Electronic Behavior

(d) a scattering center, since its positive charge produces Coulomb scattering of free carriers;

(e) an electron trap, since it may temporarily capture a free electron, with thermal reexcitation of the electron to the conduction band at a later time;

(f) a coactivator, since its presence may be necessary for the existence of a compensated acceptor suitable for radiative recombination, either because the existence of the acceptor depends on the presence of the coactivator through solubility or charge-compensation processes, or because the compensation of the acceptor depends on the presence of the coactivator donors, or because the coactivator pairs with the compensated acceptors to form centers capable of radiative recombination;

(g) a recombination center, since recombination between electrons captured at the $Cl_S^{\cdot}$ center ($Cl_S^{\cdot} + e' \to Cl_S^{\times}$) and free holes ($Cl_S^{\times} + h^{\cdot} \to Cl_S^{\cdot}$) may occur.

The electronic behavior of V'_{Cd} can similarly be described as that of

(a) an ionized acceptor, since it represents an imperfection that has given up a hole either to the valence band or to donors;

(b) a compensated acceptor, since the V'_{Cd} acceptors are compensated by the $Cl_S^{\cdot}$ donors;

(c) an optical absorption center, since excitation of an electron from the center to the conduction band is possible,

$$V'_{Cd} \xrightarrow{\hbar\omega} V^{\times}_{Cd} + e' \qquad (9.17)$$

(excitation of an electron from the valence band corresponds to excitation to the V''_{Cd} level, according to

$$V'_{Cd} + e' \xrightarrow{\hbar\omega} V''_{Cd}) \qquad (9.18)$$

(d) a scattering center, since its negative charge will produce Coulombic scattering of free carriers;

(e) a hole trap, since it may temporarily capture a free hole with subsequent thermal release of the hole to the valence band;

(f) a recombination center, since recombination between holes captured at the V'_{Cd} centers ($V'_{Cd} + h^{\cdot} \to V^{\times}_{Cd}$) and free electrons ($V^{\times}_{Cd} + e' \to V'_{Cd}$) may occur;

(g) an activator, since the recombination process just mentioned may require dissipation of the energy of the excited electron through emission of a photon;

(h) a sensitizing center for photoconductivity, since the capture of an electron by a neutral V_{Cd}^x formed by hole capture by a V'_{Cd} may be improbable, thus leading to long electron lifetimes for photoexcited electrons, and hence to sensitive photoconductivity behavior.

9.5 Theory of Shallow Imperfection Energy Levels[†]

If the ionization energy of a donor or acceptor imperfection is sufficiently small, the associated energy levels can be calculated from a perturbation calculation. The result is that the imperfection can be considered to be a hydrogenic system embedded in the crystal, with suitable corrections for the dielectric constant of the crystal and for the effective mass of the carriers involved.

Consider a crystal with a single conduction band minimum at $\mathbf{k} = 0$. In the absence of imperfections, the Hamiltonian for a free electron is

$$\mathbf{H}_0 = -\frac{\hbar^2}{2m} \nabla^2 + V(\mathbf{r}) \tag{9.19}$$

where $V(\mathbf{r})$ is the periodic potential of the crystal lattice. The eigenfunctions of this Hamiltonian may be written as Bloch waves,

$$\psi_{\mathbf{k}l}(\mathbf{r}) = \frac{1}{\mathscr{V}^{1/2}} u_{\mathbf{k}l}(r) e^{i\mathbf{k}\cdot\mathbf{r}} \tag{9.20}$$

where $\mathscr{V} = L^3$, the volume, and l is the index of the conduction band in the reduced zone of E versus $\mathbf{k}$,

$$\mathbf{H}_0 \psi_{\mathbf{k}l}(\mathbf{r}) = E_l(k) \psi_{\mathbf{k}l}(\mathbf{r}) \tag{9.21}$$

If one of the atoms of the crystal is replaced by a positively singly charged impurity ion, for example, with an additional electron for total charge neutrality, the Hamiltonian for the extra electron is

$$\mathbf{H} = \mathbf{H}_0 + U(\mathbf{r}) \tag{9.22}$$

where at large distances from the charged impurity,

$$U(\mathbf{r}) = -\frac{e^2}{\varepsilon r} \tag{9.23}$$

[†] W. Kohn, *Solid State Physics* **5**, 257 (1957).

9.5 Theory of Shallow Imperfection Energy Levels

Although the relationship is more complicated close to the impurity ion, Eq. (9.23) may be used as an approximation for all r.

The motion of the electron bound to the impurity ion neighborhood is described by

$$\mathbf{H}\Psi = (\mathbf{H}_0 + U)\Psi = E\Psi \tag{9.24}$$

For the wavefunctions of the perturbed system, Ψ, we use as usual a linear combination of the particular eigenfunctions of the unperturbed system,

$$\Psi = \sum_{\mathbf{k},l} A_l(\mathbf{k})\psi_{\mathbf{k}l} \tag{9.25}$$

Substitution of Ψ from Eq. (9.25) into Eq. (9.24), followed by multiplication through by $\psi_{\mathbf{k}'l'}^*$ and integration over all coordinate space leads to

$$[E_l(\mathbf{k}) - E] A_l(\mathbf{k}) + \sum_{\mathbf{k}',l'} U_{\mathbf{k}l,\mathbf{k}'l'} A_{l'}(\mathbf{k}') = 0 \tag{9.26}$$

where the matrix element $U_{\mathbf{k}l,\mathbf{k}'l'}$ is given by

$$U_{\mathbf{k}l,\mathbf{k}'l'} = \int \psi_{\mathbf{k}l}^* \left(-\frac{e^2}{\varepsilon r}\right) \psi_{\mathbf{k}'l'} \, d\mathbf{r}$$

$$= \frac{1}{\mathcal{V}} \int u_{\mathbf{k}l}^* u_{\mathbf{k}'l'} \, e^{i(\mathbf{k}'-\mathbf{k})\cdot\mathbf{r}} \left(-\frac{e^2}{\varepsilon r}\right) d\mathbf{r} \tag{9.27}$$

The product $u_{\mathbf{k}l}^* u_{\mathbf{k}'l'}$ is a periodic function and can be expanded in terms of the reciprocal lattice vectors $\mathbf{b}_j$:

$$u_{\mathbf{k}l}^* u_{\mathbf{k}'l'} = \sum_j C_{\mathbf{k}l,\mathbf{k}'l'}^j \, e^{i\mathbf{b}_j \cdot \mathbf{r}} \tag{9.28}$$

Substituting this expansion into the matrix element of Eq. (9.27) gives

$$U_{\mathbf{k}l,\mathbf{k}'l'} = \frac{1}{\mathcal{V}} \int \sum_j C_{\mathbf{k}l,\mathbf{k}'l'}^j \, e^{i(\mathbf{b}_j + \mathbf{k}' - \mathbf{k})\cdot\mathbf{r}} \left(-\frac{e^2}{\varepsilon r}\right) d\mathbf{r} \tag{9.29}$$

The summation over j can be removed from the integration, which then becomes

$$\int_0^{2\pi}\int_0^{\pi}\int_0^{\infty} \left(-\frac{e^2}{\varepsilon r}\right) e^{i(\mathbf{b}_j+\mathbf{k}'-\mathbf{k})\cdot\mathbf{r}} \, r^2 \sin\theta \, dr \, d\theta \, d\phi = -\frac{4\pi e^2}{\varepsilon \, |\mathbf{b}_j + \mathbf{k}' - \mathbf{k}|^2} \tag{9.30}$$

since $\int_0^\infty r \, e^{ar} = 1/a^2$. Therefore the matrix element is

$$U_{\mathbf{k}l,\mathbf{k}'l'} = \frac{1}{\mathcal{V}} \sum_j C_{\mathbf{k}l,\mathbf{k}'l'}^j \left(-\frac{4e^2}{\varepsilon \, |\mathbf{b}_j + \mathbf{k}' - \mathbf{k}|^2}\right) \tag{9.31}$$

Now we may note that

$$\int \psi_{\mathbf{k}l}^* \psi_{\mathbf{k}'l'} \, d\mathbf{r} = \frac{1}{\mathscr{V}} \int u_{\mathbf{k}l}^* u_{\mathbf{k}'l'} \, e^{i(\mathbf{k}'-\mathbf{k})\cdot\mathbf{r}} \, d\mathbf{r}$$

becomes, for $\mathbf{k} = \mathbf{k}' = 0$,

$$\delta_{l,l'} = \frac{1}{\mathscr{V}} \int u_{0l}^* u_{0l'} \, d\mathbf{r} \tag{9.32}$$

so that comparison with Eq. (9.28) gives

$$C_{0l,0l'}^0 = \delta_{l,l'} \tag{9.33}$$

For weak binding of the extra electron, it is expected that the wave function Ψ approaches closely to a Bloch wave near the conduction-band minimum. Thus we expect that Eqs. (9.26) have solutions in which $A_l(\mathbf{k})$ is negligible except when $l = 0$ and $\mathbf{k}$ is small. Setting $A_l(k) = 0$ for $l \neq 0$ transforms Eq. (9.26) to

$$[E_0(\mathbf{k}) - E] A_0(\mathbf{k}) + \sum_{\mathbf{k}'} U_{\mathbf{k}0,\mathbf{k}'0} A_0(\mathbf{k}') = 0 \tag{9.34}$$

To carry this approximation further, we substitute in Eq. (9.34) the following appropriate values for $E_0(\mathbf{k})$ and $U_{\mathbf{k}0,\mathbf{k}'0}$ for small $\mathbf{k}$ and $\mathbf{k}'$:

$$E_0(\mathbf{k}) \approx \frac{\hbar^2}{2m_e^*} |\mathbf{k}|^2 \tag{9.35}$$

$$U_{\mathbf{k}0,\mathbf{k}'0} \approx -\frac{4\pi e^2}{\mathscr{V}\varepsilon |\mathbf{k}' - \mathbf{k}|^2} \tag{9.36}$$

The result is

$$\left[\frac{\hbar^2}{2m_e^*} |\mathbf{k}|^2 - E \right] A_0(\mathbf{k}) - \frac{4\pi e^2}{\mathscr{V}\varepsilon} \sum_{\mathbf{k}'} \frac{1}{|\mathbf{k} - \mathbf{k}'|^2} A_0(\mathbf{k}') = 0 \tag{9.37}$$

In order to cast Eq. (9.37) into a more familiar and interpretable form, introduce the Fourier transform of $A_0(\mathbf{k})$,

$$F(\mathbf{r}) = \frac{1}{\mathscr{V}^{1/2}} \sum_{\mathbf{k}} A_0(\mathbf{k}) \, e^{i\mathbf{k}\cdot\mathbf{r}} \tag{9.38}$$

Now multiply Eq. (9.37) by $e^{i\mathbf{k}\cdot\mathbf{r}}$ and sum over $\mathbf{k}$ to obtain

$$\left[-\frac{\hbar^2}{2m_e^*} \nabla^2 - \frac{e^2}{\varepsilon r} \right] F(\mathbf{r}) = E \, F(\mathbf{r}) \tag{9.39}$$

9.5 Theory of Shallow Imperfection Energy Levels

since

$$-\frac{\hbar^2}{2m_e^*} \nabla^2 F(\mathbf{r}) = \sum_{\mathbf{k}} \frac{\hbar^2}{2\mathcal{V}^{1/2}m_e^*} |\mathbf{k}|^2 A_0(\mathbf{k}) e^{i\mathbf{k}\cdot\mathbf{r}}$$

and

$$-E F(\mathbf{r}) = -\frac{E}{\mathcal{V}^{1/2}} \sum_{k} A_0(\mathbf{k}) e^{i\mathbf{k}\cdot\mathbf{r}}$$

and since the Fourier transform of $1/r$ is

$$\frac{1}{r} = \frac{4\pi}{\mathcal{V}^{1/2}} \sum_{\mathbf{k}} \frac{1}{|\mathbf{k}|^2} e^{i\mathbf{k}\cdot\mathbf{r}}$$

This last result may be seen by considering the second term of Eq. (9.37) after multiplication by $e^{i\mathbf{k}\cdot\mathbf{r}}$ and summing over $\mathbf{k}$,

$$\frac{4\pi e^2}{\varepsilon \mathcal{V}} \sum_{\mathbf{k}} \sum_{\mathbf{k}'} \frac{1}{|\mathbf{k}-\mathbf{k}'|^2} A_0(\mathbf{k}') e^{i\mathbf{k}\cdot\mathbf{r}}$$

If we let $\mathbf{k}'' = \mathbf{k} - \mathbf{k}'$, this term becomes

$$\frac{4\pi e^2}{\varepsilon \mathcal{V}} \sum_{\mathbf{k}'} \sum_{\mathbf{k}''} \frac{1}{|\mathbf{k}''|^2} A_0(\mathbf{k}') e^{i(\mathbf{k}'+\mathbf{k}'')\cdot\mathbf{r}}$$

$$= \frac{4\pi e^2}{\varepsilon \mathcal{V}} \left[\sum_{\mathbf{k}'} A_0(\mathbf{k}') e^{i\mathbf{k}'\cdot\mathbf{r}}\right] \left[\sum_{\mathbf{k}''} \frac{1}{|\mathbf{k}''|^2} e^{i\mathbf{k}''\cdot\mathbf{r}}\right]$$

$$= \frac{4\pi e^2}{\varepsilon \mathcal{V}} [\mathcal{V}^{1/2} F(\mathbf{r})] \left[\frac{\mathcal{V}^{1/2}}{4\pi} \frac{1}{r}\right]$$

$$= \frac{e^2}{\varepsilon r} F(\mathbf{r})$$

Equation (9.39) is a hydrogenic Schroedinger equation with the free-electron mass replaced by the effective electron mass. The energy levels appropriate for the electron weakly bound to a positively charged impurity ion are those of a hydrogenic center where the mass of the bound electron is the effective mass of free electrons in the conduction band. These energy levels are given by

$$E = -\frac{1}{n^2} \frac{e^4 m^*}{2\hbar^2 \varepsilon^2} \tag{9.40}$$

where $n = 1, 2, \ldots$. The total wavefunction of the impurity electron in the present approximation is given by

$$\Psi = \sum_{\mathbf{k}} A_0(\mathbf{k}) \, \psi_{\mathbf{k}0}(\mathbf{r})$$

$$= \frac{1}{\mathcal{V}^{1/2}} \sum_{\mathbf{k}} A_0(\mathbf{k}) \, u_{\mathbf{k}0}(\mathbf{r}) \, e^{i\mathbf{k}\cdot\mathbf{r}}$$

$$\approx u_{00}(\mathbf{r}) \, F(\mathbf{r}) \tag{9.41}$$

If the effective mass is a tensor and not a scalar, additional complexities are introduced and the effective-mass equation relevant to this case is

$$\left(-\frac{\hbar^2}{2m_1^*} \frac{\partial^2}{\partial x^2} - \frac{\hbar^2}{2m_2^*} \frac{\partial^2}{\partial y^2} - \frac{\hbar^2}{2m_3^*} \frac{\partial^2}{\partial z^2} - \frac{e^2}{\varepsilon r} \right) F(\mathbf{r}) = E \, F(\mathbf{r}) \tag{9.42}$$

In many cases such equations can be partially separated and solved approximately. The actual solutions may depart appreciably from hydrogenic

TABLE 9.1 IONIZATION ENERGIES FOR SHALLOW IMPURITIES IN Ge AND Si

	E_I, hydrogenic value (eV)	Impurity	E_I, measured (eV)
		Donor	
Ge	0.0066	P	0.0120
		As	0.0127
		Sb	0.0096
Si	0.026	Li	0.033
		P	0.044
		As	0.049
		Sb	0.039
		Bi	0.069
		Acceptor	
Ge	0.016	B	0.0104
		Al	0.0102
		Ga	0.0108
		In	0.0112
Si	0.052	B	0.045
		Al	0.057
		Ga	0.065
		In	0.16

9.5 Theory of Shallow Imperfection Energy Levels

values if the effective-mass values in different directions are markedly dissimilar. Ionization energies for typical shallow impurities in Ge and Si are given in Table 9.1. These cubic materials do exhibit a large ratio of longitudinal to transverse effective mass and hence do show departures from the simple hydrogenic model.

Acceptor states may be treated in a way similar to that derived above for donor states. Shallow acceptor states may be described in terms of a sum of wavefunctions chosen from the top of the valence band. Because the valence bands in many materials, such as Ge and Si, are degenerate, however, the actual calculation is somewhat more complicated.

Donor and acceptor energies in typical large-bandgap semiconductors, such as the II–VI materials, are often too large to be adequately treated by the hydrogenic effective-mass approximation. These materials exhibit

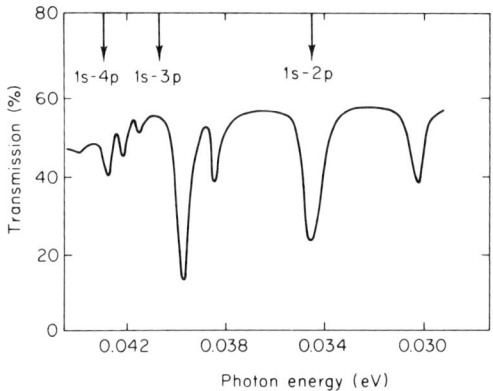

Fig. 9.7 Optical absorption spectrum for transitions related to boron hydrogenic-like impurity in silicon. [After H. J. Hrostowski and R. H. Kaiser, *J. Phys. Chem. Solids* **4**, 148 (1958).]

an electron effective mass of the order of 0.1 to $0.2m$, and a dielectric constant of about 10, which gives a hydrogenic ionization energy of about 0.026 eV. Shallow donors in these compounds (Group III or Group VII substitutional impurities) do have energies of this order of magnitude. The ionization energy of acceptors is much larger, however, varying from 0.2 eV in tellurides to over 1 eV in sulfides.

The excited states of the shallow impurity described by Eq. (9.40) have been observed in absorption spectra in Si. Figure 9.7 shows experimental data for B acceptors in Si. The strongest absorptions agree rather well with the predictions of a hydrogenic model, but more absorption peaks are found than are predicted by this simple model.

Example 9.2 Estimate how many atoms are enclosed by the "orbit" of an electron bound to a hydrogenic impurity in a crystal with dielectric constant 20, effective electron mass $0.05m$, and lattice constant of 3Å.

The ionization energy of such an electron is reduced by the factor $0.05/(20)^2$ from the ionization energy of a hydrogen atom, i.e., the ionization energy would be 0.0017 eV.

Since the relationship between the ionization energy and the radius of the ground-state a_0 is $E = e^2/2a_0$ (see, for example, Table 3.2), the radius of the hydrogenic orbit in the material is $[(20)^2/0.05]a_0$, or 4.2×10^{-5} cm.

If the lattice constant is 3Å, the radius of the orbit corresponds to about 1400 atoms, and therefore the volume included by the orbit corresponds to about 8×10^6 atoms of the crystal.

Each of these hydrogenic impurity centers is isolated on the average until the density of such impurities approaches a value of about 10^{15} to 10^{16} cm^{-3}. For higher densities there would be appreciable overlap of the wavefunctions for such electrons bound to impurities, and effects of interaction between impurities would be expected.

9.6 Thermal-Equilibrium Fermi Level in Semiconductors and Insulators

The role that a given imperfection plays in affecting the electronic behavior of a semiconductor depends to a large extent on the electronic occupancy (valence state, effective charge, ionization state). In Section 4.5 we considered how the occupancy of an electronic energy level in a metal can be described in terms of the Fermi–Dirac distribution function. In metals,

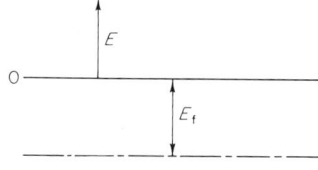

Fig. 9.8 Definition of energies for calculation of the total density of free electrons in a semiconductor.

9.6 Thermal-Equilibrium Fermi Level

the Fermi energy corresponds to that energy possessed by the most energetic electrons at 0°K, or to that energy for which the probability of occupancy is $\frac{1}{2}$ for temperatures greater than 0°K. It proves convenient to discuss the occupancy of imperfection levels in semiconductors and insulators in terms of the *Fermi level* concept also. The Fermi level is a hypothetical line drawn on an energy-band diagram at the energy of the Fermi energy. The occupancy of a given electronic level is describable in terms of the energy separation between its level and the Fermi level.

How should this Fermi level be defined for a semiconductor? In order to see the answer to this question, calculate the total number of electrons in the conduction band of a nondegenerate semiconductor or insulator, using the notation shown in Fig. 9.8:

$$n = \int_0^\infty N(E) f(E) \, dE$$
$$= \int_0^\infty \frac{1}{2\pi^2} \left(\frac{2m_e^*}{\hbar^2}\right)^{3/2} E^{1/2} e^{-(E+E_f)/kT} \, dE \qquad (9.43)$$

where we have used the density of states for spherical equal-energy surfaces, and have replaced the Fermi distribution function by the Boltzmann distribution approximation, under the assumption that the Fermi level lies several kT away from the bottom of the conduction band in the nondegenerate material being considered. The integral in Eq. (9.43) is

$$\int_0^\infty E^{1/2} e^{-(E/kT)} \, dE = \Gamma\left(\frac{3}{2}\right)(kT)^{3/2} = \frac{1}{2} \pi^{1/2} (kT)^{3/2} \qquad (9.44)$$

The result for the total density of electrons in the conduction band is

$$n = 2\left(\frac{2\pi m_e^* kT}{h^2}\right)^{3/2} e^{-E_f/kT} \qquad (9.45)$$

We can give the following interpretation of this result. By definition, the occupation of a level (in a band) lying E_f above the Fermi level (for $E_f \gg kT$) is given by

$$\frac{n}{N} = e^{-E_f/kT} \qquad (9.46)$$

if n is the density of occupied levels and N is the total density of levels. If Eq. (9.45) is compared to Eq. (9.46), we are led to the convenient definition of an "effective density of states" N_c for the conduction band,

$$N_c = 2\left(\frac{2\pi m_e^* kT}{h^2}\right)^{3/2} \qquad (9.47)$$

so that

$$n = N_c e^{-E_f/kT} \tag{9.48}$$

Thus N_c can be considered to be the effective density of levels at the bottom of the conduction band (within about kT of the bottom) which is appropriate for a Fermi distribution description *if* all of the electrons in the conduction band were located within kT of the bottom of the band. Once N_c is known for a particular material (which requires only a knowledge of the effective mass), the location of the Fermi level is specified by giving the free-electron concentration n at a specific temperature T.

Example 9.3 In a nondegenerate semiconductor, how far above the bottom of the conduction band does the density of occupied levels have its maximum?

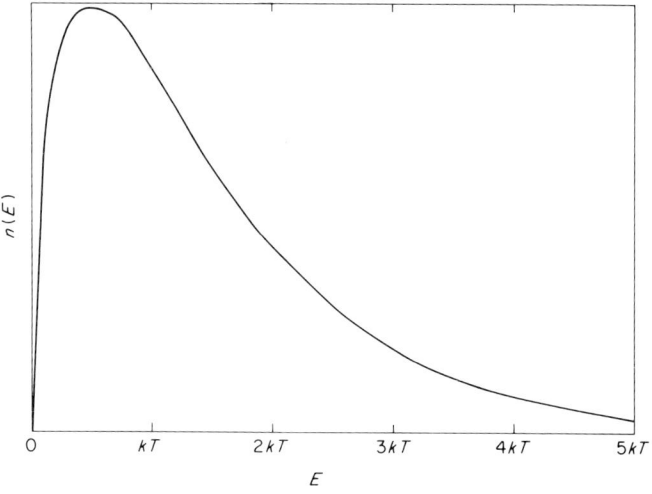

Fig. 9.9 Density of occupied states in the conduction band of a semiconductor as a function of energy from the bottom of the band.

The density of occupied levels $n(E)$ is given by the integrand of Eq. (9.43). Thus the maximum value of $n(E)$ is given by the maximum value of $E^{1/2} e^{-E/kT}$, which is readily seen to occur for $E = kT/2$. A plot of $n(E)$ versus E is given in Fig. 9.9. About 40 per cent of the electrons in the conduction band lie within kT of the bottom edge.

An identical analysis can be carried out for holes in terms of the distance of the Fermi level from the top of the valence band, an energy given in terms of our present symbols by $(E_G - E_f)$, where E_G is the bandgap, and

9.6 Thermal-Equilibrium Fermi Level

E_f is, as before, the positive energy difference between the Fermi level and the bottom of the conduction band. The calculation for holes leads to

$$p = N_v\, e^{-(E_G - E_f)/kT} \tag{9.49}$$

where N_v, the effective density of valence band states, is given by

$$N_v = 2\left(\frac{2\pi m_h^* kT}{h^2}\right)^{3/2} \tag{9.50}$$

In the case of cubic materials with ellipsoidal equal-energy surfaces like Ge and Si, the effective density of states must be calculated as

$$N_c = 2M_e\left(\frac{2\pi kT}{h^2}\right)^{3/2} m_\ell^{*1/2} m_t^* \tag{9.51}$$

for the conduction band (see Problem 6.11), where M_e is the number of equivalent minima, and as

$$N_v = 2\left(\frac{2\pi kT}{h^2}\right)^{3/2} (m_1^{*3/2} + m_2^{*3/2}) \tag{9.52}$$

for the valence band, where m_1^* and m_2^* are the effective masses of holes in the two valence bands.

The product of n and p from Eqs. (9.48) and (9.49) is

$$\begin{aligned} np &= N_c N_v\, e^{-E_G/kT} \\ &= 4\left(\frac{2\pi kT}{h^2}\right)^3 (m_e^* m_h^*)^{3/2}\, e^{-E_G/kT} \end{aligned} \tag{9.53}$$

In thermal equilibrium, the product of n and p is always a constant at a given temperature, depending only on the bandgap and the effective masses. Under conditions such that $n = p = n_i$, a material is called *intrinsic*; the density of free carriers, either electrons or holes, in an intrinsic material is given by

$$n_i = 2\left(\frac{2\pi kT}{h^2}\right)^{3/2} (m_e^* m_h^*)^{3/4}\, e^{-E_G/2kT} \tag{9.54}$$

By writing Eq. (9.48) for n_i,

$$n_i = N_c\, e^{-E_{fi}/kT} \tag{9.55}$$

and comparing with Eq. (9.54), shows that if $m_e^* = m_h^*$, $E_{fi} = E_G/2$.

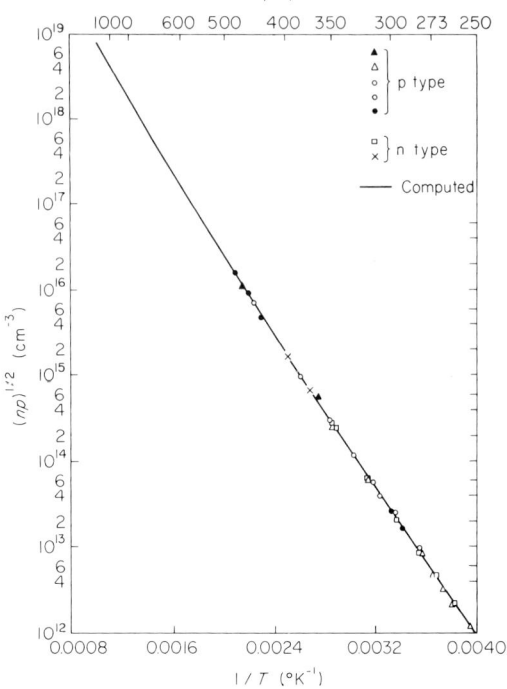
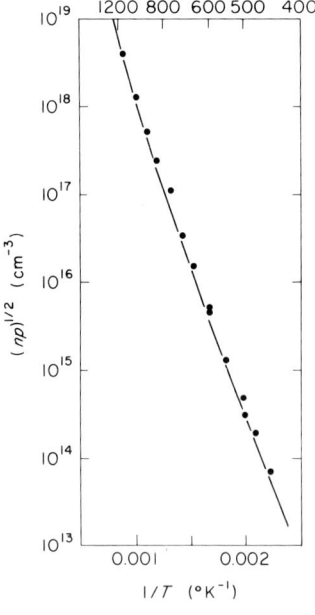

9.7 Fermi-Level Description of Electrical Conductivity

Figure 9.10 shows the variation of $(np)^{1/2}$ vs. T in the intrinsic range for Ge and Si.

OCCUPANCY OF IMPERFECTIONS

Once the location of the Fermi level is known, as by a measurement of the value of n or p in thermal equilibrium at a given temperature, then the occupancies of all other levels in thermal equilibrium are also known. A level lying E_1 below the conduction band, for example, has an occupancy given by

$$\frac{n_1}{N_1} = \frac{1}{(1/g) \, e^{(E_f - E_1)/kT} + 1} \qquad (9.56)$$

The additional factor of $(1/g)$ arises because of the possibility of effective degeneracy of the imperfection. An imperfection that can take on one electron, for example, may be able to take on an electron with either spin, thus giving a degeneracy of 2 to the level, which ought to be considered in the calculation of the Fermi distribution appropriate for such a level.

The inclusion of imperfection degeneracy effects can be effected in the following way. In this case the expression for the number of distinguishable permutations given in Eq. (4.60) should be replaced by

$$w_i = g^{n_i} \frac{G_i!}{n_i!(G_i - n_i)!} \qquad (9.57)$$

This expression results from the fact that if each imperfection is g-fold degenerate, permutations among the n_i occupied states results in g^{n_i} times as many distinguishable permutations as if g were equal to unity. If the rest of the calculation is carried out exactly the same as in Section 4.5, but with Eq. (9.57) used instead of Eq. (4.60) for the starting point, the result is the Fermi distribution used in Eq. (9.56).

9.7 Fermi-Level Description of Electrical Conductivity

The use of the Fermi level can be illustrated effectively by applying it to a few problems in electrical conductivity. Suppose that the conductivity

---

Fig. 9.10 Temperature dependence of the intrinsic carrier concentration $(np)^{1/2}$ in (a) germanium [after F. J. Morin and J. P. Maita, *Phys. Rev.* **94**, 1527 (1954)] and (b) silicon [after F. J. Morin and J. P. Maita, *Phys. Rev.* **96**, 28 (1954)].

is dominated by electrons, so that the conductivity may be expressed as

$$\sigma = ne\mu$$
$$= N_c e\mu \, e^{-E_f/kT} \quad (9.58)$$

Since the mobility is also in general a function of temperature, as well as N_c, we may rewrite Eq. (9.58) as

$$\sigma = C(T) \, e^{-E_f/kT} \quad (9.59)$$

In the special case of scattering by LA phonons, $\mu \propto T^{-3/2}$, and $C(T)$ is simply a constant independent of T, since $N_c \propto T^{+3/2}$. If in general we divide through by $C(T)$ in Eq. (9.59), take the logarithm of both sides, and then differentiate with respect to $(1/T)$, we obtain

$$\frac{d \ln[\sigma/C(T)]}{d(1/T)} = -\frac{E_f}{k} + \frac{T}{k}\frac{dE_f}{dT} \quad (9.60)$$

This means that a plot of $\ln[\sigma/C(T)]$ vs. $1/T$ yields a straight line if

$$E_f = E_{f0} + aT \quad (9.61)$$

where a is a constant, with a slope

$$\frac{d \ln[\sigma/C(T)]}{d(1/T)} = -\frac{E_{f0}}{k} \quad (9.62)$$

We now apply these concepts to the actual processes of imperfection ionization in a crystal. Consider as a basic example the case of a crystal containing N_D donors and no acceptors, the donors having an energy

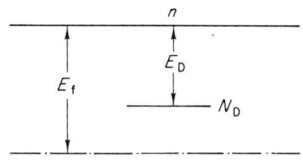

Fig. 9.11 Energy-level diagram for N_D donors with ionization energy E_D.

9.7 Fermi-Level Description of Electrical Conductivity

level lying E_D below the conduction band, as illustrated in Fig. 9.11. There are two equivalent approaches to determine the location of the Fermi level as a function of temperature: (a) the neutrality condition approach, and (b) the thermal equilibrium rate approach. Since both methods contain concepts of value for future discussions, we consider both in some detail.

NEUTRALITY CONDITION

The neutrality condition for the problem of N_D donors with an ionization energy E_D is

$$e' = N_D^{\bullet} \tag{9.63}$$

which in terms of the variables becomes

$$n = (N_D - n_D) \tag{9.64}$$

where n_D is the density of electron-occupied (un-ionized) donors, $N_D^{\times}$, i.e.,

$$N_D^{\bullet} = N_D^{(\text{total})} - N_D^{\times} \tag{9.65}$$

Equation (9.63) and the subsequent expressions neglect the thermal excitation of electrons from valence to conduction band. If this additional process is included, Eq. (9.64) becomes

$$n = (N_D - n_D) + p \tag{9.66}$$

For most materials for which $E_D < E_G/2$, neglect of p in Eq. (9.66) is a reasonable approximation.

Equation (9.64) may be rewritten as follows:

$$N_c e^{-E_f/kT} = N_D - \frac{N_D}{(1/g) e^{(E_f - E_D)/kT} + 1}$$

$$= N_D \frac{1}{1 + g e^{(E_D - E_f)/kT}} \tag{9.67}$$

Substituting $z = e^{-E_f/kT}$ and $w = e^{-E_D/kT}$, Eq. (9.67) becomes

$$z^2 + \frac{wz}{g} - \frac{N_D}{gN_c} w = 0 \tag{9.68}$$

The roots of this equation are

$$z = -\frac{w}{2g} \pm \frac{w}{2g} \left[1 + 4\left(\frac{N_D g}{N_c w}\right)\right]^{1/2} \tag{9.69}$$

Only the positive root has physical significance, and there are two limiting cases.

For $(N_D g/N_c w) \ll 1$,

$$z = \frac{N_D}{N_c} \tag{9.70}$$

or

$$E_f = kT \ln\left(\frac{N_c}{N_D}\right) \tag{9.71}$$

This is the condition that is obtained at high temperatures where the donors are completely ionized and $n = N_D$.

For $(N_D g/N_c w) \gg 1$,

$$z = \left(\frac{N_D w}{N_c g}\right)^{1/2} \tag{9.72}$$

since the first term on the right of Eq. (9.69) can be neglected in this limiting case, or

$$E_f = \frac{E_D}{2} + \frac{kT}{2} \ln\left(\frac{N_c g}{N_D}\right) \tag{9.73}$$

This is the condition obtained for low temperatures where $n \ll N_D$. Thus a plot of $\ln[\sigma/C(T)]$ vs. $1/T$ yields a straight line with slope $-E_D/2k$ terminating in a saturation of $\ln[\sigma/C(T)]$ with increasing T at a value of $\sigma = N_D e\mu$.

This approach to the problem via the neutrality condition may also be solved graphically. In this case it is easy to include the total neutrality condition of Eq. (9.66), which may be rewritten as

$$N_c e^{-E_f/kT} = \frac{N_D}{1 + g\, e^{(E_D - E_f)/kT}} + N_v\, e^{-(E_G - E_f)/kT} \tag{9.74}$$

Knowledge of N_c, N_v, N_D, E_D, and E_G permits the graphical plot shown in Fig. 9.12 for three temperatures. For sufficiently high temperatures, the graphical solution again yields E_f at $n = N_D$, and for sufficiently low temperatures E_f lies between E_D and $E_D/2$ below the conduction band and $n \ll N_D$. The rationale behind neglecting the p term in the neutrality condition is seen from Fig. 9.12 for all cases except those involving very high temperatures or very-small-bandgap semiconductors.

THERMAL EQUILIBRIUM RATES

For thermal equilibrium, the rate of thermal excitation from the donors to the conduction band is equal to the rate of capture of free electrons

9.7 Fermi-Level Description of Electrical Conductivity

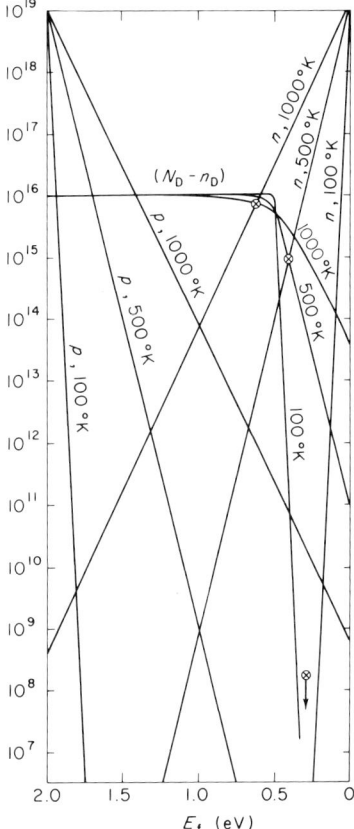

Fig. 9.12 Graphical solution of Eq. (9.74) for the determination of the location of the Fermi level in a semiconductor containing N_D donors with ionization energy E_D. Calculated for $N_c = N_v = 10^{19}$ cm^{-3}, $N_D = 10^{16}$ cm^{-3}, $E_D = 0.5$ eV, and $E_G = 2.0$ eV at $T = 100$, 500, and 1000°K. Note that the p term is negligible except for large values of E_f, e.g., even at 1000°K, it can be neglected if $E_f < 1.2$ eV. The intersections marking the solution values of E_f are represented by the symbol $\otimes$. The value of g has been taken equal to unity.

by empty donors. We may write this mathematically as

$$(N_D - n)\nu \, e^{-E_D/kT} = n \, S\langle v \rangle \, n \tag{9.75}$$

Here the quantity ν is usually referred to as the *attempt-to-escape* frequency; it is approximately the number of times per second an electron can absorb energy from the lattice vibrations, multiplied by the transition probability from the bound level to the conduction band. Fortunately we can relate it to other constants of direct interest in thermal equilibrium, as will be shown.

The Boltzmann factor $e^{-E_D/kT}$ expresses the probability of an electron having sufficient energy to be freed from the bound donor level.

The quantity S on the right of Eq. (9.75) is the *capture cross section* of an ionized donor for a free electron. The capture cross section describes the probability of a free electron being captured at that donor. The concept is quite similar to that of a scattering cross section, except that the capture cross section describes the probability of a carrier with a charge of a given sign being *captured* at an imperfection with a charge of a given sign. Thus the scattering cross section for an electron is about the same for a positively charged and a negatively charged scattering imperfection, but the capture cross section is much larger for the positively charged than for the negatively charged capturing imperfection. The rate of capture of n free electrons by N available imperfections can be written as

$$\text{Capture rate} = nN(Sv) \tag{9.76}$$

The *capture probability* is $N(Sv)$, and the *capture coefficient* is (Sv). The capture coefficient is composed of a quantity with the dimensions of area, S, the capture cross section, and v, the thermal velocity of a free carrier. The product (Sv) is the volume swept out per unit time by one imperfection in relative motion between the imperfection and a free carrier. Since the capture coefficient actually is a function of the energy of the free carrier, an average value $\langle (Sv) \rangle$ should be used. It has become customary to write this average as

$$\langle (Sv) \rangle = S \langle v \rangle \tag{9.77}$$

where $\langle v \rangle$ is the average thermal velocity of a carrier. In most of the future discussion, we indicate the average thermal velocity of electrons and holes by v_e and v_h respectively. Magnitudes of S cover a wide range depending on the nature of the imperfection and the mechanism for the capture process, experimental values having been found between 10^{-11} cm^2 for a Coulomb-attractive capture process at low temperature, to less than 10^{-22} cm^2 for a Coulomb-repulsive center at low temperatures. We consider theoretical ways of calculating values of capture cross section for various processes in Chapter 12. Experimentally measured extremes for electron capture with Coulomb attraction at a donor in Ge, and with Coulomb repulsion at a double acceptor in Si, are illustrated in Fig. 9.13.

Returning to Eq. (9.75), we see that $(N_D - n)$ is the density of electron-occupied (un-ionized) donors in view of the neutrality condition of Eq. (9.64), and that n is the density of unoccupied (ionized) donors. Thus the left side of Eq. (9.75) expresses the rate of thermal excitation into the con-

9.7 Fermi-Level Description of Electrical Conductivity

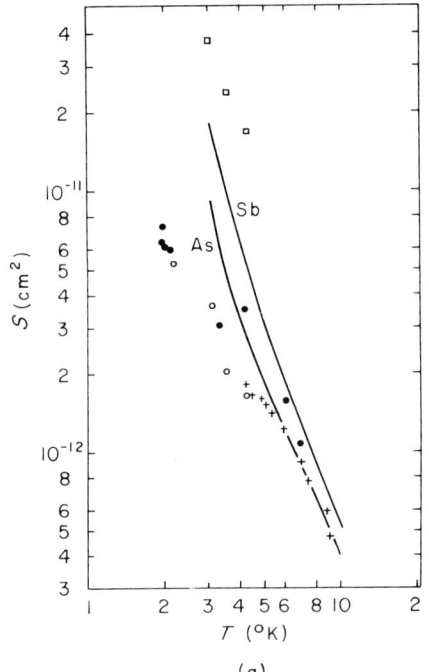

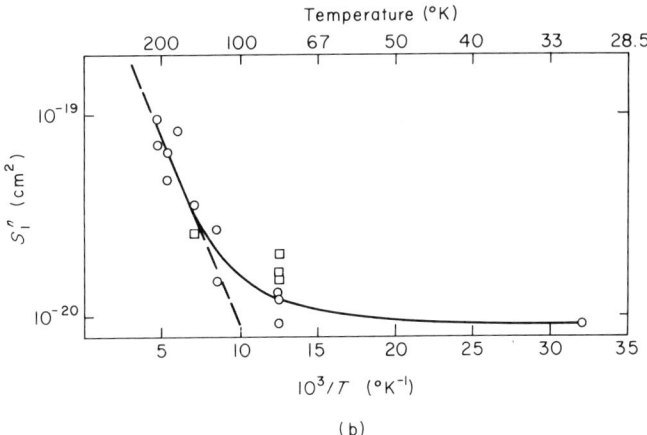

Fig. 9.13 Examples of extremes in the values of capture cross section for free electrons by imperfections. (a) Capture by positively charged donor impurities As or Sb in Ge. [After G. Ascarelli and S. Rodriguez, *Phys. Rev.* **124**, 1321 (1961).] (b) Capture by negatively charged acceptor impurity Zn in Si. [After A. Sklensky and R. H. Bube, *Phys. Rev.* **6**, 1328 (1972).]

duction band when there are $(N_D - n)$ occupied donors, and the right side expresses the rate of recombination between n free electrons and n unoccupied donors.

As mentioned above, the attempt-to-escape frequency ν can be related to other constants of the system. To see this, express Eq. (9.75) in its general form,

$$n_D \nu\, e^{-E_D/kT} = nSv_e(N_D - n_D) \tag{9.78}$$

and then express the concentrations n_D and $(N_D - n_D)$ in terms of the location of the Fermi level,

$$\frac{N_D}{(1/g)\, e^{(E_f-E_D)/kT} + 1}\, \nu\, e^{-E_D/kT} = nSv_e \frac{N_D}{1 + g\, e^{(E_D-E_f)/kT}} \tag{9.79}$$

If we replace n in Eq. (9.79) by Eq. (9.48) and cross multiply, we obtain

$$\nu = \frac{N_c S v_e}{g} \tag{9.80}$$

Thus the ratio of the attempt-to-escape frequency and the capture cross section is a constant at a given temperature, and given as a function of temperature by

$$\frac{\nu}{S} = \left(\frac{8}{g}\right) m_e^* \left(\frac{\pi}{h^2}\right)^{3/2} (kT)^2 \tag{9.81}$$

Using Eq. (9.80) for ν, Eq. (9.75) becomes

$$(N_D - n)N_c\, e^{-E_D/kT} = gn^2 \tag{9.82}$$

Setting $z = n/N_c$ and $w = e^{-E_D/kT}$, as before, we find that Eq. (9.82) is identical with Eq. (9.68) previously derived and solved for the neutrality-condition approach.

For partial ionization of donors, it may be seen directly from Eq. (9.82), as well as from the previous analysis, that when $n \ll N_D$,

$$n = \left(\frac{N_c N_D}{g}\right)^{1/2} e^{-E_D/2kT} \tag{9.83}$$

and

$$\sigma = C(T) \left(\frac{N_D}{gN_c}\right)^{1/2} e^{-E_D/2kT} \tag{9.84}$$

according to Eq. (9.59). Now it is always possible that the donor ionization energy E_D is itself a function of temperature, e.g., $E_D = E_{D0} + bT$. In this case,

$$\sigma = C(T) \left(\frac{N_D}{gN_c}\right)^{1/2} e^{-b/2k}\, e^{-E_{D0}/2kT} \tag{9.85}$$

9.7 Fermi-Level Description of Electrical Conductivity

If the ionization energy itself, therefore, is a linear function of temperature, the value computed from a plot of $\ln[\sigma/C(T)]$ vs. $1/T$ is E_{D0}.

Example 9.4 Consider the more realistic example of a crystal with both donors and acceptors present, with $N_D > N_A$, as illustrated in Fig. 9.14, and calculate the temperature dependence of the Fermi level by both the neutrality condition and the thermal equilibrium rate analysis.

Neutrality Condition. The N_A acceptors can be considered to be completely ionized (compensated, filled with electrons). The neutrality condition now is

$$n + N_A = (N_D - n_D) + p$$

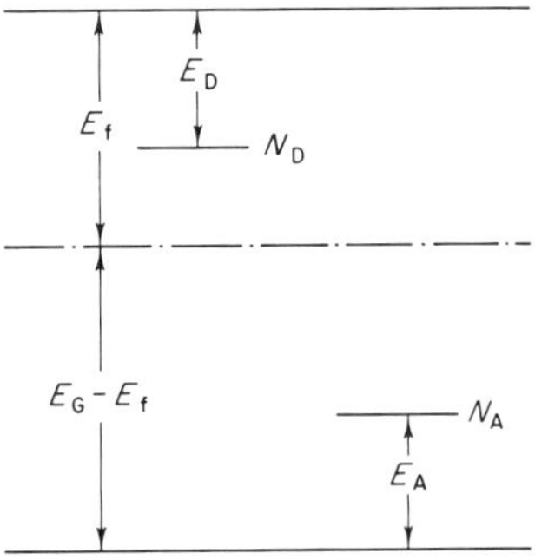

Fig. 9.14 Energy level diagram for N_D donors with ionization energy E_D, and N_A acceptors with ionization energy of E_A.

If we neglect the p in the neutrality condition because $N_D > N_A$, we may rewrite it as before as

$$N_c e^{-E_f/kT} + N_A = \frac{N_D}{1 + g\, e^{(E_D - E_f)/kT}}$$

or in terms of the previous notation,

$$z^2 + \left(\frac{w}{g} + \frac{N_A}{N_c}\right)z + \frac{w}{gN_c}(N_A - N_D) = 0$$

with the solution

$$z = -\frac{1}{2}\left(\frac{w}{g} + \frac{N_A}{N_c}\right) + \frac{1}{2}\left[\left(\frac{w}{g} + \frac{N_A}{N_c}\right)^2 + 4\frac{w}{gN_c}(N_D - N_A)\right]$$

There are three ranges of interest.

For very low free-electron concentrations, i.e., only slight ionization of the uncompensated donors, $w/g \ll N_A/N_c$ and $(N_A/N_c)^2 \gg 4(w/gN_c)(N_D - N_A)$. We obtain

$$z = \frac{w}{g} \frac{N_D - N_A}{N_A}$$

or

$$E_f = E_D + kT \ln\left(\frac{gN_A}{N_D - N_A}\right)$$

A plot of $\ln[\sigma/C(T)]$ vs. $1/T$ gives a straight line with slope $-E_D/k$,

$$\sigma = C(T)\left(\frac{N_D - N_A}{gN_A}\right) e^{-E_D/kT}$$

The effect of compensation is therefore in this range to change the activation energy of dark conductivity from $E_D/2$ to E_D.

For intermediate free-electron concentrations, $w/g \ll N_A/N_c$ and $(N_A/N_c)^2 \ll 4(w/gN_c)(N_D - N_A)$. We then obtain

$$z = \left[\frac{w(N_D - N_A)}{gN_c}\right]^{1/2}$$

and

$$E = \frac{E_D}{2} + \frac{kT}{2} \ln\left(\frac{gN_c}{N_D - N_A}\right)$$

The activation energy changes back to $E_D/2$ as in the uncompensated case.

Finally, for very high electron concentrations, when the donors are all ionized, $w/g \gg N_A/N_c$ and $(w/g)^2 \gg 4(w/gN_c)(N_A - N_D)$. We obtain

$$z = \frac{N_D - N_A}{N_c}$$

and

$$E_f = kT \ln\left(\frac{N_c}{N_D - N_A}\right)$$

This result corresponds to $n = N_D - N_A$.

Thermal Equilibrium Rates. The thermal equilibrium rate equation is

$$(N_D - N_A - n)N_c S v_e e^{-E_D/kT} = gnSv_e(n + N_A)$$

which may be rewritten as

$$(N_D - N_A)N_c w - N_c^2 zw = gN_c^2 z^2 + gN_c N_A z$$

9.7 Fermi-Level Description of Electrical Conductivity

Dividing through by gN_c^2 and rearranging terms leads to the same basic quadratic equation in z just discussed in its various limiting cases. The three ranges described correspond to $n \ll N_A$, $(N_D - N_A)$; $N_A < n < (N_D - N_A)$; and $n \approx (N_D - N_A)$.

The Intrinsic Case

We may apply the same kind of approach to conductivity under intrinsic conditions. The neutrality condition is simply

$$n = p \tag{9.86}$$

From the equality

$$N_c\, e^{-E_f/kT} = N_v\, e^{-(E_G - E_f)/kT} \tag{9.87}$$

one obtains directly

$$E_f = \frac{E_G}{2} + \frac{kT}{2} \ln\left(\frac{N_c}{N_v}\right) \tag{9.88}$$

or

$$E_f = \frac{E_G}{2} + \frac{3kT}{4} \ln\left(\frac{m_e^*}{m_h^*}\right) \tag{9.89}$$

This is the result pointed out earlier in connection with Eq. (9.55).

The intrinsic case may also be treated by the thermal equilibrium rate approach,

$$(N_v - n)N_c\, e^{-E_G/kT} = np = n^2 \tag{9.90}$$

yielding the same result of Eq. (9.89).

General Treatment

The previous treatments for extrinsic cases have been approximations, for not only has p been consistently neglected (probably justified in an n-type material) in the neutrality equation, but also N_A, the total density of acceptors, has been used as an approximation to n_A, the density of electron-occupied acceptors. The complete neutrality condition including free electrons, free holes, one type of donor, and one type of acceptor is

$$n + n_A = p + (N_D - n_D) \tag{9.91}$$

or, using the previous notation and with $G \equiv e^{-E_G/kT}$ and $x \equiv G\, e^{E_A/kT}$,

$$N_c z + \frac{N_A}{1 + x/gz} = \frac{N_v G}{z} + \frac{N_D}{1 + gz/w} \tag{9.92}$$

We can determine the Fermi level graphically as in Fig. 9.12 for specific varies of the variables, or we can attempt to simplify the quartic equation in z represented by Eq. (9.92),

$$z^4 + \left(\frac{w+x}{g} + \frac{N_A}{N_c}\right)z^3 + \left[w\left(\frac{x}{g^2} + \frac{N_A - N_D}{gN_c}\right) - G\frac{N_v}{N_c}\right]z^2$$

$$- \left[\frac{N_v G(w+x)}{gN_c} + \frac{wN_D x}{g^2 N_c}\right]z - \frac{N_v Gwx}{g^2 N_c} = 0 \qquad (9.93)$$

Certain simplifications can be made with reasonable generality. For example, for a donor in the top half of the bandgap and an acceptor in the bottom half of the bandgap, it is always true that $w \gg x$, and therefore the x can be neglected in those terms involving $(w + x)$. The result is the only slightly less general quartic,

$$z^4 + \left(\frac{w}{g} + \frac{N_A}{N_c}\right)z^3 + \left[\frac{w}{g}\left(\frac{x}{g} + \frac{N_A - N_D}{N_c}\right) - G\frac{N_v}{N_c}\right]z^2$$

$$- \frac{w}{gN_c}\left(N_v G + \frac{N_D x}{g}\right)z - \frac{N_v Gwx}{g^2 N_c} = 0 \qquad (9.94)$$

Now the terms involving the factor G appear only because of the retention of the p term in the neutrality condition of Eq. (9.91). In an even slightly n-type material, it is likely that the terms involving G present only a minor influence on the behavior. If it is possible to omit them, Eq. (9.94) can be reduced to a cubic in z,

$$z^3 + \left(\frac{w}{g} + \frac{N_A}{N_c}\right)z^2 + \frac{w}{g}\left(\frac{x}{g} + \frac{N_A - N_D}{N_c}\right)z - \frac{wN_D x}{g^2 N_c} = 0 \qquad (9.95)$$

(Of course, in the case of slightly p-type material, a similar reduction can be made by neglecting the contribution of terms involving n in the neutrality condition.)

Now if x/g is small, so that it can be neglected in the third term of Eq. (9.95), and so that the fourth term in Eq. (9.95) can be neglected, Eq. (9.95) reduces to exactly the same equation previously treated in Example 9.4 for compensated donors. This calculation corresponds to the situation where the compensation is not very close. The dark conductivity as a function of temperature gives an activation energy of either E_D or $E_D/2$, depending on the temperature range.

If, on the other hand, x/g is large, corresponding to the case for close compensation between donors and acceptors, new possibilities occur. Eq.

9.7 Fermi-Level Description of Electrical Conductivity

(9.95) becomes

$$z^3 + \left(\frac{w}{g} + \frac{N_A}{N_c}\right)z^2 + \frac{wx}{g^2}z - \frac{wN_Dx}{g^2N_c} = 0 \qquad (9.96)$$

For this case of close compensation, $z < w$. For any case where $N_A \leq N_D$, $x < z$. Incorporating these considerations reduces Eq. (9.96) to

$$z^2\left(\frac{w}{g} + \frac{N_A}{N_c}\right) = \frac{wN_Dx}{g^2N_c} \qquad (9.97)$$

There are two limiting cases; if $w/g \ll N_A/N_c$,

$$z = \left(\frac{wN_Dx}{g^2N_A}\right)^{1/2} \qquad (9.98)$$

or

$$E_f = \frac{E_G - E_A + E_D}{2} + \frac{kT}{2}\ln\left(\frac{g^2N_A}{N_D}\right) \qquad (9.99)$$

This means that under these conditions of close compensation, the activation energy measured from a plot of $\ln[\sigma/C(T)]$ vs. $1/T$ is neither E_D nor $E_D/2$, but is rather $E_D + (E_G - E_A - E_D)/2$.

For $w/g \gg N_A/N_c$,

$$z = \left(\frac{N_Dx}{gN_c}\right)^{1/2} \qquad (9.100)$$

or

$$E_f = \frac{E_G - E_A}{2} + \frac{kT}{2}\ln\left(\frac{gN_c}{N_D}\right) \qquad (9.101)$$

Under these conditions the primary supply of free electrons is by thermal excitation from the acceptors.

Example 9.5 Some of the energy levels formed when Cu impurity is substitutionally incorporated for Zn in ZnSe crystals are shown in Fig. 9.15. As usual the solid level line indicates the energy of an electron which may be raised to the conduction band if sufficient energy is supplied. The dashed energy levels represent the energy level to which an electron must be replaced.

The electronic transitions represented by arrows are:

(1) $Cu'_{Zn} \xrightarrow{1.98 \text{ eV}} Cu^\times_{Zn} + e'$

(2) $Cu^\times_{Zn} \xrightarrow{0.72 \text{ eV}} Cu'_{Zn} + h^\cdot$

(3) $Cu^{\cdot}_{Zn} + e' \rightarrow Cu^\times_{Zn} + 2.35 \text{ eV}$

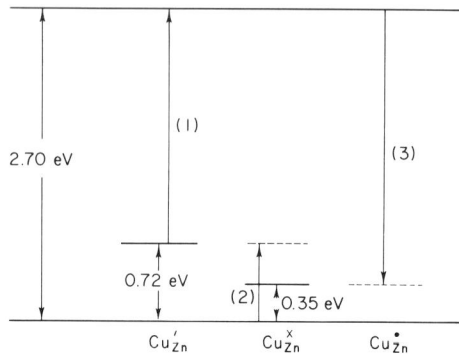

Fig. 9.15 Possible levels for Cu impurity in ZnSe.

Describe the relative concentrations of Cu'_{Zn}, Cu^x_{Zn} and $Cu^{\cdot}_{Zn}$ centers for various locations of the Fermi level throughout the bandgap.

If the concentration of some very shallow donors (e.g., interstitial zinc) is much larger than the copper concentration, then all the Cu is in the Cu'_{Zn} state. If the Cu concentration is N_{Cu},

$$[Cu'_{Zn}] = N_{Cu} \quad \text{and} \quad [Cu^x_{Zn}] = [Cu^{\cdot}_{Zn}] = 0$$

for this case. The brackets around the Cu'_{Zn} signify "concentration of."

If the concentration of some very shallow acceptors (no real ones known) is sufficient to bring the Fermi level down to something like 0.05 eV above the valence band, all of the Cu is in the $Cu^{\cdot}_{Zn}$ states, and

$$[Cu^{\cdot}_{Zn}] = N_{Cu} \quad \text{and} \quad [Cu'_{Zn}] = [Cu^x_{Zn}] = 0$$

If only copper impurity is present, and all other donors and acceptors are absent, the Fermi level must lie halfway between the Cu'_{Zn} level and the $Cu^{\cdot}_{Zn}$ level in order to produce charge neutrality,

$$[Cu'_{Zn}] + n = [Cu^{\cdot}_{Zn}] + p$$

where the densities of the free carriers can be neglected because of the energies involved. Thus the Fermi level lies about 0.53 eV above the valence band, giving

$$[Cu'_{Zn}] = [Cu^{\cdot}_{Zn}] = \frac{N_{Cu}}{e^{(0.19\,eV)/kT} + 1} = 10^{-3} N_{Cu}$$

$$[Cu^x_{Zn}] = N_{Cu}(1 - 2 \times 10^{-3}) = 0.998 N_{Cu}$$

Most of the Cu is in the neutral state (with respect to the crystal).

9.8 Imperfection Interactions

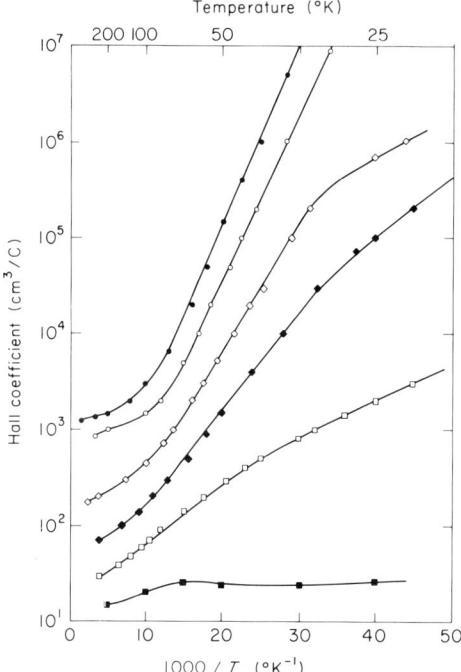

Fig. 9.16 Temperature dependence of the Hall coefficient for the samples of CdS:Ga for which the temperature dependence of mobility is given in Fig. 8.10. The Hall coefficient (see Chapter 10) is given approximately by $(1/ne)$ and is therefore inversely proportional to the electron density. In order of increasing Hall coefficient the density of Ga donors are 6×10^{17} cm^{-3}, 2×10^{17} cm^{-3}, 2×10^{16} cm^{-3}, and no donors were intentionally added for the three highest curves. The curves show the exponential variation of electron density with temperature and the decrease in electron ionization energy with increasing Ga density. The curves with the lowest donor densities show saturation of electron density at high temperatures corresponding to donors empty. (After W. W. Piper and R. E. Halsted, "Proc. Intern. Conf. Semiconductor Physics, Prague, 1960," Czechoslovak Academy of Sciences, Prague, 1961, pp. 1046–1048.)

Typical variation of carrier density with temperature and impurity density is shown in Fig. 9.16 for the CdS:Ga samples for which the temperature dependence of mobility was shown in Fig. 8.10.

9.8 Imperfection Interactions

As long as the density of imperfections in a crystal is sufficiently low, interactions between imperfections are minimized. As the concentration of imperfections increases, however, a variety of interaction effects may be expected.

Decrease in Ionization Energy of Donors and Acceptors

The ionization energy of a donor or acceptor imperfection decreases as the concentration of the imperfection increases into the range where imperfection interaction is possible. One contribution to this effect is the screening of bound carriers in large "orbits" around imperfection centers from the charge of the center by the free carriers in the system. A second contribution is the effect of the beginning of the formation of *impurity bands*.

Impurity Bands

When the concentration of imperfections becomes sufficiently high that there is appreciable overlap of the wavefunctions of the bound carriers at neighboring impurity sites, interaction effects give rise to impurity bands in much the same way that the application of the tight-binding LCAO approach gave energy bands for interacting atoms (see Section 5.5). When impurity banding occurs, charge transport directly from imperfection to imperfection becomes possible, the necessity for ionization of the imperfection to obtain conductivity disappears, and the total conductivity is the sum of the conductivity in the conduction band and in the impurity band. Since the impurity band is likely to be a fairly narrow band compared to the regular bands of the periodic lattice, the imperfection-band mobilities are considerably smaller than those of the free carriers and may be characterized by different physical processes such as hopping. For shallow impurities, the impurity band may overlap the nearest crystal band to produce a degenerate material.

Coulombic Interactions

A very simple model describing the Coulombic interaction effects between a negatively charged acceptor and a positively charged donor with

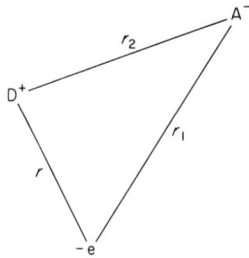

Fig. 9.17 Diagram for estimating the effects of an ionized acceptor on the binding energy of a donor imperfection for an electron.

9.8 Imperfection Interactions

its electron in a large "orbit" may be formulated. Given the situation shown in Fig. 9.17, the ionization energy of the electron may be written

$$|E_\text{I}| = \frac{e^2}{\varepsilon}\left(\frac{1}{r} - \frac{1}{r_1}\right) \tag{9.102}$$

As long as the acceptor can be considered to be relatively far from the donor center, compared with the size of the electron "orbit," then $r_1 \approx r_2 \approx N_\text{I}^{-1/3}$, where N_I is the concentration of donors or acceptors, if these are about equal in a nearly compensated material. Setting $e^2/(\varepsilon r) = E_{\text{I}0}$, the ionization energy in the absence of the interacting acceptor, we obtain

$$|E_\text{I}| = |E_{\text{I}0}| - \frac{e^2}{\varepsilon} N_\text{I}^{1/3} \tag{9.103}$$

Many of the experimentally observed concentration-dependent ionization energies follow a relation that is at least formally similar to Eq. (9.103).

Pair Formation

Two imperfections with opposite charges, such as a positively charged donor and a negatively charged acceptor, find it energetically favorable to exist as pairs in the crystal. Evidence for the existence of such pairs has grown appreciably in recent years. If E_D is the ionization energy of the donor member of the pair and E_A is the ionization energy of the acceptor member, the energy separation between donor and acceptor levels (such as would be detected in terms of the photon energy of radiation emitted in a transition from the electron-occupied donor to the hole-occupied acceptor), is given by

$$\varDelta E = E_\text{G} - \left(E_\text{D} + E_\text{A} - \frac{e^2}{\varepsilon r}\right) \tag{9.104}$$

where r is the spacing between the donor and acceptor imperfections in the crystal. The closer the two members of the pair involved in the transition are to each other, the larger is the value of $\varDelta E$. Since the transition probability decreases exponentially with spacing r, it follows that the light emitted in the recombination process is of higher energy for higher intensity excitation (as the more distant pairs become saturated), and that the light emitted in the recombination process at long times of decay after stopping excitation of the crystal is of lower energy than at short times (as the more closely spaced centers recombine first). Since r must vary by discrete increments because of the crystal lattice structure, a series of emis-

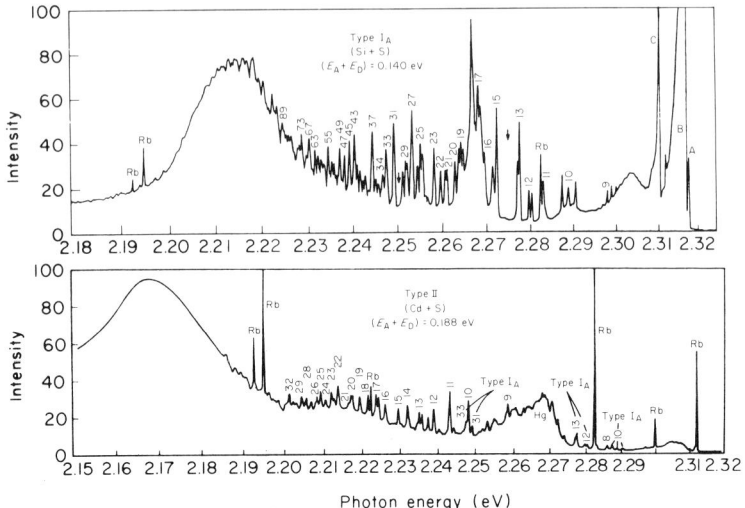

Fig. 9.18 Examples of pair-emission spectra for Si–S and Zn–S pairs in GaP at 1.6°K. Lines marked Rb are rubidium calibration lines. A, B, and C are bound-exciton lines. The small numbers indicate the indices of the successive values of r in Eq. (9.104) as determined by the lattice. [After D. G. Thomas, M. Gershenzon, and F. A. Trumbore, *Phys. Rev.* **133**, A269 (1964).]

sion *lines* may be observed under suitable conditions, which may be indexed in terms of the values of r provided by the crystal structure. (Further discussion of radiative recombination processes is found in Chapter 12.) Discrete line spectra for pair recombination processes in GaP are shown in Fig. 9.18.

9.9 Device Applications Describable by the Band Picture of Imperfect Semiconductors

There is unfortunately not enough space in a book such as this to treat in any detail all of the many significant device applications that have followed from an understanding of the band picture in imperfect semiconductors. It would be an unforgivable neglect, however, if no word was made of some of these applications, which illustrate the close relationship between theory and practice in the solid state field.

The first and most striking success of the energy-band picture was in describing the differences in electrical properties of insulators, semiconductors, and metals in terms of a single simple model. The whole field of solid-state research supplies other examples of successes. In semiconductor research, for example, the understanding of p-n junctions, optical absorp-

9.9 Device Applications

tion, transistor behavior, photoconductivity, photovoltaic effects, Gunn effect, solid-state light emitters and lasers, to mention only a few of the most common, has been based on the band picture of solids. We select here just a few of these possibilities which embody in a unique way a utilization of the energy-band picture in imperfect materials for their description, and offer a purely qualitative description.

p–n Junctions

It is no exaggeration to say that the p–n junction is one of the most significant and versatile structures to be developed in recent decades. When an n-type form of a given material is produced (i.e., the material with donor imperfections which provide free electrons) in contact with a p-type form of the same material (or even of a different material) (i.e., the material with acceptor imperfections which provide free holes), electrons flow from the n-type to the p-type side and holes flow from the p-type to the n-type side until the potential barrier that is formed just counterbalances this flow. The result is the p–n junction shown in various applications in Fig. 9.19. Such a p–n junction can be used to rectify, amplify, detect light, convert energy, and emit light, depending on the specific application.

Figure 9.19a illustrates the use of a p–n junction as a rectifier. When such a junction is biased by the application of an electric field in the reverse direction (n-side positive), there is a high resistance in the dark, since the p-type region cannot supply free electrons to the n-type region, and the n-type region cannot supply free holes to the p-type region. For bias in the forward direction, however, (p-side positive) there is enhanced injection of electrons from the n-type side into the p-type side, and enhanced injection of holes from the p-type side into the n-type side. The result of applying an alternating electric field to a p–n junction is therefore to produce rectification according to the typical diode rectifier current versus voltage plot shown in Fig. 9.20.

If an additional semiconductor region is added to a p–n junction to produce either an n–p–n or a p–n–p junction, as shown in Fig. 9.19b, amplification is possible. Consider the n–p–n junction for the following description. If an excess density of holes is injected into the p-type region by optical absorption or by application of a suitable electric field, the presence of these additional holes reduces the height of the barrier at the forward-biased p-n_1 junction and permits an increase of electrons to enter the p-type region from the n_1 region. This increased flow of electrons rhrough the p region continues until the excess positive charge either tecombines with these electrons or diffuses out over the n_1–p junction.

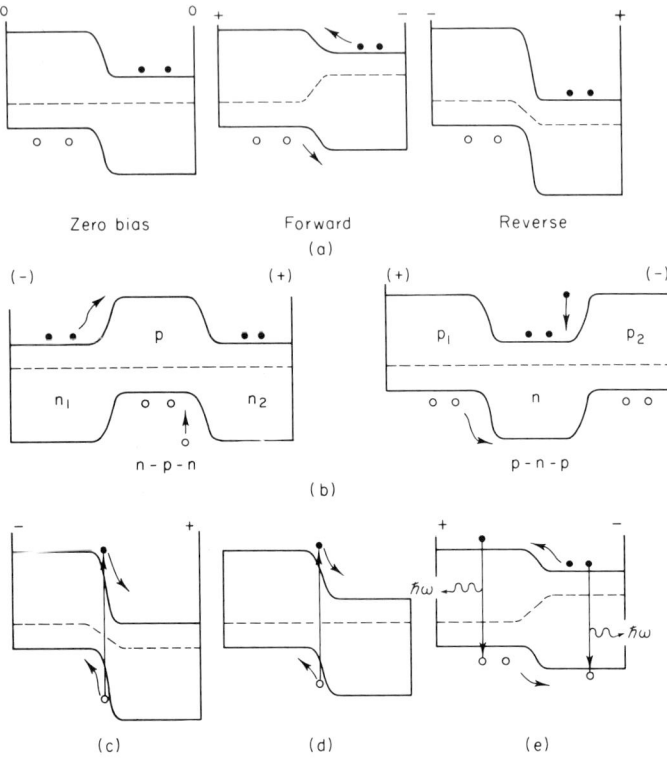

Fig. 9.19 Applications of the p–n junction to (a) diode rectification, (b) amplification, (c) radiation detection, (d) energy conversion, and (e) light emission in either incoherent or coherent (laser) form.

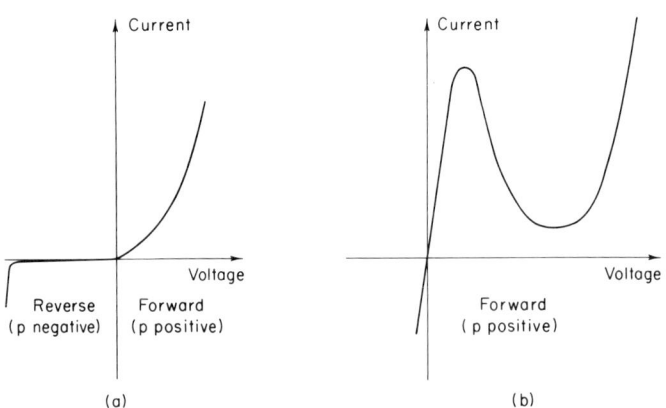

Fig. 9.20 Comparison of current versus voltage characteristics for (a) a normal diode and (b) a tunnel diode.

9.9 Device Applications

Since many electrons may flow through the device per hole injected, amplification is achieved.

Figure 9.19c shows the p–n junction in use as a detector of radiation. Electron–hole pairs created by optical radiation within a diffusion length of the junction are separated by the junction and are collected by a reverse bias. One electronic charge crosses the junction for each photon absorbed. The n–p–n or p–n–p junctions can also be used as radiation detectors, as suggested above; because of the amplification factor of this structure, more than one electronic charge is collected for each photon absorbed.

Figure 9.19d shows the p–n junction in use as an energy converter, converting radiation energy into electrical energy. This is the type of device commonly used in solar cells to convert energy from the sun's radiation into electricity. No electric field is applied externally to the junction. Absorption of light creates free carriers which are separated by the junction. If the device is run in short-circuit configuration (load resistance less than the cell internal resistance), a steady short-circuit current can be drawn. If the device is run in open-circuit configuration (load resistance much larger than cell resistance), an open-circuit voltage is built up, the maximum value of which is equal to the difference between the Fermi levels of the separated n- and p-type material. The polarity of this open-circuit voltage is the same as for forward bias.

Figure 9.19e shows the p–n junction in use as a radiation emitter. In this application the junction is forward biased so that electrons and holes flow across the junction in opposite directions. If recombination of these electrons and holes occurs via transitions in which the energy of the excited carrier is given off as photons, luminescence results. Frequently this luminescence occurs as the result of recombination of electrons and holes at particular imperfections on either the p-type or n-type sides (or both), or by direct recombination of free electrons and free holes. Normally this recombination radiation is incoherent (i.e., with random phases), but if the injection level is high enough to cause appreciable population inversion in the recombination regions, stimulated emission becomes a dominant process and coherent-radiation-emitting lasers result.

Tunnel Diode[†]

A tunnel diode is a particular kind of p–n junction capable of giving a negative-resistance region, a characteristic useful in making oscillators. It is a device in which band properties play a dominant role.

[†] L. Esaki, *Phys. Rev.* **109**, 603 (1958).

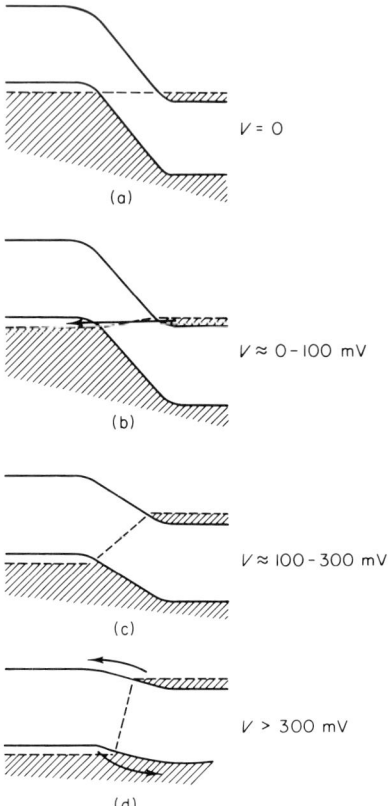

Fig. 9.21 Description of the operation of a tunnel diode with increasing forward bias voltage applied in terms of the energy-band diagram of the junction. (a) No tunneling; zero current. (b) Strong tunneling; current reaches maximum. (c) No available empty states for tunneling; current decreases. (d) Normal forward current dominates.

The current–voltage characteristic of a normal p–n junction is compared with that for a typical tunnel diode in Fig. 9.20. The tunnel diode is a p–n junction formed between degenerate p-type material and degenerate n-type material, as indicated in Fig. 9.21a for no applied external electric field. The form of the observed current–voltage characteristic can be explained in terms of the band picture as follows.

For zero applied voltage, the current is also zero. When a voltage is applied such that the p side is positive, the possibility of current flow begins. Current can flow either by the motion of electrons from the n-type side over the potential barrier or by the motion of holes from the p-type side over the potential barrier, both very small contributions, or by the

9.9 Device Applications

tunneling of electrons from the conduction band of the n-type side into empty states at the top of the valence band in the p-type side. The maximum current flows when the occupied states in the n-type material are energetically exactly opposite (the same as) the unoccupied states in the p-type material, as shown in Fig. 9.21b. If the applied voltage is raised still further, the occupied states in the n-type side now are energetically the same as the forbidden gap in the p-type side, and the current decreases as shown in Fig. 9.21c, since there are now no available states into which to tunnel. Finally, if the applied voltage is raised still further, the normal forward p–n junction current caused by electrons and holes passing over the greatly reduced potential barrier dominates, as in Fig. 9.21d.

Gunn Effect[†]

A negative-resistance characteristic is also found when moderately high electric fields are applied to a crystal with the appropriate band structure. In GaAs, for example, the effect depends upon the presence of a high-effective-mass conduction band minimum lying about 0.36 eV above a low-effective-mass conduction band minimum, as shown in Fig. 9.22. When the field is raised sufficiently, electrons are scattered from the lower-lying low-mass band into the higher-lying high-mass band with a consequent decrease in conductivity because of a decrease in carrier mobility.

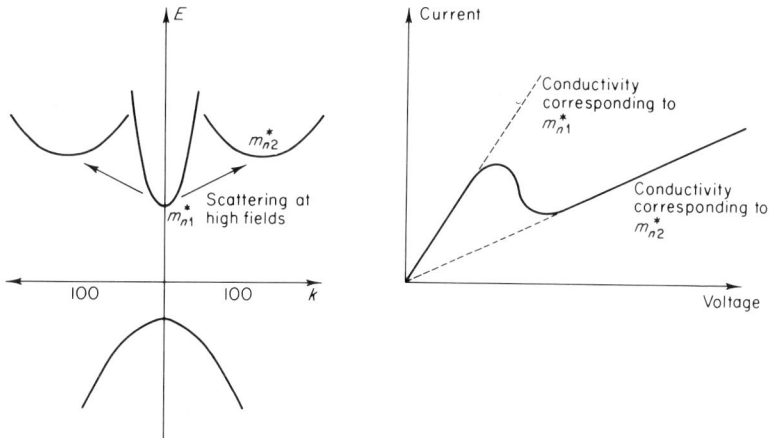

Fig. 9.22 Gunn effect in GaAs. Scattering of electrons from the low-mass m_{n1}^* conduction band to the higher-mass m_{n2}^* conduction band for elevated electric fields results in negative resistance.

[†] J. B. Gunn, *IBM Journal* **8**, 141 (1964).

Néel Switches

Materials like V_2O_3 show a very large change in conductivity by over eight orders of magnitude for a very small temperature change at the so-called Néel temperature. Figure 9.23 shows typical data for a single crystal specimen. A phase transition occurs at the critical temperature, accompanied by a large volume change and rearrangement of the crystal structure. As a result of the rearrangement, the material changes from a semiconductor below the Néel temperature to a metal (or semimetal) above the Néel temperature.

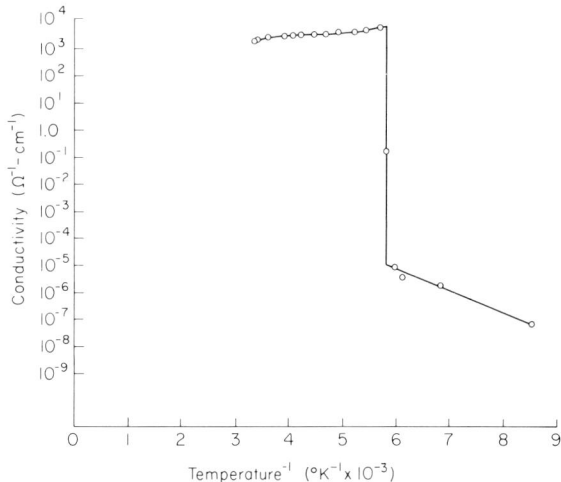

Fig. 9.23 Temperature dependence of conductivity through the phase transition at the Néel temperature of 168°K for V_2O_3. [After G. Goodman, *Phys. Rev. Letters* **9**, 305 (1962).]

Amplification via Negative-Effective-Mass Effects: A Failure

Failures are sometimes as interesting as successes. It was pointed out in connection with Fig. 6.13 that the reentrant type of energy contours found for heavy holes in Ge and Si results in the existence of negative transverse effective masses within a cone close to the 100 axis. Motivated by a search for negative-resistance effects, the suggestion was made[†] that amplification of a radio-frequency electric field applied perpendicular to the 100 direction could be achieved if the hole population could be significantly shifted into

---

[†] H. Kroemer, *Phys. Rev.* **109**, 1856 (1958).

9.10 Summary 353

the negative-mass cone by applying a strong dc electric field bias in the 100 direction. This ingenious attempt to utilize distinctly band characteristics of solids for device purposes failed because unfavorable scattering processes prevented concentration of the heavy holes in the negative-mass cone.

9.10 Summary

Many of the scientifically and technologically significant properties of solids depend on the existence and properties of imperfections. Often the concentration of these imperfections, either intrinsic defects or extrinsic impurities, needed for desired effects is very small, and the practical achievement of the desired properties is a complex problem is materials research.

The description of imperfection incorporation, properties, and interactions is facilitated by the choice of a notation language for describing such imperfections. We have chosen a notation that describes the charge of an imperfection in relation to the charge of an atom at that site in the perfect crystal. Imperfections can be characterized in terms of the categories of identity, location, and behavior. A full investigation of the way in which the density of imperfections depends on the preparation conditions is beyond the scope of this book, but is a well-treated problem in physical chemistry of defects in solids. References in the Bibliography will guide further study in this particular area.

The concept of an imperfection *center* is a key idea in the description of imperfection properties. This means that many times it is not simply the electronic properties of the imperfection itself that must be considered, but the effect of the presence of the imperfection on the neighboring atoms of the crystal. Often the volume covered by the wavefunction of the electronic state associated with an imperfection extends over many hundreds of atoms in the crystal. This is particularly true for imperfection centers with a small ionization energy, which can be treated quantum mechanically fairly directly and which can be represented by a hydrogenic model embedded in the crystal with the mass of the bound carrier replaced by the effective mass of the carrier in the adjacent band. More highly localized imperfection states, corresponding to larger ionization energies, are still the subject of active theoretical research.

Knowledge of the occupancy of various imperfection levels in a crystal is necessary for a systematic description of the electronic properties. A convenient concept for the definition of the occupancy is the thermal-equilibrium Fermi level, the location of which can be defined from a knowledge only of the density of free carriers at a given temperature,

$$E_f = kT \ln\left(\frac{N_c}{n}\right)$$

or

$$(E_G - E_f) = kT \ln\left(\frac{N_v}{p}\right)$$

where E_f is the positive energy separation between the Fermi level and the bottom of the conduction band and E_G is the width of the bandgap. Once the location of the Fermi level is known, then the occupancy of all other levels and of the conduction and valence bands is also known. One significant consequence of the investigation of the properties of the Fermi level is that the product of the densities of free electrons and free holes is always constant in a semiconductor in thermal equilibrium at a fixed temperature.

The description of the temperature dependence of electrical conductivity in a semiconductor or insulator is a natural application of the Fermi-level concept. We have shown how the temperature dependence of the Fermi level, and hence the corresponding activation energy for dark conductivity as a function of temperature, can be calculated for several representative systems of donor and acceptor levels by invoking either the principle of charge neutrality or of equal thermal equilibrium excitation-and-capture rates. In materials not closely compensated, the activation energy for dark conductivity is found to be equal to, or equal to one-half of, the ionization energy of the dominant imperfection. In closely compensated materials, however, a third possibility exists in which the activation energy is not related simply to that of the dominant imperfection.

When the density of imperfections becomes sufficiently large, interaction effects between different imperfections may play an important role in the resulting properties. One of the most prominent of these interaction effects is the formation of imperfection bands that makes it possible for conductivity to occur via imperfection levels only, without it being necessary for thermal excitation to the bands of the perfect crystal to take place. There is considerable evidence that electronic properties associated with imperfections in large-bandgap materials are more often than not associated with a complex of more than one imperfection, forming a specific type of active center. One of the most common of these complexes is a pair of imperfections of opposite charge.

In conclusion we reviewed in a qualitative way some examples of the success of a band picture of imperfect semiconductors in describing effects that have proved to be of considerable technological significance.

Problems

9.1 Write all possible incorporation reactions, using both the ionic and the covalent picture, for substituting Li and O impurities in BN.

9.2 Show that the expression for the number of distinguishable permutations for imperfection levels with degeneracy g given in Eq. (9.57) leads to the distribution given in Eq. (9.56).

9.3 Calculate the intrinsic conductivity of pure GaAs at 300°K. Use the bandgap of 1.4 eV, an electron mobility of 7000 cm^2/V-sec, a hole mobility of 300 cm^2/V-sec, an electron effective mass of $0.07m$, and a hole effective mass of $0.5m$.

10^{15} cm^{-3} donors are incorporated into the intrinsic GaAs with an ionization energy of 0.01 eV. What is the concentration of holes in this impure sample at 300°K? What is the location of the Fermi level?

9×10^{14} cm^{-3} acceptors are incorporated in addition to the 10^{15} donors. What is the concentration of holes in this sample at 300°K? What is the location of the Fermi level?

9.4 The temperature coefficient of the bandgap in GaAs is -4.5×10^{-4} eV/°K. Plot the location of the Fermi level with respect to the bottom of the conduction band as a function of temperature between 0 and 300°K.

9.5 Two possible energy-level schemes for a Group VII donor level and for cation vacancy levels in a II–VI compound are given in Fig. 9.24. If the density of donors is 10^{17} cm^{-3} and the density of cation vacancies is 5×10^{16} cm^{-3}, calculate the location of the Fermi level in each case at 100°K.

9.6 Calculate the slope to be expected in a plot of ln(conductivity) vs. $1/T$ for a *uniform quasi-continuous* distribution of donors with depth. (Sug-

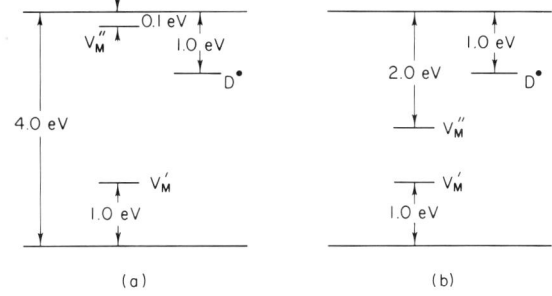

Fig. 9.24 Two possible energy-level schemes for a Group VII donor level and fo cation vacancy levels in a II–VI compound, as described in Problem 9.5.

gestion: Express the free-electron density in terms of the donor density and the location of the Fermi level *approximately*.)

9.7 Make Problem 8.4 more quantitative to consider the following problem. By photoexcitation at 100°K, the mobility in a crystal is increased from 4000 to 6000 cm^2/V-sec by the removal of the charge from a charged impurity. Neglecting other complicating factors and considering the scattering due to a neutral impurity to be negligible, determine whether the effect of photoexcitation has been to (a) make a singly charged impurity neutral, (b) make a doubly charged impurity neutral, or (c) make a doubly charged impurity singly charged. Assume an effective mass for the charged carriers of $0.5m$, a dielectric constant of 10, and a total density of charged impurities of 2×10^{15} cm^{-3}.

9.8 A crystal has a conduction band with a scalar effective mass of $0.01m$. The impurity content of this crystal is adjusted so that there are 10^{18} cm^{-3} free electrons in the conduction band. Where is the Fermi level? What is the temperature dependence of mobility for LA phonon scattering?

9.9 The data of Fig. 9.25 represent the variation of dark conductivity with temperature for a crystal containing partially compensated acceptors. Given that the effective mass for holes is a scalar, $0.2m$, determine from the presented data: (a) the acceptor ionization energy, (b) the density of acceptors, (c) the density of donors, (d) the hole mobility at 100°K. Assume that scattering by LA phonons is dominant throughout.

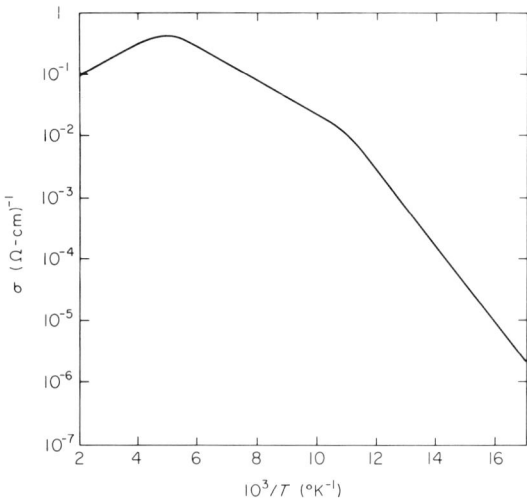

Fig. 9.25 Temperature dependence of electrical conductivity in a crystal with partially compensated acceptors, as described in Problem 9.9.

Chapter 10

Magnetic-Field Effects

When a magnetic field is applied to a solid in which an electric field and/or a thermal gradient exists, a whole new set of interactions become possible. Sometimes a general expression is written for the current density in a material in terms of a power series in electric and magnetic field strengths and tensor "conductivities" as coefficients

$$J_i = \sum_j \sigma^*_{ij}\mathscr{E}_j + \sum_{j,l} \sigma^*_{ijl}\mathscr{E}_j B_l + \sum_{j,l,m} \sigma^*_{ijlm}\mathscr{E}_j B_l B_m \qquad (10.1)$$

The first term in this series represents the normal electrical conductivity and σ^*_{ij} is the electrical conductivity tensor. The second term represents the interaction between an electric field and a magnetic field, and σ^*_{ijl} is the corresponding conductivity tensor; when $\mathscr{E} \cdot \mathbf{B} = 0$, this term describes the *Hall effect*. The third term represents the interaction between an electric field and a second-order magnetic field, and σ^*_{ijlm} is the corresponding conductivity tensor; the variety of possible contributions from this interaction are classed in the category of *magnetoresistance* phenomena. If thermal gradients are also included in Eq. (10.1) it is evident that a large number of possible effects occur. In this chapter we consider particularly the Hall effect, magnetoresistance effects, and briefly magnetothermal effects. Thermoelectric effects were considered in Section 7.9.

Magnetic-field effects can be separated into three principal ranges depending on the magnitude of the magnetic field and the scattering processes in the material. First there are the low-field range and the high-field range. The distinction between these ranges is defined in terms of a basic parameter of a free or quasi-free carrier in a magnetic field. If a free electron, for example, with effective mass m_e^* is acted on only by interaction with a magnetic field **B**, the electron moves in a circular orbit in a plane perpendicular to **B** with an angular frequency ω_c, called the *cyclotron frequency*, determined by equating the centrifugal and the magnetic force on the electron

$$\frac{m_e^* v^2}{r} = ev\frac{B}{c} \tag{10.2}$$

This gives

$$\omega_c = \frac{v}{r} = \frac{eB}{m_e^* c} \tag{10.3}$$

The low-field case corresponds to the situation where $\omega_c \tau \ll 1$, where τ is the relaxation time for the dominant scattering mechanism, where such an expression is appropriate. This condition corresponds physically to the situation where the electron passes through only a small arc of the circular orbit described above before scattering removes it from this orbit into another. The magnetic field can appropriately be treated as a small perturbation and the relaxation-time solution of the Boltzmann equation may be appropriate. The high-field case corresponds to the situation where $\omega_c \tau \gg 1$, corresponding to the case where the circular orbit is well defined and the magnetic field can no longer be treated as a perturbation. In the high-field range, quantum effects due to the magnetic field occur, and the solution can be obtained only by starting with the appropriate solution of the Schroedinger equation.

Another distinction can be made between solutions in the low-field case, depending on whether the magnetic field is small enough to introduce disturbances that can be treated in a linear approximation or large enough that a nonlinear approach must be used.

Measurements of electrical conductivity yield information about the product of carrier density and mobility. The Hall effect, measured simultaneously with conductivity, makes it possible to determine the sign of the charge carriers, their density, and their mobility as separate quantities. The Hall effect and the thermoelectric effect (see Section 7.9) are among the most effective ways for obtaining this kind of information. In measurements in which the temperature dependence of carrier density and of mobility are required separately, therefore, the Hall effect has proved to be of great value (see again, for example, Figs. 8.10 and 9.16). Since there is no *a priori* reason why the Hall effect must be applicable only to equilibrium situations, its use in separating charge density and mobility changes

10.1 Low Magnetic Fields in Linear Approximation

under nonequilibrium situations is also possible; photo–Hall measurements to detect the effect of photoexcitation on carrier density and mobility have been widely used.

Measurements of magnetoresistance have played a role of some significance in attempts to obtain experimental insight into the shape of equal-energy surfaces in the electronic band structure of a material.

The quantum effects produced by large enough magnetic fields to bring about the high-field case are varied and of great interest for developing an understanding of the electronic structure of materials. In semiconductors the phenomena of cyclotron resonance and of magnetoabsorption between bands come into being; these represent some of the most accurate methods available for the determination of effective-mass tensors. In metals wide use has been made of the de Haas–van Alphen effect, Az'bel–Kaner cyclotron resonance, and high-field magnetoresistance and Hall effects to map out the topology of the Fermi surface.

10.1 Low Magnetic Fields in the Linear Approximation

In order to obtain a better feeling for the physical processes involved in the simplest type of Hall effect, we consider first the small perturbation effects in the low-field range. Since this approximation is linear in magnetic field, no magnetoresistance effects are predicted.

If a magnetic field is applied at right angles to an applied electric field, the interaction between the magnetic field and the moving free carriers produces a potential difference in the third mutually orthogonal direction. For an initial treatment of the effect, we assume the simplifying assumptions of spherical equal-energy surfaces and a relaxation time that is independent of electron energy. Consider then an infinite semiconductor with a current density j_x in the x direction, and a magnetic field B_z in the z direction, as indicated in Fig. 10.1.

The current density for electrons is

$$\mathbf{j}_e = -ne\mathbf{v}_e = \sigma_e \mathscr{E} \tag{10.4}$$

where

$$\mathbf{v}_e = -\mu_e \mathscr{E} \tag{10.5}$$

and the current density for holes is

$$\mathbf{j}_h = +pe\mathbf{v}_h = \sigma_h \mathscr{E} \tag{10.6}$$

where
$$\mathbf{v}_h = +\mu_h \mathscr{E} \tag{10.7}$$

The Lorentz force, for either electron or hole, moving in an electric and magnetic field, is given by

$$\mathbf{F} = e\left(\mathscr{E} + \frac{\mathbf{v} \times \mathbf{B}}{c}\right) \tag{10.8}$$

ONE-CARRIER EFFECT

Consider first the Hall effect in terms of Fig. 10.1 if all of the current density j_x is due to free electrons. Then for $\mathbf{F} = 0$,

$$e\mathscr{E} = -\frac{e}{c}\mathbf{v}_e \times \mathbf{B} = +\frac{1}{nc}\mathbf{j}_e \times \mathbf{B} \tag{10.9}$$

which gives

$$\mathscr{E} = +\frac{1}{ne}\frac{\mathbf{j}_e \times \mathbf{B}}{c} \tag{10.10}$$

or

$$\mathscr{E}_y = -\frac{1}{ne}\frac{j_x B_z}{c} \tag{10.11}$$

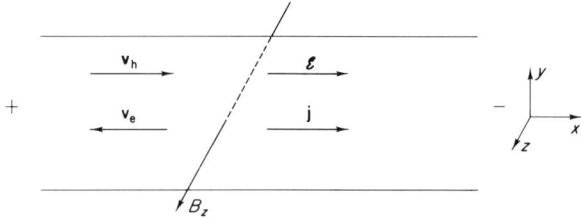

Fig. 10.1 Quantities of interest for the calculation of the Hall effect.

This means physically that the force on moving electrons due to the magnetic field tends to displace electrons until an electric field is set up that is just sufficient to maintain the total force $\mathbf{F} = 0$. If the current density is associated with electrons and is j_x, then the presence of a magnetic field B_z produces a Hall field $\mathscr{E}_y$ that is proportional to the product $(j_x B_z/c)$ through the proportionality or *Hall constant* R_e,

$$R_e = -\frac{1}{ne} \tag{10.12}$$

for electrons.

10.1 Low Magnetic Fields in Linear Approximation

Alternatively if the Hall effect is calculated for a current density j_x due only to holes,

$$e\mathscr{E} = -\frac{e}{c}\mathbf{v}_h \times \mathbf{B} = -\frac{1}{pc}\mathbf{j}_h \times \mathbf{B} \tag{10.13}$$

which gives

$$\mathscr{E} = -\frac{1}{pe}\frac{\mathbf{j}_h \times \mathbf{B}}{c} \tag{10.14}$$

or

$$\mathscr{E}_y = +\frac{1}{pe}\frac{j_x B_z}{c} \tag{10.15}$$

with the Hall constant for holes being given by

$$R_h = +\frac{1}{pe} \tag{10.16}$$

Thus the polarity of the Hall field indicates whether conductivity is due to electrons or to holes, and the carrier density can be determined from the magnitude of the Hall constant.

Because the electric field intensity initially is $\mathscr{E}_x$ in the absence of the magnetic field, but is $\mathscr{E}_x + \mathscr{E}_y$ in the presence of the magnetic field, the effect of the magnetic field is to cause a rotation of the electric field through a specific angle θ, called the *Hall angle*. In view of Eq. (10.11), for electrons,

$$\tan\theta_e = -\frac{\mathscr{E}_y}{\mathscr{E}_x} = -\mu_e\frac{B_z}{c} \tag{10.17}$$

since $\mathscr{E}_x = j_x/(ne\mu_e)$. Likewise in view of Eq. (10.15), for holes,

$$\tan\theta_h = +\mu_h\frac{B_z}{c} \tag{10.18}$$

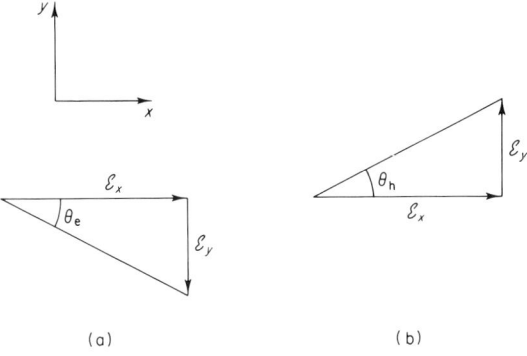

Fig. 10.2 Hall angles for (a) electrons and (b) holes.

The Hall angles are directly proportional to the product of mobility and magnetic field; the relationships are shown in Fig. 10.2.

From Eqs. (10.12) and (10.16) for the Hall constant, it is also evident that

$$-\mu_e = \sigma_e R_e \quad \text{for electrons} \qquad (10.19)$$

$$\mu_h = \sigma_h R_h \quad \text{for holes} \qquad (10.20)$$

In more general situations than we are considering in this section, R and the product σR may not have the same simple form as given in the above equations. It is generally accepted, however, to refer to

$$R \equiv \frac{\mathscr{E}_y}{j_x B_z/c} \qquad (10.21)$$

as the Hall constant, and to

$$\mu_H \equiv \sigma R \qquad (10.22)$$

as the Hall mobility.

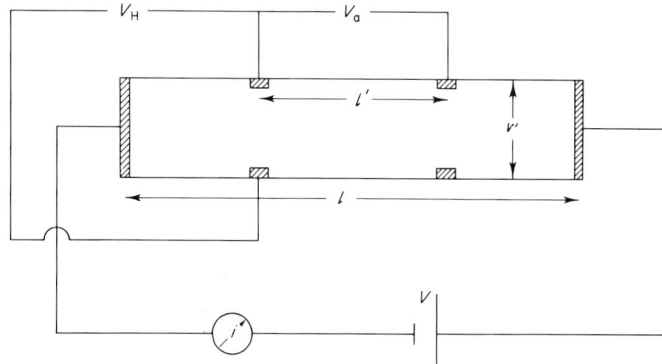

Fig. 10.3 Typical experimental arrangement for the measurement of Hall effect.

In actual laboratory measurements, the Hall effect is usually measured by determining the voltage developed across a width w, small compared to the length l of the crystal, as indicated in Fig. 10.3. This voltage V_H, the voltage drop V_a across the potential probes separated by l', the current i, and the magnetic field B, are the measured parameters. In terms of these parameters,

$$\sigma = \frac{i}{V_a} \frac{l'}{wd} \qquad (10.23)$$

10.1 Low Magnetic Fields in Linear Approximation

if d is the thickness of the crystal. In laboratory units,

$$R = 10^8 \frac{V_H d}{Bi} \tag{10.24}$$

with R in cm³/coulomb, V_H in volts, i in amperes, d in cm, and B in gauss. The Hall mobility is therefore

$$\mu_H = \frac{10^8}{B} \frac{V_H}{V_a} \frac{l'}{w} \text{ cm}^2/\text{V-sec} \tag{10.25}$$

Operationally, therefore, measurements can be extended to materials with smaller values of Hall mobility by increasing B and/or V_a, or by decreasing the lower limit of detectable V_H. It is of some practical interest to note from Eq. (10.25) that for fixed B and geometry,

$$V_H \propto \mu_H V_a \tag{10.26}$$

and that therefore the magnitude of the Hall voltage does not depend on the free-carrier density explicitly. In laboratory units, the carrier density is

$$n = \frac{10^{-8} Bi}{edV_H} \text{ cm}^{-3} \tag{10.27}$$

The geometry of the Hall sample can play an important role in determining the magnitude of the Hall voltage measured. Figure 10.4 indicates the correction factor that must be applied to the theoretical V_H if either

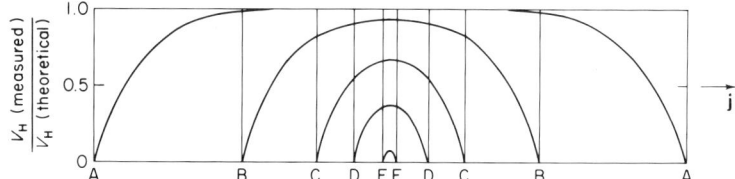

Fig. 10.4 Ratio of the experimentally measured Hall voltage to the Hall voltage calculated theoretically for an ideal configuration as a function of the geometry of the experimental configuration. The solid vertical lines represent the ends of five different experimental samples, AA to EE, corresponding to progressively smaller ratios of length to width from 4/1 for AA to 1/8 for EE. Horizontal distance corresponds to the placement of the Hall probes on the sample. For example, for a sample with the length/width ratio of AA, Hall probes must be at least one fourth the length of the crystal in from the ends, in order for the true theoretical Hall voltage to be measured. For length/width ratios of 2/1 or less, no position of the Hall probes gives the theoretical Hall voltage. [After J. Volger, *Phys. Rev.* **79**, 1023 (1950).]

the ratio l/w is not sufficiently large, or if the Hall probes are placed too close to the end of the sample.

TWO-CARRIER EFFECT

If both electrons and holes contribute to the conductivity, we need to establish the conditions for which the total current flow in the y direction is zero. The situation is illustrated in Fig. 10.5.

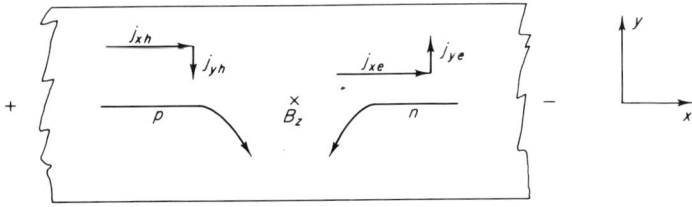

Fig. 10.5 Quantities of interest for the determination of the Hall effect in the case of two-carrier conductivity for low magnetic fields.

The total current density in the y direction is

$$j_y = j_{ye} + j_{yh} \tag{10.28}$$

with

$$j_{ye} = \frac{j_{xe}\mu_e B_z}{c} = \frac{\sigma_e \mathscr{E}_x \mu_e B_z}{c} \tag{10.29}$$

and

$$j_{yh} = -\frac{j_{xh}\mu_h B_z}{c} = -\frac{\sigma_h \mathscr{E}_x \mu_h B_z}{c} \tag{10.30}$$

If we substitute explicitly for σ_e and σ_h in terms of the carrier density and mobility, and write $\mathscr{E}_x = j_x/\sigma$, substitution of Eqs. (10.29) and (10.30) into Eq. (10.28) gives

$$\begin{aligned} j_y &= ne\mu_e \frac{j_x}{\sigma} \frac{\mu_e B_z}{c} - pe\mu_h \frac{j_x}{\sigma} \frac{\mu_h B_z}{c} \\ &= \frac{eB_z j_x}{\sigma c}(n\mu_e^2 - p\mu_h^2) \end{aligned} \tag{10.31}$$

The Hall field required to reduce this transverse current to zero is

$$\mathscr{E}_y = -\frac{j_y}{\sigma} = \frac{e j_x B_z}{\sigma^2 c}(p\mu_h^2 - n\mu_e^2) \tag{10.32}$$

10.1 Low Magnetic Fields in Linear Approximation

The Hall constant for the case in which both electrons and holes contribute to the conductivity is therefore

$$R = \frac{\mathscr{E}_y}{j_x B_z/c} = \frac{e}{\sigma^2}(p\mu_h^2 - n\mu_e^2)$$

$$= \frac{1}{e} \frac{p\mu_h^2 - n\mu_e^2}{(p\mu_h + n\mu_e)^2} \tag{10.33}$$

The corresponding Hall mobility is

$$\mu_H = \sigma R = \frac{p\mu_h^2 - n\mu_e^2}{p\mu_h + n\mu_e} \tag{10.34}$$

These results show that the Hall constant and the Hall mobility go to zero when $p\mu_h^2 = n\mu_e^2$, corresponding to a cancellation of the Hall voltage due to electrons by the Hall voltage due to holes. Examples of the variation of Hall mobility with the ratio p/n are given in Fig. 10.6.

Example 10.1 A particular Hall apparatus is equipped with a magnetic field of 5000 G and a voltage detector whose limit is 0.1 mV. Specimens to be used in the apparatus measure $10 \times 5 \times 1$ mm³, and have a conductivity of 1 (ohm-cm)⁻¹. If the temperature of the sample cannot be maintained constant if Joule heating exceeds 100 mW, what is the minimum mobility that can be determined?

A conductivity of 1 (ohm-cm)⁻¹ gives a current according to Eq. (10.23) of $0.05 V_a$ A. The power limitation means that iV_a must not exceed 0.1 W. It follows that the maximum value of V_a that can be safely used is 1.4 V.

Using Eq. (10.25) to calculate the smallest value of μ_H detectable under these experimental constraints gives

$$\mu_H = \frac{10^8}{5000} \frac{10^{-4}}{1.4} \frac{1}{0.5}$$

$$= 3 \text{ cm}^2/\text{V-sec}$$

Such equipment would be quite sensitive for the measurement of the Hall effect in normal semiconductors. If the conductivity were due to a two-carrier effect rather than a one-carrier effect, and if $\mu_e = 10^3$ cm²/V-sec and $\mu_h = 10^2$ cm²/V-sec, Eq. (10.34) shows that the measurable range is described by

$$3 \leq \left| \frac{10^4(p/n) - 10^6}{10^2(p/n) + 10^3} \right|$$

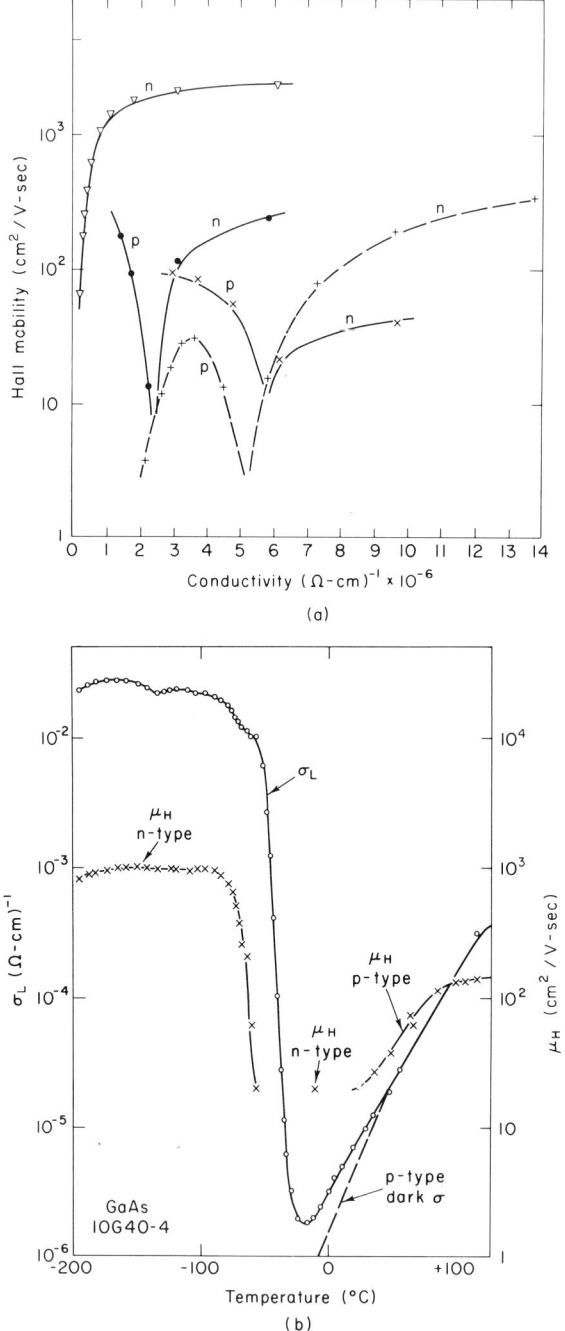

10.2 Types of Mobility

In other words the only range excludable from measurement possibilities is the range $97 < (p/n) < 103$.

10.2 Types of Mobility

In general, four different kinds of mobility enter into common discussions.

1. *Microscopic mobility.* This is the actual velocity per unit electric field of a free carrier in a crystal. It cannot be measured directly.

2. *Conductivity mobility.* This is the mobility that enters into the conductivity expression, $\sigma_e = ne\mu_e$. This mobility involves an averaged relaxation time $\langle \tau \rangle$ dependent on the nature of the scattering process, and in crystals with nonspherical equal-energy surfaces the conductivity mobility also involves a combined effective mass.

3. *Hall mobility.* This mobility is the product of the measured conductivity and the measured Hall constant. In general the Hall mobility differs from the conductivity mobility by a constant factor, as discussed in Section 10.4; this factor is of order unity and depends on the type of scattering and on the band structure. Under certain special conditions, such as inhomogeneity in a crystal, the measured Hall mobility may be considerably different from the conductivity mobility.

4. *Drift mobility.* This is the velocity or drift per unit field for a carrier moving in an electric field. Experimentally a pulse of carriers is injected (optically or electrically) into a sample at one point as shown in Fig. 10.7. The time t is measured that it takes for the carriers to drift in a field to another point a distance d away, and the drift mobility is defined as

$$\mu_{\text{drift}} = \frac{d^2}{Vt} \quad (10.35)$$

if the electric field is V/d. If trapping centers are present, so that the actual drift process is not simply motion through the conduction band, but in-

---

Fig. 10.6 Examples of two-carrier Hall effects. (a) Variation of Hall mobility with conductivity, as varied by increasing photoexcitation on the material, for crystals of high-resistivity GaAs. The much larger electron than hole mobility in GaAs causes equal densities of electrons and holes to give a strongly n-type Hall effect. [After J. Blanc, R. H. Bube and H. E. MacDonald, *J. Appl. Phys.* **32**, 1666 (1961).] (b) Temperature dependence of photoconductivity (for steady photoexcitation), and photo–Hall mobility for GaAs:Si:Cu crystal. The photo–Hall mobility becomes n-type at low temperatures, whereas it is p-type at room temperature, and two-carrier conductivity is prominent between -50 and $+50°C$ in the light but not in the dark. [After R. H. Bube and H. E. MacDonald, *Phys. Rev.* **128**, 2062 (1962).]

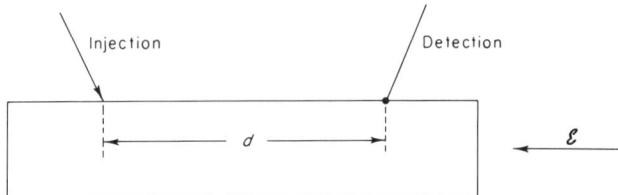

Fig. 10.7 Experimental arrangement for the measurement of the drift velocity of electrons in an electric field.

volves a series of trapping and untrapping processes, the drift mobility can be much less than the conductivity mobility. If one argues that all the electrons available for drift, i.e., the sum of the actually free electrons n and the electrons in traps at any particular instant n_t, contribute to the conductivity via the drift mobility the same as all the free carriers contribute via the conductivity mobility, an expression for the drift mobility can be obtained:

$$(n + n_t)\mu_{\text{drift}} = n\mu \qquad (10.36)$$

giving

$$\mu_{\text{drift}} = \left(\frac{n}{n + n_t}\right)\mu \qquad (10.37)$$

If the traps active in this process have a discrete energy level E_t below the conduction band, and if $n_t \gg n$,

$$\mu_{\text{drift}} = \left(\frac{N_c}{gN_t} e^{-E_t/kT}\right)\mu \qquad (10.38)$$

if the occupancy of both the conduction band and the N_t traps can be approximately given by a Boltzmann function. Therefore the drift mobility has a temperature activation energy under these circumstances, corresponding to the trap depth.

The four mobilities are all equal,

$$\mu = \mu_{\text{con}} = \mu_{\text{Hall}} = \mu_{\text{drift}}$$

only when the following three conditions are met: (a) spherical equal-energy surfaces with extremum at $\mathbf{k} = 0$, (b) relaxation time independent of electron energy, and (c) negligible trapping effects, i.e., $n \gg n_t$.

10.3 General Treatment of the Low-Magnetic-Field Range

Within the low-field range, when the magnetic field becomes sufficiently large that a linear approximation is no longer valid, a more general treatment is required. It is perhaps worthwhile to indicate how this more general treatment can be carried out both in terms of a semiclassical consideration of the equations of motion for carriers in the presence of both electric and magnetic fields, and in terms of a solution of the Boltzmann equation.

Equations of Motion

The equations of motion for electrons in the presence of an electric field are

$$\frac{dv_x}{dt} = -\frac{e}{m_e^*}\mathscr{E}_x - \omega_c v_y \tag{10.39}$$

$$\frac{dv_y}{dt} = -\frac{e}{m_e^*}\mathscr{E}_y + \omega_c v_x \tag{10.40}$$

We desire to solve these two simultaneous equations for v_x and v_y so that we can calculate $j_x = -en\bar{v}_x$ and $j_y = -en\bar{v}_y$ and proceed with the calculation of the Hall effect.

One method for obtaining a solution of Eqs. (10.39) and (10.40) is to introduce the complex quantity

$$Z \equiv v_x + iv_y \tag{10.41}$$

Then

$$\frac{dZ}{dt} = \frac{dv_x}{dt} + i\frac{dv_y}{dt} = -\frac{e}{m_e^*}(\mathscr{E}_x + i\mathscr{E}_y) + i\omega_c Z \tag{10.42}$$

This differential equation in Z has the familiar general solution,

$$Z = Z_0 + \frac{1}{i\omega_c}\frac{e}{m_e^*}(\mathscr{E}_x + i\mathscr{E}_y)(1 - e^{i\omega_c t}) \tag{10.43}$$

with $Z_0 = Z(t=0) = v_{x0} + iv_{y0}$.

If the carriers are moving under the effects of a scattering process with relaxation time τ, the average value of Z over all collisions is given by

$$\bar{Z} = \frac{1}{\tau}\int_0^\infty Z\, e^{-t/\tau}\, dt \tag{10.44}$$

The result is

$$\bar{Z} = \frac{e}{m_e^*} (\mathscr{E}_x + i\mathscr{E}_y) \frac{1}{i\omega_c \tau} \left\{ \int_0^\infty e^{-t/\tau} dt - \int_0^\infty \exp\left[\left(-\frac{1}{\tau} - i\omega_c\right)t\right] dt \right\}$$

$$= \frac{e}{m_e^*} (\mathscr{E}_x + i\mathscr{E}_y) \frac{\tau}{i\omega_c \tau - 1} \quad (10.45)$$

Since $\bar{Z} = \bar{v}_x + i\bar{v}_y$, separation of the results of Eq. (10.45) into real and imaginary parts by multiplication by $(1 + i\omega_c\tau)/(1 + i\omega_c\tau)$ enables us to express $\bar{v}_x$ and hence j_x, $\bar{v}_y$ and hence j_y. The results are

$$j_x = \frac{ne^2}{m_e^*} \left(\frac{\tau \mathscr{E}_x}{1 + \omega_c^2 \tau^2} - \frac{\omega_c \tau^2 \mathscr{E}_y}{1 + \omega_c^2 \tau^2} \right) \quad (10.46)$$

$$j_y = \frac{ne^2}{m_e^*} \left(\frac{\tau \mathscr{E}_y}{1 + \omega_c^2 \tau^2} + \frac{\omega_c \tau^2 \mathscr{E}_x}{1 + \omega_c^2 \tau^2} \right) \quad (10.47)$$

These are fairly general expressions for the components of the current density in the presence of crossed electric and magnetic fields. Generalization to energy-dependent relaxation times (see Section 10.4) is accomplished by averaging the quantities entering the above equations over energy, and generalization to nonspherical equal-energy surfaces is accomplished by using the appropriate effective masses.

Example 10.2 Do these general results differ from the previous linear approximation when only electrons or only holes contribute to the conductivity?

Setting $j_y = 0$ in Eq. (10.47) permits the calculation of

$$\tan \theta_e = \frac{\mathscr{E}_y}{\mathscr{E}_x} = -\omega_c \tau = -\frac{eB_z \tau}{m_e^* c} = -\mu_e \frac{B_z}{c}$$

in agreement with Eq. (10.17). For one-carrier conductivity the linear approximation gives the same results.

Also if we calculate the value of j_x corresponding to the condition of $j_y = 0$, by substituting $\mathscr{E}_y = -\omega_c \tau \mathscr{E}_x$ into Eq. (10.46), we obtain

$$j_x = \frac{ne^2}{m_e^*} \tau \mathscr{E}_x = ne\mu_e \mathscr{E}_x = \sigma_e \mathscr{E}_x$$

In the particular case of spherical equal-energy surfaces, a relaxation time independent of energy, and one-carrier conductivity, there is no change of resistance in the initial direction of current flow because of the presence of the magnetic field, i.e., no magnetoresistance.

10.3 General Treatment of Low-Magnetic-Field Range

The Hall constant R for the condition that $j_y = 0$ is

$$R = \frac{\mathscr{E}_y}{j_x B_z/c} = -\frac{\omega_c \tau \mathscr{E}_x}{ne\mu_c \mathscr{E}_x B_z/c} = -\frac{1}{ne}$$

which is the same as Eq. (10.12). For the case of one-carrier conductivity, the Hall effect calculated by the more generalized method yields the same results as the simple condition that the net force $F = 0$ on an electron in crossed electric and magnetic fields.

A similar calculation can be demonstrated for holes by the substitution of $-\omega_c$ for ω_c in the above equations.

SOLUTION OF THE BOLTZMANN EQUATION

The general solution of the Boltzmann equation for spherical equal-energy surfaces in the relaxation-time approximation in the presence of a magnetic field and a thermal gradient is given in Eqs. (7.78) and (7.79). In the present case we set the thermal gradient equal to zero and obtain

$$\boldsymbol{\theta} = \frac{(\hbar/m^*)\tau[e\mathscr{E} - (e\tau/m^*c)\mathbf{B} \times e\mathscr{E} + (e\tau/m^*c)^2\mathbf{B}(\mathbf{B} \cdot e\mathscr{E})]}{1 + (e\tau B/m^*c)^2} \quad (10.48)$$

For the magnetic field orthogonal to the electric field, i.e., $\mathbf{B}$ in the z direction and $\mathscr{E}$ in the xy plane, we write the x and y components of $\boldsymbol{\theta}$, using the definition of the cyclotron frequency from Eq. (10.3):

$$\theta_x = \frac{\hbar \tau e}{m^*} \frac{\mathscr{E}_x + \omega_c \tau \mathscr{E}_y}{1 + (\omega_c \tau)^2} \quad (10.49)$$

$$\theta_y = \frac{\hbar \tau e}{m^*} \frac{\mathscr{E}_y - \omega_c \tau \mathscr{E}_x}{1 + (\omega_c \tau)^2} \quad (10.50)$$

The current density $\mathbf{j}$ is given by

$$\mathbf{j} = -\frac{e}{4\pi^3} \int (\mathbf{k} \cdot \boldsymbol{\theta}) \mathbf{v} \frac{\partial f_0}{\partial E} d\mathbf{k} \quad (10.51)$$

as in Eq. (7.81). Writing the x and y components of the integrand of Eq. (10.51) using Eqs. (10.49) and (10.50), we can then express the x and y components of $\mathbf{j}$ in terms of transport integrals. Proceeding in a manner similar to that used in Section 7.6, we arrive at the same expressions for j_x and j_y as are given in Eqs. (10.46) and (10.47).

HALL EFFECT FOR TWO-CARRIER CONDUCTIVITY

Now we apply the more general formulation to the problem of two-carrier conductivity, i.e., when both electrons and holes are contributing to the electrical conductivity. Here the generalized result is different from the simple result previously derived on the assumption of small magnetic fields. As before, we must add the contribution of electrons and holes to obtain j_x and j_y, and we assume that the relaxation times for electrons and holes are given respectively by τ_e and τ_h.

Since the algebraic expressions become somewhat unwieldy, it is convenient to resort to a somewhat simpler notation. Equations (10.46) and (10.47) for electrons may be written as

$$j_x = C_{xx}\mathscr{E}_x - C_{xy}\mathscr{E}_y \qquad (10.52)$$

$$j_y = C_{yx}\mathscr{E}_x + C_{yy}\mathscr{E}_y \qquad (10.53)$$

where

$$C_{xx} = C_{yy} \equiv C_{1e} = \frac{ne^2}{m_e^*}\frac{\tau_e}{1+\omega_{ce}^2\tau_e^2} = \frac{\sigma_e}{1+\sigma_e^2 R_e^2(B_z/c)^2} \qquad (10.54)$$

using

$$\mu_e = \frac{e}{m_e^*}\tau_e, \qquad \omega_{ce} = \frac{eB_z}{m_e^* c}, \qquad \text{and} \qquad \mu_e = -\sigma_e R_e$$

and where

$$C_{xy} = C_{yx} \equiv C_{2e} = \frac{ne^2}{m_e^*}\frac{\omega_{ce}\tau_e^2}{1+\omega_{ce}^2\tau_e^2} = -\frac{R_e \sigma_e^2 B_z/c}{1+\sigma_e^2 R_e^2(B_z/c)^2} \qquad (10.55)$$

In terms of Example 10.2, for example, for $j_y = 0$,

$$\tan\theta_e = \frac{\mathscr{E}_y}{\mathscr{E}_x} = -\frac{C_{2e}}{C_{1e}} = \sigma_e R_e \frac{B_z}{c} = -\mu_e \frac{B_z}{c}$$

Equations for holes similar to Eqs. (10.52) and (10.53) are

$$j_x = C_{1h}\mathscr{E}_x + C_{2h}\mathscr{E}_y \qquad (10.56)$$

$$j_y = -C_{2h}\mathscr{E}_x + C_{1h}\mathscr{E}_y \qquad (10.57)$$

where

$$C_{1h} = \frac{\sigma_h}{1+\sigma_h^2 R_h^2(B_z/c)^2} \qquad (10.58)$$

$$C_{2h} = \frac{\sigma_h^2 R_h B_z/c}{1+\sigma_h^2 R_h^2(B_z/c)^2} \qquad (10.59)$$

10.3 General Treatment of Low-Magnetic-Field Range

When both electrons and holes are contributing to the conductivity,

$$j_x = (C_{1e} + C_{1h})\mathscr{E}_x - (C_{2e} - C_{2h})\mathscr{E}_y \qquad (10.60)$$

$$j_y = (C_{2e} - C_{2h})\mathscr{E}_x + (C_{1e} + C_{1h})\mathscr{E}_y \qquad (10.61)$$

For $j_y = 0$,

$$\mathscr{E}_x = -\left(\frac{C_{1e} + C_{1h}}{C_{2e} - C_{2h}}\right)\mathscr{E}_y \qquad (10.62)$$

If this value for $\mathscr{E}_x$ is substituted into Eq. (10.60),

$$j_x = -\frac{(C_{1e} + C_{1h})^2 + (C_{2e} - C_{2h})^2}{C_{2e} - C_{2h}}\mathscr{E}_y \qquad (10.63)$$

Therefore we obtain, for the Hall constant $R = \mathscr{E}_y/(j_x B_z/c)$,

$$R = -\frac{(C_{2e} - C_{2h})}{(C_{1e} + C_{1h})^2 + (C_{2e} - C_{2h})^2}\frac{1}{B_z/c} \qquad (10.64)$$

The algebra following substitution for the C's from Eqs. (10.54), (10.55), (10.58), and (10.59) can be simplified by defining a quantity

$$X_i = \left[1 + \sigma_i^2 R_i^2\left(\frac{B_z}{c}\right)^2\right], \qquad i = e, h \qquad (10.65)$$

We have

$$R = \frac{\sigma_e^2 R_e X_h + \sigma_h^2 R_h X_e}{\left(\dfrac{\sigma_e}{X_e} + \dfrac{\sigma_h}{X_h}\right)^2 + \left(\dfrac{\sigma_e^2 R_e B_z/c}{X_e} + \dfrac{\sigma_h^2 R_h B_z/c}{X_h}\right)^2 X_e X_h} \qquad (10.66)$$

The results of simplification can be written finally as

$$R = \frac{\sigma_e^2 R_e + \sigma_h^2 R_h + \sigma_e^2 \sigma_h^2 R_e R_h (R_e + R_h)(B_z/c)^2}{(\sigma_e + \sigma_h)^2 + \sigma_e^2 \sigma_h^2 (R_e + R_h)^2 (B_z/c)^2} \qquad (10.67)$$

For small magnetic fields, Eq. (10.67) reduces directly to Eq. (10.33) upon substitution of $\sigma_e = ne\mu_e$, $\sigma_h = pe\mu_h$, $R_e = -1/ne$, and $R_h = 1/pe$. In most actual cases, the dependence of R on B_z is small. In an impurity-dominated conductivity, usually either σ_e or σ_h is essentially zero, and R is not a function of magnetic field. In a pure intrinsic material, $R_e = -R_h$, and again R is not a function of magnetic field. The strongest dependence of R on magnetic field occurs near the point where $p\mu_h^2 = n\mu_e^2$, and the value of p/n for which $R = 0$ varies with magnetic field. Data taken with InSb and given in Fig. 10.8 illustrate this point.

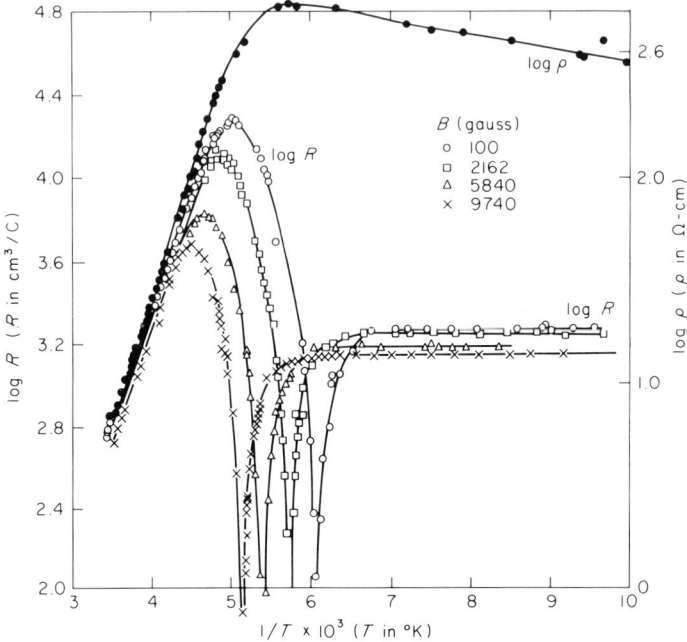

Fig. 10.8 Temperature dependence of the Hall coefficient of a p-type InSb crystal for different magnetic fields, showing the variation of the $(n\mu_n^2 = p\mu_p^2)$ point as a function of magnetic field. [After D. Howarth, R. Jones and E. Putley, *Proc. Phys. Soc. (London)* **B70**, 124 (1957).]

For large values of magnetic field, Eq. (10.67) simplifies appreciably to

$$R_\infty = \frac{R_e R_h}{R_e + R_h} = \frac{1}{(p-n)e} \qquad (10.68)$$

Typical data for InSb are given in Fig. 10.9.

MAGNETORESISTANCE FOR TWO-CARRIER CONDUCTIVITY

Example 10.2 for one-carrier conductivity showed that the presence of a magnetic field did not affect the resistance in the current-flow direction for spherical equal-energy surfaces and a constant relaxation time. This is no longer true if any one of these three simplifying conditions is violated. Whenever the Hall-effect electric field cannot cancel out j_y for *all* the carriers, there is an effective lengthening of the drift path length for some carriers, and a *magnetoresistance* effect results.

A simple situation in which to demonstrate the existence of a magneto-

10.3 General Treatment of Low-Magnetic-Field Range

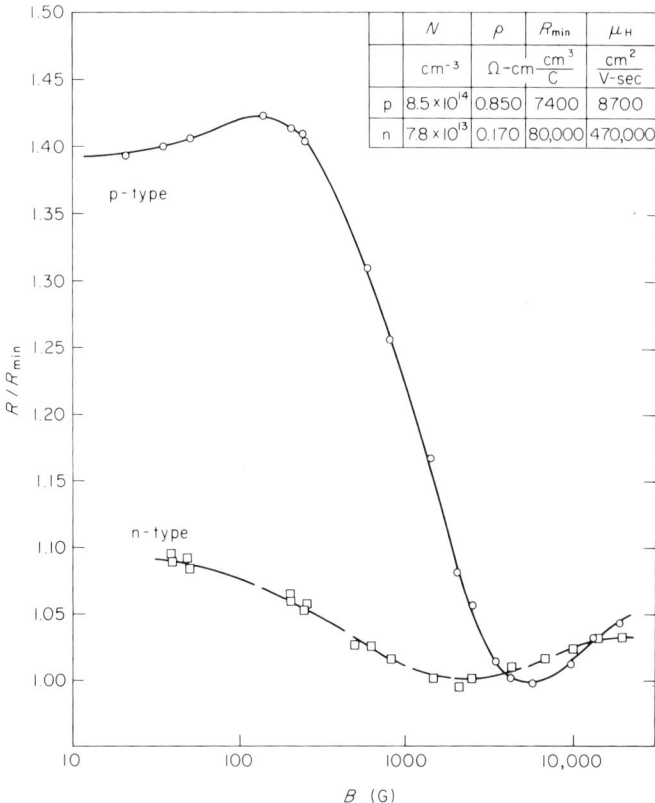

Fig. 10.9 Variation of Hall coefficient, normalized to its minimum value, as a function of magnetic field for n- and p-type InSb. (Willardson and Beer, unpublished; cited in A. C. Beer, "Galvanomagnetic Effects in Semiconductors," Academic Press, New York, 1963, p. 173.)

resistance effect is that of spherical equal-energy surfaces and a constant relaxation time, but with two-carrier rather than one-carrier conductivity. From Eq. (10.64) we have

$$R \frac{B_z}{c} = \frac{\mathscr{E}_y}{j_x} = -\frac{(C_{2e} - C_{2h})}{(C_{1e} + C_{1h})^2 + (C_{2e} - C_{2h})^2} \quad (10.69)$$

Combining this result with Eq. (10.61) gives

$$\sigma = \frac{j_x}{\mathscr{E}_x} = \frac{(C_{1e} + C_{1h})^2 + (C_{2e} - C_{2h})^2}{(C_{1e} + C_{1h})} \quad (10.70)$$

Substitution of Eqs. (10.54), (10.55), (10.58), and (10.59) permits the determination of the conductivity σ. The operation can be made easier if the product σR is calculated,

$$\sigma R = -\frac{C_{2e} - C_{2h}}{C_{1e} + C_{1h}} \frac{1}{B_z/c} \tag{10.71}$$

and then evaluated,

$$\sigma R = \frac{\sigma_e^2 R_e^2[1 + \sigma_h^2 R_h^2 (B_z/c)^2] + \sigma_h^2 R_h^2[1 + \sigma_e^2 R_e^2 (B_z/c)^2]}{(\sigma_e + \sigma_h) + \sigma_e \sigma_h (B_z/c)^2 (\sigma_e R_e^2 + \sigma_h R_h^2)} \tag{10.72}$$

Equation (10.72) gives the Hall mobility for two-carrier conductivity. If the value of the Hall constant from Eq. (10.67) is combined with Eq. (10.72), an expression for the conductivity is obtained,

$$\sigma = \frac{(\sigma_e + \sigma_h)^2 + \sigma_e^2 \sigma_h^2 (B_z/c)^2 (R_e + R_h)^2}{(\sigma_e + \sigma_h) + \sigma_e \sigma_h (B_z/c)^2 (\sigma_e R_e^2 + \sigma_h R_h^2)} \tag{10.73}$$

If we now let $\sigma_0 = (\sigma_e + \sigma_h)$, represent the conductivity for $B_z = 0$, and let $\Delta\sigma$ represent the change in σ when $B_z \neq 0$, we can calculate $\Delta\sigma$ by subtracting $(\sigma_e + \sigma_h)$. Neglecting the term in B_z^2 in the denominator, we obtain

$$\frac{\Delta\sigma}{\sigma_0 (B_z/c)^2} \approx \frac{\sigma_e^2 \sigma_h^2 (R_e + R_h)^2}{\sigma_0^2} - \frac{\sigma_e \sigma_h (\sigma_e R_e^2 + \sigma_h R_h^2)}{\sigma_0} \tag{10.74}$$

If we substitute $\sigma_e = ne\mu_e$, $\sigma_h = pe\mu_h$, $R_e = -1/ne$, and $R_h = 1/pe$, Eq. (10.74) is transformed to

$$-\frac{\Delta\sigma}{\sigma_0 (B_z/c)^2} = \frac{\Delta\varrho}{\varrho_0 (B_z/c)^2} = \frac{np\mu_e\mu_h(\mu_e + \mu_h)^2}{(n\mu_e + p\mu_h)^2} \tag{10.75}$$

where ϱ is the resistivity. Using the value for R_0 given in Eq. (10.33),

$$-\frac{\Delta\sigma}{\sigma_0} = \frac{\Delta\varrho}{\varrho_0} = \xi R_0^2 \sigma_0^2 \left(\frac{B_z}{c}\right)^2 \tag{10.76}$$

where the coefficient

$$\xi = \frac{np\mu_e\mu_h(\mu_e + \mu_h)^2}{(p\mu_h^2 - n\mu_e^2)^2} \tag{10.77}$$

is called the *transverse magnetoresistance coefficient*. In terms of the mobility ratio $b = \mu_e/\mu_h$,

$$\xi = \frac{npb(1 + b)^2}{(p - nb^2)^2} \tag{10.78}$$

10.4 Effects of Scattering Mechanisms

As expected in this model, $\xi = 0$ for an extrinsic semiconductor where either n or p is essentially zero.

Example 10.3 According to the model for magnetoresistance in the case of two-carrier conductivity in a crystal with constant relaxation time and spherical equal-energy surfaces, how large must the mobility be in order to see a 10-per-cent increase in resistivity under the most favorable circumstances?

The most favorable circumstances are determined by the conditions that give maximum magnitude to the expression of Eq. (10.75). The initial reaction that the maximum value is found for $p\mu_h^2 = n\mu_e^2$, since this gives an infinite value of ξ in Eq. (10.77), is not correct, because the same condition gives a zero value for R_0 in Eq. (10.76). Actually the maximum value of the expression of Eq. (10.75) is obtained for $n\mu_e = p\mu_h$. Under this most favorable condition,

$$\frac{\Delta\varrho}{\varrho_0} = \frac{(\mu_e + \mu_h)^2}{4}\left(\frac{B_z}{c}\right)^2$$

Again the most favorable circumstances are controlled by the largest value of B_z obtainable. Let us suppose that this is of the order of 10^5 G. Then, in order for $\Delta\varrho = 0.1\varrho_0$, $(\mu_e + \mu_h) = 630$ cm^2/V-sec. In an ordinary magnet with about 10^3 G, the resistivity would change by 10 per cent only if the mobility sum $(\mu_e + \mu_h)$ were 63,000 cm^2/V-sec.

10.4 Effects of Scattering Mechanisms

When the relaxation-time approximation is appropriate and the relaxation time itself is a function of carrier energy, the necessary account of this dependence can be taken through suitable averages of the relaxation time over energy, similar to that discussed earlier in connection with electrical conductivity in Sections 7.6 and 7.7. In order to separate the effects of scattering mechanism and band structure, we consider in this section spherical equal-energy surfaces only. In the following section we consider the additional corrections made necessary by nonspherical equal-energy surfaces.

HALL EFFECT

The correction due to the energy dependence of the relaxation time for scattering can be introduced into the Hall coefficient as a multiplicative factor K,

$$R = \frac{K}{ne} \tag{10.79}$$

or

$$\mu_H = \sigma R = K\mu \tag{10.80}$$

The starting point is the two equations (10.46) and (10.47) for the two components of the current density. When the relaxation time is a function of energy, these components become, for electron conductivity,

$$j_x = \frac{ne^2}{m_e^*} \left(\left\langle \frac{\tau}{1 + \omega_c^2 \tau^2} \right\rangle \mathscr{E}_x - \left\langle \frac{\omega_c \tau^2}{1 + \omega_c^2 \tau^2} \right\rangle \mathscr{E}_y \right) \tag{10.81}$$

$$j_y = \frac{ne^2}{m_e^*} \left(\left\langle \frac{\tau}{1 + \omega_c^2 \tau^2} \right\rangle \mathscr{E}_y + \left\langle \frac{\omega_c \tau^2}{1 + \omega_c^2 \tau^2} \right\rangle \mathscr{E}_x \right) \tag{10.82}$$

For $j_y = 0$ in order to calculate the Hall effect, we solve Eq. (10.82) for $\mathscr{E}_x$ and then substitute into Eq. (10.81) to obtain an equation relating j_x and $\mathscr{E}_y$, thus making it possible to evaluate

$$R = \frac{\mathscr{E}_y}{j_x(B_z/c)} = -\frac{1}{ne} \frac{\left\langle \frac{\tau^2}{1 + \omega_c^2 \tau^2} \right\rangle}{\left\langle \frac{\tau}{1 + \omega_c^2 \tau^2} \right\rangle^2 + \omega_c^2 \left\langle \frac{\tau^2}{1 + \omega_c^2 \tau^2} \right\rangle^2} \tag{10.83}$$

The correction constant K is given by

$$K = \frac{\left\langle \frac{\tau^2}{1 + \omega_c^2 \tau^2} \right\rangle}{\left\langle \frac{\tau}{1 + \omega_c^2 \tau^2} \right\rangle^2 + \omega_c^2 \left\langle \frac{\tau^2}{1 + \omega_c^2 \tau^2} \right\rangle^2} \tag{10.84}$$

There are two limiting cases. When $(\omega_c \tau)^2 \ll 1$, corresponding to a sufficiently strong scattering that a carrier is prevented from undergoing an appreciable portion of its circular orbit in the magnetic field before it is scattered out of this orbit, we neglect all terms involving ω_c^2, and conclude that

$$K = \frac{\langle \tau^2 \rangle}{\langle \tau \rangle^2} \tag{10.85}$$

If, for a particular scattering process, $\tau = aE^{-s}$, we have calculated previously [Eq. (8.63) and Example 8.3]

$$\langle \tau \rangle = a(kT)^{-s} \frac{\Gamma(\frac{5}{2} - s)}{\Gamma(\frac{5}{2})}$$

10.4 Effects of Scattering Mechanisms

So also

$$\langle \tau^2 \rangle = a^2 \frac{\int_0^\infty E^{3/2-2s} e^{-E/kT} dE}{\int_0^\infty E^{3/2} e^{-E/kT} dE}$$

$$= a^2 (kT)^{-2s} \frac{\Gamma(\frac{5}{2} - 2s)}{\Gamma(\frac{5}{2})} \quad (10.86)$$

The value of K given by Eq. (10.85) is therefore

$$K = \frac{\Gamma(\frac{5}{2})\Gamma(\frac{5}{2} - 2s)}{[\Gamma(\frac{5}{2} - s)]^2} \quad (10.87)$$

For lattice scattering by LA phonons with $s = \frac{1}{2}$, $K = 3\pi/8 = 1.18$. For impurity scattering with $s = -\frac{3}{2}$, $K = 315\pi/512 = 1.93$. A plot of K as a function of the proportion of impurity scattering in a crystal in which both lattice and impurity scattering are present is shown in Fig. 10.10.

For scattering processes for which the relaxation time approximation is not appropriate, it may still be possible to take account of the process through an expression like that of Eq. (10.79). For optical-mode lattice scattering, for example, it is found that the correction factor K is a function of T. Values are given in Fig. 10.11.

For a degenerate semiconductor or metal,

$$\langle \tau^2 \rangle = [\tau(E_F)]^2 = \langle \tau \rangle^2$$

and $K = 1$.

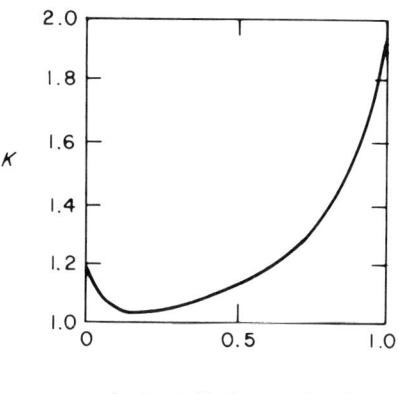

$(1/\mu_I)/(1/\mu_I + 1/\mu_L)$

Fig. 10.10 Magnitude of the correction factor K depending on the nature of the scattering process. K is plotted as a function of the fraction of the resistance due to impurity scattering in a situation where both LA scattering (L) and charged-impurity scattering (I) are present. [After V. A. Johnson and K. Lark-Horovitz, *Phys. Rev.* **82**, 977 (1951).]

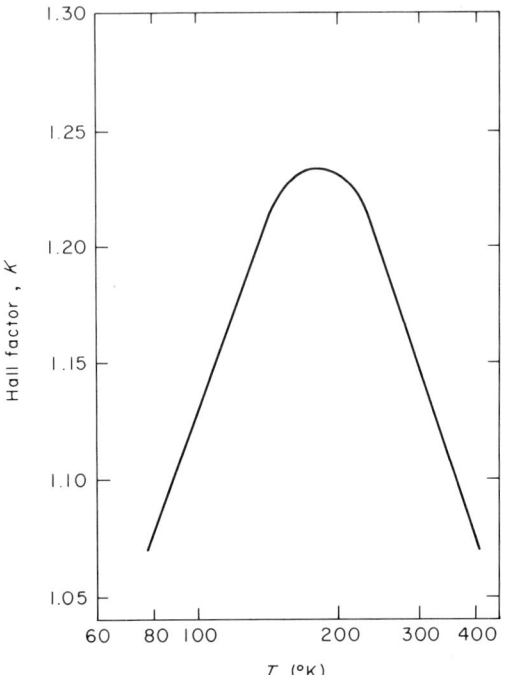

Fig. 10.11 Magnitude of the correction factor K as a function of temperature for optical-mode lattice scattering. (After S. S. Devlin, *in* "Physics and Chemistry of II–VI Compounds" (M. Aven and J. Prener, eds.), Wiley, New York, 1967, p. 561.)

When $(\omega_c\tau)^2$ is not less than unity, on the other hand, so that the full expression of Eq. (10.84) must be used for K, large values of ω_c, i.e., large values of magnetic field, lead again to $K = 1$.

MAGNETORESISTANCE

We consider the case of one-carrier conductivity, and start with Eqs. (10.46) and (10.47) under the assumption that $\omega_c\tau$ is small but not negligible. Under this assumption we have, for j_x,

$$j_x = \frac{ne^2}{m_e^*}\left[\tau\mathscr{E}_x(1 - \omega_c^2\tau^2) - \omega_c\tau^2\mathscr{E}_y(1 - \omega_c^2\tau^2)\right]$$

$$\approx \frac{ne^2}{m_e^*}\left[\tau\mathscr{E}_x - \omega_c\tau^2\mathscr{E}_y - \omega_c^2\tau^3\mathscr{E}_x\right] \qquad (10.88)$$

retaining no terms in B_z greater than second order. In writing j_y for this

10.4 Effects of Scattering Mechanisms

example, it is sufficient to neglect $\omega_c \tau$ with respect to unity completely:

$$j_y = \frac{ne^2}{m_e^*}[\tau \mathscr{E}_y + \omega_c \tau^2 \mathscr{E}_x] \qquad (10.89)$$

For $j_y = 0$, $\mathscr{E}_y = -\omega_c \mathscr{E}_x \langle \tau^2 \rangle / \langle \tau \rangle$, if we take the averages over energy at this point in the calculation. Using this value to eliminate $\mathscr{E}_y$ from Eq. (10.88),

$$j_x = \frac{ne^2}{m_e^*}\left[\mathscr{E}_x\langle\tau\rangle - \omega_c^2\langle\tau^3\rangle + \omega_c^2 \frac{\langle\tau^2\rangle^2}{\langle\tau\rangle}\right] \qquad (10.90)$$

This permits the calculation of $\sigma = j_x/\mathscr{E}_x$:

$$\sigma = \frac{ne^2}{m_e^*}\langle\tau\rangle - \frac{ne^2}{m_e^*}\omega_c^2\left[\langle\tau^3\rangle - \frac{\langle\tau^2\rangle^2}{\langle\tau\rangle}\right]$$

$$= \sigma_0 - \frac{e^2(B_z/c)^2}{m_e^{*2}}\sigma_0\left[\frac{\langle\tau^3\rangle\langle\tau\rangle - \langle\tau^2\rangle^2}{\langle\tau\rangle^2}\right] \qquad (10.91)$$

From this we conclude that

$$-\frac{\Delta\sigma}{\sigma_0} = \frac{\Delta\varrho}{\varrho_0} = \frac{e^2(B_z/c)^2}{m_e^{*2}}\left[\frac{\langle\tau^3\rangle\langle\tau\rangle - \langle\tau^2\rangle^2}{\langle\tau\rangle^2}\right] \qquad (10.92)$$

In view of the definitions of σ_0 and R_0, i.e.,

$$\sigma_0 = \frac{ne^2}{m_e^*}\langle\tau\rangle$$

$$R_0 = -\frac{1}{ne}\frac{\langle\tau^2\rangle}{\langle\tau\rangle^2}$$

we rewrite Eq. (10.92) as

$$-\frac{\Delta\sigma}{\sigma_0} = \frac{\Delta\varrho}{\varrho_0} = \zeta R_0^2 \sigma_0^2 \left(\frac{B_z}{c}\right)^2 \qquad (10.93)$$

with

$$\zeta + 1 = \frac{\langle\tau^3\rangle\langle\tau\rangle}{\langle\tau^2\rangle^2} \qquad (10.94)$$

If the relaxation time is related to energy by $\tau = aE^{-s}$,

$$\zeta + 1 = \frac{\Gamma(\frac{5}{2} - 3s)\Gamma(\frac{5}{2} - s)}{[\Gamma(\frac{5}{2} - 2s)]^2} \qquad (10.95)$$

For lattice scattering with $s = \frac{1}{2}$, $\zeta + 1 = 4/\pi = 1.275$. For impurity scattering, with $s = -\frac{3}{2}$, $\zeta + 1 = 1.57$. If τ is a constant, $\zeta = 0$, giving consistent results with previous calculations.

Another region of interest is that of high magnetic fields. In this range, Eqs. (10.46) and (10.47) reduce to

$$j_x = \frac{ne^2}{m_e^*}\left(-\frac{\mathscr{E}_y}{\omega_c}\right) \tag{10.96}$$

$$j_y = \frac{ne^2}{\omega_c^2 m_e^*}\left(\mathscr{E}_y\left\langle\frac{1}{\tau}\right\rangle + \omega_c \mathscr{E}_x\right) \tag{10.97}$$

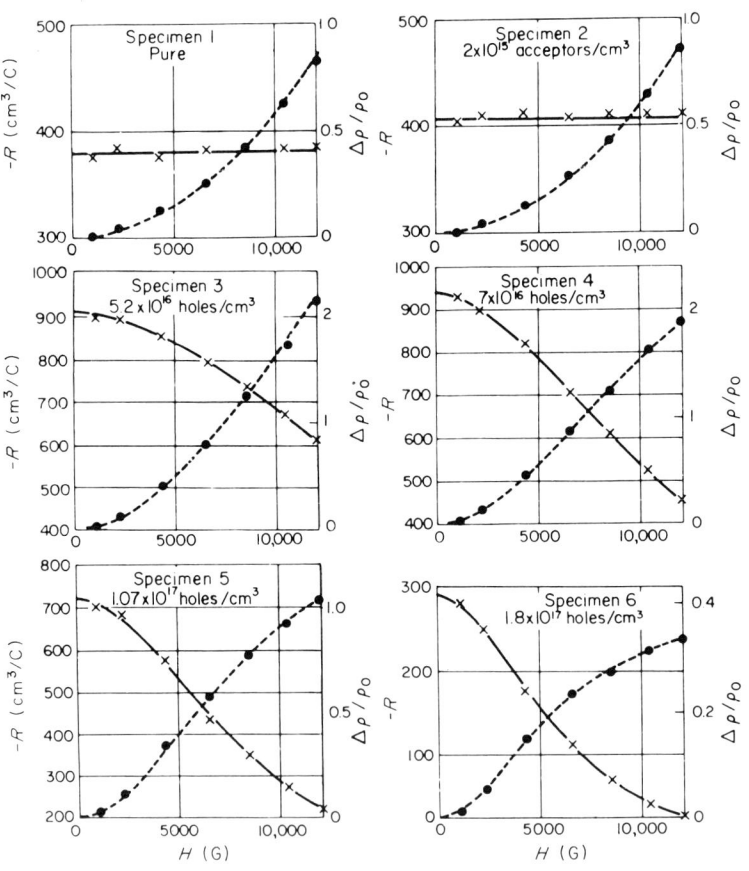

Fig. 10.12 Variation of the Hall coefficient (solid curves) and transverse magnetoresistance (dashed curves) in p-type InSb with magnetic field, for samples with different impurity concentrations. [After C. Hilsum and R. Barrie, *Proc. Phys. Soc. (London)* **71**, 676 (1958).]

10.5 Effects of Band Structure

For $j_y = 0$, $\mathscr{E}_x = -\mathscr{E}_y \langle 1/\tau \rangle / \omega_c$. Using this relation to eliminate $\mathscr{E}_y$ from Eq. (10.112) gives

$$j_x = \frac{ne^2}{m_c^*} \frac{1}{\langle 1/\tau \rangle} \mathscr{E}_x \qquad (10.98)$$

If σ_∞ represents the high-magnetic-field conductivity, as σ_0 represents the low-magnetic-field conductivity,

$$\frac{\sigma_0}{\sigma_\infty} = \frac{\varrho_\infty}{\varrho_0} = \left\langle \frac{1}{\tau} \right\rangle \langle \tau \rangle \qquad (10.99)$$

As the magnetic field is increased, therefore, the conductivity decreases to a saturation value σ_∞, where

$$\frac{\sigma_0}{\sigma_\infty} = \frac{\Gamma(\frac{5}{2} + s)\Gamma(\frac{5}{2} - s)}{[\Gamma(\frac{5}{2})]^2} \qquad (10.100)$$

if $\tau = aE^{-s}$. For lattice scattering, $\sigma_0/\sigma_\infty = 32/9\pi = 1.13$. For charged-impurity scattering, $\sigma_0/\sigma_\infty = 32/3\pi = 3.4$.

Typical variations of Hall coefficient and magnetoresistance for InSb with different concentrations of acceptors are given in Fig. 10.12.

10.5 Effects of Band Structure

Considerations to this point have deliberately been limited to spherical equal-energy surfaces. The contributions of nonspherical surfaces may also be included.

HALL EFFECT

If the bands involved in conductivity do not have spherical equal-energy surfaces, another factor involving the actual band structure enters the proportionality constant between the Hall coefficient and the reciprocal one-carrier density, or between the Hall mobility and the conductivity mobility. We call this factor due to the band structure M, and rewrite Eq. (10.80) as

$$\mu_H = \sigma R = KM\mu \qquad (10.101)$$

Consider the case of ellipsoidal equal-energy surfaces with effective masses in the x, y, and z directions of m_1^*, m_2^*, and m_3^* respectively. Assume that the relaxation time itself is isotropic and depends only on the electron energy. The equations of motion are

$$\frac{dv_x}{dt} = -\frac{e}{m_1^*}\mathscr{E}_x - \omega_{c1}v_y \qquad (10.102)$$

$$\frac{dv_y}{dt} = -\frac{e}{m_2^*}\mathscr{E}_y + \omega_{c2}v_x \qquad (10.103)$$

The procedure to be followed is identical to that used in the previous section, and the results for the components of current density *for a single ellipsoid* are

$$-e\bar{v}_x = e^2\left[\frac{\tau}{m_1^*(1+\omega_{c1}\omega_{c2}\tau^2)}\mathscr{E}_x - \frac{\omega_{c1}\tau^2}{m_2^*(1+\omega_{c1}\omega_{c2}\tau^2)}\mathscr{E}_y\right] \qquad (10.104)$$

$$-e\bar{v}_y = e^2\left[\frac{\tau}{m_2^*(1+\omega_{c1}\omega_{c2}\tau^2)}\mathscr{E}_y + \frac{\omega_{c2}\tau^2}{m_1^*(1+\omega_{c1}\omega_{c2}\tau^2)}\mathscr{E}_x\right] \qquad (10.105)$$

If the several equivalent ellipsoids are oriented along the x, y, and z axes, the sum over all ellipsoids must be included in the calculation of the total current density. In making this sum and in the following calculations, we make the simplifying assumption that $\omega_{c1}\omega_{c2}\tau^2 \ll 1$.

$$\begin{aligned}j_x &= -ne\sum \bar{v}_x \\ &= \frac{ne^2}{3}\left[\langle\tau\rangle\mathscr{E}_x\left(\frac{1}{m_1^*}+\frac{1}{m_2^*}+\frac{1}{m_3^*}\right)\right. \\ &\quad \left. - e\frac{B_z}{c}\langle\tau^2\rangle\mathscr{E}_y\left(\frac{1}{m_1^*m_2^*}+\frac{1}{m_2^*m_3^*}+\frac{1}{m_3^*m_1^*}\right)\right] \qquad (10.106)\end{aligned}$$

The expression for $j_y = -ne\sum \bar{v}_y$ is similar. If we solve the equation for j_y for $\mathscr{E}_y/\mathscr{E}_x$ and then use this to eliminate $\mathscr{E}_x$ from Eq. (10.106) in the same kind of calculation used to determine K in the previous section, we find that the M factor of Eq. (10.101) is given by

$$M = \frac{3(m_1^* + m_2^* + m_3^*)}{m_1^* m_2^* m_3^* \left(\frac{1}{m_1^*}+\frac{1}{m_2^*}+\frac{1}{m_3^*}\right)^2} \qquad (10.107)$$

if factors in B_z^2 are neglected. If the equal-energy ellipsoids are ellipsoids of revolution so that $m_2^* = m_3^* = m_t^*$ and $m_1^* = m_\ell^*$, and if $\gamma \equiv m_\ell^*/m_t^*$,

$$M = \frac{3\gamma(\gamma+2)}{(2\gamma+1)^2} \qquad (10.108)$$

10.5 Effects of Band Structure

Values indicating the order of magnitude of the M factor are given in Table 10.1 for Ge and Si.

TABLE 10.1 Values of M Factor

	γ	M
Ge	15.8	0.795
Si	5.1	0.87

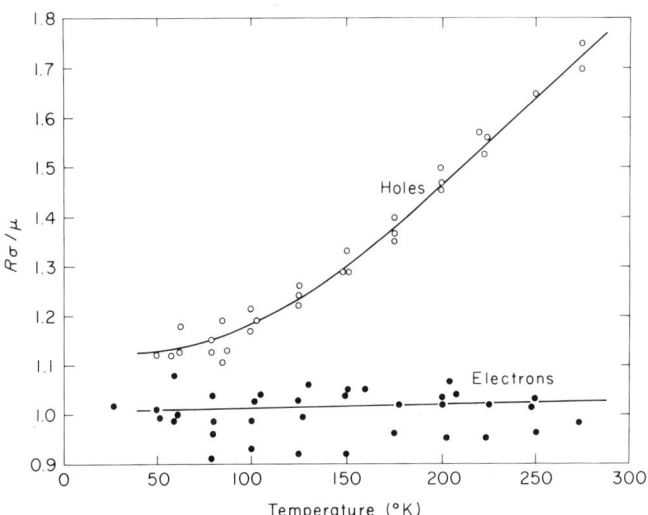

Fig. 10.13 Variation of the ratio of Hall mobility to conductivity mobility in n- and p-type Ge as a function of temperature. The behavior of the holes requires account to be taken of the degenerate hole bands making up the valence band in Ge, and of the fact that the light-hole band is in the nonlinear range at low temperatures while the heavy-hole band is still in the linear range. [After F. J. Morin, *Phys. Rev.* **93**, 62 (1954).]

Since the value of $M < 1$, and the value of $K > 1$, the two correction factors tend to cancel each other out. For lattice scattering of electrons in Si for example, the product $KM = 1.02$. Variation of KM with temperature for electrons and holes in Ge is shown in Fig. 10.13.

Magnetoresistance

The calculation of magnetoresistance coefficients can become quite a tedious assignment as the complexity of the band structure increases. Since we have repeated the kind of calculation needed to extend the previous calculations to ellipsoidal equal-energy surfaces, both in the discussion of the Hall effect for ellipsoidal equal-energy surfaces and in the discussion of magnetoresistance to this point, we attempt here to give only an indication of the nature of the results.

Suppose that the equal-energy surfaces consist of a number of ellipsoids of revolution with $m_1^* = m_\ell^*$, and $m_2^* = m_3^* = m_t^*$. Then, if $\gamma \equiv m_\ell^*/m_t^*$,

$$-\frac{\Delta\sigma}{\sigma_0} = [(\zeta + 1) F(\gamma) - 1]R_0^2\sigma_0^2\left(\frac{B_z}{c}\right)^2 \qquad (10.109)$$

where $(\zeta + 1)$ is given by Eq. (10.94) and

$$F(\gamma) = \frac{(\gamma^2 + \gamma + 1)(2\gamma + 1)}{\gamma(\gamma + 2)^2} \qquad (10.110)$$

There is an evident analogy between the Hall effect K and M factors, and the magnetoresistance ζ and $F(\gamma)$ factors, respectively.

The magnetoresistance coefficient is usually written in general with a directional superscript indicating the direction of the magnetic field and a directional subscript indicating the direction of the current. Previous calculations with the magnetic field in the z direction and the current in the x direction therefore correspond to a coefficient $\mathscr{M}_{100}^{001}$, i.e.,

$$\mathscr{M}_{100}^{001} = [(\zeta + 1) F(\gamma) - 1] \qquad (10.111)$$

In analyzing the energy-band structure of materials, it is useful to make magnetoresistance measurements with different arrangements of magnetic-field and electric-current orientation. For example,

$$\mathscr{M}_{110}^{110} = (\zeta + 1)\frac{(\gamma - 1)^2(2\gamma + 1)}{2\gamma(\gamma + 2)^2} \qquad (10.112)$$

i.e., $B_x = B_y$ and $B_z = 0$; $j_x = j_y$ and $j_z = 0$.

A detailed measurement of the magnetoresistance coefficients $\mathscr{M}_{100}^{001}$, $\mathscr{M}_{110}^{110}$, and $\mathscr{M}_{100}^{100}$ permit a determination of the orientation of ellipsoidal equal-energy surfaces.

10.6 Magnetothermal Effects

Interactions between magnetic-field and thermal-gradient effects produce a number of so-called magnetothermal effects.

Ettingshausen Effect

When perpendicular electric and magnetic fields exist in a crystal, not only does a transverse potential difference result, the Hall voltage, but also a transverse temperature difference which produces a thermoelectric voltage that adds to the Hall voltage. Usually the error in the Hall voltage caused by neglecting the Ettingshausen effect in semiconductors, however, is less than 1 per cent.

Physically the effect can be understood as resulting from the fact that the force on a carrier in a magnetic field is proportional to its velocity. Thus the carriers that are deflected most are those with the higher energies and the side of the crystal where the carriers accumulate becomes the hot side.

Nernst Effect

The Nernst effect involves perpendicular temperature gradient and magnetic field, producing a potential difference in the third mutually perpendicular direction. It is equivalent to the Hall effect of carriers moving under the thermal gradient, rather than in an electric field.

Righi–Leduc Effect

When carriers moving in a thermal gradient are deflected by a magnetic field to produce the Nernst effect, a temperature gradient in the transverse direction is also set up. Thus the Righi–Leduc effect is related to the Nernst effect (carrier movement under a thermal gradient) in an analogous way to that in which the Ettingshausen effect is related to the Hall effect (carrier movement in an electric field gradient).

All of these thermomagnetic effects become more significant in more highly conducting materials, since they increase in magnitude with the magnitude of electrical conductivity, and also with the relative contribution of electrons to the thermal conductivity, as compared to the contribution of lattice vibrations.

10.7 High-Magnetic-Field Effects

When the magnetic field becomes large enough, all the previous treatments of this chapter in which the magnetic field is treated approximately as a perturbation as far as the energy levels of the system are concerned, are inadequate. It is necessary to return to the Schroedinger equation and to incorporate the magnetic field in the Hamiltonian.

If $\mathbf{P} = -i\hbar\nabla$ is the normal momentum operator in the absence of a magnetic field, so that

$$\frac{\mathbf{P}^2}{2m} = -\frac{\hbar^2}{2m}\nabla^2$$

this $\mathbf{P}$ becomes, in the presence of a magnetic field,

$$\mathbf{P} = -i\hbar\nabla - \frac{e\mathbf{A}}{c} \quad (10.113)$$

where $\mathbf{A}$ is the *vector potential* associated with the magnetic field such that $\mathbf{H} = \nabla \times \mathbf{A}$, and $\nabla \cdot \mathbf{A} = 0$.

The wave equation for an electron with scalar effective mass m_e^* in the presence of a magnetic field is therefore

$$\frac{1}{2m_e^*}\left(i\hbar\nabla + \frac{e\mathbf{A}}{c}\right)^2 F(\mathbf{r}) = E\, F(\mathbf{r}) \quad (10.114)$$

Consider a magnetic field along the z axis, and take $\mathbf{A} = (0, Hx, 0)$, which satisfies the relation $\mathbf{H} = \nabla \times \mathbf{A}$. Expanding Eq. (10.114), substituting this explicit value for $\mathbf{A}$, and dividing through by $\hbar^2$, yields

$$\left(\nabla^2 - \frac{2ieHx}{\hbar c}\frac{\partial}{\partial y} - \frac{e^2H^2x^2}{\hbar^2 c^2}\right)F(\mathbf{r}) + \frac{2m_e^*}{\hbar^2}E\, F(\mathbf{r}) = 0 \quad (10.115)$$

or

$$\frac{\partial^2 F}{\partial x^2} + \frac{\partial^2 F}{\partial z^2} + \left(\frac{\partial}{\partial y} - \frac{ieHx}{\hbar c}\right)^2 F + \frac{2m_e^*}{\hbar^2} EF = 0 \quad (10.116)$$

Separation of variables in this equation is possible by taking

$$F = \phi(x)\, e^{i(k_y y + k_z z)} \quad (10.117)$$

Substitution of Eq. (10.117) into Eq. (10.116) yields

$$\frac{\partial^2 \phi}{\partial x^2} + \left[\frac{2m_e^*}{\hbar^2}E - k_z^2 - \left(k_y - \frac{eHx}{c\hbar}\right)^2\right] F = 0 \quad (10.118)$$

The following transformation of variables clarifies the significance of this result.

$$E' = E - \frac{\hbar^2 k_z^2}{2m_e^*} \quad (10.119)$$

$$x' = x - \frac{\hbar k_y c}{eH} \quad (10.120)$$

10.7 High-Magnetic-Field Effects

The result is

$$\frac{\partial^2 \phi}{\partial x'^2} + \left\{ \frac{2m_e^*}{\hbar^2} \left[E' - \frac{1}{2} m_e^* \left(\frac{eH}{m_e^* c} \right)^2 x'^2 \right] \right\} \phi = 0 \quad (10.121)$$

This is the wave equation for a harmonic oscillator with frequency $\omega_c = eH/m_e^* c$, the cyclotron resonance frequency. The allowed values of E' are $(n + \frac{1}{2})\hbar\omega^*$ or the allowed values of the energy E are

$$E = (n + \tfrac{1}{2})\hbar\omega_c + \frac{\hbar^2 k_z^2}{2m_e^*}$$

$$= (n + \tfrac{1}{2}) \frac{\hbar eH}{m_e^* c} \frac{\hbar^2 k_z^2}{2m_e^*} \quad (10.122)$$

This means that the effect of the magnetic field is to split the conduction and valence bands into a series of levels, as indicated in Fig. 10.14. These levels are frequently called Landau levels.

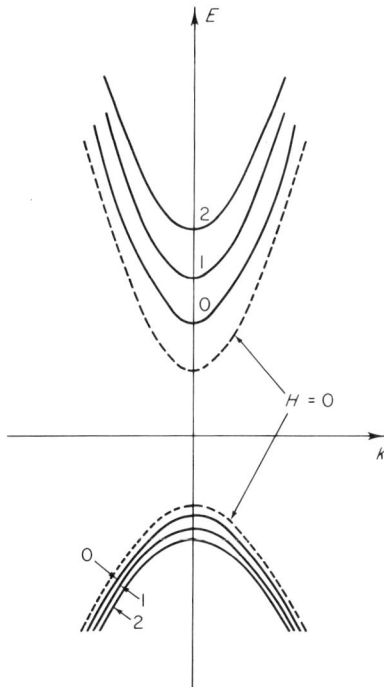

Fig. 10.14 Splitting of the E versus k curves in a high magnetic field to produce Landau levels. The spacing between levels in a given band is given by $\hbar\omega_c$.

Cyclotron Resonance

Transitions between two neighboring levels of the same band can be optically caused by absorption of photons of energy $\hbar\omega_c$. This is the phenomenon of *cyclotron resonance* and is usually measured by observing the resonance absorption corresponding to an applied rf electric-field frequency equal to the cyclotron frequency of the free carriers. In order to observe cyclotron resonance experimentally, there must be sufficient density of free carriers and there must be a sufficiently large value of $\omega_c\tau > 1$ that the carrier remains in a magnetic-field-induced orbit for the major portion of a period without being scattered out of that orbit. These conditions mean that cyclotron resonance is possible in general only in relatively pure materials at low temperatures, sometimes with optical excitation to produce a sufficient density of free carriers.

Because of the direct dependence of ω_c on the effective mass of the carriers, cyclotron resonance is one of the most useful methods for determining the effective masses of carriers in solids. If the carriers are present in an energy band with spherical equal-energy surfaces, a single absorption resonance is expected. For materials with nonspherical equal-energy surfaces, an arbitrary orientation of magnetic field with respect to the axes of the equal-energy surfaces can produce a number of resonance peaks. Typical cyclotron-resonance curves for such an arbitrary orientation for Ge and Si are shown in Fig. 10.15. For the ellipsoidal equal-energy surfaces in these materials, the resonance effective-mass values m^* observed in experiment are related to m_t^* and m_ℓ^* by

$$m^* = m_t^*\left[\frac{m_\ell^*}{m_\ell^* c^2 + m_t^*(1-c^2)}\right] \qquad (10.123)$$

where c is the direction cosine between the direction of the magnetic field and the principal axis of the ellipsoid. Resonance peaks for electrons and holes can be separated by using either a right-hand circularly polarized electric field or a left-hand circularly polarized electric field respectively.

Straightforward measurement of cyclotron resonance is not possible in metals, since the condition for $\omega_c T > 1$ calls for electric-field frequencies so large that the field cannot appreciably penetrate the metal because of the skin effect (see Section 11.1). In the Az'bel–Kaner effect, the magnetic fields is applied *parallel* to the surface of the metal so that the electron orbit in the field intersects the penetration depth of the electric field once in each period. Resonant absorption occurs for an electric-field frequency equal to the cyclotron frequency and also for all subharmonics of the electric-field frequency, ω_c/n, where n is an integer.

10.7 High-Magnetic-Field Effects

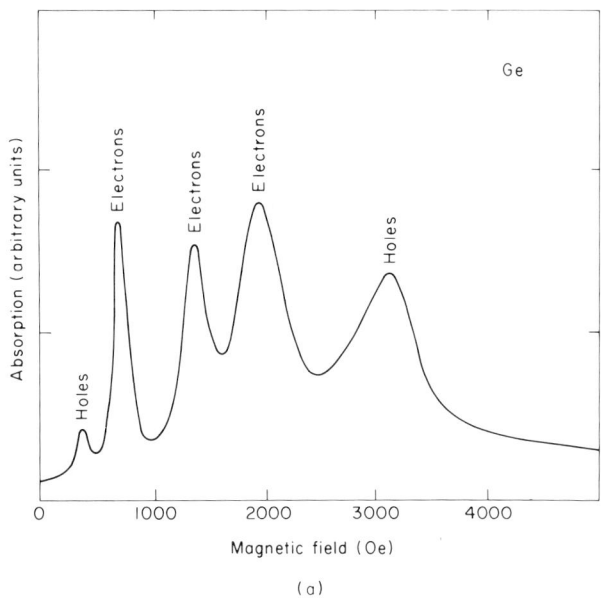

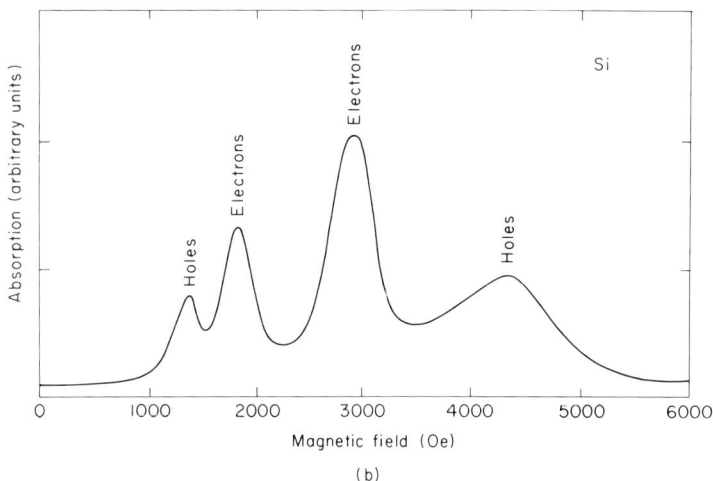

Fig. 10.15 Cyclotron-resonance peaks for carriers in (a) Ge and (b) Si, measured with an arbitrary orientation of the magnetic field (in a (110) plane at 60° from a [100] axis for Ge, and at 30° from a [100] axis in Si) to reveal the number of different resonance peaks occurring for spheroidal equal-energy surfaces with a number of equivalent minima. [After G. Dresselhaus, A. Kip, and C. Kittel, *Phys. Rev.* **98**, 368 (1955).]

MAGNETOABSORPTION EDGE EFFECT

The extrema of the bands shift with increasing magnetic field as

$$E_c = E_c \Big]_{H=0} + \frac{\hbar e}{2m^*c} H \qquad (10.124)$$

which means that the optical absorption edge shifts to higher photon energies with increasing magnetic field. The shift of the bandgap is given by

$$E_G = E_G \Big]_{H=0} + \frac{\hbar e}{2m_r^*c} H \qquad (10.125)$$

For a reduced mass $m_r^* = m$, the coefficient of H in Eq. (10.125) has a value of 3.3×10^{-8} eV/G. Some of the first reported evidence for this effect in InSb is shown in Fig. 10.16.

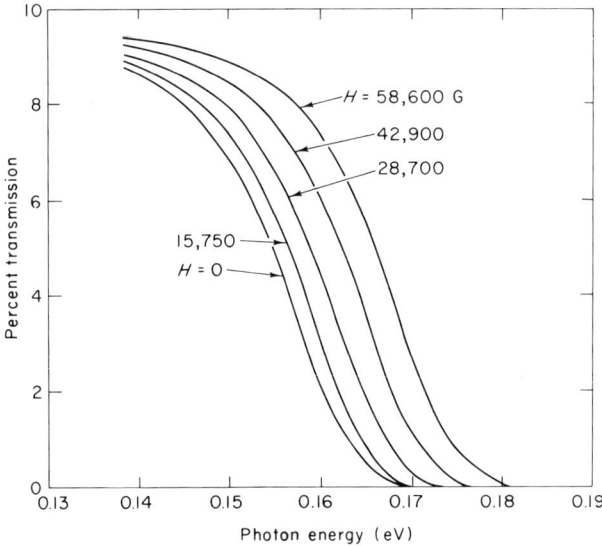

Fig. 10.16 Variation of optical absorption edge (measuring the bandgap) in InSb with magnitude of magnetic field. [After E. Burstein, G. Picus, H. Gebbie, and F. Platt, *Phys. Rev.* **103**, 826 (1956).]

OSCILLATORY MAGNETOABSORPTION

Optical transitions between levels in the valence band and conduction band may occur. Such transitions give rise to an absorption that oscillates with magnetic field as illustrated in Fig. 10.17. A plot of the energy of any

10.7 High-Magnetic-Field Effects

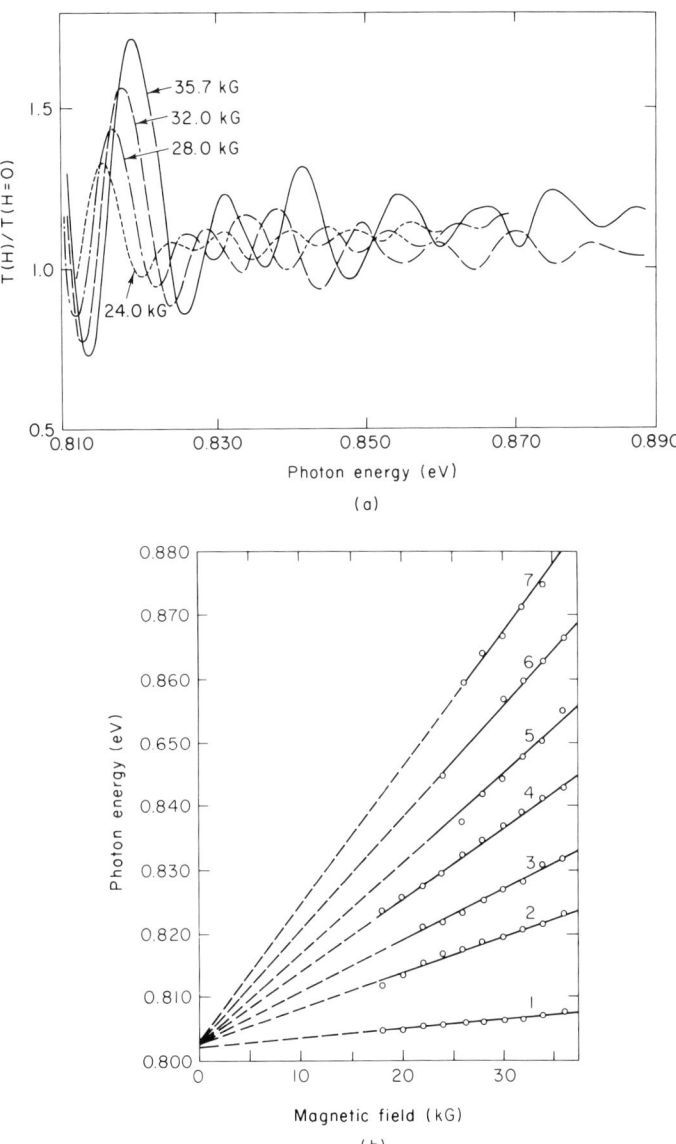

Fig. 10.17 (a) Oscillatory magnetoabsorption in Ge. [After S. Zwerdling, B. Lax, and W. Roth, *Phys. Rev.* **108**, 1402 (1957).] (b) Energy values of transmission minima as a function of magnetic field for successive minima in a measurement of oscillatory magnetoabsorption. All lines converge to give a bandgap of 0.803 ± 0.001 eV at 300°K at $k = 0$. From the slope of the lines and an independent knowledge of the hole effective mass from cyclotron-resonance measurements, the value $m_e^* = (0.037 \pm 0.0001)m$ is obtained. [After B. Lax and S. Zwerdling, *Progress in Semiconductors*, **5**, 227 (1961).]

particular absorption peak as a function of H can be used to get an accurate value of bandgap at $\mathbf{k} = 0$ from the $H = 0$ intercept, and of m_r^* at $\mathbf{k} = 0$ from the slope, according to Eq. (10.122). The selection rule for such optical transitions is $\Delta l = 0$, if l is the index of the Landau level as indicated in Fig. 10.14. Figure 10.17 also shows a plot of transmission minima as a function of magnetic field for several different minima for Ge.

DE HAAS–VAN ALPHEN EFFECT

The de Haas–van Alphen effect is the oscillatory behavior of the magnetic moment with varying magnetic field in a metal at low temperatures. The period of the oscillations (which occur at equal intervals in $1/H$) in units of $1/H$ is inversely proportional to the extremal area of the Fermi surface normal to the direction of $\mathbf{H}$; measurements for different directions of $\mathbf{H}$ permit the mapping of the size and shape of the Fermi surface, and the effect has been one of the most important for the determination of these parameters in metals. The effect arises from a change in the distribution of states for a free-electron gas in the presence of a high magnetic field, these states no longer being described by the components k_i of the wave vector $\mathbf{k}$ but instead by the index number l of a quantized state with degeneracy proportional to H.

OTHER HIGH-MAGNETIC-FIELD EFFECTS IN METALS

Useful insight into the nature of the Fermi surface in metals can also be obtained by Hall-effect and magnetoresistance measurements at high magnetic fields. These insights are usually associated with anisotropy effects and are gained by making the measurements in various crystallographic orientations. One example is the failure of magnetoresistance to saturate at high magnetic fields if it is possible for an open orbit to exist carrying current normal to the applied electric and magnetic fields (an open orbit is one in which the $\mathbf{k}$-space trajectory in the extended-zone scheme passes from one zone to an adjacent zone without ever returning to the starting point).

10.8 Summary

Magnetic-field effects in a material can be considered either as perturbations on the system, which do not change the existing electronic energy levels (the low-field range), or as making major quantization effects in the

10.8 Summary

allowed energies for free carriers (the high-field range). Physically the former range corresponds to the situation in which a carrier in a magnetic field does not complete a major portion of a cycle of its circular orbit before being scattered out of it; the latter range corresponds to a well-established circular orbit in the magnetic field. The angular frequency of a free carrier in a magnetic field is given by the cyclotron frequency ω_c, which is proportional to the magnetic field and inversely proportional to the effective mass of the carrier. If the scattering process can be characterized by a relaxation time τ, quantization effects due to the magnetic field can be observed only if $\omega_c \tau \gg 1$.

If electric and magnetic fields are present in perpendicular orientation in a material, a potential difference in the third mutually perpendicular direction occurs which is known as the Hall voltage. The Hall voltage exists to prevent net current flow in this third direction. The magnitude of the Hall voltage is directly proportional to the Hall mobility, the sign of the Hall voltage distinguishes between electrons and holes, and the combination of the Hall voltage with a conductivity measurement permits the determination of both carrier density and mobility. In order to correlate the Hall mobility measured in this way with the mobility of significance for the calculation of electrical conductivity, it is necessary to take account of the specific scattering process, the degeneracy of the material, and the specific band structure.

Except for the single special case of an energy-independent relaxation time for scattering, one-carrier conductivity, and spherical equal-energy surfaces, the presence of a magnetic field also affects the conductivity in the material. Whenever the transverse current of carriers deflected by the magnetic field cannot be cancelled out for all the carriers participating, the path length of some of the carriers is effectively lengthened, and the corresponding resistivity is increased. The effect is known as magnetoresistance and can prove useful in determining the nature and orientation of the equal-energy surfaces of a semiconductor, as well as the size and shape of the Fermi surface of a metal.

High magnetic fields giving rise to quantization effects make possible another whole range of measurements of significance for materials characterization. Cyclotron resonance in which the optical excitation of carriers from one quantized level to the next highest occurs with photons of energy $\hbar\omega_c$ permit the determination of effective mass. Magnetoabsorption effects from quantized levels in one "band" to quantized levels in the next higher "band" make possible an accurate determination of bandgap and effective mass. A variety of magnetic-field-associated phenomena in metals have been explored to provide a wealth of information about the shape and size of the Fermi surface.

General transport involving electric and magnetic fields should also consider the effect of a simultaneous temperature gradient. The number of different kinds of effect produced by combinations and permutations of these three external influences is large indeed. Some of the properties of electric fields and thermal gradients have been discussed previously. The Hall-effect experiment produces not only the transverse Hall voltage but also a transverse Ettingshausen temperature gradient, which in turn produces a thermoelectric voltage that adds slightly to the Hall voltage. Similarly if carrier flow (but not current flow) is occurring in a simple temperature gradient, a magnetic field produces a Nernst transverse voltage, and also a transverse Righi–Leduc temperature gradient, which in turn produces a thermoelectric voltage that adds slightly ot the Nernst voltage. These thermomagnetic effects are of significant magnitude only in high-conductivity semiconductors or metals.

Problems

10.1 A cubic compound has a conduction band with equivalent ellipsoidal equal-energy surfaces in the (100) direction with longitudinal effective mass of $0.64m$, transverse effective mass of $0.2m$, a valence band with spherical equal-energy surfaces with maximum at $\mathbf{k} = 0$, and scalar effective mass of $0.8m$. The bandgap is 2.0 eV at 300°K. The electron mobility at 300°K is 1000 cm^2/V-sec, and the Hall constant at 300°K is zero for low values of the magnetic field. Calculate the conductivity at 300°K.

10.2 At 300°K the Hall mobility in an intrinsic cubic covalent semiconductor is 500 cm^2/V-sec with a sign appropriate for electrons. The transverse effective mass of electrons is $0.1m$, and the longitudinal effective mass for electrons is $0.8m$. The effective mass for holes is a scalar but unknown. Calculate the hole mobility, if it is known that the electron mobility is 1000 cm^2/V-sec.

10.3 Shallow donor impurities with a density of 10^{16} cm^{-3}, essentially completely ionized at 300°K, are diffused into the crystal of Problem 10.2. The density of free electrons is calculated from the Hall constant measured in the presence of a magnetic field of 10^4 G. Calculate the error introduced by magnetoresistance effects if the current density used in calculating n is measured in the presence of the magnetic field rather than in the absence of the magnetic field.

10.4 At low temperatures for the crystal of Problem 10.1, a scattering process dominates for which $\tau = aE^{+1/2}$. If, over these same temperatures, one-carrier conductivity by holes dominates, calculate: (a) $m_h^* \mu_h/e$, (b) $R\sigma/\mu_h$, (c) $\Delta\varrho/[\varrho_0 R_0^2 \sigma_0^2 (B/c)^2]$, and (d) ϱ_∞/ϱ_0.

10.5 A Hall mobility of 10 cm²/V-sec is measured for a crystal at 300°K. What value of electrical conductivity is consistent with the interpretation that this low Hall mobility is the result of a two-carrier conductivity process, if the electron mobility is 1000 cm²/V-sec, the hole mobility is 500 cm²/V-sec, the bandgap is 1.0 eV, the electron and hole effective masses are scalars, $0.2m$ and $0.5m$ respectively, and LA scattering dominates?

10.6 Another application of the Hall effect is in use as a wattmeter, with the magnitude of the Hall voltage being proportional to the electrical power being expended in some external test circuit. Indicate how the Hall effect can be used in this way, and discuss the limitations of the apparatus described in Example 10.1 if used in a wattmeter application.

10.7 For two-carrier conductivity there are four unknowns: n, p, μ_e, and μ_h. If the other properties of the material are known, e.g., its bandgap, effective masses, and so on, what measurements can be made to provide four equations from which the four unknowns above can be determined? Proceed with the solution.

10.8 In several practical applications, a large-area polycrystalline film is used, which consists of an array of grains (n_1, μ_1) surrounded by intergranular material (n_2, μ_2). Hall-effect measurements on such films can produce unusual results. Using the model shown in Fig. 10.18, with $n_2 \ll n_1$, $\mu_2 = \mu_1$, and $l_2 \ll l_1$, calculate the carrier density n_H and Hall mobility μ_H, as determined from a Hall-effect measurement on the film, in terms of the actual parameters n_1, n_2, μ_1, μ_2, l_1, and l_2.

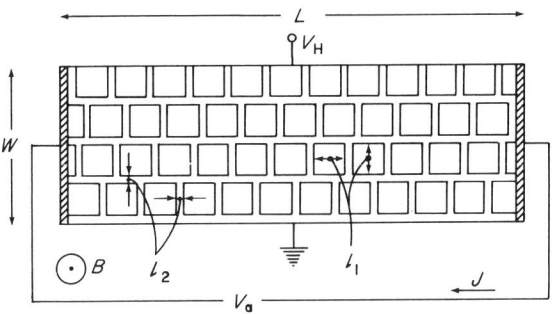

Fig. 10.18 A simple model for an inhomogeneous material consisting of grains of width l_1 and intergranular regions with width l_2.

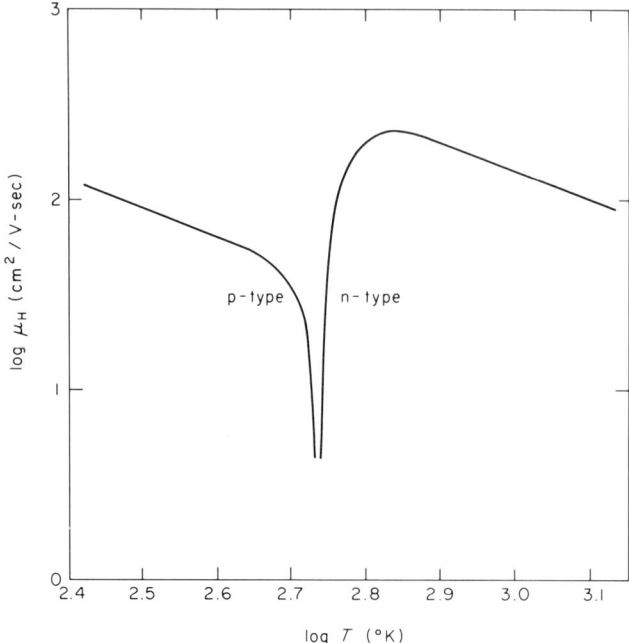

Fig. 10.19 Variation of Hall mobility with temperature; see Problem 10.9.

10.9 A material with a bandgap of 1.0 eV (neglect temperature dependence) has N_A acceptors with ionization energy of 0.05 eV. As the temperature is raised, intrinsic conductivity becomes important and the variation of Hall mobility with temperature is as given in the data of Fig. 10.19. Given that the effective density of states for both electrons and holes is 1.76×10^{18} cm^{-3} at 300°K, and that scattering by LA phonons dominates over the whole temperature range of interest, determine, for a material with spherical equal-energy surfaces, (a) the electron mobility at 500°K, (b) the location of the Fermi level at the temperature at which the Hall mobility goes to zero, and (c) the acceptor density N_A.

10.10 In a cubic nondegenerate semiconductor with equivalent spheroidal equal-energy surfaces in (100) directions ($m_t^* = 0.1m$ and $m_\ell^* = 0.5m$), the relaxation time for complete dominance of scattering by charged-impurity scattering, such as is found in the dark, is 10^{-12} sec. When high-intensity light is shined on this material, charged impurities are effectively neutralized, and scattering becomes dominated only by longitudinal acoustical phonon scattering. If this photoexcitation produces an *increase* of 10^4 cm^2/V-sec in the carrier mobility (one-carrier conductivity), what is the *exact* predicted *change* in the measured Hall mobility as a result of photo-excitation?

10.11 Consider a semiconductor with spherical equal-energy surfaces and scattering by LA phonons dominant, with a sufficiently high density of donor impurities that an impurity band is formed from these donors. At high temperatures, all donors are ionized, the impurity band is empty, and all conductivity takes place in the conduction band as in a nondegenerate semiconductor. As the temperature decreases, some of the donors become occupied by electrons, and thus there are electrons available for transport in the impurity band as well as in the conduction band.

(a) If N is the density of donors (no acceptors), show that the Hall

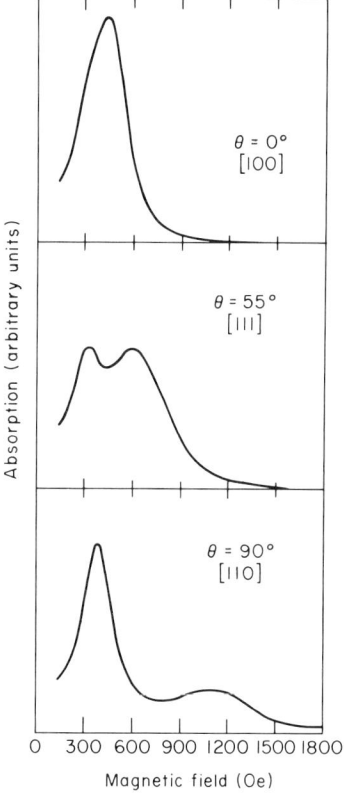

Fig. 10.20 Cyclotron-resonance peaks for electrons in n-type Ge at 4.2°K, measured in a (110) plane as a function of the angle θ between the direction of the magnetic field and the [100] direction. [After B. Lax, H. Zeiger, R. Dexter, and E. Rosenblum, *Phys. Rev.* **93**, 1418 (1954).]

coefficient is given by $R_H = -K_e/Ne$ at any temperature if the conduction band and impurity band mobilities are equal, and at very high temperatures or very low temperatures even if the conduction-band mobility is greater than the impurity-band mobility.

(b) If the conduction-band mobility is greater than the impurity-band mobility, show that R_H exhibits a maximum at some intermediate temperature.

(c) If the conduction-band mobility is ten times the impurity-band mobility, what is the ratio of electrons in the conduction band to electrons in the impurity band at the maximum R_H?

10.12 The ratio of Hall mobility to drift mobility in a trap-free cubic semiconductor with equivalent spheroidal equal-energy surfaces in (100) directions is 1.85. For an electron density of 10^{15} cm^{-3}, the measured thermoelectric power at 300°K is -1.13 mV/°K. If the longitudinal effective mass is twice the transverse effective mass, determine the values of these effective masses.

10.13 Figure 10.20 shows cyclotron-resonance curves for free electrons in n-type germanium at 4.2°K. The measurements were made in a (110) plane as a function of the angle θ between the direction of the magnetic field and the [100] direction. Explain the structure found and determine the longitudinal and transverse effective masses in Ge from these data.

Chapter 11

Optical Absorption

A number of different optical phenomena may be associated with the incidence of light on a crystalline solid, including reflection, refraction, absorption, and transmission. A variety of electronic processes may contribute to these observed phenomena. Figure 11.1 shows a schematic representation of optical absorption as a function of wavelength of the incident electromagnetic radiation. From higher to lower energies the processes involved are optical excitation (1) to and from energy bands with greater energy separation than the uppermost occupied band and the next-higher-lying empty band, (2) of excitons, (3) from the uppermost occupied levels to the next-higher-lying empty band, (4) from an imperfection to a band, from a band to an imperfection, or within the energy levels of an imperfection, (5) of free carriers to a higher energy state in the same band, and (6) of a Reststrahlen mode of the lattice vibrations (Section 1.4).

Electronic optical absorption processes are called *direct transitions* if only an optical photon is involved in the process, producing negligible change in the wavevector of the electron. Direct transitions are pictured as vertical transitions on an $E(\mathbf{k})$ vs. $\mathbf{k}$ diagram. Electronic absorption processes are called *indirect transitions* if a photon and one or more phonons are involved in the process, producing a change in the wavevector by the wavevector of the phonon. Indirect transitions are pictured as nonvertical transitions on an $E(\mathbf{k})$ vs. $\mathbf{k}$ diagram. Direct transitions are first-order processes; indirect transitions are second-order processes. For most electronic transitions in crystalline solids conservation of energy and electron wavevector between initial and final states is required.

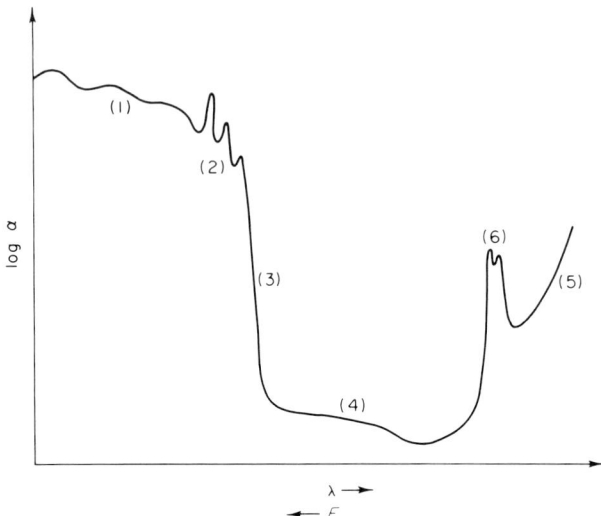

Fig. 11.1 Schematic display of different types of optical absorption typically found in crystals. (1) Transitions to high-lying bands, (2) excitons, (3) absorption edge for valence-band-to-conduction-band transition, (4) imperfection absorption, (5) free-carrier absorption, and (6) Reststrahlen absorption.

Metals are distinguished by their high reflectivity and opacity except in very thin films. These optical properties are associated primarily with the absorption of electromagnetic radiation by the free electrons in the metal. Indirect optical transitions are involved in such absorption processes. These transitions may be reasonably adequately described by a quasi-classical approach, although a quantum approach is required for a more general treatment. We confine our discussion here to that of the quasi-classical view in the interest of simplicity and brevity.

Semiconductors may also be highly reflecting and opaque over certain ranges of wavelength, and yet appreciably transmitting over others. All of the absorption processes pictured in Fig. 11.1 can be detected in semiconductors, including the free-electron absorption characteristic of metals provided that the free-carrier density is sufficiently large. The dominant absorption process in semiconductors is usually the excitation of an electron from the valence band across the forbidden gap to the conduction band. This process can occur via either a direct or an indirect transition. The dependence of the absorption constant on the photon energy can be calculated for these two kinds of process by the use of time-dependent perturbation theory.

11.1 Free-Carrier Absorption

The quasi-classical treatment of free-carrier absorption starts with the electromagnetic wave equation derived from Maxwell's equations, as given in Table 1.6. When this wave equation is written for plane electromagnetic waves in the y–z plane, wave equations can be written for each of the components of the electric and magnetic field: $\mathscr{E}_y$, $\mathscr{E}_z$, H_y, and H_z. The wave equation for $\mathscr{E}_y$, for example, is

$$\frac{\partial^2 \mathscr{E}_y}{\partial x^2} = \frac{\varepsilon}{c^2} \frac{\partial^2 \mathscr{E}_y}{\partial t^2} + \frac{4\pi\sigma}{c^2} \frac{\partial \mathscr{E}_y}{\partial t} \tag{11.1}$$

for a material with dielectric constant ε, conductivity σ, and magnetic permeability $\mu = 1$. A harmonic wave of the form $\mathscr{E}_y = A\, e^{i(kx-\omega t)}$ is a solution of Eq. (11.1) provided that the complex velocity v^* is given by

$$\frac{1}{v^{*2}} = \frac{k^2}{\omega^2} = \frac{\varepsilon}{c^2} + i\,\frac{4\pi\sigma}{\omega c^2} \tag{11.2}$$

The physical significance of a complex wave velocity is that an absorption process is involved, as may be shown as follows.

Define a complex index of refraction in terms of the complex velocity,

$$r^* \equiv \frac{c}{v^*} \tag{11.3}$$

Then from Eq. (11.2),

$$r^{*2} = \varepsilon + i\,\frac{4\pi\sigma}{\omega} \tag{11.4}$$

Also define a relationship between the complex index of refraction r^* and the real index r by

$$r^* \equiv r(1 + i\gamma) \tag{11.5}$$

where γ is called the *absorption index*. Substitution of $k^* = \omega/v^*$ into $\mathscr{E}_y = A\, e^{i(k^* x - \omega t)}$ shows that for a complex velocity,

$$\mathscr{E}_y = A\,[e^{-xr\omega\gamma/c}]\, e^{i(kx-\omega t)} \tag{11.6}$$

The factor $e^{-xr\omega\gamma/c}$ is an *attenuation factor* acting on the wave as it proceeds through the material. The magnitude of this factor depends on the magnitude of γ, which can in turn be related to the magnitude of σ, and hence ultimately to the free-electron density.

To relate γ to σ, square Eq. (11.5) and compare with Eq. (11.4) to obtain

$$r^2(1 - \gamma^2) = \varepsilon \tag{11.7}$$

$$\frac{\omega r^2 \gamma}{2\pi} = \sigma \tag{11.8}$$

Explicit expressions for $r(\sigma)$ and $\gamma(\sigma)$ can be obtained by solving these two equations

$$r^2 = \frac{1}{2}\left[\varepsilon + \left(\varepsilon^2 + \frac{16\pi^2\sigma^2}{\omega^2}\right)^{1/2}\right] \tag{11.9}$$

$$\gamma^2 = 1 - \frac{\varepsilon}{\frac{1}{2}\left[\varepsilon + \left(\varepsilon^2 + \frac{16\pi^2\sigma^2}{\omega^2}\right)^{1/2}\right]} \tag{11.10}$$

The extent of the deviation from $r = \varepsilon^{1/2}$, as found for a material without absorption, depends on the magnitude of the ratio σ/ω. In order for the refractive index to depart appreciably from $\varepsilon^{1/2}$ for 5000-Å light, for example by 10 per cent, $\sigma/\omega \approx 1$ if ε is of the order of 10–20. This means that σ must be equal to or greater than 4×10^3 (Ω-cm)$^{-1}$ before the absorption affects the index of refraction, i.e., a value characteristic of a metal or semimetal.

Optical absorption is usually defined in terms of an *absorption constant* α defined operationally by the relationship

$$I_t = (I_0 - I_R)\, e^{-\alpha d} \tag{11.11}$$

where I_0 is the intensity of the light incident on a crystal (as measured, for example in W cm^{-2} sec^{-1}), I_R is the reflected intensity, so that $(I_0 - I_R)$ is the intensity actually propagated into the crystal (with second-order corrections for reflection from the back face, and from repetitive crossings of the crystal as a result), I_t is the intensity of the transmitted light, and d is the thickness of the crystal. The quantity $d^* = 1/\alpha$ is called the *penetration depth*; it represents the path length in the material necessary to reduce the intensity by a factor e. Since the energy of an electromagnetic wave is proportional to the product of $\mathscr{E}$ and $\mathbf{H}$, and since both $\mathscr{E}$ and $\mathbf{H}$ contain similar exponential terms, comparison of Eq. (11.11) with Eq. (11.6) gives

$$\alpha = \frac{2\omega r}{c}\gamma = \frac{4\pi}{cr}\sigma \tag{11.12}$$

11.1 Free-Carrier Absorption

OPTICAL REFLECTION

Metals, with strong absorption by free electrons, also show high reflectivity. The relationship between absorption index and reflectivity can be seen by considering the interaction of an electromagnetic wave at an interface between two materials.

Consider an electromagnetic wave moving in the $+x$ direction and impinging on an interface between a material with ε_1 and σ_1 for $x < 0$, and a material with ε_2 and σ_2 for $x > 0$. We need to consider incident, reflected and transmitted electric and magnetic wave components, and the effect of required continuity in the tangential components of electric and magnetic field at the interface.

Incident waves:
$$\mathscr{E}_y^{(i)} = A\, e^{i\omega(xr_1^*/c - t)} \tag{11.13}$$

$$H_z^{(i)} = A\, r_1^*\, e^{i\omega(xr_1^*/c - t)} \tag{11.14}$$

since $\mathscr{E}_y$ is related to H_z in this way through Maxwell's equations. In the reflected wave, $\mathscr{E}_y$ and H_z must have opposite signs to satisfy Maxwell's equations.

Reflected waves:
$$\mathscr{E}_y^{(r)} = -A'\, e^{-i\omega(xr_1^*/c + t)} \tag{11.15}$$

$$H_z^{(r)} = A'r_1^*\, e^{-i\omega(xr_1^*/c + t)} \tag{11.16}$$

Transmitted waves:
$$\mathscr{E}_y^{(t)} = A''\, e^{i\omega(xr_2^*/c - t)} \tag{11.17}$$

$$H_z^{(t)} = A''r_2^*\, e^{i\omega(xr_2^*/c - t)} \tag{11.18}$$

Continuity of $\mathscr{E}_y$ and H_z at the interface gives

$$\frac{A'}{A} = \frac{r_2^* - r_1^*}{r_2^* + r_1^*} \tag{11.19}$$

The reflection coefficient R is given by $A'A'^*/AA^* = |A'|^2/|A|^2$, i.e., the square of the magnitude of the reflected radiation divided by the square of the magnitude of the incident radiation. Substituting $r_j^* = r_j(1 + i\gamma_j)$ into Eq. (11.19) and multiplying by the complex conjugate gives

$$R = \frac{(r_2 - r_1)^2 + (r_2\gamma_2 - r_1\gamma_1)^2}{(r_2 + r_1)^2 + (r_2\gamma_2 + r_1\gamma_1)^2} \tag{11.20}$$

For $\sigma_j = 0$ and $r_1 = 1$, i.e., an interface between vacuum and a non-conducting material,

$$R = \frac{(r_2 - 1)^2}{(r_2 + 1)^2} \tag{11.21}$$

which is the familiar expression for reflection from glass, for example. For the interface between vacuum and a conducting medium,

$$R = \frac{(r_2 - 1)^2 + r_2^2 \gamma_2^2}{(r_2 + 1)^2 + r_2^2 \gamma_2^2} \tag{11.22}$$

As the terms involving the absorption index exceed the terms involving the index of refraction, the reflection coefficient approaches unity.

For a metal, for example, Eq. (11.9) becomes

$$r = \left(\frac{2\pi\sigma}{\omega}\right)^{1/2} \tag{11.23}$$

Also $\gamma \approx 1$, and Eq. (11.22) gives

$$R \approx 1 - \frac{2}{(2\pi\sigma/\omega)^{1/2}} \tag{11.24}$$

For light in the visible portion of the spectrum, a metal with $\sigma = 10^6$ $(\Omega\text{-cm})^{-1}$ has a reflection coefficient $R \approx 0.93$. In order for R to be as large as 0.5, the conductivity σ must be greater than 10^1 $(\Omega\text{-cm})^{-1}$.

The general results derived here for absorption and reflection are more general than the specific absorption process due to free carriers. An absorption index γ can be assigned to *any* optical absorption process with an absorption constant α through Eq. (11.12). Then the same results describing the dependence of the index of refraction and optical reflection coefficient on γ can be used to describe the effects of this other absorption process.

FREQUENCY DEPENDENCE

The relationship between absorption constant α and electrical conductivity σ in Eq. (11.12) holds for some particular frequency of the radiation. If the conductivity is a function of electric-field frequency, then the appropriate conductivity for the frequency involved must be used in these calculations. Insight into the frequency dependence of conductivity can also be obtained in a simple quasi-classical manner.

Consider the equation of motion for an electron moving in the presence of an electric field in the x direction with frequency ω, in a material with definable scattering relaxation time τ.

$$\frac{dv_x}{dt} + \frac{v_x}{\tau} = -\frac{e}{m_e^*} \mathscr{E}_x e^{-i\omega t} \tag{11.25}$$

11.1 Free-Carrier Absorption

With $\mathscr{E}_x = 0$, $v_x = v_{x0} e^{-t/\tau}$. If the electric field is proportional to $e^{-i\omega t}$, v_x is also proportional to $e^{-i\omega t}$, and Eq. (11.25) becomes

$$\left(-i\omega + \frac{1}{\tau}\right)v_x = -\frac{e}{m_e^*}\mathscr{E}_x e^{-i\omega t} \tag{11.26}$$

The current density j_x can be calculated as $-nev_x$, and after clearing complex terms from the denominators, we have

$$j_x = \frac{ne^2}{m_e^*}\left(\frac{\tau}{1+\omega^2\tau^2} + i\omega\frac{\tau^2}{1+\omega^2\tau^2}\right)\mathscr{E}_x e^{-i\omega t} \tag{11.27}$$

Equation (11.4) for the complex index of refraction can be considered as an expression for a complex dielectric constant,

$$\varepsilon^* = r^{*2} \tag{11.28}$$

A complex conductivity σ^* can also be defined such that $j_x = \sigma^*(\mathscr{E}_x e^{-i\omega t})$,

$$\sigma^* = -i\frac{\omega}{4\pi}\varepsilon_e^* = \sigma - i\frac{\omega}{4\pi}\varepsilon_e \tag{11.29}$$

where the subscript e has been added to ε to emphasize that we are here concerned with the electronic contribution to the dielectric constant. There is also an ionic contribution to the dielectric constant from the polarization of the lattice ions, ε_L, so that the total dielectric constant is given by $\varepsilon = \varepsilon_L + \varepsilon_e$. Comparison of Eq. (11.29) with Eq. (11.27) gives

$$\sigma = \frac{ne^2}{m_e^*}\left\langle\frac{\tau}{1+\omega^2\tau^2}\right\rangle = \sigma_0\frac{\left\langle\frac{\tau}{1+\omega^2\tau^2}\right\rangle}{\langle\tau\rangle} \tag{11.30}$$

$$\varepsilon = \varepsilon_L - \frac{4\pi ne^2}{m_e^*}\left\langle\frac{\tau^2}{1+\omega^2\tau^2}\right\rangle = \varepsilon_L - 4\pi\sigma_0\langle\tau\rangle\frac{\left\langle\frac{\tau^2}{1+\omega^2\tau^2}\right\rangle}{\langle\tau\rangle^2} \tag{11.31}$$

The indicated averages are over the electron energy to account for $\tau(E)$, and the subscript 0 indicates the dc or low-frequency value of conductivity. The characteristics of the phenomena encountered depend upon the magnitudes of frequency ω and conductivity σ_0, and several physically significant ranges can be delineated.

In the low-frequency, low-conductivity range ($\omega\tau \ll 1$; $r = \varepsilon^{1/2}$), the $\omega^2\tau^2$ term can be neglected completely with respect to unity, and $\sigma = \sigma_0$.

A second-order solution can be obtained by replacing $(1 + \omega^2\tau^2)^{-1}$ by $(1 - \omega^2\tau^2)$, giving

$$\sigma = \sigma_0\left(1 - \frac{\omega^2\langle\tau^3\rangle}{\langle\tau\rangle}\right) \tag{11.32}$$

To first order the absorption constant for this range is simply

$$\alpha = \frac{4\pi\sigma_0}{\varepsilon^{1/2}c} \tag{11.33}$$

In the low-frequency, high-conductivity range ($\omega\tau \ll 1$; $r > \varepsilon^{1/2}$), such as would be encountered for long-wavelength electromagnetic radiation in metals, the index of refraction is given by Eq. (11.23) with $\sigma = \sigma_0$, and the absorption constant becomes

$$\alpha = \left(\frac{8\pi}{c^2}\right)^{1/2}(\sigma_0\omega)^{1/2} \tag{11.34}$$

The absorption constant in this case increases not only with the conductivity but also with the frequency, i.e., penetration of the material is proportional to $\omega^{-1/2}$. The higher the frequency of the electric field (within the range of the approximation), the more the field is restricted to a thin region near the surface. This phenomenon is called the *skin effect*.

In the high-frequency range ($\omega\tau \gg 1$), Eq. (11.30) becomes

$$\sigma = \frac{ne^2}{m_e^*\omega^2}\left\langle\frac{1}{\tau}\right\rangle = \frac{\eta\sigma_0}{\omega^2\langle\tau\rangle^2} = \frac{\eta e^2\sigma_0}{\omega^2 m_e^{*2}\mu^2} \tag{11.35}$$

where $\eta = \langle\tau\rangle/\langle 1/\tau\rangle$, and $\mu = (e/m_e^*)\langle\tau\rangle$. Then the absorption constant is

$$\alpha = \frac{4\pi\eta e^2\sigma_0}{\omega^2 m_e^{*2}\mu^2 rc}$$

$$= n\lambda^2 \frac{\eta e^3}{\pi m_e^{*2}\mu rc^3} \tag{11.36}$$

where λ is the wavelength of the electromagnetic radiation. The absorption constant in this range is proportional both to the carrier density and to the square of the wavelength, the proportionality constant also including the effective mass of the carriers. Figure 11.2a shows experimental curves for free-carrier absorption by electrons and holes in Ge crystals. The absorption due to electrons varies approximately as λ^2, in good agreement with

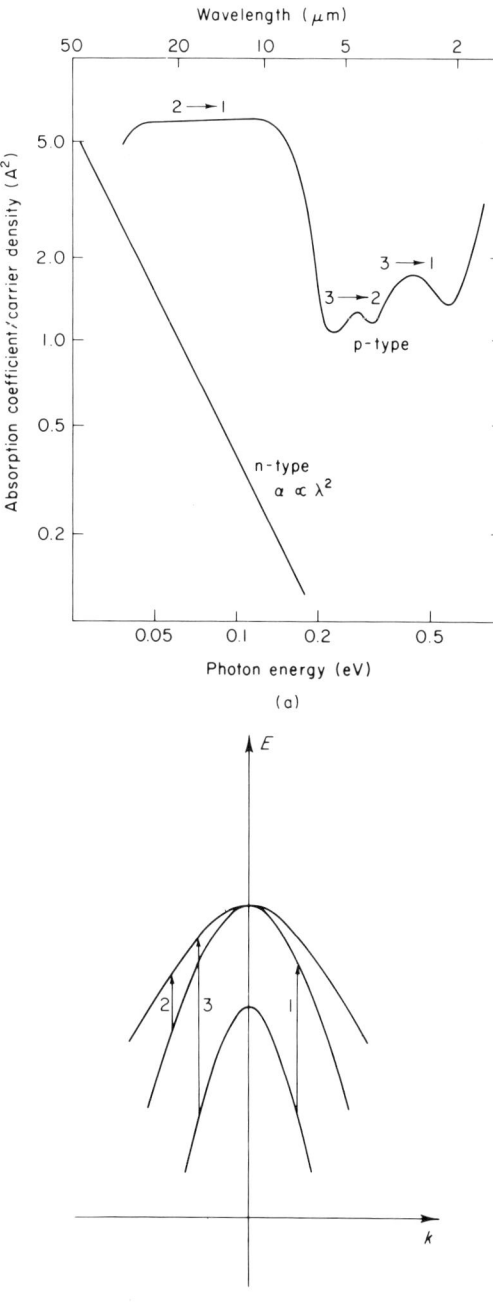

Fig. 11.2 (a) Free-carrier absorption due to electrons in n-type Ge and to holes in p-type Ge [After W. Kaiser, R. Collins, and H. Y. Fan, *Phys. Rev.* **91**, 1380 (1953)]. (b) Typical absorption transitions between the three valence bands of Ge.

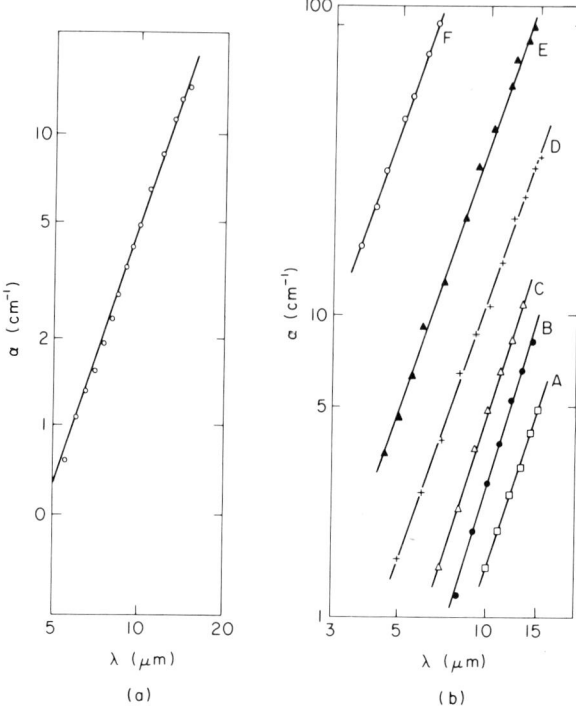

Fig. 11.3 (a) Free-electron absorption in nonstoichiometric CdSe with conductivity of 2 $(\Omega\text{-cm})^{-1}$ and mobility of 600 cm^2/V-sec (after A. L. Robinson, Ph.D. Thesis, Stanford University). (b) Free-electron absorption in n-type InAs for different values of the free-electron density in 10^{17} cm^{-3}. (A) 0.28, (B) 0.85, (C) 1.4, (D) 2.5, (E) 7.8, (F) 39. (After J. R. Dixon, "Proc. Intern. Conf. Semiconductor Phys., Prague, 1960," Czechoslovak Academy of Sciences, Prague, and Academic Press, New York, 1961, p. 366.)

the prediction of Eq. (11.36), but the absorption due to holes has a completely different structure. This structure for the free-hole absorption in p-type Ge is the result of the multiple valence-band structure; the transitions indicated in Fig. 11.2a are illustrated in Fig. 11.2b in a representative energy-band diagram for the valence band. If the scattering process is considered in more detail than in the above quasi-classical treatment, the wavelength dependence of the absorption constant may be different from that reported above; for optical-mode scattering, for example, a variation of $\alpha \propto \lambda^{2.5}$ is predicted and for charged-impurity scattering, $\alpha \propto \lambda^3$; Fig. 11.3 shows absorption by free electrons in an n-type CdSe crystal and in n-type InAs.

Another phenomenon occurring when $\omega\tau \gg 1$ can be described in terms

11.1 Free-Carrier Absorption

of the effect on the dielectric constant. In this range, Eq. (11.31) reduces to

$$\varepsilon = \varepsilon_L - \frac{4\pi\sigma_0}{\omega^2\langle\tau\rangle}$$

$$= \varepsilon_L - \frac{4\pi n e^2}{m_e^* \omega^2} \quad (11.37)$$

If $\varepsilon = 0$, then $r = 0$, and the absorption constant becomes very large. The frequency at which this phenomenon occurs is known as the *plasma frequency*, and is given by

$$\omega_p = \left(\frac{4\pi n e^2}{m_e^* \varepsilon_L}\right)^{1/2} \quad (11.38)$$

It is the frequency at which an undamped plasma of free electrons can oscillate as a whole. Actually the resonance is not very sharp unless $\omega\tau$ exceeds unity by an order of magnitude or so, and a second-order calculation shows that even when the two terms of Eq. (11.37) cancel,

$$\varepsilon = \frac{4\pi n e^2}{m_e^* \omega^4} \frac{1}{\langle\tau^2\rangle} \quad (11.39)$$

Since the plasma frequency is proportional to the square-root of the free-carrier density, it occurs at quite different values of frequency for metals and semiconductors. For the alkali metals, for example, the plasma frequency corresponds to the ultraviolet region of the spectrum, and the transmission found at $\omega > \omega_p$, corresponding to a real index of refraction, gives rise to the so-called "ultraviolet transparency" of these metals. The plasma frequency for semiconductors is commonly in the infrared in the 10–100-μm range because of the smaller value of the free-carrier densities encountered. Figure 11.4 shows measurements of the plasma edge in InSb for several different free-electron densities; for wavelengths longer than the edge, the index of refraction is complex and high reflectivity is observed, whereas for wavelengths shorter than the edge, the index of refraction is real and transmission is observed.

The interpretation of ω_p as the plasma frequency for the collective oscillation of the electron gas can be understood by considering the motion of the electron gas with respect to the fixed positive ions of the lattice. A displacement of the gas by ξ produces an electric field $\mathscr{E} = 4\pi n e \xi$, which constitutes a restoring force in the equation of motion,

$$nm\frac{d^2\xi}{dt^2} = -ne\mathscr{E} = -4\pi n^2 e^2 \xi \quad (11.40)$$

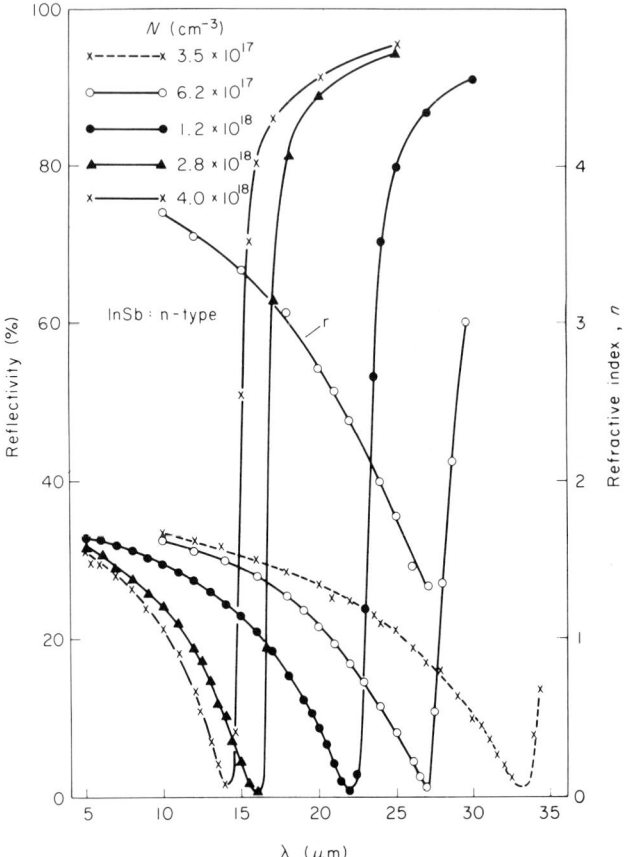

Fig. 11.4 Reflectivity spectra for n-type InSb samples, showing the variation of the plasma edge with the density of free electrons. The curve for index of refraction (r) is for the sample with 6.2×10^{17} cm^{-3} electrons. [After W. G. Spitzer and H. Y. Fan, *Phys. Rev.* **106**, 882 (1957).]

But this is imply the equation of an oscillator with the frequency ω_p,

$$\frac{d^2\xi}{dt^2} + \omega_p^2 \xi = 0 \qquad (11.41)$$

In the quantum development of this problem, such an oscillating system is of course quantized, and the corresponding quantum $\hbar\omega_p$ is called a *plasmon*. This plasmon quantization can be detected experimentally by measurements of the energy lost by energetic electrons passing through thin metallic films.

11.2 Optical Transitions between Bands

A direct optical transition involving the absorption of a photon only is pictured in Fig. 11.5a. The energy of the photon absorbed must be equal to the bandgap energy for a transition from the top of the valence band to the bottom of the conduction band, and there is almost no change in **k** upon making the transition. Actually, of course, there is a finite change in **k** because the photon does have a finite momentum. How large a change in **k** results from the momentum of the photon can be determined by recognizing that

$$P_{\text{photon}} = h\nu \frac{r}{c} \qquad (11.42)$$

where ν is the frequency of the light wave, and (c/r) is the velocity of the wave in the crystal with index of refraction r. Since Δk for the optical transition is given by

$$\hbar \Delta k = P_{\text{photon}} \qquad (11.43)$$

the ratio of the Δk caused by the photon momentum to the value of k_{max}

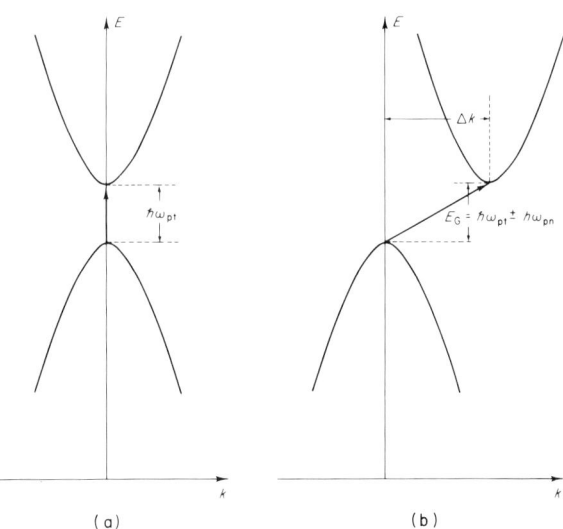

Fig. 11.5 (a) Direct optical transition between two bands with extrema at **k** = 0. (b) Indirect optical transition from the top of the valence band at **k** = 0 to the bottom of the conduction band at some finite value of **k**.

at the edge of the zone is given by

$$\frac{\Delta k}{k_{\max}} = \frac{2\pi \nu r/c}{\pi/a} = 2ra\frac{E_G}{hc} \quad (11.44)$$

For typical values such as $r = 4$, $a = 2$ Å, and $E = 2$ eV, the above ratio is about 1/500. Therefore the Δk due to the photon momentum alone can ordinarily be neglected, and a direct optical transition is represented by a vertical line on an E vs. $\mathbf{k}$ plot.

An indirect optical transition involving the absorption of a photon and either the absorption or the emission of a phonon is pictured in Fig. 11.5b. Conservation of energy and of wave vector under this condition requires

$$E_{\text{photon}} \pm E_{\text{phonon}} = E_{G(\text{indirect})} \quad (11.45)$$

$$\Delta k = K_{\text{phonon}} \quad (11.46)$$

Since a direct transition involves absorption of a photon only, the optical transition involved is a first-order process. On the other hand, an indirect transition, involving both photon and phonon, is a second-order process.

11.3 Direct Intrinsic Transition

For the sake of the calculation of the dependence of the absorption constant on the wavelength of the incident radiation in the case of a direct optical transition between the valence and conduction bands of a material, we assume nondegenerate bands and a scalar effective mass. As shown in Fig. 11.6, we take the top of the valence band at $\mathbf{k} = 0$, so that the energy corresponding to wave vector $\mathbf{k}'$ in the valence band is given by

$$E_v(\mathbf{k}') = -E_G - \frac{\hbar^2 k'^2}{2m_h^*} \quad (11.47)$$

The corresponding dependence of energy in the conduction band on wave vector $\mathbf{k}''$ is given by

$$E_c(\mathbf{k}'') = \frac{\hbar^2 k''^2}{2m_e^*} \quad (11.48)$$

E_G is the separation between valence-band and conduction-band extrema at $\mathbf{k} = 0$, and the zero of energy is taken at the bottom of the conduction band.

For an optical transition from a state with wave vector $\mathbf{k}'$ in the valence

11.3 Direct Intrinsic Transition

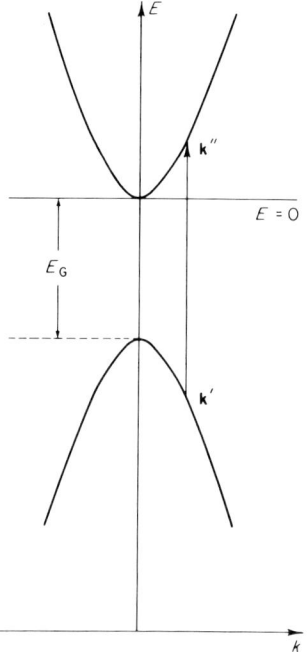

Fig. 11.6 Direct optical transition from a state with wave vector $\mathbf{k}'$ in the valence band to a state with wave vector $\mathbf{k}''$ in the conduction band.

band to a state with wave vector $\mathbf{k}''$ in the conduction band, conservation of energy requires that

$$E_{\text{photon}} = \hbar\omega = E_c(\mathbf{k}'') - E_v(\mathbf{k}') \tag{11.49}$$

Optical absorption creates an electron in the conduction band with velocity $\hbar\mathbf{k}''/m_e^*$ and a hole in the valence band with velocity $\hbar\mathbf{k}'/m_h^*$.

PERTURBATION POTENTIAL FOR THE TRANSITION

From Eq. (8.31) recall that the transition probability is proportional to the matrix element for the transition,

$$H_{m0} = \frac{1}{2} \int \psi_m^* \mathbf{H}_0' \psi_0 \, d\mathbf{r} \tag{11.50}$$

if the total perturbation is

$$\mathbf{H}'(\mathbf{r}, t) = \frac{1}{2} \mathbf{H}_0'(\mathbf{r})(e^{i\omega t} + e^{-i\omega t}) \tag{11.51}$$

11 Optical Absorption

The wavefunction for the initial state ψ_0 is given by

$$\psi_0 \equiv \psi_{\mathbf{k}'} = \frac{1}{N^{1/2}} u_v(\mathbf{r}, \mathbf{k}') e^{i\mathbf{k}' \cdot \mathbf{r}} \tag{11.52}$$

where N as usual is the number of unit cells per crystal and normalizes the wavefunction over the crystal. The wavefunction for the final state ψ_m is given by

$$\psi_m \equiv \psi_{\mathbf{k}''} = \frac{1}{N^{1/2}} u_c(\mathbf{r}, \mathbf{k}'') e^{i\mathbf{k}'' \cdot \mathbf{r}} \tag{11.53}$$

For the case of an optically induced transition, the perturbation $\mathbf{H}'(\mathbf{r}, t)$ must be associated with the electromagnetic radiation causing the transition. With the momentum operator in the form given in Eq. (10.113), we have

$$\frac{\mathbf{P}^2}{2m} = \frac{1}{2m} \left(i\hbar \nabla + \frac{e\mathbf{A}}{c} \right)^2$$

$$= -\frac{\hbar^2}{2m} \nabla^2 + \frac{i\hbar e}{mc} \mathbf{A} \cdot \nabla + \frac{e^2 |\mathbf{A}|^2}{2mc^2} \tag{11.54}$$

On the assumption that terms involving $\mathbf{A}$ can be treated as a small perturbation, the final term in $|\mathbf{A}|^2$ in Eq. (11.54) can be neglected completely. Thus the perturbation potential is

$$\mathbf{H}'(\mathbf{r}, t) = \frac{ie\hbar}{mc} \mathbf{A} \cdot \nabla \tag{11.55}$$

Now the vector potential $\mathbf{A}$ can be written as

$$\mathbf{A} = \frac{1}{2} A \mathbf{a}_0 e^{i(\mathbf{K} \cdot \mathbf{r} - \omega t)} + \text{c.c.} \tag{11.56}$$

where $\mathbf{a}_0$ is a unit polarization vector, and $\mathbf{K}$ is the wave vector for the propagation of plane-polarized light waves. Then $E_{\text{photon}} = \hbar\omega$, $\mathbf{a}_0 \cdot \mathbf{K} = 0$, and $|\mathbf{K}| = \omega r/c$. The perturbation potential of Eq. (11.55) becomes

$$\mathbf{H}'(\mathbf{r}, t) = \frac{ie\hbar}{2mc} A e^{i(\mathbf{K} \cdot \mathbf{r} - \omega t)} \mathbf{a}_0 \cdot \nabla + \text{c.c.} \tag{11.57}$$

And the quantity $\mathbf{H}_0'$ entering the matrix element of Eq. (11.50) is

$$\mathbf{H}_0' = \frac{ie\hbar}{2mc} e^{i\mathbf{K} \cdot \mathbf{r}} \mathbf{a}_0 \cdot \nabla \tag{11.58}$$

1.13 Direct Intrinsic Transition

MATRIX ELEMENT FOR THE TRANSITION

From Eq. (8.33), the transition probability is given by

$$|A_{k''}|^2 = \frac{4|H_{k''k'}|^2 \sin^2[(E_{k''} - E_{k'} - \hbar\omega)t/2\hbar]}{(E_{k''} - E_{k'} - \hbar\omega)^2} \quad (11.59)$$

where the matrix element $H_{k''k'}$ is given by

$$H_{k''k'} = \frac{ie\hbar A}{2Nmc} \int u_c^*(r, k') \, e^{-i\mathbf{k}''\cdot\mathbf{r}} (e^{i\mathbf{K}\cdot\mathbf{r}} \mathbf{a}_0 \cdot \nabla) \, u_v(\mathbf{r}, \mathbf{k}') \, e^{i\mathbf{k}'\cdot\mathbf{r}} \, d\mathbf{r} \quad (11.60)$$

This is the total contribution to the direct optical transition, for only the $e^{-i\omega t}$ term in Eq. (11.57) for $\mathbf{H}'(\mathbf{r}, t)$ is able to satisfy the conservation of energy conditions since $E_{k''} > E_{k'}$.

Since the perturbation Hamiltonian is a differential operator, the matrix element contains two terms,

$$H_{k''k'} = \frac{ie\hbar A}{2Nmc} \int u_c^*(\mathbf{r}, \mathbf{k}'')[\mathbf{a}_0 \cdot \nabla \, u_v(\mathbf{r}, \mathbf{k}') + i(\mathbf{a}_0 \cdot \mathbf{k}') u_v(\mathbf{r}, \mathbf{k}')]$$
$$\times [e^{i(\mathbf{k}'+\mathbf{K}-\mathbf{k}'')\cdot\mathbf{r}}] \, d\mathbf{r} \quad (11.61)$$

Because of the periodicity of the u functions, this integral over the whole crystal can be expressed as a sum of integrals over a unit cell,

$$H_{k''k'} = \sum_j e^{i(\mathbf{k}'+\mathbf{K}-\mathbf{k}'')\cdot\mathbf{R}_j} \int_{\text{cell}} \phi(\mathbf{r}, \mathbf{k}', \mathbf{k}'') \, d\mathbf{r} \quad (11.62)$$

where $\mathbf{R}_j$ is the lattice vector defining the jth cell, and the ϕ function is an abbreviation for the expression remaining in Eq. (11.61). For small values of the electron wave vectors we have seen in Eq. (8.43) that the above sum is nonzero only if

$$\mathbf{k}'' - \mathbf{k}' = \mathbf{K} \approx 0 \quad (11.63)$$

in which case the sum is simply N. Thus the matrix element is

$$H_{k''k'} = \frac{ie\hbar A}{2mc} \int_{\text{cell}} u_c^*(\mathbf{r}, \mathbf{k}'')[\mathbf{a}_0 \cdot \nabla \, u_v(\mathbf{r}, \mathbf{k}') + i(\mathbf{a}_0 \cdot \mathbf{k}') u_v(\mathbf{r}, \mathbf{k}')] \, d\mathbf{r} \quad (11.64)$$

The two contributions to the matrix element are of greatly different magnitude, so much so that the large contribution due to the term involving ∇u_v is commonly referred to as the *allowed* direct transition, whereas the small contribution due to the other term is commonly referred to as the *forbidden* transition.

The matrix element for the forbidden transition is

$$H^{f}_{k''k'} = -\frac{e\hbar A}{2mc}(\mathbf{a}_0 \cdot \mathbf{k}') \int_{\text{cell}} u_c^*(\mathbf{r}, \mathbf{k}'') u_v(\mathbf{r}, \mathbf{k}')\, d\mathbf{r} \quad (11.65)$$

which would be identically zero because of the orthogonality of the u functions if $\mathbf{k}'' = \mathbf{k}'$. The finite magnitude of $H^{f}_{k''k'}$ depends therefore on the small change in $\mathbf{k}$ due to the contribution from the photon $\mathbf{K}$, and is generally much less than the matrix element for the allowed transition,

$$H^{a}_{k''k'} = \frac{ie\hbar A}{2mc} \int_{\text{cell}} u_c^*(\mathbf{r}, \mathbf{k}'')\, \mathbf{a}_0 \cdot \nabla\, u_v(r, k')\, d\mathbf{r} \quad (11.66)$$

ALLOWED TRANSITION

The notation may be simplified by expressing $H^{a}_{k''k'}$ in terms of the matrix element of the momentum operator,

$$P_{k''k'} = -i\hbar \int_{\text{cell}} u_c^*(\mathbf{r}, \mathbf{k}'')\, \nabla\, u_v(\mathbf{r}, \mathbf{k}')\, d\mathbf{r} \quad (11.67)$$

so that

$$H^{a}_{k''k'} = -\frac{eA}{2mc}\, \mathbf{a}_0 \cdot P_{k''k'} \quad (11.68)$$

In calculating the total probability of a transition, we assume monochromatic radiation and perform the calculation by summing over all allowed values of $\mathbf{k}'$. We assume in addition that the valence band is fully occupied and that the density of states from which transitions may be made is given by $1/(4\pi^3)$ per unit volume of $\mathbf{k}$ space.

If $P(E)$ is the transition probability per unit volume per unit time,

$$P(E)\, t = \frac{1}{4\pi^3} |A_{k''}|^2$$

$$= \frac{e^2 A^2}{4\pi^3 m^2 c^2} \int \frac{|\mathbf{a}_0 \cdot P_{k''k'}|^2 \sin^2[(E_{k''} - E_{k'} - \hbar\omega)t/2\hbar]}{(E_{k''} - E_{k'} - \hbar\omega)^2}\, d\mathbf{k}' \quad (11.69)$$

combining Eqs. (11.59) and (11.68). Since $(E_{k''} - E_{k'} - \hbar\omega)$ is a function only of k', the angular portion of this integration may be taken care of separately. Recognizing that $d\mathbf{k}' = k'^2\, dk'\, d\Omega$, and setting

$$\int |\mathbf{a}_0 \cdot P_{k''k'}|^2\, d\Omega = 4\pi \bar{P}^2_{k''k'} \quad (11.70)$$

11.3 Direct Intrinsic Transition

allows us to rewrite Eq. (11.69) as

$$P(E)t = \frac{e^2 A^2}{\pi^2 m^2 c^2} \int \frac{\bar{P}^2_{k''k'} \sin^2[(E_{k''} - E_{k'} - \hbar\omega)t/2\hbar]k'^2\, dk'}{(E_{k''} - E_{k'} - \hbar\omega)^2} \qquad (11.71)$$

Now the integrand is large only under the conditions for the conservation of energy, i.e., only if $(E_{k''} - E_{k'}) \approx \hbar\omega$. We are therefore justified in taking $k'\bar{P}^2_{k''k'}$ outside the integral, and setting this k' equal to k_0, where k_0 is defined by the conservation-of-energy relationship, $(E_{k''} - E_{k_0}) = \hbar\omega$. Since

$$\hbar\omega = E_G + \frac{\hbar^2 k_0^2}{2m_r^*} \qquad (11.72)$$

where the reduced mass $m_r^* = m_e^* m_h^*/(m_e^* + m_h^*)$, according to Eqs. (11.47) and (11.48), it follows that

$$k_0 = \frac{(2m_r^*)^{1/2}}{\hbar}(\hbar\omega - E_G)^{1/2} \qquad (11.73)$$

This provides the energy dependence of the transition probability.

The reason for taking one factor of k' out of the integrand and leaving the other factor of k' in the integrand is that this remaining integral can be simply integrated. Let $x = (E_{k''} - E_{k'} - \hbar\omega)t/2\hbar$, so that

$$dx = \frac{t}{2\hbar} d(E_{k''} - E_{k'}) = \frac{t}{2\hbar} d\left(\frac{\hbar^2 k'^2}{2m_r^*}\right)$$

$$= \frac{\hbar k' t}{2m_r^*} dk' \qquad (11.74)$$

Then

$$P(E)t = \frac{e^2 A^2}{\pi^2 m^2 c^2} k_0 \bar{P}^2_{k''k'} \frac{t^2}{4\hbar^2} \frac{2m_r^*}{\hbar t} \int_{-\infty}^{+\infty} \frac{\sin^2 x}{x^2} dx \qquad (11.75)$$

Since the value of the definite integral of Eq. (11.75) is π, we have

$$P(E) = \frac{e^2 A^2 (2m_r^*)^{1/2} \bar{P}^2_{k''k'}}{4\pi m^2 c^2 \hbar^4}(\hbar\omega - E_G)^{1/2} \qquad (11.76)$$

using Eq. (11.73).

For a direct allowed optical transition from the valence band to the conduction band, the transition probability varies as the square root of the difference between the photon energy and the actual bandgap.

The absorption constant α can be calculated from the transition prob-

420   11 Optical Absorption

ability. The incident intensity is

$$I_0 = \frac{|\mathbf{S}|}{\hbar\omega} \tag{11.77}$$

where $\mathbf{S}$ is the Poynting vector of the electromagnetic radiation, the energy crossing unit area per unit time. The transmitted intensity (neglecting reflection losses) is given by

$$I = \frac{|\mathbf{S}|}{\hbar\omega} - P(E)\,d \tag{11.78}$$

where the second term represents the number of quanta per unit time absorbed in a thin strip of thickness d and unit area normal to the light, we have

$$\frac{|\mathbf{S}|}{\hbar\omega} - P(E)\,d = \frac{|\mathbf{S}|}{\hbar\omega} e^{-\alpha d}$$

$$\approx \frac{|\mathbf{S}|}{\hbar\omega}(1 - \alpha d) \tag{11.79}$$

so that

$$\alpha = \frac{P(E)\,\hbar\omega}{|\mathbf{S}|} \tag{11.80}$$

under conditions permitting the approximation used in Eq. (11.79), i.e., small αd. Since $\mathbf{S} = (c/4\pi)\mathscr{E} \times \mathbf{H}$, and we can express $\mathscr{E}$ and $\mathbf{H}$ in terms of $\mathbf{A}$ using Eq. (11.56), the average value of the Poynting vector over a period is

$$|\bar{\mathbf{S}}| = \frac{A^2 K \omega}{8\pi} \tag{11.81}$$

Inserting appropriate units, the final result for the absorption constant for an allowed direct transition can be conveniently written as

$$\alpha_{\mathrm{da}} = 2.7 \times 10^5\, r^{-1}(2m_\mathrm{r}^*/m)^{3/2}\,f(\hbar\omega - E_\mathrm{G})^{1/2}\ \mathrm{cm}^{-1} \tag{11.82}$$

if $\hbar\omega$ and E_G are both given in eV. The quantity f is called the *oscillator strength* for the transition and is given by

$$f = \bar{f}_{k''k'}; \qquad f_{k''k'} = \frac{2\,|P_{k''k'}|^2}{m\hbar\omega} \tag{11.83}$$

The oscillator strength has a value near unity, and is given approximately

11.3 Direct Intrinsic Transition

by[†]

$$f \approx 1 + \frac{m}{m_\mathrm{h}{}^*} \tag{11.84}$$

FORBIDDEN TRANSITION

The matrix element for the forbidden transition is given by Eq. (11.65). Define a quantity

$$f' \equiv \left| \int_\mathrm{cell} u_\mathrm{c}{}^*(\mathbf{r}, \mathbf{k}'') u_\mathrm{v}(\mathbf{r}, \mathbf{k}') \, d\mathbf{r} \right|^2 \tag{11.85}$$

such that $f' \ll 1$. If we proceed in this case in the same way used above for the allowed transition, setting $(\mathbf{a}_0 \cdot \mathbf{k}') = k' \cos \theta$ and taking the average value $\tfrac{1}{3}$ for $\cos^2 \theta$, we obtain

$$\begin{aligned} P(E) &= \frac{e^2 A^2 (2m_\mathrm{r}{}^*)^{3/2} f' k_0{}^2 (\hbar\omega - E_\mathrm{G})^{1/2}}{12\pi m^2 \hbar^2} \\ &= \frac{e^2 A^2 (2m_\mathrm{r}{}^*)^{5/2} f' (\hbar\omega - E_\mathrm{G})^{3/2}}{12\pi m^2 \hbar^4} \end{aligned} \tag{11.86}$$

using Eq. (11.73) for k_0. For a forbidden direct transition the transition probability varies as the $\tfrac{3}{2}$ power of the difference between the photon energy and the bandgap. The expression for the absorption constant in conventional units for this case is

$$\alpha_\mathrm{df} = 1.8 \times 10^5 \, r^{-1} \left(\frac{2m_\mathrm{r}{}^*}{m} \right)^{5/2} f' \, \frac{(\hbar\omega - E_\mathrm{G})^{3/2}}{\hbar\omega} \quad \mathrm{cm}^{-1} \tag{11.87}$$

for $\hbar\omega$ and E_G in eV.

Typical orders of magnitude for the absorption constant are as follows. For a direct allowed transition, with $2m_\mathrm{r}{}^* = m$, $r = 4$, and $f = 1$, $\alpha_\mathrm{da} = 6.7 \times 10^3 \, \mathrm{cm}^{-1}$ for $(\hbar\omega - E_\mathrm{G}) = 0.01$ eV. For a direct forbidden transition, with the same constants and $f' = 0.1$ and $\hbar\omega = 1$ eV, $\alpha_\mathrm{df} = 4.5 \, \mathrm{cm}^{-1}$ for $(\hbar\omega - E_\mathrm{G}) = 0.01$ eV.

Some idea concerning the magnitude of f' for the forbidden transition can be obtained by evaluating it with $\mathbf{k}'' = \mathbf{k}' + \mathbf{K}$, expressing the deviation of $\mathbf{k}''$ from $\mathbf{k}'$ in terms of an expansion,

$$\int u_\mathrm{c}{}^*(\mathbf{r}, \mathbf{k}'') u_\mathrm{v}(\mathbf{r}, \mathbf{k}') \, d\mathbf{r}$$
$$= \int u_\mathrm{v}(\mathbf{r}, \mathbf{k}') \left[1 + \mathbf{K} \cdot \nabla_{\mathbf{k}'} + \frac{1}{2} (\mathbf{K} \cdot \nabla_{\mathbf{k}'})^2 + \cdots \right] u_\mathrm{c}{}^*(\mathbf{r}, \mathbf{k}') \, d\mathbf{r} \tag{11.88}$$

[†] See, for example, R. A. Smith, "Wave Mechanics of Crystalline Solids," Wiley, New York, 1961, pp. 409, 464.

The first term in the expansion on the right of Eq. (11.88) is zero because of the orthogonality of u_c and u_v. The second term is proportional to $\bar{P}_{k''k'}$ which is equal to zero if the allowed transition has zero matrix element, as would normally be the case when considering the forbidden transition. The third term is proportional to K^2, and hence is also small.

Experimental evidence for a direct intrinsic transition in GaAs is shown in Fig. 11.7.

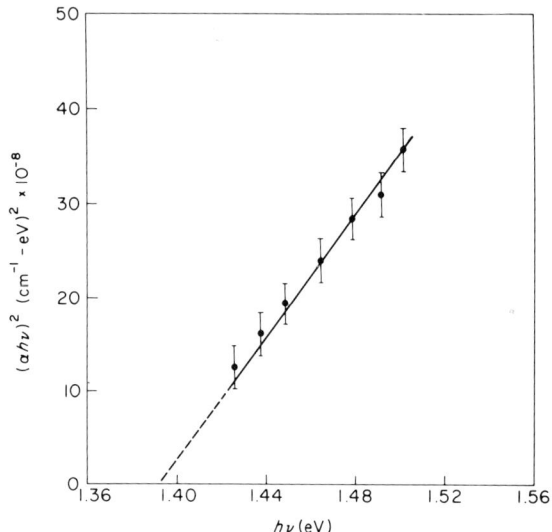

Fig. 11.7 Analysis of the absorption edge of p-type GaAs showing the presence of a direct optical transition corresponding to a bandgap of 1.39 eV. [After I. Kudman and T. Seidel, *J. Appl. Phys.* **33**, 771 (1962).]

GENERAL COMMENTS

The following qualifying general comments about optical absorption may be made.

(1) For small values of $(\hbar\omega - E_G)$, the derived expressions do not hold. The Coulomb interaction between electron and hole created by the photo-excitation has been neglected, and hence absorption involving the formation of excitons has not been considerd.

(2) For large values of $(\hbar\omega - E_G)$, the absorption constant does not continue to increase, for the point is reached where the quadratic relationship between E and **k** no longer holds. Transitions to higher bands also contribute to the absorption constant. Usually the absorption constant

11.3 Direct Intrinsic Transition

reaches a value of 10^5–10^6 cm^{-1} with fluctuations due to the structure of higher-lying bands for large values of $(\hbar\omega - E_G)$.

(3) If the effective mass is not a scalar, then the reduced mass m_r^* must be replaced by an appropriately averaged effective mass.

(4) If the valence band is degenerate, then the absorption due to various bands must be added.

(5) If the Fermi level lies in the conduction band, then the optical absorption "gap" is larger than the true bandgap, and the optical absorption is a function of the free-carrier density. This happens most commonly in materials with a small conduction-band effective mass, and hence a small density of states N_c. Suppose that the Fermi level lies an energy E' above the bottom of the conduction band. The energy of the level in the conduction band to which an optical transition is made, above the bottom of the conduction band, is

$$\delta E = \frac{m_r^*}{m_e^*} (\hbar\omega - E_G) \qquad (11.89)$$

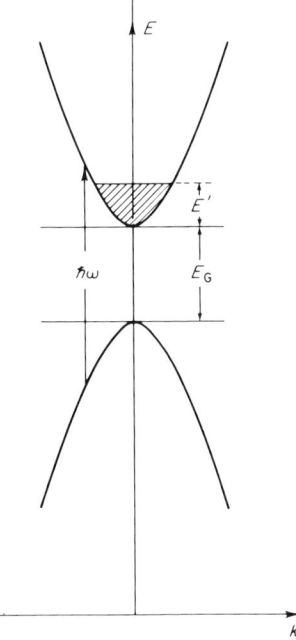

Fig. 11.8 The minimum energy for optical absorption is larger than E_G in a degenerate material, where the Fermi level lies in the conduction band.

as indicated in Fig. 11.8. The probability that this level is unoccupied is

$$\frac{1}{1 + \exp\{-[m_r^*(\hbar\omega - E_G) - m_e^*E']/m_e^*kT\}} \quad (11.90)$$

Therefore optical absorption does not become strong until

$$\hbar\omega = E_G + \frac{m_e^*}{m_r^*} E' \quad (11.91)$$

E' is a function of the free-carrier density, and so the apparent optical absorption edge of the material is also a function of the free-carrier density. Variation of the absorption edge of InSb with increasing degeneracy is shown in Fig. 11.9.

(6) When degeneracy is not important, the absorption constant varies with temperature through the dependence of E_G on temperature. Over

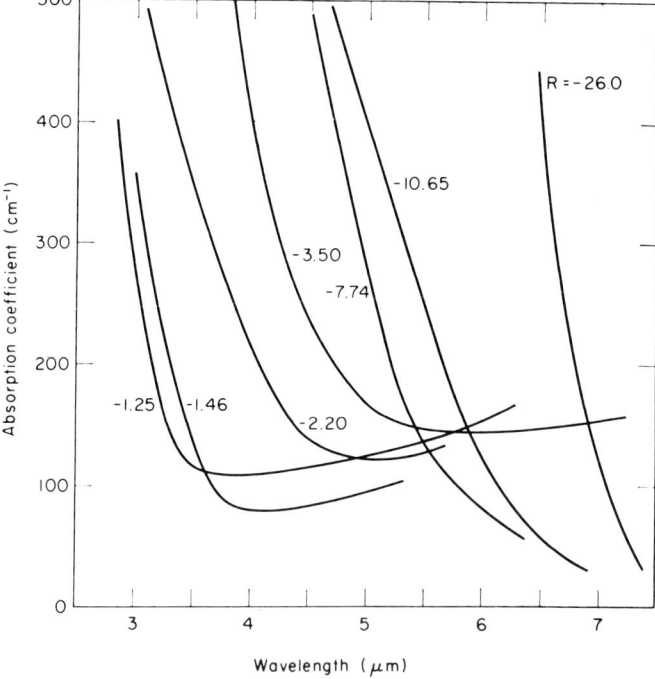

Fig. 11.9 Variation of the absorption edge in InSb with electron density, given in terms of the value of the Hall constant in cm³/C. [After R. Barrie and J. T. Edmond, *J. Electronics* **1**, 161 (1955).]

11.3 Direct Intrinsic Transition

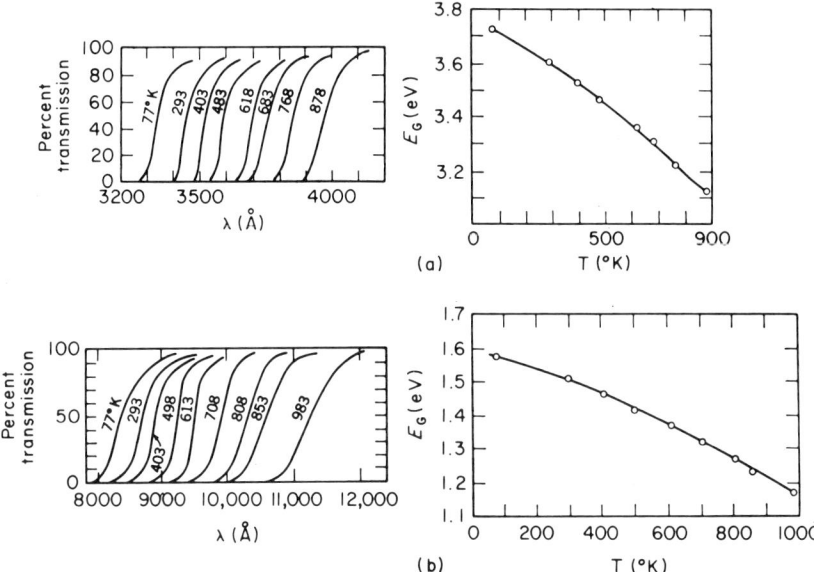

Fig. 11.10 Temperature dependence of the absorption spectra of (a) ZnS and (b) CdTe, and temperature dependence of the bandgap as determined from the absorption spectra. [After C. Z. Van Doorn, *Physica* **20**, 1155 (1954), and C. Z. Van Doorn and D. de Nobel, *Physica* **22**, 338 (1956).]

fairly wide temperature ranges, not at low temperatures, it is empirically found that

$$E_G = E_{G0} + \beta T \tag{11.92}$$

where the temperature coefficient $|\beta|$ has values in the range of several 10^{-4} eV/°K. Typical data for ZnS and CdTe are given in Fig. 11.10.

A part of the temperature dependence of the bandgap results from expansion or contraction of the lattice. The effect of such change in lattice constant on the bandgap can be tested by measurements of the effect of pressure on the bandgap. The overall dependence of bandgap on temperature and pressure may be expressed

$$\Delta E_G = \left(\frac{\partial E_G}{\partial P}\right)_T \Delta P + \left(\frac{\partial E_G}{\partial T}\right)_P \Delta T \tag{11.93}$$

or

$$\Delta E_G = \left(\frac{\partial E_G}{\partial P}\right)_T \Delta P + \left[\left(\frac{\partial E_G}{\partial T}\right)_{\mathscr{V}} - \frac{C}{\chi}\left(\frac{\partial E_G}{\partial P}\right)_T\right] \Delta T \tag{11.94}$$

where C is the thermal coefficient of volume expansion, and χ is the compressibility. The coefficients of bandgap dependence are usually defined as

$$\gamma = \left(\frac{\partial E_G}{\partial P}\right)_T \qquad (11.95)$$

$$\beta = \left(\frac{\partial E_G}{\partial T}\right)_{\mathscr{P}} - \frac{C}{\chi}\gamma \qquad (11.96)$$

For most materials β is negative, i.e., the bandgap decreases with increasing temperature, but for a few, such as PbS, PbSe, and PbTe, β is positive. The pressure coefficient γ is typically of the order of several 10^{-6} eV/atm, and may have either sign or even change sign for a given material with increasing pressure. Characteristically this comes about because of different signs of γ for different energy bands. A good example is that for Ge, for which the data and the shift of the energy bands with pressure are illustrated in Fig. 11.11.

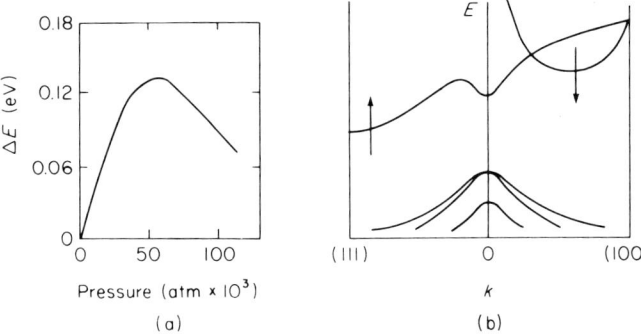

Fig. 11.11 Pressure dependence of the absorption edge in Ge [After T. E. Slykhouse and H. G. Drickamer, *J. Phys. Chem. Solids* **7**, 210 (1958)]; and energy-band diagram to describe the nature of the results in terms of a different pressure dependence of the two bands shown.

(7) In some materials the location of the absorption edge depends on the polarization of the light used. CdS or CdSe, with hexagonal crystal structure, are good examples. Energy-band structures for these materials are given in Fig. 6.16. Selection rules for optical absorption are such that transitions from both V_1 and V_2 bands to the conduction band are allowed if the light is polarized with $\mathscr{E}$ perpendicular to the c axis of the crystal, but only transitions from V_2 are allowed if the light is polarized with $\mathscr{E}$ parallel to the c axis. Representative absorption curves are shown in Fig. 11.12.

11.4 Indirect Intrinsic Transition

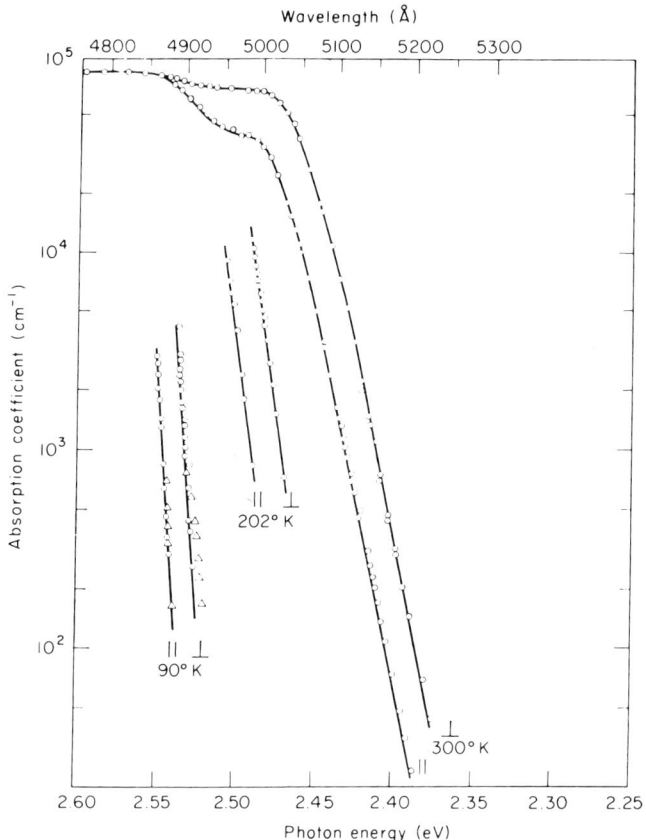

Fig. 11.12 Absorption edge of CdS at 300 and 90°K, illustrating the dependence of the measured edge on the polarization of the light because of the selection rules for the two uppermost valence bands. [After D. Dutton, *Phys. Rev.* **112**, 785 (1958).]

11.4 Indirect Intrinsic Transition

In an indirect transition from valence to conduction band, conservation of wavevector can be achieved only if a phonon is absorbed or emitted in the process of photon absorption. Multiple-phonon processes are much less probable than those involving only a single phonon and are not be considered here.

Conservation of wavevector requires that

$$\mathbf{k}'' \pm \mathbf{K}_{\mathrm{pn}} = \mathbf{k}' + \mathbf{K}_{\mathrm{pt}} \tag{11.97}$$

where $\mathbf{k}''$ is the electron wavevector for the final state, $\mathbf{K}_{pn}$ is the wavevector of the phonon, $\mathbf{k}'$ is the electron wavevector for the initial state, and $\mathbf{K}_{pt}$ is the photon wavevector. Neglecting the small contribution from the photon, the conservation of wavevector requires that

$$\mathbf{k}'' - \mathbf{k}' = \pm \mathbf{K}_{pn} \qquad (11.98)$$

Conservation of energy requires that

$$E_c(\mathbf{k}'') - E_v(\mathbf{k}') = \hbar\omega_{pt} \pm \hbar\omega_{pn} \qquad (11.99)$$

One way to consider an indirect transition is in terms of a short-lived virtual state, as indicated in Fig. 11.13. The indirect transition from the

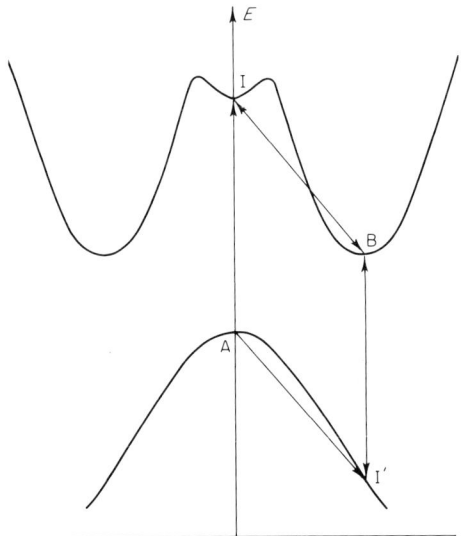

Fig. 11.13 Treatment of the indirect transition in terms of direct transitions to (I) or from (I') a virtual state, with simultaneous absorption or emission of a phonon to scatter from I to B, or from I' to A.

initial state with energy $E_v(\mathbf{k}')$ to the final state with energy $E_c(\mathbf{k}'')$ proceeds by one of two possible paths: (a) a direct optical transition from A to I with conservation of $\mathbf{k}$ but not of E, followed by a phonon scattering from I to B with conservation of $\mathbf{k}$ but not of E, in such a way that the transition from A to B conserves both $\mathbf{k}$ and E; or (b) a direct optical transition from I' to B with conservation of $\mathbf{k}$ but not of E, followed by a phonon scattering from I' to A with conservation of $\mathbf{k}$ but not of E, in

11.4 Indirect Intrinsic Transition

such a way that the transition from A to B again conserves both **k** and E. The short-lived nature of the virtual state permits the breakdown of conservation of energy in the intermediate processes, as long as it is conserved in the final accounting.

Transitions involving such virtual states may be described in terms of second-order time-dependent perturbation theory as summarized below. It might be expected that the transition probability for such a second-order process would be very small, much smaller for example than for a direct transition. The indirect transition does have a smaller probability, but the difference is partially counterbalanced by an increase in the number of possible states from which a transition can be made and to which a transition can be made. This effect is illustrated in Fig. 11.14. For a direct transition, conservation of **k** requires a vertical transition and conservation of energy fixes the states from which transitions can be made to a narrow range. For an indirect transition, however, the energy of the initial state may lie between 0 and $[\hbar(\omega_{pt} \pm \omega_{pn}) - E_G]$ below the top of the valence band, and the energy of the final state may lie between 0 and $[\hbar(\omega_{pt} \pm \omega_{pn}) - E_G]$ above the bottom of the conduction band, where E_G is here the indirect bandgap. The total probability is obtained by a sum over the possible states.

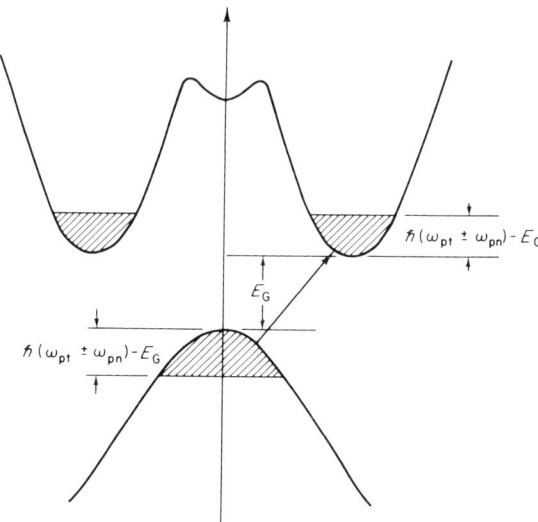

Fig. 11.14 The magnitude of an indirect optical transition is larger than expected for a second-order process because transitions can be made from a range of states in the valence band to a range of states in the conduction band, as long as each particular transition conserves energy between initial and final states.

Second-Order Time-Dependent Perturbation Theory

Instead of the first-order approximation of Eq. (8.27), used to extend the expression for dA_m/dt to times $t > 0$, assume now the utility of second-order terms as well and write in general

$$A_0 = 1 + A_0' + A_0''$$
$$A_m = A_m' + A_m'' \qquad (11.100)$$

Equation (8.23) may be rewritten as

$$i\hbar \frac{dA_m}{dt} = A_0 H_{m0} e^{i(E_m - E_0)t/\hbar} + \sum_{n \neq 0} A_n H_{mn} e^{i(E_m - E_n)t/\hbar} \qquad (11.101)$$

Substitution from Eq. (11.100), retaining only terms of first or second order ($A_m'' H_{mn}$ is a third-order term), gives

$$i\hbar \frac{dA_m'}{dt} + i\hbar \frac{dA_m''}{dt} = (1 + A_0') H_{m0} e^{i(E_m - E_0)t/\hbar} + \sum_{n \neq 0} A_n' H_{mn} e^{i(E_m - E_n)t/\hbar} \qquad (11.102)$$

From Eqs. (8.26) and (8.27) it follows that the first term on the left of Eq. (11.102) cancels with the first term on the right, leaving

$$i\hbar \frac{dA_m''}{dt} = \sum_{n \neq 0} A_n' H_{mn} e^{i(E_m - E_n)t/\hbar} \qquad (11.103)$$

For a periodic perturbation like that given in Eq. (8.29), Eq. (11.103) becomes

$$i\hbar \frac{dA_m''}{dt} = \sum_{n \neq 0} A_n' H_{mn} \; e^{i(E_m - E_n + \hbar\omega)t/\hbar} + e^{i(E_m - E_n - \hbar\omega)t/\hbar} \qquad (11.104)$$

An expression for A_n' is obtainable from Eq. (8.30) and subsequent integration yields A_m''. The resulting general expression is cumbersome to write down, containing terms with denominators involving $(E_m - E_0 \pm \hbar\omega)$ and $(E_m - E_n \pm \hbar\omega)$. It is possible to simplify this expression by consideration of the relevant physical conditions for indirect transitions.

The expression for $A_m''(t)$ does not contain the matrix element H_{m0} for a transition directly from the initial state to the final state, but does contain the matrix elements H_{n0} and H_{mn} for transitions via the intermediate state with energy E_n. Therefore, even if $H_{m0} = 0$ so that $A_m'(t) = 0$, a finite

11.4 Indirect Intrinsic Transition

transition probability corresponding to $A_m''(t)$ still may exist, with a magnitude depending on the product of matrix elements $H_{n0}H_{mn}$. Furthermore only those terms in A_m'' with $(E_m - E_0 \pm \hbar\omega)$ in the denominator will contribute to a large transition probability increasing linearly with time, since energy is not conserved in the transition from E_0 to E_n, or from E_n to E_m.

The Perturbation Hamiltonian

The perturbation Hamiltonian for indirect transitions contains two contributions, one related to the perturbation caused by the light as given in Eq. (11.57),

$$\mathbf{H}'_{pt} = \frac{ie\hbar}{2mc} e^{i(\mathbf{K}_{pt}\cdot\mathbf{r}-\omega_{pt}t)} \mathbf{a}_0 \cdot \nabla + \text{c.c.} \qquad (11.105)$$

and the other related to the perturbation caused by the lattice waves as given in Eq. (8.37),

$$\mathbf{H}'_{pn} = \frac{1}{2}(\mathbf{A}_{pn} \cdot \nabla V_0) e^{i(\mathbf{K}_{pn}\cdot\mathbf{r}-\omega_{pn}t)} + \text{c.c.} \qquad (11.106)$$

Terms in $\pm\omega_{pn}$, corresponding to phonon absorption and emission, are both retained in the final calculation, but only the term in $-\omega_{pt}$, corresponding to photon absorption, is retained, since the energy of the final state is greater than that of the initial state.

The Perturbation Calculation

For ease in notation, express

$$\begin{aligned}\mathbf{H}' &= \mathbf{H}'_{pt} + \mathbf{H}'_{pn} \\ &= \mathbf{H}'_p\, e^{-i\omega_{pt}t} + \mathbf{H}'_{p'}\, e^{-i\omega_{pn}t} + \text{c.c.}\end{aligned} \qquad (11.107)$$

The corresponding matrix elements are

$$H^e_{mn} = \int \psi_m^* \mathbf{H}'_{p'} \psi_n\, d\mathbf{r} \qquad (11.108)$$

$$H^p_{mn} = \int \psi_m^* \mathbf{H}'_p \psi_n\, d\mathbf{r} \qquad (11.109)$$

The expression for the first-order coefficients A_n' are obtained by integrating as in Eq. (8.30) to obtain

$$A_n' = -H_{n0}^{\text{p}}\left[\frac{\exp[i(E_n - E_0 + \hbar\omega_{\text{pt}})t/\hbar] - 1}{E_n - E_0 + \hbar\omega_{\text{pt}}}\right.$$
$$\left.+ \frac{\exp[i(E_n - E_0 - \hbar\omega_{\text{pt}})t/\hbar] - 1}{E_n - E_0 - \hbar\omega_{\text{pt}}}\right]$$
$$- H_{n0}^{e}\left[\frac{\exp[i(E_n - E_0 + \hbar\omega_{\text{pn}})t/\hbar] - 1}{E_n - E_0 + \hbar\omega_{\text{pn}}}\right.$$
$$\left.+ \frac{\exp[i(E_n - E_0 - \hbar\omega_{\text{pn}})t/\hbar] - 1}{E_n - E_0 - \hbar\omega_{\text{pn}}}\right] \quad (11.110)$$

For the present case, Eq. (11.104) can be written as

$$i\hbar \frac{dA_m''}{dt} = \sum_{n \neq 0} A_n'(H_{mn}^{\text{p}}\{\exp[i(E_m - E_n + \hbar\omega_{\text{pt}})t/\hbar]$$
$$+ \exp[i(E_m - E_n - \hbar\omega_{\text{pt}})t/\hbar]\}$$
$$+ H_{mn}^{e}\{\exp[i(E_m - E_n + \hbar\omega_{\text{pn}})t/\hbar]$$
$$+ \exp[i(E_m - E_n - \hbar\omega_{\text{pn}})t/\hbar]\}) \quad (11.111)$$

If we substitute Eq. (11.110) into Eq. (11.111) and integrate again, we obtain a complex expression with denominators containing various combinations of $(E_n - E_0)$, $(E_m - E_n)$, $\hbar\omega_{\text{pt}}$, and $\hbar\omega_{\text{pn}}$. Fortunately for our present calculation, only two of these combinations need to be considered as significant in this case, since they alone lead to a transition probability increasing linearly with time, i.e., $[(E_m - E_0) - \hbar\omega_{\text{pt}} \pm \hbar\omega_{\text{pn}}]$, expressing the conservation of energy between the initial and final states.

PERTINENT MATRIX ELEMENTS

Altogether there are six different forms of matrix element, corresponding to transitions between valence and conduction bands, and transitions within valence and conduction bands, with the aid of photons and phonons. These may be summarized as follows.

(1) Optical absorption by free electrons in the conduction band

$$H'_{\text{p,cc}} = \int \psi_c^*(\mathbf{k}, \mathbf{r}) \, \mathbf{H}_{\text{p}}' \, \psi_c(\mathbf{k}', \mathbf{r}) \, d\mathbf{r} \quad (11.112)$$

These transitions are not included in the present calculation.

11.4 Indirect Intrinsic Transition

(2) Optical absorption by electrons in the valence band

$$H'_{p,vv} = \int \psi_v^*(\mathbf{k}, \mathbf{r}) \, \mathbf{H}_p' \, \psi_v(\mathbf{k}', \mathbf{r}) \, d\mathbf{r} \tag{11.113}$$

These transitions are also excluded from the present calculation.

(3) Optical absorption causing a transition from the valence band to the conduction band.

$$H'_{p,cv} = \int \psi_c^*(\mathbf{k}, \mathbf{r}) \, \mathbf{H}_p' \, \psi_v(\mathbf{k}', \mathbf{r}) \, d\mathbf{r} \tag{11.114}$$

This is the principal optical transition of the present calculation.

(4) Phonon-scattering transition within the conduction band.

$$H'_{\ell,cc} = \int \psi_c^*(\mathbf{k}, \mathbf{r}) \, \mathbf{H}_\ell' \, \psi_c(\mathbf{k}', \mathbf{r}) \, d\mathbf{r} \tag{11.115}$$

This is one of the processes used to obtain conservation of $\mathbf{k}$ in an indirect transition.

(5) Phonon-scattering transition within the valence band.

$$H'_{\ell,vv} = \int \psi_v^*(\mathbf{k}, \mathbf{r}) \, \mathbf{H}_\ell' \, \psi_v(\mathbf{k}', \mathbf{r}) \, d\mathbf{r} \tag{11.116}$$

This is the other process used to obtain conservation of $\mathbf{k}$.

(6) Thermal excitation from valence to conduction band.

$$H'_{\ell,cv} = \int \psi_c^*(\mathbf{k}, \mathbf{r}) \, \mathbf{H}_\ell' \, \psi_v(\mathbf{k}', \mathbf{r}) \, d\mathbf{r} \tag{11.117}$$

In the present calculation the bandgap is assumed to be sufficiently large that thermal excitation across the gap can be neglected compared to optical excitation.

Only three of the possible forms of the matrix element need to be considered: $H'_{p,cv}$, $H'_{\ell,cc}$, and $H'_{\ell,vv}$. We choose for the initial state $\psi_v(\mathbf{k}_0, \mathbf{r})$ where $\mathbf{k}_0 \approx 0$, and for the final state $\psi_c(\mathbf{k}_f, \mathbf{r})$ where $\mathbf{k}_f \approx \mathbf{k}_m$, and then finally we sum over allowed values of $\mathbf{k}_0$ and $\mathbf{k}_f$ near the extrema of the bands. The matrix element of type $H'_{p,cv}$ is either

$$H'_{p,cv}(\mathbf{k}_0, \mathbf{k}_0) \equiv H_0^p \tag{11.118}$$

for an optical absorption at $\mathbf{k}_0$, or

$$H'_{p,cv}(\mathbf{k}_f, \mathbf{k}_f) \equiv H_f^p \tag{11.119}$$

434　　　　　　　　　　　　　　　　　　　　　　　　11　Optical Absorption

for an optical absorption at $\mathbf{k}_f$. The other pertinent matrix elements are

$$H'_{\ell,cc}(\mathbf{k}_0, \mathbf{k}_f) \equiv H_c^\ell \tag{11.120}$$

for electron scattering from $\mathbf{k}_0$ to $\mathbf{k}_f$ in the conduction band, and

$$H'_{\ell,vv}(\mathbf{k}_0, \mathbf{k}_f) \equiv H_v^\ell \tag{11.121}$$

for scattering from $\mathbf{k}_0$ to $\mathbf{k}_f$ in the valence band. The relevant energy differences are shown in Fig. 11.15, and are given by

$$\hbar\omega_{f0} = E_c(\mathbf{k}_f) - E_v(\mathbf{k}_0) \tag{11.122}$$

$$\hbar\omega_{i0} = E_c(\mathbf{k}_0) - E_v(\mathbf{k}_0) \tag{11.123}$$

$$\hbar\omega'_{i0} = E_c(\mathbf{k}_f) - E_v(\mathbf{k}_f) \tag{11.124}$$

THE TRANSITION PROBABILITY

If we substitute the pertinent matrix elements into Eq. (11.111), carry out the integration, and retain only those terms with denominators that can go to zero and hence can contribute to a probability increasing linearly

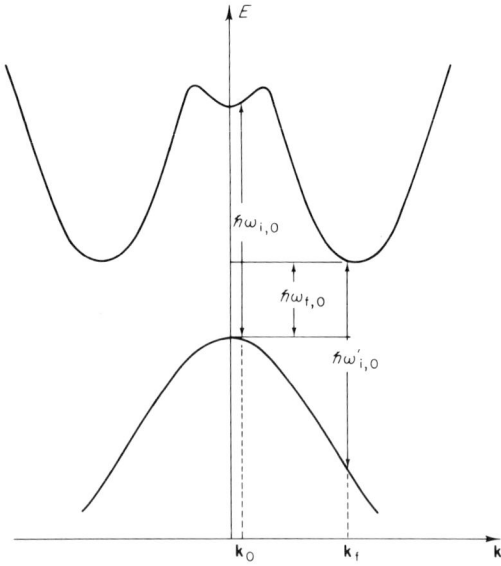

Fig. 11.15 Definition of the energies involved in the indirect transition calculation.

11.4 Indirect Intrinsic Transition

with time, we obtain

$$A_f''(t) = \frac{1}{\hbar^2}\left(\frac{H_0^p H_c}{\omega_{i0} - \omega_{pt}} + \frac{H_f^p H_v}{\omega_{i0}' - \omega_{pt}}\right)$$

$$\left(\frac{\exp[i(\omega_{f0} - \omega_{pt} - \omega_{pn})t] - 1}{\omega_{f0} - \omega_{pt} - \omega_{pn}} + \frac{\exp[i(\omega_{f0} - \omega_{pt} + \omega_{pn})t] - 1}{\omega_{f0} - \omega_{pt} + \omega_{pn}}\right)$$

(11.125)

Upon squaring $A_f''(t)$, we neglect cross-product terms, since they do not lead to a probability increasing linearly with time. The final result is just four terms,

$$|A_f''|^2 = \frac{4|H_0^p|^2 |H_c^e|^2 \sin^2[(\omega_{f0} - \omega_{pt} - \omega_{pn})t/2]}{\hbar^4 (\omega_{i0} - \omega_{pt})^2 (\omega_{f0} - \omega_{pt} - \omega_{pn})^2}$$

$$+ \frac{4|H_0^p|^2 |H_c^e|^2 \sin^2[(\omega_{f0} - \omega_{pt} + \omega_{pn})t/2]}{\hbar^4 (\omega_{i0} - \omega_{pt})^2 (\omega_{f0} - \omega_{pt} + \omega_{pn})^2}$$

$$+ \frac{4|H_f^p|^2 |H_v^e|^2 \sin^2[(\omega_{f0} - \omega_{pt} - \omega_{pn})t/2]}{\hbar^4 (\omega_{i0}' - \omega_{pt})^2 (\omega_{f0} - \omega_{pt} - \omega_{pn})^2}$$

$$+ \frac{4|H_f^p|^2 |H_v^e|^2 \sin^2[(\omega_{f0} - \omega_{pt} + \omega_{pn})t/2]}{\hbar^4 (\omega_{i0}' - \omega_{pt})^2 (\omega_{f0} - \omega_{pt} + \omega_{pn})^2} \quad (11.126)$$

The first term in Eq. (11.126) represents optical excitation from $\mathbf{k}_0$ followed by scattering of the electron to $\mathbf{k}_f$ by absorption of a phonon. The second term represents the same process but with emission of a phonon. The third term represents optical excitation to the conduction band at $\mathbf{k}_f$ followed by scattering of an electron from $\mathbf{k}_0$ to $\mathbf{k}_f$ in the valence band by absorption of a phonon. The fourth term represents the same process but with emission of a phonon.

Allowed Direct Transition

As a typical example of the procedure involved in calculating the transition probability, consider the first term of Eq. (11.126) only, and for the case of an allowed direct transition.

The first step is to sum over all allowed values of the final wave vector $\mathbf{k}_f$, keeping the initial wave vector fixed at $\mathbf{k}_0$. When this integration is carried out, we assign to other terms their value for $\omega_{f0} = \omega_{pt} + \omega_{pn}$ (this being the condition for a probability increasing linearly with time) and take them outside the integral, as done previously in similar calculations. The result gives for P_{c0}, the transition probability from $\mathbf{k}_0$ to a range of states with final wave vector $\mathbf{k}_f$, per unit time,

$$P_{c0} = \frac{2\pi M \langle |H_0^p|^2\rangle_{av} |H_c^\ell|^2 N_c(E')}{\hbar^4(\omega_{i0} - \omega_{pt})^2} \delta(\omega_{f0} - \omega_{pt} - \omega_{pn}) \quad (11.127)$$

where M is the number of equivalent minima, and $E' = E_c(\mathbf{k}_f) - E_c(\mathbf{k}_m)$. $N_c(E')$ is the density of states at an energy E' above the minimum of the conduction band. The average over $|H_0^p|^2$ indicates an angular average for this quantity.

In order to perform the second sum over the allowed initial states, we must take account of the requirement of conservation of energy that

$$E_c(\mathbf{k}_f) = \hbar\omega_{pt} + \hbar\omega_{pn} + E_v(\mathbf{k}_0)$$
$$\approx \hbar\omega_{pt} + \hbar\omega_m + E_v(\mathbf{k}_0) \quad (11.128)$$

assuming that $\hbar\omega_{pn} \approx \hbar\omega_m$ at the conduction-band minimum. In view of the definition of E',

$$E' = \hbar\omega_p + \hbar\omega_m + E_v(\mathbf{k}_0) - E_c(\mathbf{k}_m) \quad (11.129)$$

Now let E'' be the depth in energy of the state corresponding to $\mathbf{k}_0$ in the valence band below the top of the valence band, i.e., $E'' = E_v(0) - E_v(\mathbf{k}_0)$. Then

$$E' = \hbar\omega_{pt} + \hbar\omega_m - E'' - E_G \quad (11.130)$$

If this value of E' is used in Eq. (11.127), there is no longer any need for the delta function. To obtain the total transition probability P_{cv} for this first term of Eq. (11.126), we must now sum over the allowed values of $\mathbf{k}_0$ for which E'' lies between 0 and $(\hbar\omega_{pt} + \hbar\omega_m - E_G)$, i.e., over the allowed values of $\mathbf{k}_0$ for which both E' and E'' are positive.

Except for $N_c(E')$, other factors in P_{c0} of Eq. (11.127) do not vary much with $\mathbf{k}_0$, and they may therefore be assigned approximate values corresponding either to $\mathbf{k} = 0$ for the valence band, or to $\mathbf{k} = \mathbf{k}_m$ for the conduction band. Then

$$P_{cv} = \frac{4\pi M \langle |H_0^p|^2\rangle_{av} |H_c^\ell|^2}{\hbar^4(\omega_{i0} - \omega_{pt})^2} \int_0^{E_m} N_v(E) N_c(E_m - E) \, dE \quad (11.131)$$

with $E_m = (\hbar\omega_{pt} + \hbar\omega_m - E_G)$, and with an additional factor of 2 to account for electron spin. The integration in the valence-band term is between the two limiting values of E'', i.e., 0 and E_m. The integration in the conduction-band term is over $N_c(E')$, where $E' = E_m - E$, the variable E going between the allowed limits of E''.

11.4 Indirect Intrinsic Transition

For nondegenerate bands, the densities of states are

$$N_v(E) = \frac{\mathcal{V}(2m_h^*)^{3/2}}{4\pi^2 \hbar^3} E^{1/2} \tag{11.132}$$

$$N_c(E) = \frac{\mathcal{V}(2m_e^*)^{3/2}}{4\pi^2 \hbar^3} E^{1/2} \tag{11.133}$$

Substituting these values into Eq. (11.131) gives

$$P_{cv} = \frac{2M\mathcal{V}^2 \langle |H_0^p|^2 \rangle_{av} |H_e^\ell|^2 (m_h^* m_e^*)^{3/2}}{\pi^3 \hbar^{10}(\omega_{i0} - \omega_{pt})^2} \int_0^{E_m} E^{1/2}(E_m - E)^{1/2} dE \tag{11.134}$$

The integral may be evaluated by letting $u = E^{1/2}$ so that

$$\int_0^{E_m} E^{1/2}(E_m - E)^{1/2} dE$$

$$= 2 \int_0^{E_m^{1/2}} u^2 (E_m - u^2)^{1/2} du$$

$$= 2 \left\{ -\frac{u}{4}(E_m - u^2)^{3/2} + \frac{E_m}{8}\left[u(E_m - u^2)^{1/2} + E_m \sin^{-1}\left(\frac{u}{E_m^{1/2}}\right)\right]\right\}_0^{E_m^{1/2}}$$

$$= \frac{\pi E_m^2}{8} = \frac{\pi(\hbar\omega_{pt} + \hbar\omega_m - E_G)^2}{8} \tag{11.135}$$

The result for the transition probability is

$$P_{cv} = \frac{M\mathcal{V}^2 \langle |H_0^p|^2 \rangle_{av} |H_e^\ell|^2 (m_h^* m_e^*)^{3/2}(\hbar\omega_{pt} + \hbar\omega_m - E_G)^2}{4\pi^2 \hbar^8 (\hbar\omega_{i0} - \hbar\omega_{pt})^2} \tag{11.136}$$

This equation holds for $\hbar\omega_{pt} > (E_G - \hbar\omega_m)$. If $\hbar\omega_{pt} < (E_G - \hbar\omega_m)$, $P_{cv} = 0$.

The matrix element indicated by $\langle |H_0^p|^2 \rangle_{av}$ can be related to $\langle |P_{m0}|^2 \rangle_{av}$ by Eq. (11.68) and hence to the oscillator strength f through Eq. (11.83). To evaluate $|H_e^\ell|$, we may make use of the relationship of Eq. (8.66), with the significant term A_ℓ^2 being given at elevated temperatures by Eq. (8.72). More generally we must replace a Boltzmann distribution by a Bose–Einstein distribution and express

$$A_\ell^2 = \frac{2}{\mathcal{M}N\omega_{pn}^2} \frac{\hbar\omega_{pn}}{e^{\hbar\omega_{pn}/kT} - 1} \tag{11.137}$$

where $\mathcal{M}$ is the mass of the atoms and N is the number of atoms. Combining Eqs. (11.136) and (11.137) gives finally an expression for the absorption constant

$$\alpha_{cv} = G_0 \frac{(\hbar\omega_{pt} + \hbar\omega_{pn} - E_G)^2}{e^{\hbar\omega_{pn}/kT} - 1} \quad (11.138)$$

where all the multiplying constants are simply expressed as G_0.

The absorption constant for the other three terms of Eq. (11.126) can be calculated in a similar way for the case of an allowed direct transition. The total absorption constant may be written as

$$\alpha_a = (G_0 + G_f)\left[\frac{(\hbar\omega_{pt} + \hbar\omega_{pn} - E_G)^2}{e^{\hbar\omega_{pn}/kT} - 1} + \frac{(\hbar\omega_{pt} - \hbar\omega_{pn} - E_G)^2 \, e^{\hbar\omega_{pn}/kT}}{e^{\hbar\omega_{pn}/kT} - 1}\right] \quad (11.139)$$

Here the constants G_0 and G_f refer to the direct transition occurring at $\mathbf{k} = 0$ and $\mathbf{k} = \mathbf{k}_f$, respectively. The first term in the brackets corresponds to a transition at $\mathbf{k} = 0$ or $\mathbf{k} = \mathbf{k}_f$ involving phonon absorption, and the second term in the brackets corresponds to a transition either at $\mathbf{k} = 0$ or $\mathbf{k} = \mathbf{k}_f$ involving phonon emission. For an indirect transition with allowed direct transition, the absorption constant varies as the square of the photon energy. A plot of $\alpha^{1/2}$ vs. $\hbar\omega_{pt}$ has two branches, one corresponding to phonon absorption and the other to phonon emission. If the intersection of these two branches with the photon-energy axis occurs at $\hbar\omega_1$ and $\hbar\omega_2$, then

$$E_G = \frac{\hbar\omega_1 + \hbar\omega_2}{2} \quad (11.140)$$

$$\hbar\omega_{pn} = \frac{\hbar\omega_2 - \hbar\omega_1}{2} \quad (11.141)$$

for $\hbar\omega_2 > \hbar\omega_1$. From an experimental measurement of the absorption constant as a function of photon energy, it is possible to determine with reasonable accuracy both the indirect bandgap and the energy of the phonon participating in the indirect optical transition. From a knowledge of the participating phonon energy and of the phonon spectrum of the material, it is possible to deduce the phonon wavevector and hence the approximate location of the energy-band extremum in the Brillouin zone. Figure 11.16 shows an analysis of the indirect absorption spectra of Ge and Si in terms of a single phonon participating; more detailed measurements indicate the involvement of both the longitudinal and the transverse acoustic phonons in these transitions. Because of the temperature dependence of the phonon density, the phonon-absorption branch of these curves becomes less pronounced and eventually disappears at sufficiently low temperatures.

11.4 Indirect Intrinsic Transition

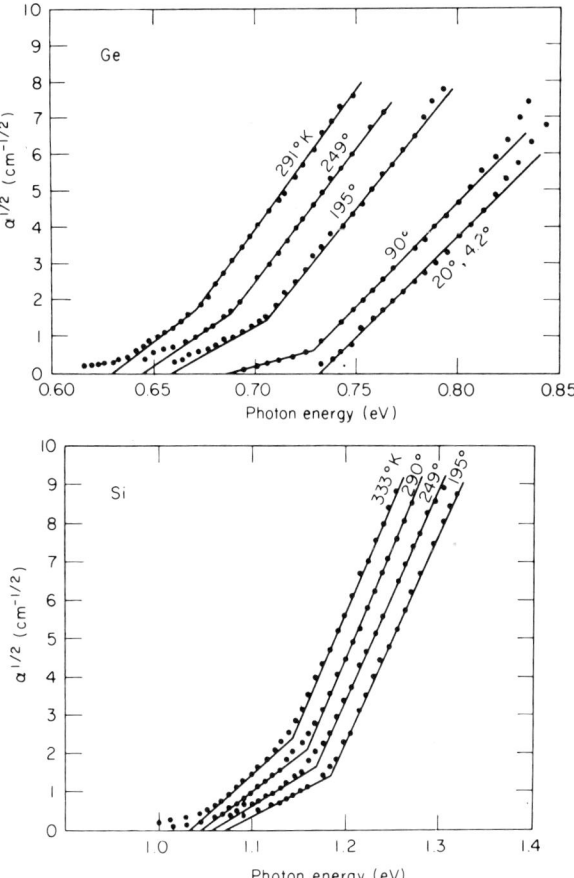

Fig. 11.16 Analysis of absorption spectra of Ge and Si with a one-phonon model to show the presence of indirect transitions. [After G. G. MacFarlane and V. Roberts, *Phys. Rev.* **97**, 1714 (1955); **98**, 1865 (1955).]

Example 11.1 Absorption data taken from the literature for Ge at 300°K are given in Fig. 11.17. Analyze the data to obtain the value of the direct bandgap, the indirect bandgap, and the phonon energy participating in the indirect transitions.

The analysis of the data is shown in Fig. 11.18, where $\alpha^{1/2}$ is plotted as a function of energy for the smaller absorption constant range from 0.2 to 50 cm^{-1}, and where α^2 is plotted as a function of energy for the larger absorption constant range from 200 to 5000 cm^{-1}.

The results give a direct bandgap of 0.81 eV, an indirect bandgap

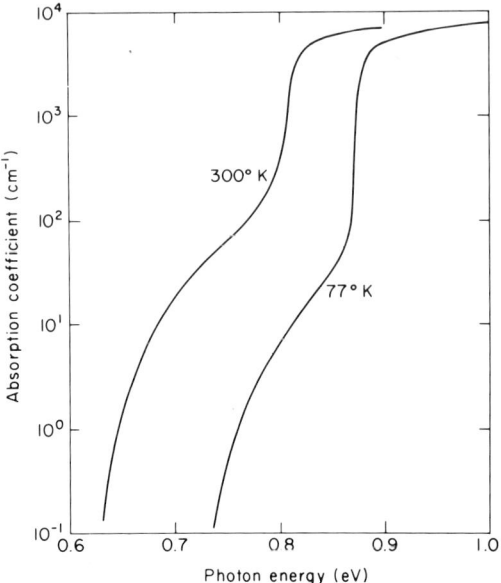

Fig. 11.17 Dependence of absorption constant on photon energy for Ge. [After W. C. Dash and R. Newman, *Phys. Rev.* **99**, 1151 (1955).]

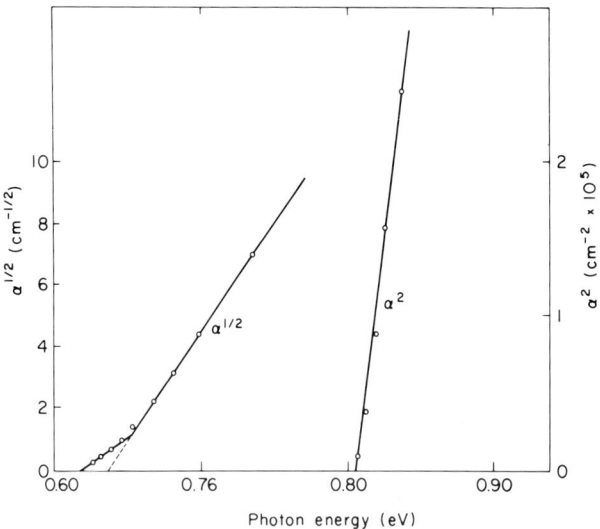

Fig. 11.18 Results of the analysis of the 300°K data of Fig. 11.17 into indirect and direct transitions.

11.4 Indirect Intrinsic Transition

of 0.63 eV, and a phonon energy of about 0.01 eV. These values are reasonably close to those obtained from more careful analyses, the phonon energy being that of the LA phonons.

FORBIDDEN DIRECT TRANSITION

If either the transition at k_0 or k_f is forbidden, then the above calculation must be adjusted. Recall in connection with the discussion of Eq. (11.86) that an additional factor of k' occurs in the matrix element in the case of the forbidden direct transition. This additional factor of k' is responsible for changing the direct transition's energy dependence of absorption constant from the $\frac{1}{2}$ to the $\frac{3}{2}$ power of the photon energy. This same factor enters in the indirect transition calculation if the direct transition is forbidden, through the matrix element $\langle | H_0^p |^2 \rangle_{av}$. The result is that the integral for P_{cv} now contains an additional factor of E under the integral,

$$P_{cv} \propto \int_0^{E_m} E^{3/2}(F_m - E)^{1/2} \, dE \tag{11.142}$$

instead of the relation given in Eq. (11.134) for an allowed direct transition. This integral can be evaluated by the substitution pictured in Fig. 11.19. Since $E = E_m \sin^2 \theta$, $dE = 2E_m \sin \theta \cos \theta \, d\theta$, and

$$\int_0^{E_m} E^{3/2}(E_m - E)^{1/2} \, dE = \int_0^{\pi/2} (E_m^{3/2} \sin^3 \theta)(E_m^{1/2} \cos \theta)(2E_m \sin \theta \cos \theta) \, d\theta$$

$$= 2E_m^3 \int_0^{\pi/2} \sin^4 \theta \cos^2 \theta \, d\theta$$

$$= 2E_m^3 \int_0^{\pi/2} \sin^4 \theta \, d\theta - 2E_m^3 \int_0^{\pi/2} \sin^6 \theta \, d\theta$$

$$= 2E_m^3 \left(\frac{3\pi}{16} + \frac{15\pi}{96} \right)$$

$$= \frac{\pi E_m^3}{16} \tag{11.143}$$

For a forbidden direct transition, the transition probability for an indirect transition varies as E_m^3 rather than as E_m^2, as it does for an allowed direct transition. The expression for the absorption constant for a forbidden direct transition is

$$\alpha_f = (G_0' + G_f') \left[\frac{(\hbar\omega_{pt} + \hbar\omega_{pn} - E_G)^3}{e^{\hbar\omega_{pn}/kT} - 1} + \frac{(\hbar\omega_{pt} - \hbar\omega_{pn} - E_G)^3 e^{\hbar\omega_{pn}/kT}}{e^{\hbar\omega_{pn}/kT} - 1} \right] \tag{11.144}$$

442   11 Optical Absorption

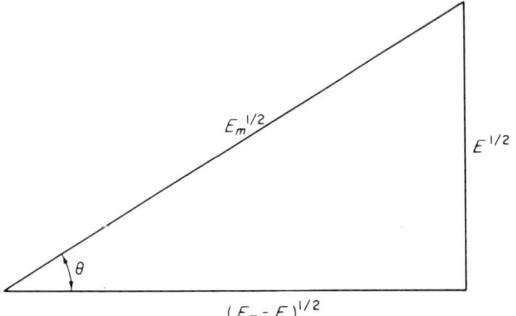

Fig. 11.19 Transformation of variables device to permit integration in the case of an indirect transition with a forbidden direct transition.

where the G' terms involve the oscillator strength f' for forbidden direct transitions.

11.5 Exciton Absorption

The optical excitation of excitons manifest itself as a series of sharp absorption lines at the low-energy side of the band edge in ionic materials. Typical data for BaO are shown in Fig. 11.20.

Excitons may be formed either by direct or indirect transitions. The total energy of an exciton is given by

$$E_{\text{ex}} = \frac{\hbar^2 K_{\text{ex}}^2}{2(m_e^* + m_h^*)} - E_{\text{ex}}^n \qquad (11.145)$$

where the first term on the right is the kinetic energy of the exciton, and the second term is the binding energy of the exciton in the nth state, as given in Eq. (6.56). For direct transitions it is required that $\mathbf{K}_{\text{ex}} \approx 0$. This condition may be understood in terms of the optical excitation process. If absorption of light creates a hole–electron pair which then becomes an electron and a hole moving in opposite directions, $\mathbf{k}$ can be conserved at the same time as a finite value of $|\hbar\mathbf{k}|$ is given to the members of the pair. When an exciton is formed, however, the electron and hole must move in the same direction. Conservation of $\mathbf{k}$ therefore requires that $\mathbf{K}_{\text{ex}} \approx 0$. A sharp line spectrum for absorption due to direct exciton excitation is observed.

Excitons may also be formed as the result of an indirect transition with the absorption or emission of a phonon. The center of gravity of the exciton

11.5 Exciton Absorption

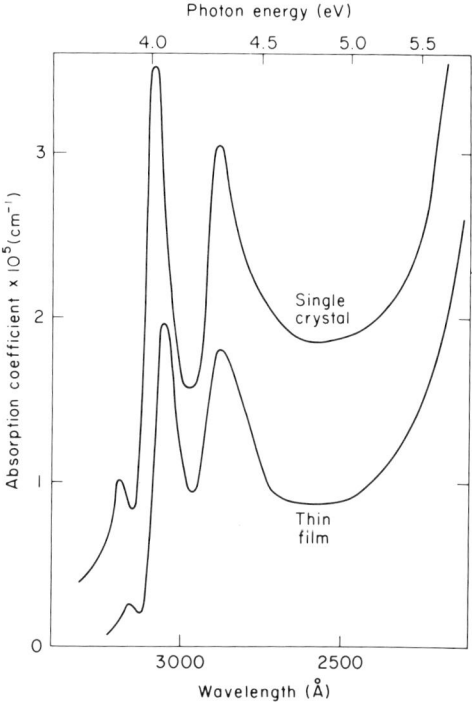

Fig. 11.20 Exciton absorption in single-crystal and thin film BaO. [After F. C. Jahoda, *Phys. Rev.* **107**, 1261 (1957).]

may now have a finite momentum $\hbar\mathbf{K}_{\text{ex}}$, since momentum can be conserved through the interacting phonon. The result is the formation of an exciton absorption band with a well-defined low-energy edge corresponding to the minimum energy required to create an exciton. If E_0 is the binding energy of the exciton in its lowest state, then the minimum value of $\hbar\omega_{\text{pt}}$ for observing such exciton absorption is given by

$$\hbar\omega_{\text{pt}} = E_G - \hbar\omega_{\text{pn}} - E_0 \qquad (11.146)$$

The energy dependence of the absorption constant for indirect exciton transition may be calculated† to show that for an allowed direct transition,

$$\alpha_{\text{ex}} \propto [\hbar\omega_{\text{pt}} - (E_G - E_0) \pm \hbar\omega_{\text{pn}}]^{1/2} \qquad (11.147)$$

† G. Dresselhaus, *J. Phys. Chem. Solids* **1**, 15 (1956); R. J. Elliott, *Phys. Rev.* **108**, 1384 (1957).

and that for a forbidden direct transition,

$$\alpha'_{ex} \propto [\hbar\omega_{pt} - (E_G - E_0) \pm \hbar\omega_{pn}]^{3/2} \qquad (11.148)$$

Since the dependence of absorption constant on energy is different for exciton absorption and band-to-band absorption, the variation of absorption constant with phonon energy gives the best indication of the existence of indirect excitons. It should also be possible to detect their presence from optical absorption measurements at very low temperatures, since no free carriers should be formed (as detectable by measurements of photoconductivity, for example) as long as

$$[(E_G - \hbar\omega_{pn}) - E_0] < \hbar\omega_{pt} < [E_G - \hbar\omega_{pn}] \qquad (11.149)$$

11.6 Summary

The optical absorption due to free carriers in metals or semiconductors manifests itself in different forms at different electric-field frequencies through the frequency dependence of the electrical conductivity. At low frequencies the absorption constant increases in direct proportion to the conductivity unless the conductivity is large enough to effect the refractive index. In this latter case the absorption constant increases as the square root of the product of the conductivity and the frequency, producing the so-called "skin effect." At high electric-field frequencies in the optical range, the absorption constant increases as a power of the wavelength of the light, the square for a simple quasi-classical calculation. At the frequency such that the electronic contribution to the dielectric constant cancels the lattice contribution, strong absorption due to excitation of free-electron plasma oscillations occurs.

Information about the band structure of a semiconducting material can be obtained if accurate measurements can be made of optical absorption for photon energies near that of the absorption edge. This is because a band structure with extrema at the same point in **k** space involves optical absorption without the emission or absorption of phonons, whereas a band structure with the conduction- and valence-band extrema at different points in **k** space requires absorption or emission of phonons in the optical absorption process. Since a direct transition, one not involving phonons, has a different dependence on photon energy from that of an indirect transition, one involving phonons, it is possible in principle to deduce which is present in a given material by optical absorption measurements.

Problems

The direct transition is a first-order process and therefore has larger values of the absorption constant than are found for the indirect transition, which is a second-order process. Since the indirect transition can involve a distribution of states in conduction and valence bands, however, the difference in absorption constant between the two types of transition is not as large as might be expected.

The energy dependences of the absorption constant are as follows:

Direct allowed transition: $(\hbar\omega_{pt} - E_G)^{1/2}$

Direct forbidden transition: $\dfrac{(\hbar\omega_{pt} - E_G)^{3/2}}{\hbar\omega_{pt}}$

Indirect transition for allowed direct transition: $\dfrac{(\hbar\omega_{pt} \pm \hbar\omega_{pn} - E_G)^2 (e^{\hbar\omega_{pn}/kT})_-}{e^{\hbar\omega_{pn}/kT} - 1}$

Indirect transition for forbidden direct transition: $\dfrac{(\hbar\omega_{pt} \pm \hbar\omega_{pn} - E_G)^3 (e^{\hbar\omega_{pn}/kT})_-}{e^{\hbar\omega_{pn}/kT} - 1}$

where the minus subscript on the $(e^{\hbar\omega_{pn}/kT})$ term in the numerator for the indirect transition expressions indicates that this term is included only when the minus sign is used with $\hbar\omega_{pn}$, i.e., only in the case of phonon emission.

Optical absorption associated with the excitation of excitons is also a possible process. Here a sharp line spectrum is found for direct transitions and a broader band spectrum for indirect transitions.

Problems

11.1 The electrical conductivity of silver is 6.12×10^5 $(\Omega\text{-cm})^{-1}$. What thickness of silver sheet is required to shield a region so that not more than 1 per cent of the intensity of a 100-m wavelength alectromagnetic field can penetrate the silver sheet? (This frequency is sufficiently low that the given dc value of conductivity can be used in the calculation.)

11.2 A given material has a high-frequency dielectric constant of 16. What must the absorption constant be for the material to have 70 per cent reflection for 5000 Å light? If this absorption arises from free carriers, what is the corresponding conductivity?

11.3 Calculate the transmission coefficient for a light wave passing through a thin metal film of thickness d, using the procedure outlined in the text for calculating the reflection coefficient from a thick piece of metal.

11.4 If the index of refraction of V_2O_3 is assumed to be of the order of 2, calculate the change in reflectivity for a wavelength causing free-carrier absorption, caused by the temperature-induced phase transition shown in Fig. 9.23.

11.5 Derive the absorption constant due to free carriers in a semiconductor at the plasma frequency when $\omega_p \tau \gg 1$.

11.6 Consider a semiconductor material for which the extrema of the valence and conduction bands are at $\mathbf{k} = 0$, and for which the bandgap is 2.4 eV. (a) What color is a pure crystal according to the normal eye by transmission? (b) What absorption constant is expected for low-intensity 1.9-eV light in a pure crystal? (c) What color is the crystal to the eye if there is strong absorption due to compensated acceptors lying 0.6 eV above the valence band? (d) Can a pure and perfect crystal of any material have a green color to the eye? Explain.

11.7 A semiconductor crystal with uniform high electron density of 10^{19} cm^{-3} is irradiated with an electron beam which introduces defects that decrease the free-electron density. Since the electron beam is not absorbed uniformly, the electron density is decreased more on the side of the crystal on which the beam of electrons is incident than on the opposite side, thus setting up a gradient in electron density across the crystal. If the total decrease in electron density is at most a factor of 10, describe a way of using a nondestructive optical measurement technique that would permit an estimate of the electron density gradient after irradiation with the electron beam. What wavelength range is involved? If $m_e^* = 0.5m$, and a strong Reststrahlen absorption occurred starting at 50 μm, what range of electron densities would be measurable?

11.8 A semiconductor crystal has parabolic energy bands with extrema at $\mathbf{k} = 0$. The effective mass of electrons is $0.01m$ and the effective mass of holes is $0.05m$. The impurity content of this crystal is adjusted so that there are 10^{18} cm^{-3} free electrons in the conduction band. Calculate the photon energy for which band-to-band absorption becomes appreciable if the bandgap of the intrinsic material is 0.20 eV.

11.9 Optical transmission measurements are made on two semiconductor crystals. The data shown in the tabulation are obtained (the data being corrected so that reflectivity losses need not be considered).

Problems

Crystal A Thickness = 100 μm		Crystal B Thickness = 1 μm	
Wavelength (Å)	Transmission (per cent)	Wavelength (μm)	Transmission (per cent)
3930	0.50	2.065	22.8
4000	1.83	2.155	25.1
4070	5.55	2.255	28.1
4130	14.1	2.360	32.0
4200	29.8	2.480	36.8
4270	48.7	2.610	44.5
4350	74.1	2.755	53.3
4425	85.2	2.915	72.8
4510	91.4		
4590	97.8		

(a) For each of these crystals determine whether the transition is direct or indirect, and whether the direct transition is allowed or forbidden. (b) Determine the bandgap of each crystal. (c) Determine the phonon energy involved in indirect transitions. (d) Make an educated guess as to the identity of the two crystals.

11.10 It is found that the absorption edge of a semiconductor crystal has the values shown in the tabulation in several samples with different impurity concentration, the free-electron density n being determined by Hall measurements.

Wavelength (μm)	Free-electron density, n (cm^{-3} × 10^{18})
4.93	0.05
4.43	0.1
3.79	0.2
2.85	0.5
2.16	1

If the electron effective mass is much smaller than the hole effective mass, (a) determine the true band gap of the material, and (b) determine the approximate electron effective mass.

11.11 Indirect optical absorption is measured in a semiconducting material with a bandgap at 0°K of 1.65 eV, with the participation of two types of phonons, one with energy of $100k$ ergs, and the other with energy of $400k$ ergs, where k is Boltzmann's constant. Draw realistically the shape of the absorption constant versus photon energy at (a) $800°K$, (b) $200°K$, and (c) $10°K$. The temperature coefficient of the bandgap is -5×10^{-4} eV/°K.

11.12 The absorption spectrum shown in Fig. 11.21 is the result of a hypothetical transition between two discrete states of the same imperfection separated by ΔE. (a) Suggest a mechanism for the observed absorption spectrum. (b) What type of phonons would be expected to be involved?

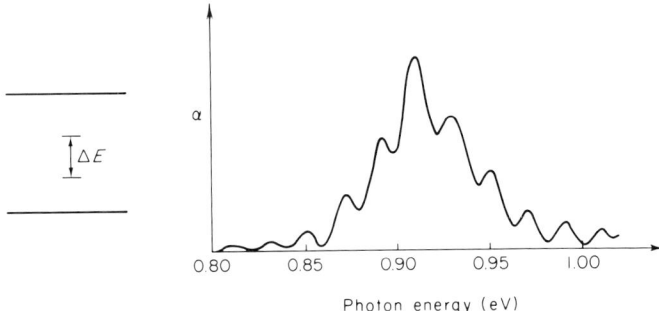

Fig. 11.21 Hypothetical absorption process and spectrum; see Problem 11.12.

(c) What is the phonon energy involved? (d) What is the conservation of energy involved in calculating the transition probability? (e) What is the value of ΔE? (f) What changes would you expect in the spectrum if remeasured at very low temperatures?

Chapter 12

Photoelectronic Effects

A variety of effects initiated by the creation of free carriers by the absorption of light in a semiconductor or insulator, and involving photoconductivity, recombination, trapping, and related phenomena, are known as *photoelectronic effects*. It is to a consideration of the most outstanding of these effects that we turn our attention in this last chapter. They are the kind of processes of importance for understanding materials used as light detectors, light emitters, or energy converters.

The exposition of many of the photoelectronic effects does not start with basic quantum mechanical considerations, but can most conveniently be undertaken with the band structure of solids as an assumed presupposition. From this point a partly phenomenological approach permits the description of observed effects in terms of a few parameters, the values of which must ultimately be calculated by quantum mechanics. Our first concern, therefore, is to define these parameters and indicate the way in which they are used. The most basic of these is the concept of lifetime of a free carrier.

Except in special circumstances, the measurement of a photoelectronic effect requires the application of electrical contacts to the material under investigation. The behavior of these contacts may be the determining factor in how a particular photoelectronic effect is manifested. Electrical processes in the material may also be influenced by carriers injected electrically from these contacts.

Photoelectronic effects commonly involve a variety of imperfection levels. In these circumstances the formal complete mathematical solution of the problem becomes intractable. By defining a steady-state Fermi level and a corollary demarcation level, to define whether or not the occupancy of a level is determined by thermal or kinetic processes, it is possible to describe a number of critical processes in a semiquantitative and conceptually helpful way.

The increase in conductivity of a material under illumination, known as *photoconductivity*, has been explained through the use of a number of different models. It is clear today that sensitive photoconductors can exist in both homogeneous-material form, and in heterogeneous-material form involving junctions or potential barriers.

The length of time that a carrier excited by light remains free is determined in part by the probability that it will recombine with a carrier of the opposite type. In this recombination process, the excess energy of the carriers must be dissipated as phonons, photons or excited carriers. In some cases the probability of recombination can be calculated by a consideration of the recombination processes.

Photovoltaic effects, in which light generates either a short-circuit current or an open-circuit voltage in the presence of a potential barrier, and photomagnetoelectric effects, in which light generates similar current or voltage in the presence of a magnetic field in a homogeneous material, are related phenomena.

12.1 General Concepts

When optical absorption by a semiconductor or insulator produces additional free carriers, the electrical conductivity of the material is increased in the phenomenon of photoconductivity. Suppose that such a material has a dark conductivity

$$\sigma_0 = n_0 e \mu_0 \tag{12.1}$$

where we consider for simplicity a material with conductivity dominated by electrons in both light and dark (e.g., the photoexcited holes are captured by imperfections much more rapidly than the electrons). We can generalize the result to two-carrier conductivity without difficulty.

In the presence of photoexcitation,

$$\sigma = ne\mu \tag{12.2}$$

where $\sigma = (\sigma_0 + \Delta\sigma)$, $n = (n_0 + \Delta n)$, and $\mu = (\mu_0 + \Delta\mu)$. If we sub-

12.1 General Concepts

stitute these definitions into Eq. (12.2), we obtain, for the photoconductivity,

$$\Delta\sigma = \sigma - \sigma_0 = e\mu_0\,\Delta n + ne\,\Delta\mu \tag{12.3}$$

Mechanisms exist for both changes in carrier density Δn, and changes in carrier mobility $\Delta\mu$ with photoexcitation.

LIFETIME

It is convenient to express the relationship between the change in carrier density Δn and the excitation intensity f per unit volume per second in terms of a lifetime,

$$\Delta n = f\tau_n \tag{12.4}$$

This relationship is in the form of a general logical statement: the steady-state quantity is equal to the generation rate times the average lifetime. It is as true of human populations as it is of electrons in crystals. The actual magnitude and variation of the lifetime can be calculated on the basis of models of photoconductivity in specific cases.

Equation (12.4) shows that the change in carrier density can come about either by a change in excitation rate f, or in lifetime τ_n, i.e.,

$$\delta(\Delta n) = \tau_n\,\delta f + f\,\delta\tau_n \tag{12.5}$$

A change in Δn with a change in f constitutes what might be called "normal" photoconductivity. A change in τ_n with a change in f must also be allowed for, since the mechanisms governing the recombination rate and hence τ_n may change with changes in f. Three ranges may be distinguished:

$$\Delta n \propto f; \qquad \tau_n \text{ constant, independent of } f \tag{12.6}$$

$$\Delta n \propto f^{<1}; \qquad \tau_n \propto f^{-a},\ \ 0 < a < 1 \tag{12.7}$$

$$\Delta n \propto f^{>1}; \qquad \tau_n \propto f^{a},\ \ a > 1 \tag{12.8}$$

Of course these power-law dependences indicated are more illustrative than exactly true. Since the behavior of Eq. (12.6) is called a *linear* variation of Δn, the behavior of Eq. (12.7) is sometimes called a *sublinear* variation, and the behavior of Eq. (12.8) a *superlinear* variation.

If we substitute Eq. (12.4) into Eq. (12.3), we obtain

$$\Delta\sigma = fe\tau_n\mu_0 + ne\,\Delta\mu \tag{12.9}$$

Since in actual materials the first term on the right of Eq. (12.9) almost always dominates, we see that the magnitude of the photoconductivity for a given excitation intensity is proportional to the product $\tau_n\mu_0$.

PHOTOEXCITATION DEPENDENCE OF MOBILITY

There are three simple processes by which the mobility can be a function of photoexcitation. The first of these is for photoexcitation to remove the charge on charged impurity centers dominating the scattering of free carriers. We have already considered this process in Problems 8.4 and 9.7. Mobility changes of a factor of 2 or less are common under photoexcitation intensities which change the free carrier density by many orders of magnitude in high-resistivity semiconductors or insulators.

A second process might involve the excitation of carriers from a low-mobility band to a high-mobility band, thus producing photoconductivity even though $\Delta n = 0$ exactly.

Finally a change in μ may be considered to result from a photoexcitation reduction of barrier heights to free-carrier flow in an inhomogeneous material. If, in a material with barriers of height E_b, we define the mobility μ_b of carriers moving through the whole material,

$$ne\mu_b = (n\, e^{-E_b/kT})e\mu \tag{12.10}$$

where n and μ are the values in the nonbarrier regions of the material, then the mobility

$$\mu_b = \mu\, e^{-E_b/kT} \tag{12.11}$$

Since E_b may be reduced by photoexcitation that produces trapping of charged carriers in or near the barriers, the mobility defined in this way becomes a strong function of excitation intensity as well as temperature.

KINDS OF LIFETIME

There are a number of different ways in which the term "lifetime" is used, depending on the particular context.

Free lifetime—the lifetime of a free carrier, not including any time spent by carriers in traps. The electron lifetime above, τ_n, is a free-carrier lifetime.

Excited lifetime—the total lifetime of an excited carrier, including both free and trapped times.

Minority-carrier lifetime—the free lifetime of a minority carrier, i.e., the carrier, electron or hole, present in lower density.

12.1 General Concepts

Majority-carrier lifetime—the free lifetime of a majority carrier, i.e., the carrier, electron or hole, present in greater density.

Pair lifetime—the free lifetime of an electron–hole pair; it is equal to the lifetime of that carrier first captured, usually the minority carrier.

Many semiconductor applications of electronic solids, transistor action for example, depend basically on the magnitude of the pair or minority-carrier lifetime. Most photoconductivity applications depend basically on the magnitude of the majority-carrier lifetime in high-resistivity semi-conductors and insulators; in these materials the majority-carrier lifetime may be much larger than the minority-carrier lifetime.

PHOTOSENSITIVITY

Three different ways of describing photosensitivity are commonly used. The first of these describes the photosensitivity as a materials property in terms of the $\tau\mu$ product, as shown in Eq. (12.9). It is comparable to considering the photosensitivity to be related to the change in conductivity per photon absorbed.

A second way of describing photosensitivity has been given the technical name *detectivity*. Detectors designed to measure small signals in the infrared are small-bandgap semiconductors with a high dark conductivity. Their utility as photodetectors depends ultimately on how small a signal they can detect, and this depends in turn on the ratio of signal to noise. The detectivity is a measure of the radiation power needed to give a signal equal to the noise. It is comparable to considering the photosensitivity to be related to the ratio of the photoconductivity to the dark conductivity.

A third way of describing photosensitivity is that of *gain*. The gain of a photoconductor is defined as the number of charge carriers passing between the electrodes per photon absorbed.

$$G = \frac{\Delta i}{eF} \quad (12.12)$$

Here $\Delta i/e$ is the number of electrons passing per second, and F is the total number of photons absorbed per second producing electron–hole pairs. Sometimes this expression is written as

$$G = \frac{\Delta i}{eqF'} \quad (12.13)$$

where q is the *quantum efficiency* giving the number of electron–hole pairs created per absorbed photon, and F' is the total number of photons absorbed per second. The gain may be expressed from a more microscopic viewpoint as the ratio of the carrier lifetime to transit time between electrodes,

$$G = \frac{\tau_n}{t_n} \tag{12.14}$$

Since the transit time is given by $t_n = l^2/\mu V$ for an electrode spacing of l,

$$G = \frac{\tau_n \mu V}{l^2} \tag{12.15}$$

The gain depends on the applied voltage and on the electrode spacing, as well as on the $\tau\mu$ product, and is thus a combination of materials and device properties.

SIMPLE KINETICS

To get a feeling for the basic types of photoconductivity kinetics, consider the special simple case of an intrinsic semiconductor with thermal and optical excitation from valence to conduction band, as indicated in Fig. 12.1. If the optical excitation rate is f, and the thermal excitation rate is g, then the rate of change of free electrons and holes is given by

$$\frac{dn}{dt} = \frac{dp}{dt} = f + g - npC_i \tag{12.16}$$

where C_i is the capture coefficient equal to Sv, as given in Eq. (9.76), S being the capture cross section for recombination between free electrons and free holes in this intrinsic material.

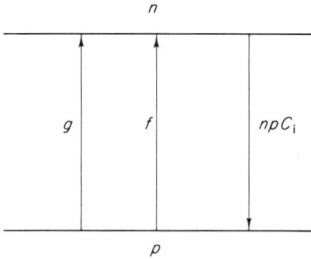

Fig. 12.1 Transitions for a simple intrinsic model. Excitation processes proceed at a rate g (thermal excitation) and/or f (optical excitation). Recombination is between a free electron and a free hole with a capture coefficient C_1.

12.1 General Concepts

In thermal equilibrium,

$$g = n_0 p_0 C_i = n_i^2 C_i \tag{12.17}$$

The addition of photoexcitation f gives

$$f + g = (n_0 + \Delta n)^2 C_i \tag{12.18}$$

since again $(n_0 + \Delta n) = (p_0 + \Delta p)$. Combining these two equations gives

$$f = \Delta n (2n_0 + \Delta n) C_i \tag{12.19}$$

Two simple cases occur. The first is that when $\Delta n \ll n_0$, and is characteristic of "semiconductor-like" behavior,

$$\tau_n = \tau_p = \frac{\Delta n}{f} = \frac{1}{2n_0 C_i} \tag{12.20}$$

In this case the lifetime is constant and depends on the dark conductivity of the material. The photoconductivity varies linearly with f. The second case is that when $\Delta n \gg n_0$, and is characteristic of "insulator-like" behavior,

$$\tau_n = \tau_p = \frac{\Delta n}{f} = \frac{1}{C_i \Delta n} = \left(\frac{1}{C_i f}\right)^{1/2} \tag{12.21}$$

In this case the lifetime varies inversely as the square root of the excitation intensity and depends on the change in carrier density. The photoconductivity varies as the square root of f. In nonintrinsic high-resistivity semiconductors or insulators, the existence of imperfection levels may cause the lifetime to depart from the relation of Eq. (12.21), as well as making electron and hole lifetimes unequal, but when the excitation intensity is high enough to make the recombination of free electrons and holes dominate, the behavior of Eq. (12.21) is found.

Spectral Response

Spectral response curves for photoconductivity in a "pure" CdS crystal are shown in Fig. 12.2, together with an interpretive diagram. The spectral response curve consists of three parts. For wavelengths much less than the absorption edge of the material, absorption is very strong and occurs near the surface of the crystal. In this range the photosensitivity is limited by the lifetime for recombination at the surface, which may be appreciably smaller than in the volume due to the greater imperfection density at the

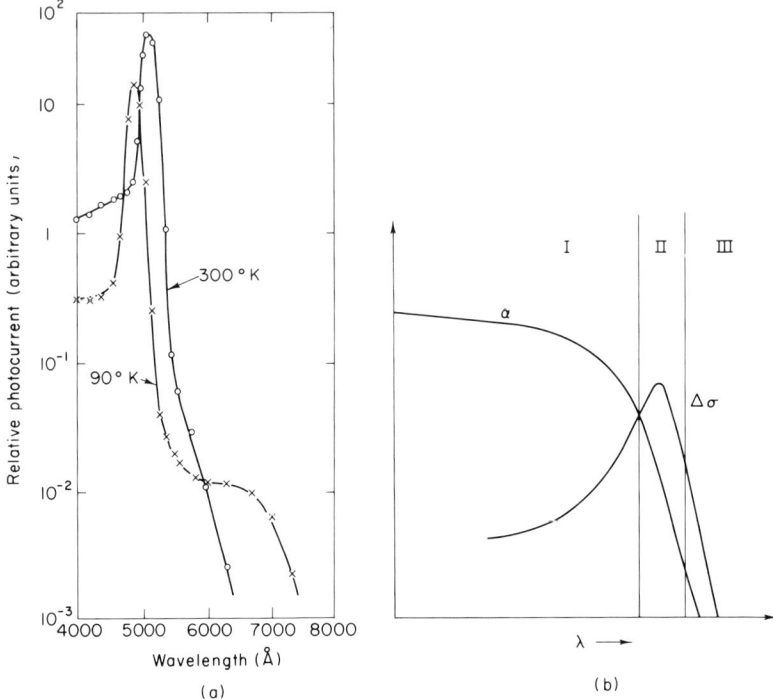

Fig. 12.2 (a) Photoconductivity excitation spectra for a photosensitive "pure" CdS crystal at 90 and 300°K [after R. H. Bube and L. A. Barton, *RCA Rev.* **20**, 564 (1959)]. (b) Illustrative comparison of the shape of the photoconductivity excitation spectrum and the absorption edge of the material, defining the three characteristic regions. This correspondence can be seen for the actual data of (a) by comparison with Fig. 11.12, giving the corresponding absorption edge.

surface. For wavelengths near the absorption edge, appreciable penetration of photoexcitation into the volume of the crystal occurs. The maximum photoconductivity usually occurs for that wavelength corresponding to an absorption constant approximately equal to the reciprocal of the thickness of the crystal, i.e., when $(1 - 1/e)$ of the incident radiation is absorbed in the crystal. For wavelengths much longer than the absorption edge, the incident radiation is only weakly absorbed and the resulting photoconductivity decreases rapidly. The shape of the spectral response curve in the neighborhood of the absorption edge depends critically on the relative magnitudes of surface and volume recombination lifetime.

One way to calculate the shape of the spectral response curve is to consider the competition between excitation at a certain depth from the surface

12.1 General Concepts

on which the radiation is incident, with diffusion to the surface followed by surface recombination, and with volume recombination.† The geometry is illustrated in Fig. 12.3. It is assumed that the radiation is incident on the crystal at right angles to the direction of the applied electric field. The rate of change of the density of excited carriers at a distance x from the surface is

$$\frac{dn(x)}{dt} = A e^{-\alpha x} - \frac{di}{dx} - \frac{n}{\tau} \tag{12.22}$$

The first term on the right represents the generation rate. If L is the total light absorbed,

$$L = \int_0^\infty A e^{-\alpha x} = \frac{A}{\alpha} \tag{12.23}$$

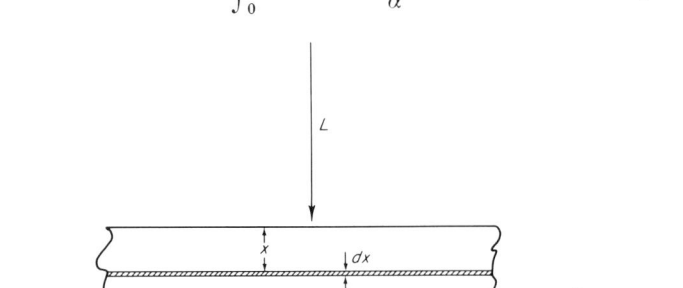

Fig. 12.3 Geometry for the calculation of the photoconductivity excitation spectrum when the excitation is incident normal to the electric-field direction.

since the region of α of interest is that for which most of the light will be absorbed in the crystal. The second term is the diffusion term with

$$i = n \frac{dx}{dt} = -D \frac{dn}{dx} \tag{12.24}$$

where D is the diffusion constant for the minority carriers. The steady-state form of Eq. (12.22) then becomes

$$\frac{d^2 n(x)}{dt^2} = \frac{n(x)}{D\tau} - \frac{L\alpha}{D} e^{-\alpha x} \tag{12.25}$$

The τ in Eqs. (12.22) and (12.25) is the pair or minority-carrier lifetime, it being assumed in this model that electrons and holes have equal lifetimes.

† H. DeVore, *Phys. Rev.* **102**, 86 (1956).

The general solution of Eq. (12.25) is

$$n(x) = C_1 e^{x/(D\tau)^{1/2}} + C_2 e^{-x/(D\tau)^{1/2}} + \frac{L\alpha\tau}{1 - \alpha^2 D\tau} e^{-\alpha x} \qquad (12.26)$$

This general solution is subject to the following two boundary conditions:

$$i = n_0 s = D \left.\frac{dn}{dx}\right]_{x=0}; \quad i = n_l s = D \left.\frac{dn}{dx}\right]_{x=l} \qquad (12.27)$$

where s is the *surface recombination velocity*. The recombination rate at the surface is approximately proportional to the concentration of excess carriers; the proportionality factor, with dimensions of velocity, is called the surface recombination velocity s.

The complete solution is obtained by applying the boundary conditions of Eq. (12.27) to the particular solution of Eq. (12.26) and then integrating:

$$n = \int_0^l n(x)\, dx \qquad (12.28)$$

The result of this calculation can be written in the following form, using dimensionless parameters to emphasize the physical processes acting.

$$\frac{n}{L\tau} = \frac{1 - e^{-Z}}{1 + R \coth(W/2)} \left\{ 1 + \frac{RW[W \coth(W/2) - Z \coth(Z/2)]}{W^2 - Z^2} \right\} \qquad (12.29)$$

Here W is the thickness parameter, given by

$$W = \frac{l}{(D\tau)^{1/2}} \qquad (12.30)$$

It represents the ratio of the thickness of the crystal to one *diffusion length*, i.e., the distance covered by diffusion in a lifetime. The parameter R is the recombination parameter,

$$R = \frac{s\tau}{(D\tau)^{1/2}} \qquad (12.31)$$

being proportional to the lifetime for volume recombination divided by the lifetime for surface recombination. The parameter Z is the absorption parameter,

$$Z = \alpha l \qquad (12.32)$$

There are certain limiting cases that illustrate the applicability of this solution and that are observable upon inspection of Eq. (12.29). If the

12.1 General Concepts

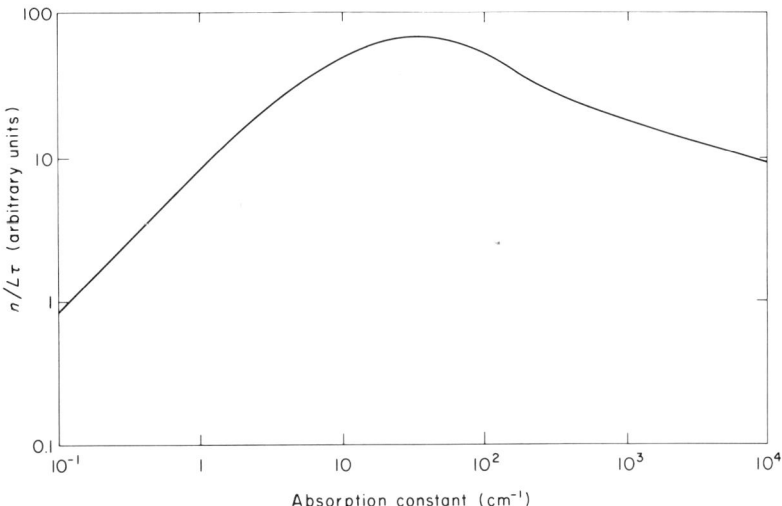

Fig. 12.4 Photoconductivity excitation spectrum calculated from Eq. (12.29) with $l = 1$ mm, $s = 10^4$ cm/sec, $\tau = 10^{-5}$ sec, and $(D\tau)^{1/2} = 10^{-2}$ cm.

absorption is very weak, Z and n both go toward zero; if the absorption is very strong, Z becomes very large, and n becomes a constant. If the material is very thin, W goes to zero and there is no maximum in the variation of n with wavelength. If the surface recombination probability is very small, R goes to zero, and there is no maximum in the n vs. λ curve. But, if R is large, corresponding to a high surface-recombination probability, then a maximum is found. A representative plot of Eq. (12.29) is given for illustration in Fig. 12.4.

Example 12.1 A crystal with electrodes 5 mm apart and with cross section of 2×1 mm^2, exhibits a photocurrent of 100 μA with an applied voltage of 10 V, for excitation by 2.0-eV photons with an intensity of 10^{15} photons cm^{-2} sec^{-1}. If the absorption constant of the crystal for 2.0-eV photons is 1 cm^{-1}, and the electron and hole mobilities are respectively 1000 and 200 cm^2/V-sec, calculate the electron and hole lifetimes, and the photoconductivity gain if electrons and holes are present in equal densities contributing to the measured photocurrent.

First we need to extend Eqs. (12.9) and (12.15) to the case of two-carrier conductivity. The equivalent equation to Eq. (12.9) is

$$\Delta\sigma = e(\mu_{e0}\,\Delta n + \mu_{h0}\,\Delta p) + e(n\,\Delta\mu_e + p\,\Delta\mu_h)$$

and the equivalent equation to Eq. (12.15) is

$$G = \frac{(\tau_n\mu_e + \tau_p\mu_h)V}{l^2}$$

if we neglect $\Delta\mu$ terms in calculating G, as we did before in the text.
The photoexcitation intensity f is given by $\alpha L = 10^{15}$ cm^{-3} sec^{-1}. The conductivity is given by

$$\Delta\sigma = \frac{100 \times 10^{-6}}{10} \frac{5 \times 10^{-1}}{2 \times 10^{-2}}$$
$$= 2.5 \times 10^{-4} \text{ } (\Omega\text{-cm})^{-1}$$

We obtain $n = p = 1.3 \times 10^{12}$ cm^{-3}, and $\tau_n = \tau_p = 1.3 \times 10^{-3}$ sec. The gain is given either by Eq. (12.12),

$$G = \frac{100 \times 10^{-6}}{1.6 \times 10^{-19} \times 10^{15} \times 10^{-2}} = 62$$

or by the two-carrier equivalent of Eq. (12.15),

$$G = \frac{(1.3 \times 10^{-3})(10^3 + 2 \times 10^2)10}{(5 \times 10^{-1})^2} = 62$$

A gain of 62 means that 62 carriers pass between the electrodes of the crystal for every photon that creates an electron–hole pair. The only way this is physically possible is for the negative electrode (the cathode) to replenish an electron for each electron that leaves the crystal under the applied electric field at the positive electrode (the anode). This crossing of the crystal and being replenished continues until the free carrier is removed by recombination.

12.2 Electrical Contacts

Most measurements of electrical conductivity require that electrical contacts be made to the material under investigation, usually with metal electrodes. There are alternatives to this procedure, as for example to measure conductivity in terms of the optical absorption of electromagnetic radiation by free carriers in an experimental arrangement with a microwave cavity, or to measure the charging and discharging of layers or films of the material. These alternative methods are helpful in certain cases, but it remains of basic importance to control and understand the nature of metallic contacts to semiconductors and insulators.

12.2 Electrical Contacts

RECTIFYING, BLOCKING, NONOHMIC CONTACTS

If a junction is made between a metal and an n-type semiconductor for which the work function of the metal is greater than that of the semiconductor, transfer of electrons from semiconductor to metal occurs, as illustrated in Fig. 12.5. The result is a *depletion layer* in the semiconductor from which all free electrons have been removed. In order to estimate the potential distribution in the depletion layer, assume that the total charge in this region comes from ionized donors with density N^+. Then

$$\nabla \cdot \mathscr{E} = \frac{4\pi N^+ e}{\varepsilon} \tag{12.33}$$

Since $\mathscr{E} = -\nabla V$,

$$\frac{\partial^2 V}{\partial x^2} = -\frac{4\pi N^+ e}{\varepsilon} \tag{12.34}$$

with the solution

$$V = -\frac{2\pi N^+ e}{\varepsilon} x^2 \tag{12.35}$$

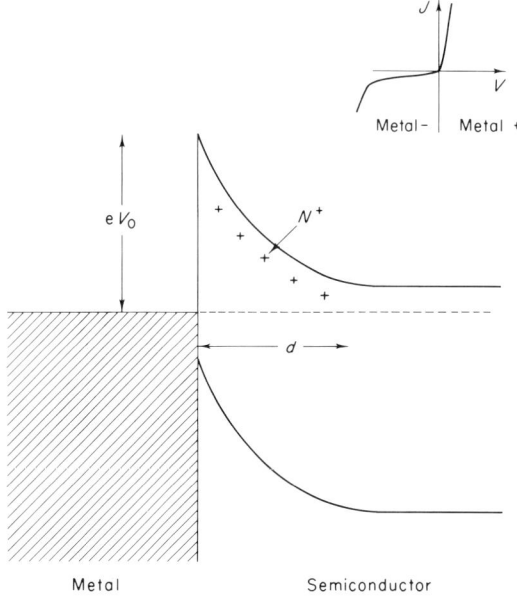

Fig. 12.5 Potential barrier at the junction of a metal and a semiconductor when a nonohmic contact is formed. The width of the depletion layer in the semiconductor is given by d. Upper insert shows the variation of current with voltage in the forward and reverse directions.

If the *barrier height* is eV_0, then the thickness of the depletion layer d is given by

$$d = \left(\frac{\varepsilon V_0}{2\pi e N^+} \right)^{1/2} \tag{12.36}$$

For a typical case of $N^+ = 10^{17}$ cm^{-3}, $\varepsilon = 10$, and $V_0 = 0.5$ V, the depletion layer thickness $d = 7.5 \times 10^{-6}$ cm.

The current–voltage characteristic of such a contact is shown in Fig. 12.5 also. When the metal contact is positive with respect to the semiconductor, the barrier for electron flow from semiconductor to metal is reduced, and large currents flow. When the metal contact is negative (the cathode), however, only very small currents can be drawn until very high voltage is applied and the barrier breaks down from some process such as tunneling from the metal into the semiconductor.

Since the capacitance associated with such a barrier and depletion layer is given approximately by the relationship $C = \varepsilon A/(4\pi d)$, with no applied bias voltage,

$$\frac{1}{C^2} = \frac{8\pi V_0}{\varepsilon A^2 e N^+} \tag{12.37}$$

or with a positive bias of V volts,

$$\frac{1}{C^2} = \frac{8\pi (V_0 - V)}{\varepsilon A^2 e N^+} \tag{12.38}$$

Thus if $1/C^2$ is plotted as a function of applied bias, a straight line is obtained with slope related to N^+ and intercept equal to V_0. Figure 12.6 shows typical plots for Au on CdS, and degenerate p-type Cu$_2$S on CdS, both of which cause depletion layers in the CdS.

If there are acceptor levels in the material, $N^+ = N_D - N_A$ and photoexcitation may result in the capture of photoexcited holes by these acceptors, and hence an increase in positive charge in the depletion layer. The existence of these acceptors is therefore detectable by a change in the slope of the $1/C^2$ vs. V curve after photoexcitation, if the measurements are done at a low enough temperature so that the captured holes are not thermally freed. From the temperature dependence of this "excess" capacitance, the ionization energy of the acceptors can be determined.

If a material is contacted with nonohmic electrodes and then used as a photoconductor, it follows that the gain can never exceed unity. The gain could well be less than unity, if photoexcited carriers were trapped or captured before crossing between electrodes, but since replenishment of

12.2 Electrical Contacts

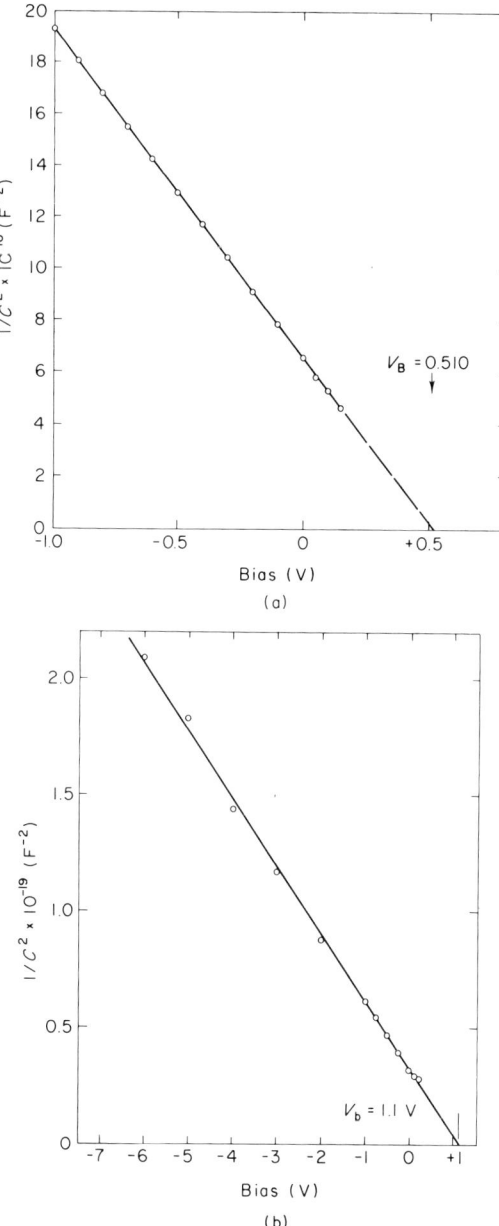

Fig. 12.6 (a) Variation of $1/C^2$ with applied bias voltage for an evaporated gold contact to CdS. [After A. M. Goodman, *J. Appl. Phys.* **35**, 573 (1964).] (b) Variation of $1/C^2$ with applied bias voltage for a layer of p-type Cu_2S on n-type CdS, producing a depletion layer in the CdS. (After W. D. Gill, Ph.D. Thesis, Stanford University, 1969.)

carriers at the electrodes is not possible, gains in excess of unity are not possible. The longest lifetime any carrier can have is its transit time.

NONRECTIFYING, NONBLOCKING, OHMIC CONTACTS

A contact that is able to replenish carriers to the semiconductor in order to maintain charge neutrality when one is drawn out at the opposite electrode by an electric field is called an ohmic contact. If a junction is made between a metal and an n-type semiconductor for which the work function of the metal is smaller than that of the semiconductor, transfer of electrons from metal to semiconductor will occur, as illustrated in Fig. 12.7. The result is an *accumulation layer* of excess charge in the semiconductor. The current–voltage characteristic of such a contact is also shown in Fig. 12.7. The behavior is ohmic over a range of voltages until a sufficiently high voltage is reached that the injected charge becomes comparable to the charge already present in the semiconductor itself. After that a space-charge-limited current flows, as described in the next section.

In order to calculate the form of the potential variation with distance in this case, recognize that the steady-state condition involves a balancing out of drift and diffusion currents to produce a net zero current. The

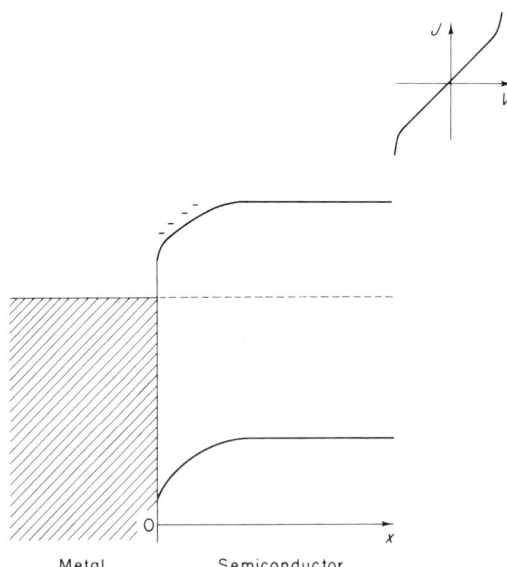

Fig. 12.7 Ohmic contact between a metal and a semiconductor, showing the accumulation layer of electrons in the semiconductor.

12.2 Electrical Contacts

governing equation is

$$[n(x)\,e\mu]\mathscr{E} - eD\frac{dn(x)}{dx} = 0 \tag{12.39}$$

The first term represents the drift current with electrons flowing toward the metal under the electric field set up at the interface, and the second term represents the diffusion current with electrons flowing into the semiconductor because of the concentration gradient. The solution of Eq. (12.39) is

$$\ln\left[\frac{n(x)}{n_0}\right] = \frac{\mu}{D}\int_0^x \mathscr{E}\,dx \tag{12.40}$$

where n_0 is the density of free electrons at the surface, $x = 0$. The potential energy of an electron $V(x) = -e\int_0^x \mathscr{E}\,dx$, so that

$$\ln\left[\frac{n(x)}{n_0}\right] = \frac{e}{kT}\int_0^x \mathscr{E}\,dx = -\frac{V(x)}{kT} \tag{12.41}$$

or

$$n(x) = n_0\,e^{-V(x)/kT} \tag{12.42}$$

We also have from Eq. (12.34),

$$\frac{\partial^2 V}{\partial x^2} = -\frac{4\pi e^2\,n(x)}{\varepsilon} = -\frac{4\pi n_0 e^2}{\varepsilon}e^{-V(x)/kT} \tag{12.43}$$

where the $V(x)$ is still the potential energy. The solution of Eq. (12.43) is

$$V(x) = 2kT\ln\left(\frac{x}{x_0} + 1\right) \tag{12.44}$$

where

$$x_0 = \left(\frac{\varepsilon kT}{2\pi n_0 e^2}\right)^{1/2} \tag{12.45}$$

so that

$$n(x) = n_0\left(\frac{x_0}{x + x_0}\right)^2 \tag{12.46}$$

In the presence of an applied electric field, the potential distribution takes the form shown in Fig. 12.8, with the accumulation layer in the semiconductor acting like a "virtual cathode" to supply charge to the semiconductor as needed to maintain charge neutrality. The behavior is analogous to that in a vacuum tube where the virtual cathode in the solid is related to the cathode of the vacuum tube, the transport of charge through

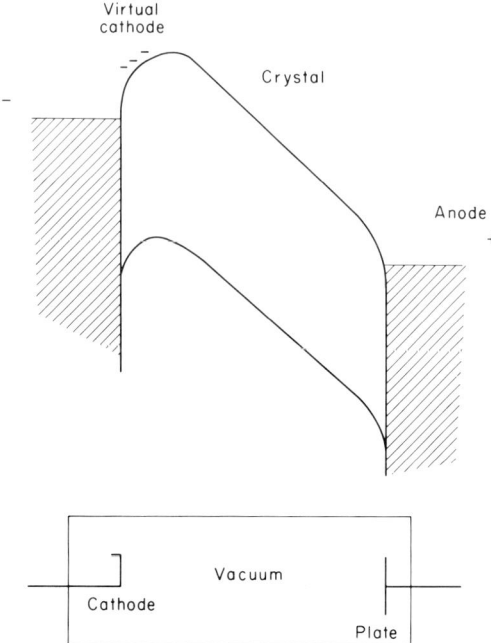

Fig. 12.8 Form of the energy bands when a potential difference is applied across a semiconductor with ohmic contacts, compared with the similar features of a vacuum-tube diode.

the solid is related to the transport of charge through the vacuum, and the collection of charge at the metal anode is related to the collection of charge at the vacuum-tube plate.

All of the above discussion has been in terms of contacts to an n-type semiconductor. Similar considerations can be applied to a p-type semiconductor. Nonohmic contacts are formed if the work function of the metal is less than that of the semiconductor, and ohmic contacts are formed if the work function of the metal is greater than that of the semiconductor. In many practical cases, however, the effect of surface states is so great that the simple comparison of work functions does not accurately predict the nature of the contact. The formation of reliable contacts in practice still requires as much art as science.

SINGLE INJECTION

When charge moves into a crystal from an ohmic contact under the effect of an applied electric field to produce a condition of non-charge-neutrality

12.2 Electrical Contacts

in the crystal, we have a case of space-charge-limited current flow in the crystal. If carriers of only one type are injected, the phenomenon is known as single injection.

The basic properties of such space-charge-limited injected current can be determined from a simple calculation. The charge injected into the crystal is given approximately by the product of its geometrical capacitance and the applied voltage. The time it takes this charge to cross the crystal is just the transit time of carriers, so that the space-charge-limited current density is given by the charge injected divided by the transit time. Such a calculation is summarized in Table 12.1 for both a semiconductor crystal and for a vacuum diode for comparison. Since the average velocity of carriers in the solid is much less than the average velocity of carriers in vacuum, the space-charge-limited current density is appreciably smaller in solids than in vacuum tubes. Nevertheless, in a trap-free crystal in which all of the injected charge is free, the space-charge-limited current has an appreciable magnitude. In most real crystals the major portion of the injected charge is held in traps; when the injected charge is sufficiently large to fill these traps, the current rises abruptly to the trap-free value. The different regimes are illustrated in Fig. 12.9.

TABLE 12.1 CALCULATION OF SPACE-CHARGE-LIMITED CURRENT DENSITY IN A CRYSTAL WITH OHMIC CONTACTS FOR THE CASE OF SINGLE INJECTION, COMPARED TO THE SIMILAR CURRENT DENSITY IN A VACUUM DIODE

	Crystal	Vacuum diode
Charge $Q \approx CV$ per unit area	$\dfrac{\varepsilon V}{4\pi L}$	$\dfrac{V}{4\pi L}$
Transit time t_n	$\dfrac{L^2}{\mu V}$	$\approx \dfrac{L}{(2eV/m)^{1/2}/2}$
Space-charge-limited current density $= Q/t_n$	$\dfrac{\varepsilon V^2 \mu}{4\pi L^3}$	$\dfrac{(2e/m)^{1/2}}{8\pi} \dfrac{V^{3/2}}{L^2}$
Value of current density for	1 mA/cm²	250 mA/cm²
$V = 100$ V		
$\mu = 100$ cm²/V-sec		
$L = 1$ mm		
$\varepsilon = 10$		

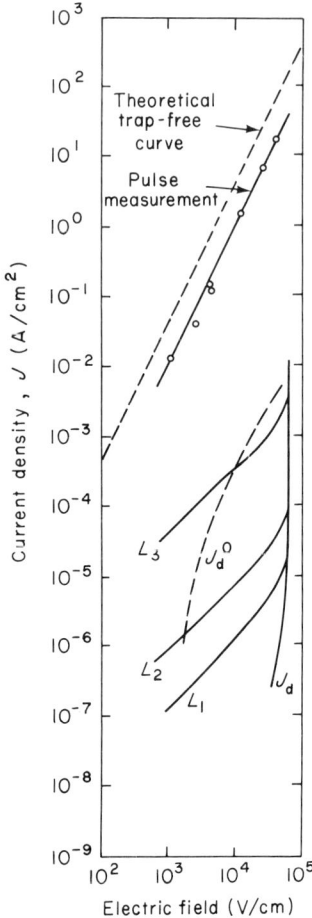

Fig. 12.9 Variation of current as a function of applied voltage with ohmic contacts in a crystal of CdS. The upper two curves show respectively the theoretical and the measured space-charge-limited current corresponding to injection of electrons without trapping. Curve J_d^0 is for the current measured for the first time after previous removal of all trapped space charge; upon trapping this current decreases to that shown in curve J_d. The abrupt rise in J_d toward the trap-free limit may correspond to that voltage for which all the traps are filled by injected electrons. Curves L_1, L_2, and L_3 are for three increasing intensities of photoexcitation, showing an ohmic range for the photoexcited current at low fields. [After R. W. Smith and A. Rose, *Phys. Rev.* **97**, 1531 (1955).]

The criterion for the importance of injected charge, as compared to the thermal-equilibrium free carriers in the crystal, can be derived in the following way. If an excess charge is placed in a crystal, the crystal redistributes its own free charge to compensate for this excess charge with a time con-

12.2 Electrical Contacts

stant given by the dielectric relaxation time (see τ_r in Table 1.6). If, however, this compensation process is sufficiently slow that the injected charge has the opportunity to pass through the crystal before compensation can take place, the injected carriers dominate the conductivity of the material. The criterion for the significance of injected current, therefore, is that the dielectric relaxation time of the material be equal to the transit time of the injected carriers, i.e.,

$$\frac{L^2}{\mu V} = \frac{\varepsilon}{4\pi\sigma} \qquad (12.47)$$

or when

$$V = \frac{4\pi\sigma L^2}{\varepsilon\mu} \qquad (12.48)$$

For typical values of $\sigma = 10^{-6}$ $(\Omega\text{-cm})^{-1}$, $L = 1$ mm, $\varepsilon = 10$, and $\mu = 10^3$ cm^3/V-sec, the critical value of V is 10 V. This calculation is also, of course, for the case of a trap-free crystal.

Double Injection

If ohmic contacts for both types of carriers are present on a crystal, application of a voltage leads to injection of both electrons from the cathode and of holes from the anode. Where the injected electron and hole densities overlap, neutralization of the injected charge becomes possible, and hence elimination of the space-charge limitation on current flow. Unneutralized electron charge is concentrated near the cathode, and unneutralized hole charge near the anode. The fraction of the charge that is uncompensated depends on the relative magnitude of the lifetime of the injected charge compared to the transit time of the injected charge, i.e., how far can it go before recombining with a carrier of opposite type?

The physical nature of double injection in a trap-free material can be illustrated by an extension of the phenomenological approach used for single injection. If Q^* represents the total uncompensated charge injected into the crystal, and if the density of injected electrons is equal to the density of injected holes,

$$Q^* = CV = Q_n^* + Q_p^*$$
$$= en\left(\frac{t_n}{\tau} + \frac{t_p}{\tau}\right) \qquad (12.49)$$

which gives

$$t_n + t_p = \frac{\varepsilon\tau V}{4\pi neL} \qquad (12.50)$$

for a unit area. The current density is given in terms of the total injected charge,

$$j = \frac{Q_n}{t_n} + \frac{Q_p}{t_p} \qquad (12.51)$$

$$= en\left(\frac{t_n + t_p}{t_n t_p}\right) \qquad (12.52)$$

Combining Eqs. (12.50) and (12.52), and substituting explicitly for t_n and t_p, gives

$$j = \frac{\mu_n \mu_p \tau \varepsilon V^3}{4\pi L^5} \qquad (12.53)$$

We expect this double injection current to be considerably larger than the single injection current of the previous section. Their ratio is

$$\frac{j_{di}}{j_{si}} \approx \frac{\mu \tau V}{L^2} \approx \frac{\tau}{t} \qquad (12.54)$$

If electron and hole mobilities are comparable, for a lifetime of 100 μsec,

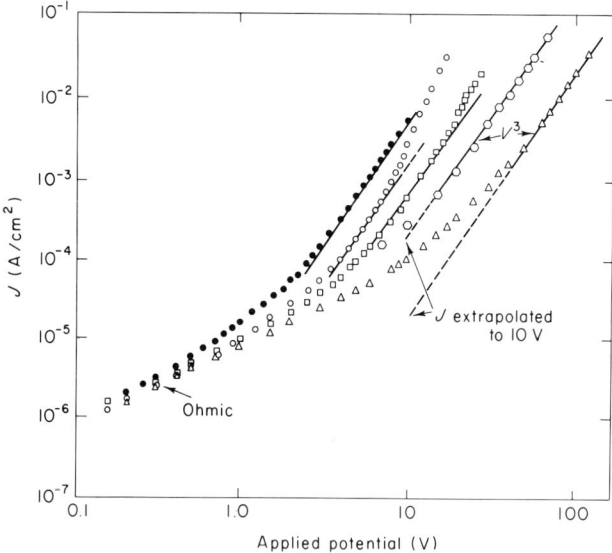

Fig. 12.10 Double-injection current density as a function of applied voltage for five p–i–n silicon junctions with different dimensions, in which electrons and holes are injected into the center high-resistivity intrinsic region, showing the predicted variation as V^3. [After J. W. Mayer, R. Baron, and O. J. Marsh, *Phys. Rev.* **137**, A286 (1965).]

$j_{di} = 10^2 j_{si}$, using the data of Table 12.1 for the calculation. Verification of the V^3 law for double injection into an insulator is shown in Fig. 12.10 for data obtained with a p–i–n silicon junction device.

12.3 Analytical Approaches

The treatment of photoelectronic effects in real crystals from a theoretical viewpoint offers two complementary approaches. Either as rigorous a mathematical treatment of all possible situations as can be managed on the basis of "first principles" is chosen, or a phenomenological viewpoint is adopted which allows the treatment of a wide variety of observed phenomena on a semiquantitative basis, relying upon experimental confirmation for vindication of the approach. In the former case, relatively simple models must be assumed in order to carry out the mathematical manipulations, with results that are likely to be fairly rigorous solutions of cases with little physical significance. In the latter case, the comfort of mathematical rigor is relinquished in order to construct at least a framework within which to view the variety of phenomena that can be observed.

GENERAL TREATMENT

Consider a model of a semiconductor involving valence and conduction bands, and imperfection levels in the forbidden gap. In order to describe the problem completely, we need the following equations.

First the continuity equations for electrons and holes. For electrons, for example,

$$\frac{\partial n}{\partial t} = f - \sum \begin{pmatrix} \text{capture from} \\ \text{conduction band} \end{pmatrix} + \sum \begin{pmatrix} \text{thermal excitation} \\ \text{to conduction band} \end{pmatrix} + \frac{\nabla \cdot \mathbf{j}_n}{e} \tag{12.55}$$

We have simply lumped all capture and thermal excitation terms together as the conduction band interacts with the variety of imperfection levels.

Second the definition of electron and hole current. For electrons, for example,

$$\mathbf{j}_n = ne\mu_n \mathscr{E} + eD_n \nabla n \tag{12.56}$$

Third, Poisson's equation,

$$\nabla \cdot \mathscr{E} = \frac{4\pi q}{\varepsilon} \tag{12.57}$$

Combining Eqs. (12.55), (12.56), and (12.57) gives for electrons

$$\frac{\partial n}{\partial t} = f - \sum (\text{capture}) + \sum (\text{thermal excitation}) + D_n \nabla^2 n$$
$$+ \mu_n \mathscr{E} \cdot \nabla n + \frac{4\pi \mu_n n q}{\varepsilon} \quad (12.58)$$

The complete solution of the problem requires the solution of Eq. (12.58) and its corresponding equation for holes, under the constraints of the appropriate boundary conditions.

The following simplifying assumptions are frequently made, one or more at a time in order to make the solution more directly obtainable.

1. $q = 0$. Neglect space charge; assume $\Delta n = \Delta p$.
2. $\nabla p = \nabla n = 0$. Neglect diffusion in the bulk of the crystal, although it is evident that strictly speaking it cannot be neglected at the contacts.

If both of these simplifying assumptions are made, and if the capture–thermal-excitation processes that just balance one another (effective thermal equilibrium) can be separated from those that do not (recombination paths), we arrive in steady state at the simple expression

$$0 = f - \frac{\Delta n}{\tau}$$

which is just Eq. (12.4). The determination of τ is still a difficult job, since it is determined by the total behavior of the imperfections. Other methods are helpful in dealing with this complex situation.

QUASI- OR STEADY-STATE FERMI LEVEL

Recall the basic expressions relating n and p to the Fermi level in a semiconductor,

$$n = N_c \, e^{-E_{fn}/kT} \quad (12.59)$$

$$p = N_v \, e^{-E_{fp}/kT} \quad (12.60)$$

where $E_{fn} = E_c - E_F$, and $E_{fp} = E_F - E_v$. The product of these two expressions gives

$$E_{fn} + E_{fp} = kT \ln\left(\frac{N_c N_v}{np}\right) \quad (12.61)$$

Since $E_G = kT \ln(N_c N_v / n_i^2)$,

$$E_{fn} + E_{fp} = E_G - kT \ln\left(\frac{np}{n_i^2}\right) \quad (12.62)$$

12.3 Analytical Approaches

In thermal equilibrium, $np = n_i^2$, and so of course $E_{fn} + E_{fp} = E_G$, as expected. Now suppose that we are dealing, not with thermal equilibrium, but with a steady-state situation in the presence of photoexcitation. Define steady-state Fermi levels using the same relations of Eqs. (12.59) and (12.60), so that the steady-state Fermi levels describe the density of free electrons and holes. They may also be expected to describe the occupation of all imperfection levels that are still essentially in thermal equilibrium with one or the other band, even in the presence of photoexcitation. Thus the occupation of levels in effective thermal equilibrium with the conduction band are given by E_{fn}^*, the occupation of levels in effective thermal equilibrium with the valence band are given by E_{fp}^*, and the occupation of levels not in effective thermal equilibrium are determined by recombination kinetics and are not given by E_{fn}^* or E_{fp}^*.

For the steady-state Fermi level situation $np \neq n_i^2$, and Eq. (12.62) shows that the two steady-state Fermi levels are separated by an energy equal to $kT \ln(np/n_i^2)$. The situation is illustrated in Fig. 12.11. The difference between E_{fn}^* and E_{fp}^* is given by

$$E_{fn}^* - E_{fp}^* = kT \ln\left[\left(\frac{m_e^*}{m_h^*}\right)^{3/2} \frac{p}{n}\right] \tag{12.63}$$

which is valid for thermal equilibrium as well as for steady-state conditions.

The relation between the steady-state and the thermal-equilibrium Fermi level is easily seen. For electrons, for example,

$$E_{fn}^* - E_{fn} = kT \ln\left(\frac{n_0}{n_0 + \Delta n}\right) \tag{12.64}$$

In an n-type semiconductor with $\Delta n \ll n_0$, photoconductivity phenomena can be described with $E_{fn}^* = E_{fn}$. In the same material, however,

$$E_{fp}^* - E_{fp} = kT \ln\left(\frac{p_0}{p_0 + \Delta p}\right) \tag{12.65}$$

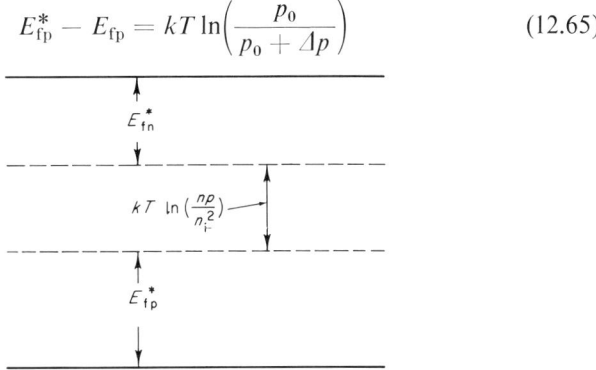

Fig. 12.11 Steady-state electron and hole Fermi levels.

and the difference between $E_{\rm fp}^*$ and $E_{\rm fp}$ may not be negligible at all. In an insulator both $E_{\rm fn}^*$ and $E_{\rm fp}^*$ are expected to differ appreciably from $E_{\rm fn}$ and $E_{\rm fp}$.

DEMARCATION LEVEL

The occupancy of levels in effective thermal equilibrium with one of the bands is given in terms of the corresponding steady-state Fermi level. The *demarcation level* defines the boundary between occupancy determined by effective thermal equilibrium with the nearest band, and occupancy determined by recombination kinetics. The location of the demarcation level depends on the specific values of capture coefficients, $C_{\rm n}$ and $C_{\rm p}$, for a particular type of imperfection; there is thus one set of demarcation levels (electron and hole) for each kind of imperfection, as defined by a particular set of capture coefficients.

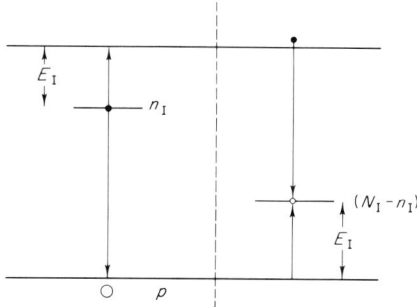

Fig. 12.12 Competing rates in the calculation of the relationship of steady-state Fermi levels and demarcation levels.

All of the Fermi levels and demarcation levels can be correlated. For example consider the situation in Fig. 12.12 in which we equate the probability of thermal excitation of electrons out of electron-occupied levels to the conduction band, to the probability of capture of free holes by these electron-occupied levels. If $n_{\rm I}$ is the density of such electron-occupied levels lying $E_{\rm I}$ below the conduction band,

$$n_{\rm I} N_{\rm c} C_{\rm n}\, e^{-E_{\rm dn}/kT} = n_{\rm I} p C_{\rm p} \qquad (12.66)$$

since for the equality $E_{\rm I} = E_{\rm dn}$; $E_{\rm dn}$ is defined consistently as the positive energy difference between the conduction-band edge and the electron demarcation level for the center with capture coefficients $C_{\rm n}$ and $C_{\rm p}$. Setting

12.3 Analytical Approaches

$N_c = n\, e^{E^*_{fn}/kT}$ in Eq. (12.66) gives

$$E_{dn} = E^*_{fn} - kT \ln\left(\frac{C_p}{C_n}\, \frac{p}{n}\right) \quad (12.67)$$

Alternatively, setting $p = N_v\, e^{-E^*_{fp}/kT}$ in Eq. (12.66) gives

$$E_{dn} = E^*_{fp} + kT \ln\left[\frac{C_n}{C_p}\left(\frac{m_e^*}{m_h^*}\right)^{3/2}\right] \quad (12.68)$$

Similarly by considering the competition between thermal excitation of holes to the valence band and electron capture by these holes at imperfections, as also shown in Fig. 12.12, gives

$$E_{dp} = E^*_{fp} - kT \ln\left(\frac{C_n}{C_p}\, \frac{n}{p}\right) \quad (12.69)$$

$$E_{dp} = E^*_{fn} + kT \ln\left[\frac{C_p}{C_n}\left(\frac{m_h^*}{m_e^*}\right)^{3/2}\right] \quad (12.70)$$

Typical sets of Fermi levels and demarcation levels in insulators and semiconductors are given in Fig. 12.13. It is always true that

$$E_{dn} + E_{dp} = E^*_{fn} + E^*_{fp} \quad (12.71)$$

Example 12.2 In a particular high-resistivity crystal it is known that a long photoexcited electron lifetime is associated with photoexcited holes being captured at imperfections with very small capture coefficient C_n. In order to be effective in this way, the occupancy of these imperfections must be determined by recombination kinetics and not

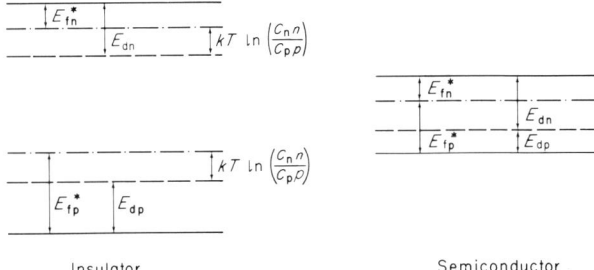

Fig. 12.13 Typical locations of steady-state Fermi levels and the demarcation levels for a particular imperfection in an insulator and in a semiconductor, i.e., respectively under conditions where $\Delta n, \Delta p \gg n_0, p_0$, and where $\Delta n, \Delta p \ll n_0, p_0$.

by thermal exchange with the valence band, i.e., it must be more probable for a captured hole to recombine with an electron than to be thermally reexcited to the valence band. What is the closest that such a level can lie to the valence band and still have the long electron lifetime present at room temperature under the following conditions? Density of the level $= 10^{16}$ cm^{-3}, excitation rate of 10^{14} cm^{-3} sec^{-1}, $C_n = 10^{-11}$ cm^3/sec, $C_p = 10^{-9}$ cm^3/sec, $m_e^* = 0.2m$, and $m_h^* = 0.5m$.

The closest that such a level can lie to the valence band is the energy separation E_{dp} corresponding to E_{fn}^* under the given conditions, as given in Eq. (12.70). If the level lies closer than E_{dp} above the valence band, its occupancy is determined by thermal interaction with the valence band and its ability to increase the electron lifetime is lost.

In order to calculate E_{fn}^*, we need to know the value of n under photoexcitation. If the lifetime is determined by recombination with the given imperfections,

$$\tau_n = \frac{1}{C_n(N_I - n_I)}$$

if the total density of imperfections is N_I and the density of empty (hole-occupied) imperfections is $(N_I - n_I)$. If essentially all of the free electrons have come from these imperfections, $(N_I - n_I) \approx n$, and $n = (f/C_n)^{1/2} = 10^{15}$ cm^{-3}. The steady-state Fermi level corresponding to this value of n is $E_{fn}^* = 0.20$ eV. Then we have

$$E_{dp} = 0.20 + kT \ln(10^5 \times 4) \text{ eV}$$
$$= 0.52 \text{ eV}$$

Therefore imperfections of the given type that result in a long electron lifetime (100 msec in the case calculated) at room temperature must lie something more than 0.5 eV above the valence band.

Note that if we write

$$E_{dp} = AkT$$

the value of A we obtain is about 21. This gives a rough order of magnitude in relating a level with a given ionization energy and the temperature at which thermal emptying or filling becomes large, and is approximately valid in many circumstances; e.g., at what temperature does a level lying 0.2 eV below the conduction band begin to empty appreciably thermally? Answer: about 110°K.

12.4 Models of Photoconductivity

The growth of interest in the mechanism of photoconductivity paralleled the growth of semiconductor research with its particular emphasis on minority-carrier and junction effects. Early models for photoconductivity therefore tended to propose some kind of repeated junction in the material as responsible for the photoconductivity behavior. Later it was shown that homogeneous materials can also show photoconductivity effects quite similar to those of junctions. In our discussion here we will start with the junction models and then move toward the homogeneous material models.

p–n JUNCTION

When an n-type form of a given material is produced (as, for example, by diffusion of a suitable impurity) in conjunction with a p-type form of the same material, electrons flow from the n-type to the p-type side and holes flow from the p-type to the n-type side until the barrier that is formed just counterbalances this flow. The result is the p–n junction discussed in Fig. 9.19. When such a junction is biased in the reverse direction (n-type side positive), there is a high resistance in the dark, since the p-type region cannot supply free electrons to the n-type region, and the n-type region cannot supply free holes to the p-type region. For photoexcitation within a diffusion length of the junction, an electron–hole pair is formed that traverses the junction. Once the excited carriers have crossed the junction, the photoconductivity process related to the absorption of the corresponding photon is at an end; the maximum gain of a p–n junction is therefore unity.

From another perspective the p–n junction can be viewed as a relatively high resistance region (the depletion-layer region) fitted with blocking contacts, the p-type region being a blocking contact for electrons and the n-type region being a blocking contact for holes. It might be expected, therefore, that some of the behavior of a p–n junction could be simulated by a high-resistivity homogeneous material fitted with metallic blocking contacts. This is indeed true, as is pointed out further below. Another variation of the p–n junction makes the similarity even closer; in the p–i–n junction, a high-resistivity intrinsic region is introduced between the p-type and n-type regions. The advantage is that the collection of photoexcited carriers can now be controlled by drift due to an applied electric field, rather than by the slower process of diffusion across the junction as in the simple p–n junction case.

n–p–n JUNCTION

Typical n–p–n and p–n–p junctions were also shown in Fig. 9.19. For the n–p–n junction, the forward-biased p–n_1 junction now plays the role of the electrode that replenishes electrons, and it is possible to have gains greater than unity. From the definition of the gain, it follows that

$$G = \frac{\text{time required for holes to diffuse out of the p-type region}}{\text{time required for electrons to diffuse across the p-type region}}$$

since the effective lifetime of an electron traversing the p-type region is the length of time that the extra positive charge remains in the p-type region.

The gain of such an n–p–n junction can be calculated as follows. Suppose that photoexcitation within a diffusion length of the p–n_2 junction causes an injection of Δp holes into the p-type region. The presence of these additional holes reduces the forward-biased p–n_1 junction by an energy ΔE. Because of this decrease in the barrier height, there is an increase of electrons entering the p-type region from the n_1 region given by

$$\Delta n = n_1 \, e^{\Delta E/kT} = \Delta p \tag{12.72}$$

Because of the decrease in the p–n_1 barrier, there is also an increased flow of holes from the p-type region to the n_1 region given by

$$\delta p = p \, e^{\Delta E/kT} \tag{12.73}$$

The diffusion time of holes can be calculated as follows. If W is the width of the p-type region, then $W \Delta p$ is the total charge per unit area. The current density for the flow of holes across the p–n_1 junction is

$$j_p = \delta p \, v = \delta p \, \frac{D_p}{L_{pn}} = \frac{\delta p \, \mu_p kT}{eL_{pn}} \tag{12.74}$$

where L_{pn} is the diffusion length of holes in n_1, i.e., $L_{pn} = (D_p \tau_p)^{1/2}$. The hole diffusion time is therefore given by

$$\frac{W \Delta p}{j_p} = \frac{WeL_{pn}}{\mu_p kT} \frac{\Delta p}{\delta p} = \frac{WeL_{pn}}{\mu_p kT} \frac{n_1}{p} \tag{12.75}$$

using Eqs. (12.72) and (12.73). The transit time for electrons by diffusion, assuming that $L_{np} > W$, is given by

$$\frac{W}{D_n/W} = \frac{eW^2}{\mu_n kT} \tag{12.76}$$

12.4 Models of Photoconductivity

Combining Eqs. (12.75) and (12.76) gives the expression for the gain,

$$G = \frac{L_{pn} n_1 \mu_n}{W p \mu_p} = \frac{L_{pn} \sigma_1}{W \sigma_p} \qquad (12.77)$$

The gain in actual n–p–n or p–n–p junctions may be as high as several hundred. This gain results, in summary, from the localization of an excess positive charge that causes an ohmic contact (the forward-biased n_1–p junction) to inject a negative charge to maintain charge neutrality. Homogeneous materials can also simulate this behavior through imperfection trapping of the photoexcited hole, as discussed further below.

Barriers

Potential barriers may exist in materials for a variety of reasons, of which p–n junctions and n–p–n junctions are just two important examples.

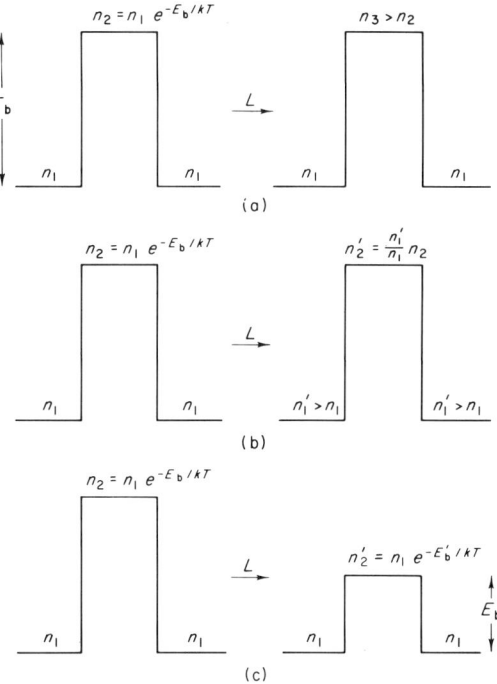

Fig. 12.14 Different ways of looking at barriers and photoconductivity-related phenomena. (a) Photoexcitation increases only the density of free carriers in the barrier region. (b) Photoexcitation increases the density of free carriers in the nonbarrier region, and hence increases the density of carriers in the barrier region as well. (c) Photoexcitation does not change the carrier densities, but decreases the barrier height by depositing suitable trapped charge in or near the barrier region.

In thin films or layers of photoconducting material, for example, there may be intergrain barriers in a polycrystalline layer. When such barriers exist, they of course affect the transport of charge through the material, and it may be possible for photoexcitation to decrease the height of the barriers through the production of a localized trapped space charge.

Figure 12.14 presents three different ways of looking at barriers in materials. Figure 12.14a shows the potential variation when two portions of the same material have a different resistivity. In this case, a common Fermi level runs throughout, and the density of carriers in the more resistive region is always related to the density of carriers in the less resistive region by an appropriate Boltzmann factor. The electrical properties of such a system can be treated on a simple resistance basis, without reference to barriers or barrier modulation. On the other hand, a choice could be made to describe the total conductivity in terms of the carrier density in the less resistive regions; then the conductivity in the more resistive regions becomes $n_1 e^{-E_b/kT} e\mu$ or $n_1 e\mu_b$, with $\mu_b = \mu e^{-E_b/kT}$. This expression has the appearance of a barrier picture, but the appearance is superficial; such a situation of the same material with two different resistivities is not essentially a barrier case.

Figure 12.14b illustrates the case of an actual barrier, where the principal effect of photoexcitation is to change the density of free carriers in the non-barrier regions and hence to increase the density of carriers passing over the barrier.

Figure 12.14c illustrates the case of an actual barrier, where the principal effect of photoexcitation is to change the barrier height and hence to increase the density of carriers passing over the barrier.

A typical barrier is shown in Fig. 12.15 limiting the flow of majority carriers through a material. If the barrier height is E_b, the diffusion current

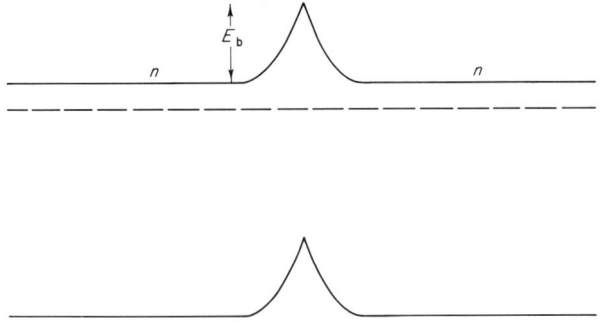

Fig. 12.15 A barrier of height E_b to majority-carrier electron flow.

12.4 Models of Photoconductivity

in one direction in the dark is given by

$$j_\mathrm{d} = \frac{1}{2} n \, e^{-E_\mathrm{b}/kT} \, e\bar{v} \tag{12.78}$$

where $\bar{v}$ is an appropriate average velocity. The same current flows across the barrier in both directions and the net current flow is zero. If a potential ΔV is applied to a particular barrier, the current densities in the two directions differ and there is a net current flow,

$$j_1 - j_2 = \frac{1}{2} n \, e^{-E_\mathrm{b}/kT} \left(e^{e\Delta V/kT} - e^{-e\Delta V/kT} \right) e\bar{v}$$

$$\approx \frac{ne^2 \Delta V \, \bar{v}}{kT} e^{-E_\mathrm{b}/kT} \tag{12.79}$$

if $e\Delta V \ll kT$, as is likely to be the case if there are many barriers distributed throughout the material. From Eq. (12.79), the conductivity is given by

$$\sigma = \frac{ne^2 \bar{v} l_\mathrm{b}}{kT} e^{-E_\mathrm{b}/kT} \tag{12.80}$$

where l_b is the average distance between barriers. If we choose to write $\sigma = ne\mu_\mathrm{b}$,

$$\mu_\mathrm{b} = \frac{e\bar{v} l_\mathrm{b}}{kT} e^{-E_\mathrm{b}/kT} \tag{12.81}$$

The conductivity changes with photoexcitation both through a change in n and through a change in E_b which affects the apparent mobility μ_b. The change in μ_b is given by

$$\Delta \mu_\mathrm{b} = - \frac{e\bar{v} l_\mathrm{b}}{(kT)^2} \Delta E_\mathrm{b} \, e^{-E_\mathrm{b}/kT} \tag{12.82}$$

In this consideration, the change in barrier height presumably results from the trapping of minority carriers in the immediate vicinity of the barrier. As the lifetime of free majority carriers determines the Δn due to photoexcitation, so under these conditions the "trapped time" of minority carriers, captured near the barrier before recombination with a majority carrier takes place, determines the $\Delta \mu_\mathrm{b}$ due to photoexcitation. Therefore, if the Δn modulation is the most important, we are most concerned about the majority-carrier lifetime in the bulk of the material, whereas if the $\Delta \mu_\mathrm{b}$ modulation is the most important, we are most concerned about the "trapped time" of minority carriers near the barrier.

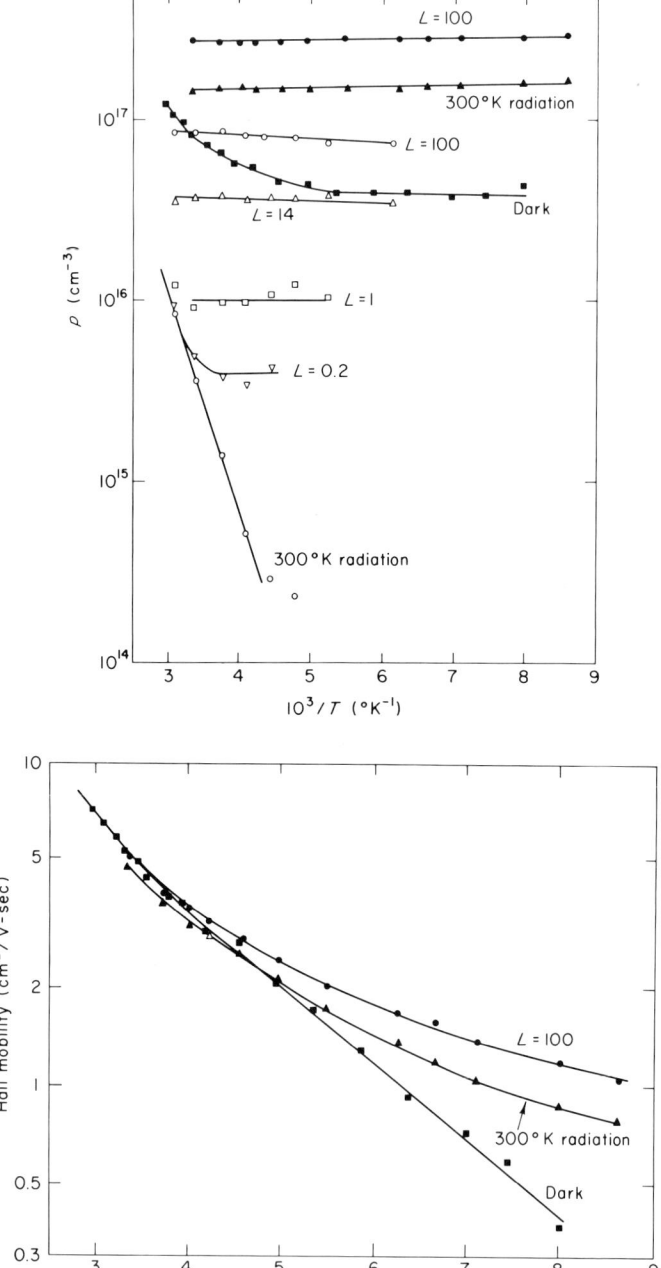

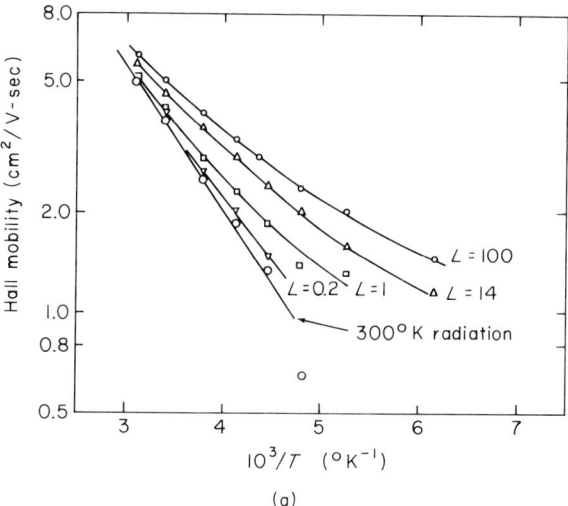

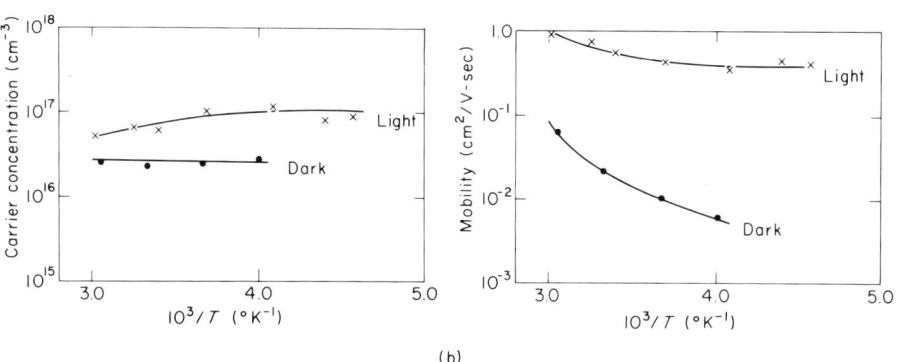

Fig. 12.16 (a) (*Opposite and top*) Effect of photoexcitation on the hole density and hole mobility as a function of temperature in two types of chemically deposited PbS layers: an initial state with appreciable adsorbed oxygen, and an oxygen-desorbed state caused by heating in vacuum. The primary effect of photoexcitation is to increase the hole density. Solid symbols indicate PbS standard material initial state; open symbols indicate PbS standard material after a 14-hour 120°C fix. [After S. Espevik, C. Wu, and R. H. Bube, *J. Appl. Phys.* **42**, 3513 (1971).] (b) (*Above*) Effect of photoexcitation on the electron density and electron mobility as a function of temperature for a chemically deposited $CdS_{0.8}Se_{0.2}$ layer. The primary effect of photoexcitation is to increase the electron mobility. [After C. Wu, R. S. Feigelson, and R. H. Bube, *J. Appl. Phys.* **43**, 756 (1972).]

Figure 12.16 shows the quite different behavior found for PbS films (large Δp and small $\Delta \mu$) and CdS–CdSe films (large $\Delta \mu$ and small Δn).

In a material with alternating n- and p-type regions, like that shown in Fig. 12.14, it is the free lifetime of minority carriers in each region that becomes important. This is because now it is the trapping of majority carriers that modulates the barrier height, and the "trapped time" of a majority carrier in a volume region can be considered equal to the free lifetime of a minority carrier in the same region.

HOMOGENEOUS MATERIALS

Five simple homogeneous photoconductor systems are illustrated in Fig. 12.17. Variations between systems depend on (a) the nature of the contact, and (b) the freedom of a carrier to move through the crystal.

In system 1, electrodes are ohmic for both electrons and holes, and both electrons and holes are free to move through the crystal. Only direct recombination between a free electron and a free hole terminates the additional conductivity resulting from photoexcitation. The gain is the sum of the electron and hole gains, and is given by the expression given in Example 12.1.

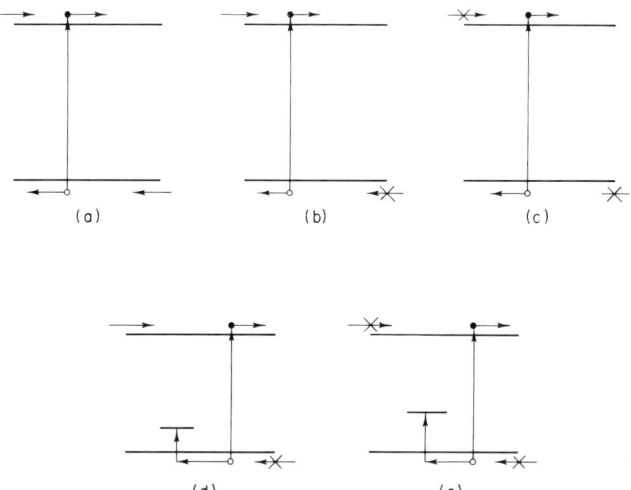

Fig. 12.17 Five models for photoconductivity in a homogeneous material. (a) Ohmic contacts for electrons and holes; both carriers free to move. (b) Ohmic contacts for electrons only; both carriers free to move. (c) Nonohmic contacts for electrons and holes; both carriers free to move. (d) Ohmic contacts for electrons only; photoexcited holes rapidly trapped and only electrons free to move. (e) Nonohmic contacts for electrons and holes; photoexcited holes rapidly trapped and only electrons free to move.

12.5 Recombination Mechanisms

In system 2, the cathode is ohmic for electron flow, but the anode is blocking for hole flow. Both electrons and holes are free to move through the crystal. Here the hole lifetime is terminated when the hole passes out of the crystal at the cathode; thus the hole lifetime is equal to the hole transit time. The electron lifetime is also equal to the hole transit time, since replenishment of electrons stops when the hole leaves the crystal. The gain is therefore given by

$$G = \frac{t_p}{t_n} + 1 = \frac{\mu_n + \mu_p}{\mu_p} \tag{12.83}$$

The gain exceeds unity, but not by as large a factor as in system 1.

In system 3, neither contact is ohmic for electrons or holes. This is the case of a homogeneous material with two blocking contacts, which is analogous to the case of the p–n junction. The maximum gain is unity. Since poor or blocking contacts were the rule experimentally before ohmic contacts were recognized, this type of system was the first to be explored historically.

In system 4, the cathode is ohmic for electron flow, but the anode is not ohmic for hole flow. In addition, only electrons are free to move through the crystal, photoexcited holes being captured at imperfections. The gain is contributed solely by electrons and is given by

$$G = \frac{\tau_n \mu_n V}{L^2} \tag{12.84}$$

The electron lifetime depends on the capture cross section of the imperfections; if this cross section is sufficiently small, large values of τ_n and G result. This system is the homogeneous analogue of the n–p–n junction case.

In system 5, the contacts are blocking and only electrons are free to move through the crystal. In this case no steady-state photoconductivity is possible, since a positive space charge builds up in the crystal to counterbalance the applied electric field. Polarization of the crystal with a resultant decrease in photoconductivity to zero was also a common experience historically with experiments being done with poor or blocking contacts.

12.5 Recombination Mechanisms

It is the energy-transfer process by which an excited carrier gives up its energy in order to recombine with a carrier of opposite type that determines the capture coefficients C_n and C_p. There are essentially three such processes for energy dissipation: (a) emission of photons (radiative recombination), (b) emission of phonons (nonradiative recombination), and (c) increase energy of another free carrier (Auger recombination). Since Auger recombination is a three-body process, its occurrence requires high densities of free carriers, and is likely to be of importance only in small-bandgap or highly imperfect semiconductors. Radiative recombination is also relatively improbable, and occurs only when some mechanism for dissipating the energy by phonon emission is absent; the probability of radiative recombination increases in general as the energy to be dissipated increases.

RADIATIVE RECOMBINATION

The simplest type of radiative recombination is that occurring between free electrons in the conduction band and free holes in the valence band. The capture coefficient for radiative recombination in this case can be calculated by equating the absorption of blackbody radiation with the emission of radiation by a material in thermal equilibrium with its surroundings.

Blackbody radiation falls on the material at a rate of

$$\frac{2\pi v^2}{c^2} \frac{1}{e^{hv/kT} - 1} \Delta v \text{ photons cm}^{-2} \text{ sec}^{-1}$$

Suppose that these photons are absorbed in a thickness d of the material. Then this same depth of semiconductor must emit photons at the same rate by the law of detailed balance. If we take $hv = E_G$, and $h\,\Delta v = kT$, the equilibrium condition is

$$n_i^2 C_r d \frac{1}{2r^2} = \frac{2\pi v^2}{hc^2} \frac{kT}{e^{E_G/kT} - 1} \tag{12.85}$$

where r is the index of refraction of the material, and $1/r^2$ is the fraction of the emission that actually leaves the material, considering effects of total reflection in a medium with index of refraction r. Substituting for n_i and solving Eq. (12.85) for C_r gives

$$C_r = \frac{r^2 h^3}{8\pi^2 c^2 d(kT)^2 (m_e^* m_h^*)^{3/2}} E_G^2 \tag{12.86}$$

if $E_G \gg kT$. For typical values of $E_G = 2$ eV, $r = 4$, $d = 1$ μm, and $m_e^* = m_h^* = m$, $C_r = 5.5 \times 10^{-12}$ cm^3/sec at 300°K, corresponding to a radiative recombination cross section of 5×10^{-19} cm^2.

Although this calculation has been specifically for the case of free–free

12.5 Recombination Mechanisms

radiative recombination, there is reason to believe that the capture coefficient is not greatly different for radiative recombination between a free electron and a captured hole, for example. Using detailed balance arguments in this situation, Blakemore[†] has shown that at room temperature the following magnitudes for the radiative cross section are predicted: 2×10^{-20} cm² for a hydrogenic imperfection, and 10^{-20} cm² for hole capture by indium acceptors in silicon. The capture of free electrons by neutral indium acceptors in silicon is reportedly[‡] totally radiative, and the measured cross section is 2×10^{-22} cm².

NONRADIATIVE RECOMBINATION

Nonradiative recombination occurs when the energy of the recombining carrier is dissipated by the emission of phonons. A relatively unlikely process for direct recombination because of the large number of phonons that must be emitted simultaneously for an appreciable energy difference, it becomes more favored in recombination at an imperfection center if the initial capture is in an excited state of the center, with subsequent cascade down to the ground state with nonsimultaneous emission of the required number of phonons.

Capture in a large excited orbit with subsequent successive decreases in energy among the excited states of a Coulomb-attractive center has been used to explain how capture cross sections as large as 10^{-11} cm² (see Fig. 9.13) can be observed at low temperatures,[§] even though the probability of capture with the simultaneous emission of sufficient phonons to dissipate the energy would be much smaller. Such processes are not expected for capture at a neutral or repulsive center, for which excited states are widely spaced or completely absent.

The maximum cross section of a Coulomb-attractive center can be estimated by asking at what radius a free electron diffuses into the center rather than away from it. A particle executing Brownian motion at a distance r from a point diffuses away from the point with an average velocity

$$\bar{v} = v \frac{2l}{3r} \qquad (12.87)$$

where l is the mean free path and v is the thermal velocity. For Coulomb

---

[†] J. S. Blakemore, *Phys. Rev.* **163**, 809 (1967).
[‡] Y. E. Pokrovsky and K. I. Svistunova, *Soviet Phys.-Solid State* **7**, 1478 (1965).
[§] M. Lax, *Phys. Rev.* **119**, 1502 (1960); G. Ascarelli and S. Rodriguez, *Phys. Rev.* **124**, 1321 (1961); **127**, 167 (1962).

attraction, the drift velocity is

$$v_d = \mu \mathscr{E} = \mu \frac{e}{\varepsilon r^2} \qquad (12.88)$$

At the critical radius, $\bar{v} = v_d$, and

$$r_c = \frac{3e\mu}{2\varepsilon v l} = \frac{3e^2}{4\varepsilon kT} \qquad (12.89)$$

substituting $\mu = (e/m^*)\tau$, $v = l/\tau$, and $kT = m^*v^2/2$. Thus the cross section is estimated to be

$$S = \pi r_c^2 = \frac{9\pi e^4}{16\varepsilon^2(kT)^2} \qquad (12.90)$$

For $\varepsilon = 10$, the value of S at $300°K$ is about 5×10^{-13} cm^2, some seven orders of magnitude larger than the cross section for radiative recombination. Since S varies as T^{-2}, a cross section of the order of 10^{-11} cm^2 is predicted from Eq. (12.90) for temperatures of about $70°K$.

AUGER RECOMBINATION

Auger recombination is the inverse of impact ionization, even as radiative recombination is the inverse of optical excitation. There are two possible variations of Auger recombination for band-to-band processes, which are illustrated in Fig. 12.18 with the complementary impact ionization processes.

For Auger recombination involving an imperfection level, the corresponding lifetime is given by

$$\frac{1}{\tau_A} = Anp + Bn^2 \qquad (12.91)$$

in n-type material, for example. The first term expresses Auger excitation of a minority carrier; the second term expresses Auger excitation of a majority carrier. In most extrinsic materials the second term dominates, and the constant B is the Auger recombination coefficient. A value for B has been calculated for a hydrogenic imperfection from a detailed balance argument equating rates of Auger recombination and impact ionization of the imperfection, with the result[†]

$$B = \frac{2.44 \times 10^{-19}}{E_I T^2} \left(\frac{m}{m^*}\right)^2 \left(\frac{1 + 0.522 \log \varepsilon}{\varepsilon}\right) \text{ cm}^6 \text{ sec}^{-1} \qquad (12.92)$$

[†] E. Burstein, G. Picus and N. Sclar, "Photoconductivity Conference," Wiley, New York, 1956, p. 353.

12.5 Recombination Mechanisms

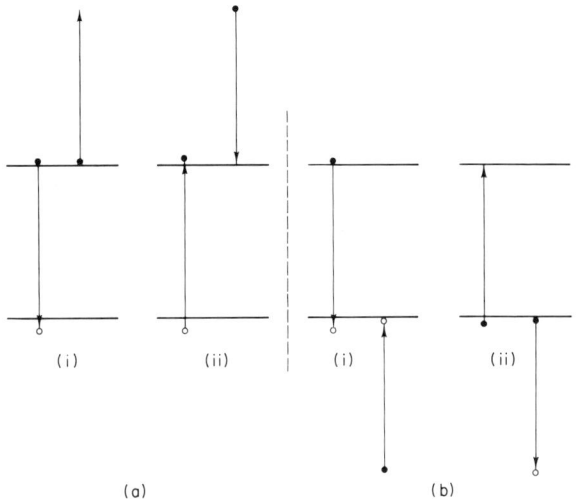

Fig. 12.18 Intrinsic Auger recombination and impact ionization processes compared. (a-i) Auger recombination with excitation of an electron; (a-ii) impact ionization of an electron–hole pair by an energetic electron; (b-i) Auger recombination with excitation of a hole; (b-ii) impact ionization of an electron–hole pair by an energetic hole.

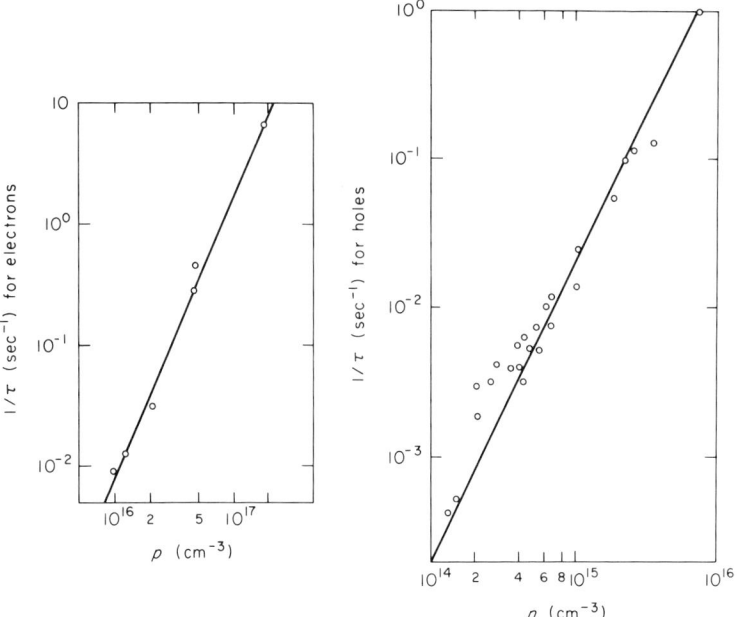

Fig. 12.19 Experimental evidence for a dependence of minority-carrier lifetime on the square of the density of the majority carriers in silicon. [After J. R. Haynes and J. A. Hornbeck, *Phys. Rev.* **100**, 606 (1955), and J. A. Hornbeck and J. R. Haynes, *Phys. Rev.* **97**, 311 (1955).]

where E_I is the ionization energy of the imperfection in eV, m^*/m is the effective mass ratio for the relevant carrier, and ε is the dielectric constant. For typical values of $E_I = 0.1$ eV, $\varepsilon = 10$, and $m^*/m = 0.1$, the value of B at room temperature is 4×10^{-22} cm^6 sec^{-1}. Data suggesting Auger recombination is silicon are given in Fig. 12.19.

12.6 Recombination Kinetics

A variety of models have been proposed to describe the variation of the free-carrier lifetime with excitation intensity, temperature, and thermal-equilibrium Fermi-level position. Such models are usually characterized as to whether they assume that one type of recombination center dominates the behavior of the photoconductivity, or whether two different types of recombination center are invoked to describe the experimental observations.

ONE-CENTER MODEL WITHOUT TRAPS

Some of the basic characteristics of recombination centers can be seen from examining a very simple and somewhat artificial model. Suppose that N_I recombination centers are located sufficiently close to the Fermi level that they are party occupied, n_I being occupied by electrons and $(N_I - n_I)$ being not occupied by electrons, or occupied by holes.

In the first case the density of free carriers is assumed to be much smaller than these densities, n_I and $(N_I - n_I)$. Thus no appreciable change in n_I or $(N_I - n_I)$ occurs under photoexcitation. The corresponding lifetimes for electrons and holes are given by

$$\tau_n = \frac{1}{C_n(N_I - n_I)} \tag{12.93}$$

$$\tau_p = \frac{1}{C_p n_I} \tag{12.94}$$

The electron and hole lifetimes are independent and unequal.

In the second case, consider that the density of free carriers is much greater than n_I or $(N_I - n_I)$. This is the kind of condition that may be brought about by high excitation intensities. For charge neutrality, $n = p$, and therefore $\tau_n = \tau_p$. If n_I^* represents the density of electron-occupied imperfections under the high excitation condition, and $(N_I - n_I)^*$ represents the density of empty centers under excitation, equating Eqs. (12.93) and (12.94) gives

$$\frac{1}{C_n(N_I - n_I)^*} = \frac{1}{C_p n_I^*} \tag{12.95}$$

12.6 Recombination Kinetics

which reduces to

$$\frac{n_1^*}{N_1} = \frac{C_n}{C_n + C_p} \quad (12.96)$$

We find therefore for the lifetime that

$$\tau_n = \tau_p = \frac{1}{N_1[C_p C_n/(C_p + C_n)]} \quad (12.97)$$

If $C_n \ll C_p$, this reduces to $1/N_1 C_n$, and if $C_p \ll C_n$, it reduces to $1/N_1 C_p$. Thus the rate of recombination is determined by the smaller of the two capture coefficients.

DETAILED ONE-CENTER MODEL WITHOUT TRAPS

A detailed analysis of a one-center model without traps in a semiconductor has been carried out by Shockley and Read.[†] The nature of the recombination center and the processes to be considered are summarized in Fig. 12.20.

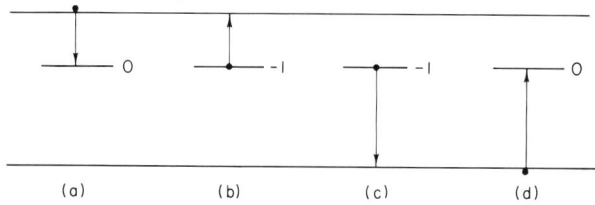

Fig. 12.20 The transitions involved in the Shockley–Read recombination model. Processes are (a) capture of an electron by a neutral center, (b) thermal excitation of an electron to the conduction band from a negatively charged center, (c) capture of a hole by a negatively charged center, and (d) thermal excitation of a hole to the valence band from a neutral center.

The center is assumed to be neutral when empty and negatively charged when occupied by an electron. The four processes to be considered are (a) electron capture by a neutral center with capture coefficient C_n, (b) thermal excitation of an electron from a negatively charged center to the conduction band, with probability $N_c C_n \, e^{-E_1/kT}$ (see Section 9.7), (c) capture of a hole by a negatively charged center with capture coefficient C_p, and (d) thermal excitation of a hole from a neutral center to the valence band with probability $N_v C_p \, e^{-(E_G - E_1)/kT}$.

[†] W. Shockley and W. T. Read, *Phys. Rev.* **87**, 835 (1952).

The net rate of electron capture by the imperfection center is given by

$$\frac{dn_\mathrm{I}}{dt} = n(N_\mathrm{I} - n_\mathrm{I})C_\mathrm{n} - n_\mathrm{I} N_\mathrm{c} C_\mathrm{n}\, e^{-E_\mathrm{I}/kT} \qquad (12.98)$$

Similarly the net rate of hole capture by the imperfection center is given by

$$\frac{d(N_\mathrm{I} - n_\mathrm{I})}{dt} = n_\mathrm{I} p C_\mathrm{p} - (N_\mathrm{I} - n_\mathrm{I}) N_\mathrm{v} C_\mathrm{p}\, e^{-(E_\mathrm{G} - E_\mathrm{I})/kT} \qquad (12.99)$$

These processes are pictured in Fig. 12.21. In the steady state, Eqs. (12.98) and (12.99) must be equal to each other, and must each be equal to the total excitation rate. If the total excitation rate is R, and if photoexcitation produces changes in carrier density of Δn and Δp, then the lifetimes are

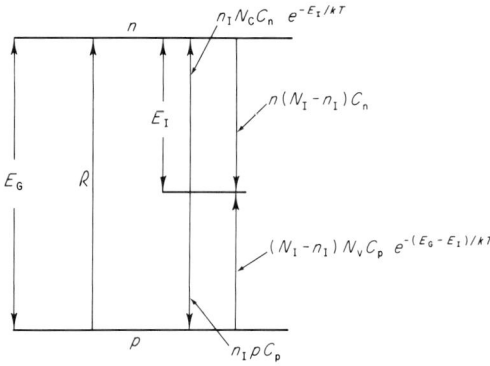

Fig. 12.21 Transitions of importance for the Shockley–Read recombination model. Recombinations between free electrons and free holes are neglected.

given by

$$\tau_\mathrm{n} = \frac{\Delta n}{R} \qquad (12.100)$$

$$\tau_\mathrm{p} = \frac{\Delta p}{R} \qquad (12.101)$$

In order to proceed, it is customary to make several simplifying definitions, as follows:

$$n_\mathrm{I} = N_\mathrm{I} f_\mathrm{I} \qquad (12.102)$$

$$(N_\mathrm{I} - n_\mathrm{I}) = N_\mathrm{I}(1 - f_\mathrm{I}) \qquad (12.103)$$

$$n' \equiv N_\mathrm{c}\, e^{-E_\mathrm{I}/kT} \qquad (12.104)$$

$$p' \equiv N_\mathrm{v}\, e^{-(E_\mathrm{G} - E_\mathrm{I})/kT} \qquad (12.105)$$

12.6 Recombination Kinetics

Here f_I is the value of the Fermi function giving the occupation of the N_I imperfections, and n' and p' are respectively the density of free electrons and free holes when the Fermi level is located at the imperfection-center level, i.e., for $E_f = E_I$. If these quantities are substituted into Eqs. (12.98) and (12.99), and these four equations set equal to each other, solution for f_I gives

$$f_I = \frac{nC_n + p'C_p}{C_n(n + n') + C_p(p + p')} \quad (12.106)$$

Substituting this value back into either Eq. (12.98) or (12.99) gives

$$R = \frac{pn - p'n'}{[(n + n')/N_I C_p] + [(p + p')/N_I C_n]} \quad (12.107)$$

Here $n'p' = n_i^2 = n_0 p_0$, $1/N_I C_p$ represents the lifetime of a hole when all the imperfections are occupied by electrons, τ_{p0}, and $1/N_I C_n$ represents the lifetime of an electron when all the imperfections are occupied by holes, τ_{n0}. In the dark $R = 0$ by definition, and the right-hand side of Eq. (12.107) equals zero because $pn = p_0 n_0$ in the dark.

The effect of photoexcitation is to change p_0 to $(p_0 + \Delta p) = p$, and to change n_0 to $(n_0 + \Delta n) = n$. Equation (12.107) becomes

$$R = \frac{n_0 \Delta p + p_0 \Delta n + \Delta n \Delta p}{\tau_{p0}(n_0 + \Delta n + n') + \tau_{n0}(p_0 + \Delta p + p')} \quad (12.108)$$

The simplest case to treat is that for which $\Delta n, \Delta p \ll n_0, p_0$, and for which $\tau_n = \tau_p$ so that $\Delta n = \Delta p$. Equation (12.08) becomes

$$R = \frac{\Delta n (n_0 + p_0)}{\tau_{p0}(n_0 + n') + \tau_{n0}(p_0 + p')} \quad (12.109)$$

The lifetime is given by

$$\tau = \frac{\Delta n}{R} = \tau_{p0} \frac{n_0 + n'}{n_0 + p_0} + \tau_{n0} \frac{p_0 + p'}{n_0 + p_0} \quad (12.110)$$

For an imperfection center lying E_I below the conduction band in the upper half of the bandgap, the variation of τ with the location of the Fermi level is illustrated in Fig. 12.22. For the Fermi level very close to the valence band, $p_0 \gg p', n_0, n'$, all the centers are empty, and $\tau = \tau_{n0}$. For the Fermi level very close to the conduction band, $n_0 \gg n', p_0, p'$, all the centers are occupied, and $\tau = \tau_{p0}$. When the Fermi level lies below the middle of the gap but further than E_I above the valence band, $n' \gg p_0 \gg p', n_0$, essentially

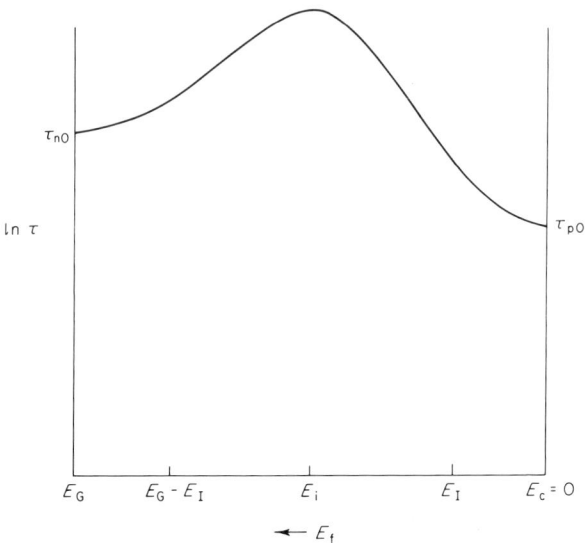

Fig. 12.22 Variation of the lifetime with the location of the Fermi level in the Shockley–Read recombination model, according to Eq. (12.110). The recombination center is assumed to lie E_I below the conduction band. As drawn, $C_n < C_p$.

all the centers are still empty but an increasing probability exists for a captured electron to be thermally reexcited to the conduction band before recombination with a free hole, and the lifetime increases with rising Fermi level as

$$\tau = \tau_{n0} + \tau_{p0} \frac{n'}{p_0} = \tau_{n0} + \tau_{p0} \frac{N_c}{N_v} e^{(E_G - E_I - E_f)/kT} \quad (12.111)$$

When the Fermi level lies above the middle of the gap but farther than E_I below the conduction band, $n' \gg n_0 \gg p_0 \gg p'$, fewer of the centers are empty and hole capture becomes more important, and the lifetime decreases with rising Fermi level as

$$\tau = \tau_{p0} \frac{n'}{n_0} + \tau_{n0} \frac{p_0}{n_0} \quad (12.112)$$

$$\approx \tau_{p0} e^{(E_f - E_I)/kT} \quad (12.113)$$

A second case of interest is that for large photoexcitation so that $\Delta n, \Delta p \gg n_0, p_0$, but still $\tau_n = \tau_p$ and $\Delta n = \Delta p$. Equation (12.108) now becomes

$$R = \frac{\Delta n \, (n_0 + p_0 + \Delta n)}{\tau_{p0}(n_0 + \Delta n + n') + \tau_{n0}(p_0 + \Delta p + p')} \quad (12.114)$$

12.6 Recombination Kinetics

giving a lifetime

$$\tau = \frac{\Delta n}{R} = \tau_{p0}\frac{n_0 + \Delta n + n'}{n_0 + p_0 + \Delta n} + \tau_{n0}\frac{p_0 + \Delta n + p'}{n_0 + p_0 + \Delta n} \quad (12.115)$$

For very large Δn,

$$\tau_\infty = \tau_{p0} + \tau_{n0} \quad (12.116)$$

If the lifetime τ from Eq. (12.110) for the small Δn case is called τ', then the lifetime from Eq. (12.115) is related to τ' by

$$\tau = \tau'\frac{1 + a\,\Delta n}{1 + b\,\Delta n} \quad (12.117)$$

where

$$a = \frac{\tau_{p0} + \tau_{n0}}{\tau_{p0}(n_0 + n') + \tau_{n0}(p_0 + p')} \quad (12.118)$$

$$b = \frac{1}{n_0 + p_0} \quad (12.119)$$

If $a > b$, then τ increases with increasing Δn, giving rise to superlinear photoconductivity; if $a < b$, then τ decreases with increasing Δn, giving rise to sublinear photoconductivity.

A third case is that for small Δn and Δp compared to n_0 and p_0, as in the first case, but now considering $\tau_n \neq \tau_p$, so that $\Delta n \neq \Delta p$. The difference between Δn and Δp must be taken up by a change in location of the Fermi level, and can be expressed as

$$\Delta p - \Delta n = N_I\,\Delta f_I \quad (12.120)$$

Equation (12.108) for this case becomes

$$R = \frac{\Delta p\, n_0 + \Delta n\, p_0}{\tau_{p0}(n_0 + n') + \tau_{n0}(p_0 + p')} \quad (12.121)$$

and we need to calculate Δf_I. Return to Eq. (12.98) and write it explicitly for the case of photoexcitation as follows:

$$\frac{dn_I}{dt} = (n_0 + \Delta n)N_I(1 - f_{I0} - \Delta f_I)\,C_n - N_I(f_{I0} + \Delta f_I)n'C_n$$

$$= \left(\frac{dn_I}{dt}\right)_0 + \Delta n\,N_I(1 - f_{I0})C_n - n_0 N_I\,\Delta f_I\,C_n - N_I\,\Delta f_I\,n'C_n$$

$$= \left(\frac{dn_I}{dt}\right)_0 + C_n[(1 - f_{I0})\,\Delta n - (n_0 + n')\,\Delta f_I] \quad (12.122)$$

Similarly from Eq. (12.99) we have

$$\frac{d(N_I - n_I)}{dt} = \left[\frac{d(N_I - n_I)}{dt}\right]_0 + C_p[f_{I0}\Delta p + (p_0 + p')\Delta f_I] \quad (12.123)$$

By equating Eqs. (12.122) and (12.123), solve for Δf_I:

$$\Delta f_I = \frac{\tau_{p0}(1 - f_{I0})\Delta n - \tau_{n0}f_{I0}\Delta p}{\tau_{n0}(p_0 + p') + \tau_{p0}(n_0 + n')} \quad (12.124)$$

In order to calculate τ_n or τ_p, it is necessary to eliminate Δn or Δp by using Eq. (12.120). For example, to calculate τ_p, substitute $\Delta n = \Delta p - N_I \Delta f_I$ into Eq. (12.124) and solve for Δf_I as a function of Δp. Then substitute $\Delta n = \Delta p - N_I \Delta f_I$ into Eq. (12.121), and calculate $\tau_p = \Delta p/R$. The results for τ_p and τ_n from such a calculation are

$$\tau_p = \frac{\tau_{n0}(p_0 + p') + \tau_{p0}[n_0 + n' + N_I(1 - f_{I0})]}{n_0 + p_0 + N_I f_{I0}(1 - f_{I0})} \quad (12.125)$$

$$\tau_n = \frac{\tau_{p0}(n_0 + n') + \tau_{n0}(p_0 + p' + N_I f_{I0})}{n_0 + p_0 + N_I f_{I0}(1 - f_{I0})} \quad (12.126)$$

The final case, the most general case, is that suitable also for large Δn and Δp, and for $\tau_n \neq \tau_p$. An exact solution is difficult to obtain, but a solution that is usually a very good approximation can be obtained; the form for τ_n, for example, is[†]

$$\tau_n = \frac{\tau_{p0}(n_0 + n' + \Delta n) + \tau_{n0}(p_0 + p' + \Delta n) + \tau_{n0}N_I\left[\dfrac{p'(n_0 + n' + \Delta n) + 2p_0\Delta n}{(n_0 + n' + \Delta n)(p_0 + p')}\right]}{(n_0 + p_0 + \Delta n) + N_I\left[\dfrac{p_0(n_0 + \Delta n)}{(n_0 + n' + \Delta n)(p_0 + p')}\right]} \quad (12.127)$$

The expression for τ_p is similar:

$$\tau_p = \frac{\tau_{p0}(n_0 + n' + \Delta n) + \tau_{n0}(p_0 + p' + \Delta p) + \tau_{p0}N_I\left[\dfrac{p_0(p_0 + p' + \Delta p) + 2p'\Delta p}{(p_0 + p')(p_0 + p' + \Delta p)}\right]}{(n_0 + p_0 + \Delta p) + N_I\left[\dfrac{p'(p_0 + \Delta p)}{(p_0 + p')(p_0 + p' + \Delta p)}\right]} \quad (12.128)$$

[†] J. S. Blakemore, "Semiconductor Statistics," Pergamon, Elmsford, New York, 1962.

12.6 Recombination Kinetics

The correspondence between the pair of Eqs. (12.127) and (12.128), and the pair of Eqs. (12.125) and (12.126), can be seen by setting the Δn or Δp terms in the former two equations equal to zero, and by recognizing that $f_{\text{I}0} = (1 + p_0/p')^{-1} = (1 + n'/n_0)^{-1}$, and that $(1 - f_{\text{I}0}) = (1 + p'/p_0)^{-1} = (1 + n_0/n')^{-1}$. Similarly Eqs. (12.127) and (12.128) reduce to Eq. (12.115) if the terms in N_{I} can be neglected.

Figure 12.23 shows the variation of minority carrier lifetime in n- and p-type Ge samples, which follows the Shockley–Read model.

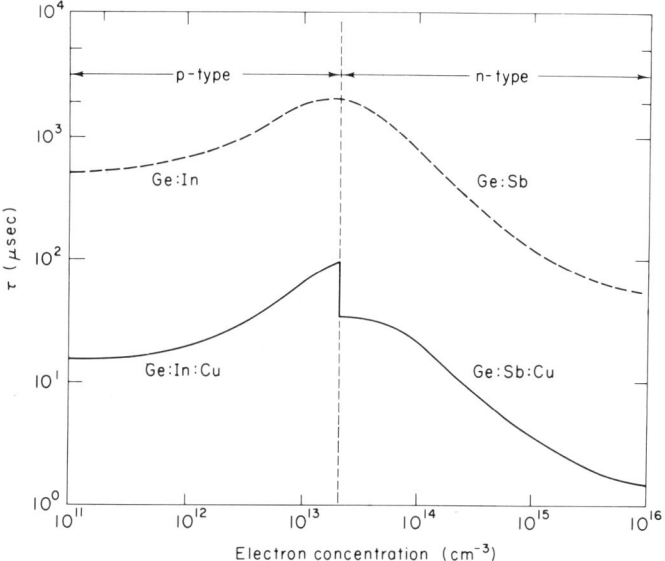

Fig. 12.23 Variation of minority-carrier lifetime with free-electron concentration in Ge with and without Cu impurity recombination centers. Experimental data coincide with curves calculated from the single-level recombination model. [After J. Burton, G. Hull, F. Morin, and J. Severiens, *J. Phys. Chem.* **57**, 853 (1953).]

Example 12.3 What kind of behavior is expected for the electron lifetime in an n-type material where the recombination centers in question lie in the middle of the bandgap and have $C_n \ll C_p$?

Since the recombination centers lie in the middle of the gap, the following inequality always holds in n-type material:

$$n_0 \gg n' \approx p' \gg p_0$$

If this inequality is applied to Eq. (12.127), the following approximations in the small-signal and large-signal cases are obtained:

$$\Delta n \ll n_0; \quad \tau_n = \frac{\tau_{p0}n_0 + \tau_{n0}(p' + N_I)}{n_0 + N_I p_0/p'}$$

$$\Delta n \gg n_0; \quad \tau_n = \frac{\tau_{p0}\Delta n + \tau_{n0}(\Delta n + N_I)}{\Delta n + N_I p_0/p'}$$

For reasonable values of the quantities involved, $p' < N_I$ in the small-signal case. If n_0 is very large, then $\tau_n = \tau_{p0}$ as expected. Otherwise we may neglect the first term in the numerator and have

$$(\tau_n)_{\text{small}} \approx \frac{\tau_{n0} N_I}{n_0 + N_I p_0/p'}$$

There are two ranges of interest. When $n_0 \ll N_I p_0/p'$,

$$(\tau_n)_{\text{small}} = \tau_{n0} p'/p_0$$

and increases with rising Fermi level. When $n_0 \gg N_I p_0/p'$,

$$(\tau_n)_{\text{small}} = \tau_{n0} \frac{N_I}{n_0} = \frac{1}{C_n n_0}$$

and decreases with rising Fermi level. A maximum lifetime occurs corresponding to

$$n_0 = N_I \frac{p_0}{p'}$$

i.e., when the density of free electrons is equal to the density of empty recombination centers.

In the large-signal case, there are three ranges of interest. When $n_0 < \Delta n < N_I p_0/p'$,

$$(\tau_n)_{\text{large}} = \tau_{n0} \frac{p'}{p_0}$$

In this range, the large-signal lifetime is the same as the small-signal lifetime cited above for the case of small n_0. When $N_I p_0/p' < \Delta n < N_I$,

$$(\tau_n)_{\text{large}} = \tau_{n0}\left(1 + \frac{N_I}{\Delta n}\right) \approx \left(\frac{\tau_{n0} N_I}{R}\right)^{1/2}$$

since $\Delta n = R\tau_n$. In this range, τ_n decreases with increasing excitation intensity, but is independent of temperature. Finally when $\Delta n > N_I$,

$$(\tau_n)_{\text{large}} = \tau_{n0}$$

12.6 Recombination Kinetics

ONE-CENTER MODELS WITH TRAPS

Models involving both one type of recombination center and also traps for the majority carrier have been developed to explain in a simple way the experimental observations on high-resistivity semiconductors that the lifetime may be constant with increasing excitation intensity while the response time decreases (the response time is either the time to rise to $(1 - 1/e)$ of the final steady-state value, or more commonly the time to decay to $1/e$ of the initial steady-state value), and that the photoconductivity is commonly found to vary as a power of the light intensity between 0.5 and 1 over extended ranges of intensity.

A collection of models proposed[†] to explain a variety of phenomena and their basic conclusions are summarized in Table 12.2. Since the majority of known sensitive high-resistivity photoconductors are n-type, the model concerns itself with free electrons, electron traps, and photoexcited holes captured at recombination centers. It is assumed that the free-hole lifetime is much less than the free-electron lifetime and that recombination takes place between a free electron and a trapped hole. Since most real materials contain traps with several different depths, the models take the mathematically simplifying form of a quasi-continuous trap distribution with depth. The particle conservation relation is

$$\Delta n + \Delta n_t = \Delta(N_I - n_I) \qquad (12.129)$$

i.e., for each free photoexcited electron and for each trapped photoexcited electron, there is a trapped photoexcited hole in a recombination center. The principle properties of these models can be summarized as follows.

1. If $\Delta n \gg \Delta n_t$, the recombination is bimolecular and $\Delta n \propto f^{1/2}$.

2. If $\Delta n \ll \Delta n_t$, the recombination is monomolecular and $\Delta n \propto f$, provided that the distribution of trap depth is reasonably uniform with depth. An exception occurs if the trapped electrons are all in shallow electron traps in effective thermal equilibrium with the conduction band; in this case $\Delta n_t \propto \Delta n$, the recombination is bimolecular, and $\Delta n \propto f^{1/2}$.

3. If $\Delta n \ll \Delta n_t$, and if the distribution of trap density with depth is sufficiently nonuniform, $\Delta n \propto f^m$ where $0.5 \leq m \leq 1$, unless of course the trapped electrons are in effective thermal equilibrium with the conduction band as in (2), in which case once again $\Delta n \propto f^{1/2}$.

4. If $\Delta n \gg \Delta n_t$, the response time is equal to the electron lifetime.

5. If $\Delta n \ll \Delta n_t$, the response time is determined by the time required

[†] A. Rose, *RCA Review* **12**, 362 (1951).

TABLE 12.2 ONE-CENTER MODELS WITH TRAPS

(a)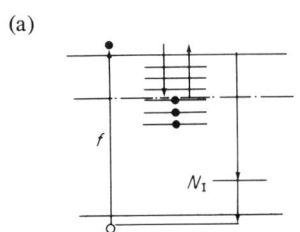

$(N_I - n_I) = N_t \approx$ constant

$$n = \frac{f}{C_n(N_I - n_I)} \; ; \quad \tau_0 = \frac{N(E)kT}{f}$$

(b)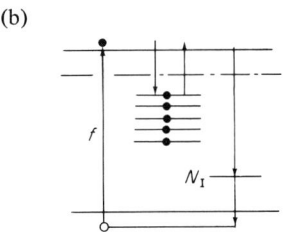

$(N_I - n_I) = N_t =$ constant

$$n = \frac{f}{C_n(N_I - n_I)} \; ; \quad \tau_0 = \tau$$

(c)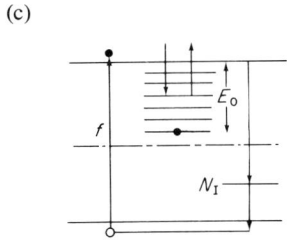

If $\left\{ \begin{array}{l} N_t \gg n \\ E_0 \gg kT \end{array} \right\}$,

$$(N_I - n_I) \approx N_t \approx \frac{kTN(E)e^{E_0/kT}}{N_c} n$$

$$\equiv \alpha(T)n$$

$$n = \left\{ \frac{f}{C_n\alpha(T)} \right\}^{1/2} \; ; \quad \tau_0 = \left\{ \frac{\alpha(T)}{C_n f} \right\}^{1/2}$$

(d)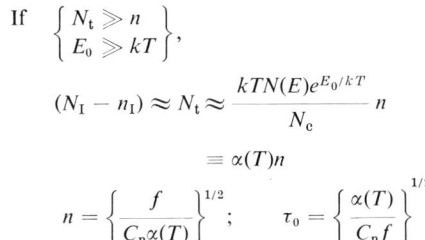

For $N_t]_{\text{below Fermi level}} > N_t]_{\text{above Fermi level}}$
$$n \propto f \; ; \quad \tau_0 = \alpha(T)\tau$$

For $N_t]_{\text{below Fermi level}} < N_t]_{\text{above Fermi level}}$
$$n \propto f^{1/2} \; ; \quad \tau_0 \propto 1/f^{1/2}$$

(e)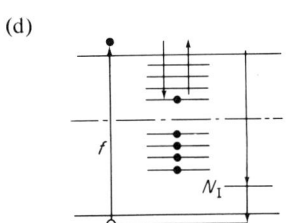

$$N(E) = N_0 e^{-E/kT^*}$$

If $N_t \gg n$, $(N_I - n_I) = N_t = kT^* N_0 e^{-E_f/kT^*}$

$$n = \left\{ \frac{f}{C_n kT^* N_0 N_c^{-T/T^*}} \right\}^{T^*/T+T^*} ;$$

$$\tau_0 = \frac{T}{T^*} \left\{ \frac{kT^* N_0 N_c^{-T/T^*}}{C_n f} \right\}^{T^*/T+T^*}$$

12.6 Recombination Kinetics

for the recombination of all free carriers, all carriers trapped above the Fermi level, and all carriers trapped within about kT below the Fermi level (since the Fermi level must drop by kT in order for the free-carrier density to drop by $1/e$). Since $\Delta n/\tau$ is the rate at which free electrons leave the conduction band, the time required to remove Δn_t carriers is $\Delta n_t/(\Delta n/\tau)$, which leads to the following relation between the response time and the lifetime:

$$\tau_0 = \tau\left(1 + \frac{\Delta n_t}{\Delta n}\right) \quad (12.130)$$

where τ_0 is the time required for the free-electron density to decrease to $1/e$ of its initial value, and Δn_t is the density of traps that must empty in order for the Fermi level to drop by kT from its steady-state position under photoexcitation.

Example 12.4 A uniform density of electron traps exists in an n-type photosensitive crystal, lying between E_1 and E_2 ($E_2 > E_1$) below the bottom of the conduction band. Recombination is dominated by centers lying below the dark Fermi level, which lies appreciably lower than E_2. Describe the dependence of photocurrent on excitation intensity.

There are three different ranges. First is the low-light range where the density of electrons trapped above the Fermi level is greater than that trapped below the Fermi level. Since in this range the trapped electrons are effectively in thermal equilibrium with the conduction band, the density of trapped electrons is proportional to the density of free electrons, and the recombination is bimolecular with $\Delta n \propto f^{1/2}$.

The intermediate-light range starts when the density of trapped electrons below the Fermi level becomes comparable to, and then greater than, the density of trapped electrons above the Fermi level. In this range monomolecular kinetics dominate if $\Delta n \ll \Delta n_t$, and $\Delta n \propto f$. Without integrating the Fermi function, we can arrive at an estimated location of the Fermi level for this change from $\Delta n \propto f^{1/2}$ to $\Delta n \propto f$. The density of trapped electrons above the Fermi level is overestimated by

$$(n_t)_{\text{above}} \approx \int_{E_1}^{E_f} N_E \, e^{(E-E_f)/kT}$$
$$\approx N_E kT$$

if $(E_f - E_1) \gg kT$. Similarly the density of trapped electrons below the Fermi level is overestimated by

$$(n_t)_{\text{below}} = \int_{E_f}^{E_2} N_E \, dE = N_E(E_2 - E_1)$$

In both expressions N_E is the uniform density of trap states. Comparison of the above equations shows that as soon as the Fermi level has risen about kT above E_2, the range of $\Delta n \propto f$ starts.

Finally for still higher light intensities, the Fermi level rises above E_1 and now each photoexcited free electron in addition has a corresponding trapped photoexcited hole in a recombination center; the recombination reverts to bimolecular, with $\Delta n \propto f^{1/2}$.

Two-Center Models

Four commonly observed phenomena in photosensitive high-resistivity photoconductors cannot be generally interpreted in terms of a one-center model. These phenomena are (1) an increase in photosensitivity resulting from the incorporation of additional recombination centers (*imperfection sensitization*), (2) a variation of photoconductivity with a power of light intensity greater than unity (*superlinearity*), (3) a rapid decrease in photosensitivity above a certain temperature (*thermal quenching*), and (4) a

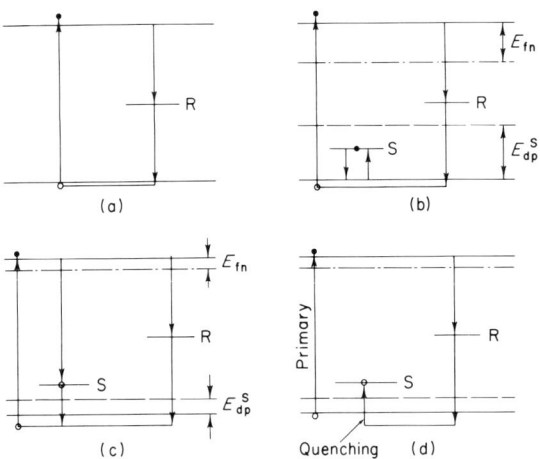

Fig. 12.24 Features of the two-center recombination model. (a) Fast recombination through R centers and short electron lifetime in a material without sensitizing centers. (b) The presence of sensitizing (S) centers does not affect the electron lifetime if the corresponding hole demarcation level lies above the S levels. (c) An increase in the electron lifetime occurs when the demarcation level of the S centers lies below the S levels. (d) In the presence of a primary excitation maintaining the position of the demarcation level, a secondary excitation that optically releases holes from S centers to the valence band where they can be captured by R centers and take part in recombination with free electrons, results in optical quenching of the photoconductivity; i.e., a decrease in the electron lifetime.

12.6 Recombination Kinetics

decrease in photoconductivity excited by a primary light when a secondary light of longer wavelength is simultaneously shined on the sample (*optical quenching*). These phenomena can all be described if two types of recombination center are proposed, the varying phenomena arising from the recombination competition between these two types of centers with different capture coefficients.

The nature of these processes is illustrated in Fig. 12.24. It is assumed that the material exists initially with certain recombination centers with relatively large capture coefficients for both electrons and holes. To this material are then added imperfection centers with a considerably different set of capture coefficients; for n-type sensitization, for example, the added *sensitizing centers* have a very small capture coefficient C_n for electrons, and a much larger coefficient C_p for holes. Such a situation occurs, for example, if the sensitizing center is a singly or doubly negatively charged acceptor imperfection.

As shown in Fig. 12.24b, if the hole demarcation level of the sensitizing centers lies above them, their occupancy is determined by effective thermal equilibrium with the valence band, they do not affect the lifetime of free electrons, and the situation is substantially unchanged from that with the sensitizing centers absent. For suitably high excitation intensities or low temperatures, however, the demarcation level lies below the level of the sensitizing centers. When this occurs in a material with $\Delta n \ll N_I$, the density of imperfections, the sensitizing centers with small C_n become largely occupied by holes, and at the same time the density of holes in the previous recombination centers with large C_n is reduced. The net effect is an increase in the electron lifetime. In actual materials such sensitizing procedures give rise to an increase in lifetime by a factor of 10^3 or more.

When the hole demarcation level is lowered through the level of the sensitizing centers by increasing light intensity at fixed temperature, the photosensitivity increases as the electron lifetime increases. The effect is superlinear photoconductivity.

When the hole demarcation level is raised through the level of the sensitizing centers by increasing the temperature at fixed excitation intensity, the photosensitivity decreases as the electron lifetime decreases. The effect is thermal quenching of photosensitivity.

When the hole demarcation level is below the level of the sensitizing centers, a second light source can excite electrons optically from the valence band to the hole-occupied sensitizing centers, releasing these holes to be captured by the recombination centers. The effect is the optical analogue of thermal quenching; it is called optical quenching.

504   12   Photoelectronic Effects

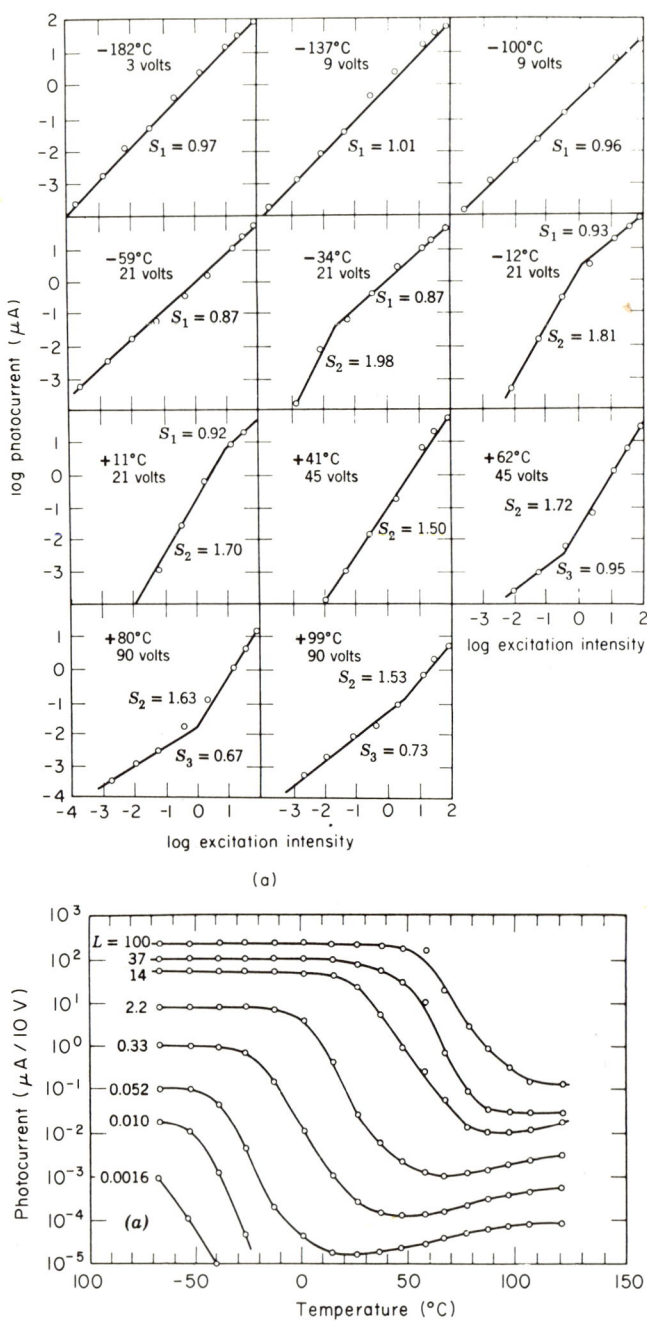

12.6 Recombination Kinetics

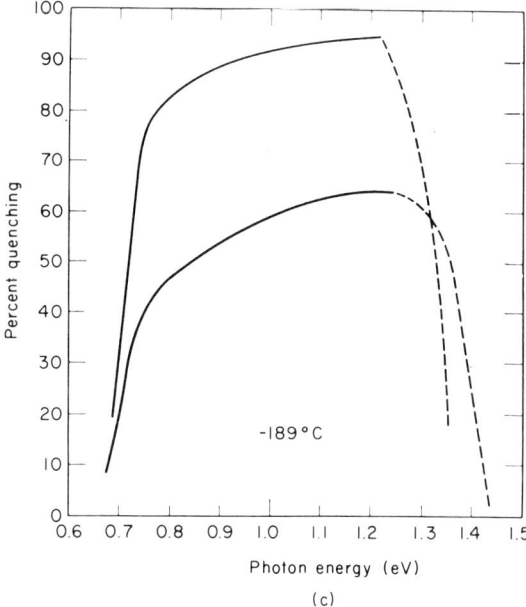

Fig. 12.25 Examples of two-center recombination processes for single-crystal CdSe. (a) Superlinear photoconductivity, developing first for low light intensities at $-34°C$, dominating the whole measured range at $+41°C$, and remaining only for high light intensities at $+99°C$. (b) Thermal quenching of photoconductivity, shown for a range of about 10^5 in steady exciting light intensities as indicated. (c) Effectiveness of optical quenching versus quenching photon energy for two different crystals, showing that the minimum energy for quenching is about 0.65 eV, the height of the sensitizing center above the top of the valence band. (After R. H. Bube, "Photoconductivity of Solids," Wiley, New York, 1960, pp. 343, 344, 350.)

The two-center model is successful in describing photosensitivity in a variety of sensitive photoconductors, including CdS, CdSe, GaAs, Ge, and Si. In the first three materials, the sensitizing centers appear to be fairly complex, involving intrinsic defects; in the case of Ge and Si, specific chemical impurities are known to be responsible, as, for example, Mn in Ge, and Zn in Si. Since the experimentally observable phenomena occur under conditions that can be correlated with the relative location of the demarcation level and the level of the sensitizing center, and since measurements of electron conductivity can be used to determine the corresponding location of the hole demarcation level by Eq. (12.70), the observation of these phenomena can be used to determine the energy level and the capture coefficients of the sensitizing centers. Illustrations of superlinearity, thermal quenching, and optical quenching in CdSe are given in Fig. 12.25.

12.7 Related Photoelectronic Effects

Of the many possible photoelectronic effects that might be considered as related to photoconductivity, we include just three: the Dember, photovoltaic, and photomagnetoelectric effects.

Dember Effect

The Dember effect is the production of a voltage difference in the direction of illumination by strongly absorbed light. It is another case of the establishment of a drift field in the presence of a diffusion field. In this case the electric field, called the Dember field, is set up to keep the total current of electrons and holes equal to zero.

In the presence of photoexcitation normal to the surface of a crystal, the particle current normal to the surface can be written for electrons and holes as

$$\mathbf{j}_n = -n\mu_n \mathscr{E}_D - D_n \frac{dn}{dx} \tag{12.131}$$

$$\mathbf{j}_p = p\mu_p \mathscr{E}_D - D_p \frac{dp}{dx} \tag{12.132}$$

where $\mathscr{E}_D$ is the Dember field. The Dember field accelerates the carriers (electrons or holes) with the smaller range and decelerates the carriers with larger range, so as to maintain $\mathbf{j}_n = \mathbf{j}_p$. Applying this condition to Eqs. (12.131) and (12.132) gives

$$\mathscr{E}_D = -\frac{(kT/2)[\mu_n(dn/dx) - \mu_p(dp/dx)]}{n\mu_n + p\mu_p} \tag{12.133}$$

With

$$f = \frac{n}{\tau_n} = \frac{p}{\tau_p} \tag{12.134}$$

Eq. (12.133) becomes

$$\mathscr{E}_D = \frac{kT}{en} \frac{dn}{dx} \left(\frac{\mu_p \tau_p - \mu_n \tau_n}{\mu_p \tau_p + \mu_n \tau_n} \right) \tag{12.135}$$

The product $\mu\tau$ is the range of a carrier; the Dember field exists only when the range of electrons and holes differs.

In order to determine the specific magnitude of the Dember field, it is necessary to evaluate $n(x)$ for specific photoexcitation conditions, and for assumptions about volume and surface recombination.

12.7 Related Photoelectronic Effects

PHOTOVOLTAIC EFFECT

Whenever photoexcitation creates free carriers in the vicinity of a barrier or junction in a material, the opportunity exists for the excitation of a photovoltaic effect. A p–n junction is commonly used in photovoltaic cells. The dark current through a p–n junction is

$$I = I_0(e^{eV/kT} - 1) \tag{12.136}$$

where I_0 is the current drawn for appreciable reverse bias on the junction and depends on the height of the p–n barrier. For a forward bias of V volts, the current is given by $I_0 e^{eV/kT}$ in the ideal case. In the presence of photo-excitation an additional current flows across the junction such that

$$I = I_L - I_0(e^{eV/kT} - 1) \tag{12.137}$$

where I is the total current, and I_L is the light-generated current. If the junction is short-circuited,

$$I_{sc} = I_L \tag{12.138}$$

If the junction is open-circuited, $I = 0$, and

$$V_{oc} = \frac{kT}{e} \ln\left(1 + \frac{I_L}{I_0}\right) \tag{12.139}$$

For small light intensities, $I_L \ll I_0$, and V_{oc} is proportional to I_L and hence to the light intensity; for high light intensities, $I_L \gg I_0$, V_{oc} is proportional to the $\ln I_L$ or ln excitation intensity. The maximum value of V_{oc} is given by the height of the junction barrier in the dark.

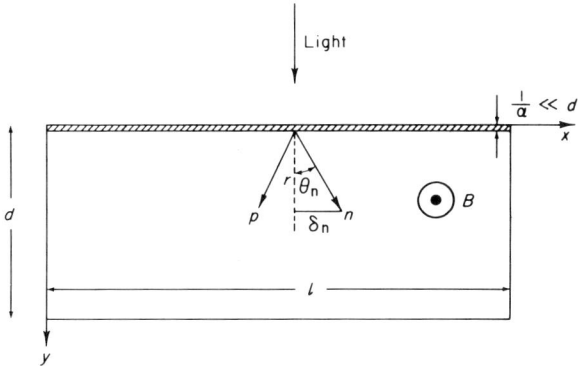

Fig. 12.26 Geometry for the photomagnetoelectric effect by strongly absorbed light incident normal to a magnetic field.

PHOTOMAGNETOELECTRIC EFFECT

In the photomagnetoelectric effect (PME effect), strongly absorbed light is incident on a material in a direction normal to that of an applied magnetic field. The Hall effect of the diffusing carriers results in a potential difference in the third orthogonal direction as the result of the setting up of a PME electric field in the material. The small-signal case can be calculated in a simple way to illustrate the nature of the effect. The geometry is given in Fig. 12.26.

For small magnetic fields the deflection of electrons, for example, is given by

$$\delta_n = \theta_n r$$
$$\approx \left(\frac{\mu_n B}{c}\right)(D_n \tau)^{1/2} \qquad (12.140)$$

where the first term represents the Hall angle (where for small θ, $\tan\theta \approx \theta$), and the second term represents the diffusion length for electrons. The charge contribution from electrons in the external circuit is therefore $e\delta_n/l$, if l is the width of the crystal. A similar contribution comes from holes, and the total short-circuited PME current is

$$I_{\text{PME}} = \frac{LeB}{cl}[\mu_n(D_n\tau)^{1/2} + \mu_p(D_p\tau)^{1/2}]$$
$$= \frac{LeB}{cl}(\mu_n^{3/2} + \mu_p^{3/2})\left(\frac{kT\tau}{e}\right)^{1/2} \qquad (12.141)$$

where L is the number of electron–hole pairs created per second per unit area, and where we have assumed that $\tau_n = \tau_p = \tau$.

Since the photocurrent measured on the same sample is given by

$$I_{\text{PC}} = \frac{eL\tau(\mu_n + \mu_p)V}{l^2} \qquad (12.142)$$

combination of Eqs. (12.141) and (12.142) gives

$$\frac{I_{\text{PC}}}{I_{\text{PME}}} = \frac{Vc}{lB}\left(\frac{e\tau}{\mu^* kT}\right)^{1/2} \qquad (12.143)$$

from which the lifetime τ can be determined without accurate determination of the absolute excitation intensity, in terms of

$$\frac{1}{\mu^{*1/2}} = \frac{\mu_n + \mu_p}{\mu_n^{3/2} + \mu_p^{3/2}} \qquad (12.144)$$

More detailed calculations take into account surface recombination and the possibility of $\tau_n \neq \tau_p$. In the latter case I_{PME} involves the minority-carrier lifetime, whereas I_{PC} involves the majority-carrier lifetime. The PME effect is one of the few that permit the determination of minority-carrier lifetime in a material in which the minority-carrier lifetime is much smaller than the majority-carrier lifetime.

12.8 Summary

When light of sufficient energy is absorbed by a crystal, free electrons and free holes are created. As long as these photoexcited carriers remain free and in the crystal, they contribute to an excess conductivity, a photoconductivity caused by the light. The length of time that a carrier stays free and contributes to the electrical conductivity is called the lifetime of that carrier.

The lifetime of a photoexcited carrier may be ended if the carrier leaves the crystal at an electrode and is not replenished at the opposite electrode, i.e., if the opposite electrode is not ohmic for carriers of that type, or if the carrier recombines with a carrier of opposite type, either at an imperfection or in a free–free recombination. The lifetime of a photoexcited carrier is interrupted if the carrier is temporarily localized at a trapping site, later to be freed thermally to the band.

The three major properties of a photoconducting material are its spectral response, its sensitivity, and its speed of response. The spectral response is determined by the bandgap of the material for intrinsic photoconductivity and by the location of imperfection energy levels for extrinsic photoconductivity. In the case of intrinsic photoconductivity, the magnitude of the response is usually a maximum near the optical absorption edge of the material. For longer wavelengths the response decreases as the absorption constant decreases, and for shorter wavelengths the response may decrease if there is a large surface recombination probability.

The sensitivity of a material is determined by the density and capture coefficients of imperfection levels, except for very high photoexcitation intensities where direct free–free recombination may become important. The sensitivity is related to the product of the lifetime and the mobility; the former varies over many orders of magnitude depending on the specific imperfections present.

The speed of response of a material is limited by the condition that the response time can never be less than the carrier lifetime. If trapping of carriers occurs, however, the response time observed in a measurement may be much longer than the carrier lifetime.

In analyzing electronic behavior under the effects of photoexcitation, it is frequently convenient to introduce the concept of a steady-state Fermi level, which describes the occupancy of levels still in effective thermal equilibrium with the nearby band, and the concept of a demarcation level for each type of imperfection, which describes whether or not a specific level is in effective thermal equilibrium with the band or not.

A variety of models can be constructed to describe the many different types of photoconductivity behavior found in experimental measurements. In each model a dominant quantity determining the ultimate lifetime of free carriers is the capture coefficient of the major recombination centers. The magnitude of this capture coefficient can be approximately calculated for the three major types of recombination: radiative recombination with the emission of a photon, nonradiative recombination with the emission of one or more phonons, and nonradiative Auger recombination with the excitation of a free carrier to a higher-energy state. The largest capture coefficients are found for Coulomb attractive centers with many excited states; free carriers can become bound in an extended orbit of this center and then drop successively down through the array of excited states to the ground state of the center with the emission of one or a few phonons with each transition. The smallest capture coefficients are expected for imperfection centers with a charge the same as that of the charge of the carrier to be captured; such centers have a capture coefficient some ten orders of magnitude smaller than that of the Coulomb-attractive type center.

A variety of other effects exist in addition to photoconductivity when light of sufficient energy is absorbed by a material under appropriate conditions. Among these are the Dember effect, the creation of an electric field in the direction of strongly absorbed light incident on a crystal surface; the photovoltaic effect, the generation of either a short-circuit current or an open-circuit voltage by photoexcitation of a junction or potential barrier in a material; and the photomagnetoelectric effect, the generation of a short-circuit current or open-circuit voltage in the third mutually orthogonal direction when strongly absorbed light is shined on a crystal surface at right angles to a magnetic field.

Problems

12.1 At $300°K$ a crystal has an index of refraction of 4, a bandgap of 2.250 eV, effective electron and hole mass ratios of 0.2 and 0.5 respectively, a minority-carrier mobility of 40 cm^2/V-sec, a free-carrier lifetime of 100 μsec, a thickness of 1 mm, and a surface recombination velocity of 10^3 cm/sec. Calculate (a) the expected variation of the photoconductivity spectral response on absorption constant, and (b) the expected variation of the photoconductivity spectral response on wavelength, assuming allowed direct optical band-to-band transitions.

12.2 A certain material is available in three forms: n-type with a carrier density of 10^{16} cm^{-3}, p-type with a carrier density of 10^{16} cm^{-3}, and near-intrinsic with both carrier densities less than 10^8 cm^{-3}, in the dark at room temperature. If it is assumed that the following values hold in all three forms of the material ($\mu_n = 10^3$ cm^2/V-sec, $\mu_p = 10^2$ cm^2/V-sec, $\tau_p = 1$ μsec $\tau_n = 10^2$ μsec), what must be the thickness of the p-type region of an n–p–n junction phototransistor made from the high-conductivity materials, to give the same photoconductivity gain as would be obtained by applying 10 V across 2 mm of the near-intrinsic high-resistivity material with ohmic contacts?

12.3 A photocurrent of 100 nA is measured for a uniform excitation of a thin photoconductor layer between two sandwich-like electrodes at the rate of 10^{13} cm^{-3} sec^{-1} when 10 mV is applied across the 0.1-mm-thick sandwich with an area of 1 cm^2. If the electron mobility is 100 cm^2/V-sec and the dielectric constant is 10, calculate the maximum voltage that can be applied before space-charge injection becomes important.

12.4 Four photoconductor cells have the following dimensions (in each case the interelectrode spacing is given first): 2 cm × 1 cm, 1 cm × 2 cm, 1 cm × 1 cm, and 2 cm × 2 cm. Each consists of identical material and each has the same voltage applied. All sit in the same illumination that shines perpendicular to the large face of the cell and may be assumed to be homogeneously absorbed. If the contacts are ohmic, give the ratio of (a) the measured photocurrent, (b) the photosensitivity, and (c) the photoconductivity gain.

12.5 It is observed that the photocurrent varies as the 0.87 power of the excitation intensity for a particular crystal at room temperature. If it is assumed that this behavior arises from an exponential trap distribution, calculate the expected variation of photocurrent with light intensity at 100 and 600°K, as well as the expected ratio of the photocurrents at these two temperatures.

12.6 It is frequently found that the onset of thermal quenching (constant excitation intensity and increasing temperature) or the onset of superlinear photoconductivity (constant temperature and decreasing excitation intensity) occurs when the hole demarcation level of the sensitizing centers lies at the level of the sensitizing centers. Following are real data measured on a crystal of CdS with electrode spacing of 1 mm and cross-sectional area of 1×0.35 mm^2, with 10 V applied. Using the data, make plots of both photocurrent versus temperature for different excitation intensities, and photocurrent versus intensity for different temperatures. Determine the hole ionization energy of sensitizing centers, and the ratio of capture coefficients for these centers. Effective-mass ratios in CdS are 0.2 and 1.0 for electrons and holes respectively, and the electron mobility may be taken as 100 cm^2/V-sec and as independent of temperature. In photosensitive crystals of CdS, the photoexcited holes are quickly trapped at sensitizing centers, and the photoconductivity is due to electrons only.

Relative excitation intensity	Current (μA)								
	298°K	314°K	327°K	337°K	348°K	355°K	362°K	371°K	380°K
100	175	200	195	190	180	150	130	94	35
47	100	110	100	100	97	82	63	28	1.2
20	57	60	56	53	48	36	20	2.0	0.12
9	31	31	28	25	21	11	2.4	0.1	0.06
4	15	15	13	11	6.5	1.4	0.046		
2	7.1	7.1	5.8	3.5	1.0	0.019			
0.8	3.7	3.3	2.2	0.94	0.015				
0.4	1.9	1.6	0.75	0.028	0.003				
0.2	0.80	0.48	0.052	0.002					

12.7 The photocurrent is measured as a function of temperature for different photoexcitation intensities through the region of thermal quenching of photoconductivity in a crystal with sensitizing centers, both for photoexcitation with photons with absorption constant considerably greater than the reciprocal of the crystal thickness, and for photoexcitation with photons with absorption constant considerably less than the reciprocal of the crystal thickness. In terms of the data available from such an experiment, show how the "effective volume" for excitation by the strongly absorbed photons can be calculated.

12.8 Consider the situation of extrinsic excitation from an imperfection level to the conduction band. Remember that in this case the rate of excitation is proportional to the density of occupied centers (rate $\approx fS_0n_I$, where S_0 is the optical cross section of the imperfection, and n_I is the density of occupied imperfections) as well as to the light intensity. Calculate the dependence of the free-electron density on the exciting light intensity over the whole range from zero to very high, and plot a typical curve.

Appendix

Units and Conversion Factors

Units and Conversion Factors

Quantity	Symbol	Laboratory	MKS	esu	emu
Capacitance	C	farad	1 $m^{-1}L^{-2}t^2q^2$	9×10^{11} εL	10^{-9} $\mu^{-1}L^{-1}t^2$
Charge	q	coulomb	1 q	3×10^9 $\varepsilon^{1/2}m^{1/2}L^{3/2}t^{-1}$	10^{-1} $\mu^{-1/2}m^{1/2}L^{1/2}$
Conductivity, volume	σ	(ohm-cm)$^{-1}$	10^2 $m^{-1}L^{-3}tq^2$	9×10^{11} εt^{-1}	10^{-9} $\mu^{-1}L^{-2}t$
Current	i	ampere	1 $t^{-1}q$	3×10^9 $\varepsilon^{1/2}m^{1/2}L^{3/2}t^{-2}$	10^{-1} $\mu^{-1/2}m^{1/2}L^{1/2}t^{-1}$
Current density	j	ampere cm^2	10^4 $L^{-2}t^{-1}q$	3×10^9 $\varepsilon^{1/2}m^{1/2}L^{-1/2}t^{-2}$	10^{-1} $\mu^{-1/2}m^{1/2}L^{-3/2}t^{-1}$
Displacement, electric	$\mathbf{D}$	volt/cm	$(36\pi \times 10^7)^{-1}$ $L^{-2}q$	$(9 \times 10^{12})^{-1}$ $\varepsilon^{1/2}m^{1/2}L^{-1/2}t^{-1}$	10^8 $\mu^{-1/2}m^{1/2}L^{-3/2}$
Field intensity, electric	$\mathscr{E}$	volt/cm	10^2 $mLt^{-2}q^{-1}$	$(300)^{-1}$ $\varepsilon^{-1/2}m^{1/2}L^{1/2}t^{-1}$	10^8 $\mu^{1/2}m^{1/2}L^{1/2}t^{-2}$
Field intensity, magnetic	$\mathbf{H}$	oersted	$(4\pi \times 10^{-3})^{-1}$ $L^{-1}t^{-1}q$	3×10^{10} $\varepsilon^{1/2}m^{1/2}L^{1/2}t^{-2}$	1 $\mu^{-1/2}m^{1/2}L^{-1/2}t^{-1}$
Flux, magnetic	Φ	maxwell	10^{-8} $mL^2t^{-1}q^{-1}$	$(3 \times 10^{10})^{-1}$ $\varepsilon^{-1/2}m^{1/2}L^{1/2}$	1 $\mu^{1/2}m^{1/2}L^{3/2}t^{-1}$
Inductance	L	henry	1 mL^2q^{-2}	$(9 \times 10^{11})^{-1}$ $\varepsilon^{-1}L^{-1}t^2$	10^9 μL

Units and Conversion Factors

Quantity	Symbol	Unit			
Induction, magnetic	B	gauss	10^{-4} $mt^{-1}q^{-1}$	$(3 \times 10^{10})^{-1}$ $\varepsilon^{-1/2}m^{1/2}L^{-3/2}$	1 $\mu^{1/2}m^{1/2}L^{-1/2}t^{-1}$
Permeability	μ	unitless	$4\pi \times 10^{-7}$ mLq^{-2}	$(9 \times 10^{20})^{-1}$ $\varepsilon^{-1}L^{-2}t^2$	1 μ
Permittivity, dielectric constant	ε	unitless	$(36\pi \times 10^9)^{-1}$ $m^{-1}L^{-3}t^2q^2$	1 ε	$(9 \times 10^{20})^{-1}$ $\mu^{-1}L^{-2}t^2$
Potential, electric	V	volt	1 $mL^2t^{-2}q^{-1}$	$(300)^{-1}$ $\varepsilon^{-1/2}m^{1/2}L^{1/2}t^{-1}$	10^8 $\mu^{-1/2}m^{1/2}L^{3/2}t^{-2}$
Potential, magnetic vector	A	maxwell/cm	10^{-6} $mLt^{-1}q^{-1}$	$(3 \times 10^{10})^{-1}$ $\varepsilon^{-1/2}m^{1/2}L^{-1/2}$	1 $\mu^{1/2}m^{1/2}L^{1/2}t^{-1}$
Resistance	R	ohm	1 $mL^2t^{-1}q^{-2}$	$(9 \times 10^{11})^{-1}$ $\varepsilon^{-1}L^{-1}t$	10^9 μLt^{-1}
Resistivity, volume	ϱ	ohm-cm	10^{-2} mL^3tq^{-2}	$(9 \times 10^{11})^{-1}$ $\varepsilon^{-1}t$	10^9 μL^2t^{-1}

Example of How to Use This Table

1 Laboratory unit of resistivity = 1 ohm-cm
$\qquad = 10^{-2}$ **MKS** units of resistivity = 10^{-2} ohm-m
$\qquad = (9 \times 10^{11})^{-1}$ esu units of resistivity = $(9 \times 10^{11})^{-1}$ esu-ohm-cm
$\qquad = 10^9$ emu units of resistivity = 10^9 emu-ohm-cm

Electronic charge = 1.6×10^{-19} coulomb (laboratory and **MKS** unit)
$\qquad = 4.8 \times 10^{-10}$ esu units
$\qquad = 1.6 \times 10^{-20}$ emu units

Bibliography

Background General Reading

Azaroff, L. V. and Brophy, J. J., "Electronic Processes in Materials," McGraw-Hill, New York, 1963.

Barrett, C. R., Tetelman, A. S., and Nix, W. D., "Introduction to Materials Science," Prentice-Hall, Englewood Cliffs, New Jersey, 1973.

Greig, D., "Electrons in Metals and Semiconductors," McGraw-Hill, New York, 1969.

Hutchison, T. S., and Baird, D. C., "The Physics of Engineering Solids," Wiley, New York, 1963.

Jain, G. C., "Properties of Electrical Engineering Materials," Harper and Row, New York, 1967.

Kittel, C., "Introduction to Solid State Physics" (3rd ed.), Wiley, New York, 1966.

Nussbaum, A., "Electronic and Magnetic Behavior of Materials," Prentice-Hall, Englewood Cliffs, New Jersey, 1967.

Ramey, R. L., "Physical Electronics," Wadsworth, 1961.

Rose, R. M., Shepard, L. A., and Wulff, J., "The Structure and Properties of Materials: Vol. IV, Electronic Properties," Wiley, New York, 1966.

Stringer, J., "An Introduction to the Electron Theory of Solids," Pergamon, Elmsford, New York, 1967.

Van der Ziel, A., "Solid State Physical Electronics," Prentice-Hall, Englewood Cliffs, New Jersey, 1957.

Wert, C. A., and Thomson, R. M., "Physics of Solids," McGraw-Hill, New York, 1964.

Properties of Waves

Brillouin, L., "Wave Propagation in Periodic Structures," Dover, New York, 1953.

Towne, D. H., "Wave Phenomena," Addison-Wesley, New York, 1969.

Bibliography

Quantum Mechanics

Pohl, H. A., "Quantum Mechanics for Science and Engineering," Prentice-Hall, Englewood Cliffs, New Jersey, 1967.
Rojansky, V. "Introductory Quantum Mechanics," Prentice-Hall, Englewood Cliffs, New Jersey, 1946.
Schiff, L. I., "Quantum Mechanics," McGraw-Hill, New York, 1955.
White, R. L., "Basic Quantum Mechanics," McGraw-Hill, New York, 1966.

Wave Mechanics of Solids

Donovan, B., "The Elementary Theory of Metals," Pergamon, Elmsford, New York, 1967.
Raimes, S., "The Wave Mechanics of Electrons in Metals," Wiley-Interscience, New York, 1961.
Smith, R. A., "Wave Mechanics of Crystalline Solids," Wiley, New York, 1961.

Energy Bands in Solids

Callaway, J., "Energy Band Theory," Academic Press, New York, 1964.
Long, D., "Energy Bands in Semiconductors," Wiley-Interscience, New York, 1968.
Slater, J. C., "Quantum Theory of Molecules and Solids: Vol. 2, Symmetry and Energy Bands in Crystals," McGraw-Hill, New York, 1965.

Transport Phenomena

Beam, W. R., "Electronics of Solids," McGraw-Hill, New York, 1965.
Beer, A. C., "Galvanomagnetic Effects in Semiconductors," Academic Press, New York, 1963.
Blatt, F. J., "Physics of Electronic Conduction in Solids," McGraw-Hill, New York, 1968.
Conwell, E. M., "High Field Transport in Semiconductors," Academic Press, New York, 1967.
Gossick, B. R., "Potential Barriers in Semiconductors," Academic Press, New York, 1964.
Grove, A. S., "Physics and Technology of Semiconductor Devices," Wiley, New York, 1967.
Harman, T. C., and Honig, J. M., "Thermoelectric and Thermomagnetic Effects and Applications," McGraw-Hill, New York, 1967.
Lampert, M. A., and Mark, P., "Current Injection in Solids," Academic Press, New York, 1970.

Lark-Horovitz, K., and Johnson, V. A. (eds.), "Methods of Experimental Physics: Vol. 6B. Electrical, Magnetic, and Optical Properties," Academic Press, New York, 1959.

Moll, J. L., "Physics of Semiconductors," McGraw-Hill, New York, 1964.

Putley, E. H., "The Hall Effect and Related Phenomena," Butterworths, London, 1960.

Shockley, W., "Electrons and Holes in Semiconductors," Van Nostrand, New York, 1950.

Slater, J. C., "Quantum Theory of Molecules and Solids: Vol. 3, Insulators, Semiconductors and Metals," McGraw-Hill, New York, 1967.

Smith, A. C., Janak, J. F., and Adler, R. B., "Electronic Conduction in Solids,"McGraw-Hill, New York, 1967.

Smith, R. A., "Semiconductors," Cambridge Univ. Press, New York, 1961.

Spenke, E., "Electronic Semiconductors," McGraw-Hill, New York, 1958.

Tredgold, R. H., "Space Charge Conduction in Solids," Elsevier, 1966.

Van Gool, W., "Principles of Defect Chemistry of Crystalline Solids," Academic Press, New York, 1966.

Wang, S., "Solid State Electronics," McGraw-Hill, New York, 1966.

Ziman, J. M., "Electrons and Phonons," Oxford Univ. Press, New York, 1960.

Ziman, J. M., "Electrons in Metals," Taylor and Francis, London, 1963.

Ziman, J. M., "Principles of the Theory of Solids," Cambridge Univ. Press, New York, 1964.

Optical and Photoelectronic Properties

Abeles, F. (ed.), "Optical Properties of Solids," North-Holland Publ., Amsterdam, 1972.

Aven, M., and Prener, J. S. (eds.), "Physics and Chemistry of II–VI Compounds," North-Holland, Amsterdam, 1967.

Blakemore, J. S., "Semiconductor Statistics," Pergamon, Elmsford, New York, 1962.

Bube, R. H., "Photoconductivity of Solids," Wiley, New York, 1960.

Henisch, H. K., "Electroluminescence," Pergamon, Elmsford, New York, 1962.

Larach, S. (ed.), "Photoelectronic Materials and Devices," Van Nostrand, New York, 1965.

Moss, T., "Optical Properties of Semiconductors," Butterworths, London, 1959.

Schulman, J. H., and Compton, W. D., "Color Centers in Solids," Macmillan, New York, 1962.

Rose, A., "Concepts in Photoconductivity and Allied Problems," Wiley-Interscience, New York, 1963.

Ryvkin, S. M., "Photoelectric Effects in Semiconductors," Consultants Bureau, 1964.

Thornton, P. R., "The Physics of Electroluminescent Devices," Spon, London, 1967.

Vavilov, V. S., "Effects of Radiation on Semiconductors," Consultants Bureau, 1965.

Willardson, R. K., and Beer, A. C. (eds.), "Semiconductors and Semimetals: Vol. 3, Optical Properties of III–V Compounds," Academic Press, New York, 1967.

Index

A

Absorption, optical, 401–445
Absorption index, 15, 403
Acceptors, 312
Accumulation layer, 464
Activation energy, 338, 341
Activator, 317
Amorphous materials, 167, 200
Amplification, 347, 352
Atomic energy levels, 75–83
Attempt-to-escape frequency, 333
Az'bel-Kaner resonance, 359, 390

B

Bandgap
 direct, 189, 401, 413–427
 indirect, 189, 401, 427–442
 pressure dependence of, 425, 426
 temperature dependence of, 424–426
Band model, 179–184
 conduction, 181
 valence, 181
Barriers, 452, 462, 479–484
Basis vectors, 34
Binding, types of, 201, 311
Bloch condition, 124
Bloch functions, 121, 122, 217, 262
Bohr magneton, 78
Bohr orbits, 68
Boltzmann equation, 211, 220–231, 371
 electric field only, 225, 226
 general solution, 228, 229, 371
 relaxation time solution, 223–229
 thermal gradient only, 225–228
Boltzmann's constant, 103
Bonds, 167, 200–206
Bose-Einstein distribution, 99, 100

Boundary conditions, 5
 periodic, 74, 92
Box
 one-dimensional, 22, 49–53
 three-dimensional, 83–90
Bra symbol, 42
Bragg reflection, 121, 148, 263
Bravais lattice, 130, 153
Brillouin zones, 128–141
 extended, 131, 133
 fundamental, 133
 reduced, 133, 139, 186
Brooks–Herring scattering 250, 281

C

Capacitance, junction, 462, 463
Capture
 coefficient, 334
 cross section, 334, 486, 488
 probability, 334
Cellular method, 124–128
Central force field, 59, 275
Charge density, 16
Chemical potential, 103
Coactivator, 317
Commutator, 40
Compensation, 313
 self, 313
Complementary observables, 40
Conductivity
 complex, 407
 effective mass, 233, 235
 electrical, 15, 225, 229–239, 329–343
 frequency dependence, 406–408
 nobility, 235
 thermal, 239–244
Contacts, 449, 460–471
 double injection, 469–471

nonohmic, 461–464
ohmic, 464–471
single injection, 466–469
Continuity equation, 471
Conwell–Weisskopf scattering, 250, 274–280
Cross sections
capture, 334, 486, 488
scattering, 260, 270, 273, 277
Crystal field splitting, 197
Crystal vibrations, 5–16
Cyclotron frequency, 358
Cyclotron resonance, 390

D

de Broglie wavelength, 22
Debye length, 281
Debye temperature, 12, 237
Defects, 306
Deformation potential, 250
Degenerate states, 36, 85
de Haas–van Alphen effect, 359, 394
Demarcation level, 450, 474–476
Dember effect, 506
Density of states, 87–90, 95–98
in bands, 171–174
effective density, 325
effective mass, 235
Depletion layer, 461, 462
Detectivity, 453
Device applications, 346–354
Dielectric constant, 15, 403, 407
Dielectric relaxation time, 469
Diffusion
constant, 457
length, 458
Dirac notation, 42
Dislocations, 286
Dispersion, 2
Donors, 312, 316
Drift
mobility, 367, 368
path length, 374
velocity, 296, 488

E

Effective mass, 127, 150, 159–161, 166, 172, 174–179

conductivity, 235
density-of-states, 235
longitudinal, 189
negative, 352, 353
tensor, 178, 191
transverse, 189
Eigenfunctions, 27, 30, 39
Eigenvalues, 27, 30
Elastic energy, 253
Electrical conductivity, 211, 229–239, 329–343
Electric field, 16, 225
high-field effects, 288–299
Electrochemical potential, 104
Electromagnetic waves, 15–17, 405
Electronegativity, 201, 202
Emission of radiation, 345
Energy bands, 121–207
calculations, general, 167–171
electrical properties, 179–184
Hall effect, 383–385
impurity bands, 344
Kronig–Penney model, 163, 164, 307
magnetoresistance, 385, 386
origin of, 121–163
overlapping bands, 153
perturbed free electrons, 141–153
properties of, 166–207
real crystals, 184–200
tight-binding approximation, 153–163
Energy conservation, 349
Energy gaps, 149
Energy levels, 85
hydrogenic, 318–324
localized, 305–354
Entropy, 103
Ettingshausen effect, 387
Excited states, 52
hydrogenic impurity, 323, 324
Excitons, 167, 199, 200, 422, 442–444
Expectation values, 38–42

F

Fermi–Dirac statistics, 75, 99–106
Fermi energy, 89
Fermi level, 306, 324–343
steady-state, 449, 472–474
Fermi sphere, 95
Fermi surface, 237, 359

Index

Field emission, 114-117
Field ionization, 289
Free electron model, 73-118
  applications of, 106-117
Free particle, 48

G

Gain, photoconductivity, 453, 479, 484, 485
God, 70
Ground state, 52
Group velocity, 4
Grüneisen relation, 242, 243
Gunn effect, 347, 351

H

Hall effect, 357, 359-374
  angle, 361
  constant, 360
  dependence on magnetic field, 373
  energy band effects, 383-385
  experimental apparatus, 362
  field, 360
  geometry of sample, 363
  mobility, 362, 363, 367, 376
  photo, 359
  scattering effects, 377-380
Hamiltonian, 27
Hartree-Fock model, 81
Hartree model for metal, 93-98
Hartree potential, 74, 76
Heat capacity, electronic, 108-110
Heat current, 240
Heisenberg uncertainty principle, 40, 215, 236
Hermite polynomials, 56
Hermitian operator, 35, 36
Hole, 166, 181-184
Hopping process, 200
Hot electrons, 289-297
Hydrogenic atom, 58-68
  angular dependence, 60-64
  radial dependence, 64, 65
Hydrogenic impurity, 306, 318-324

I

Impact ionization, 289

Imperfections, 305
  bands, 344
  centers, 311
  covalent model, 315, 316
  electronic behavior, 316-318
  excited states, 323, 324
  hydrogenic, 318-324
  incorporation, 310-316
  interactions, 343-346
  ionic model, 311-315
  ionization energy, 323
  occupancy of, 329
  pairs, 345, 346
  photoelectronic effects, 449
  sensitization, 502, 505
  terminology, 310
Impurities, 306
Indeterminacy, 40
Injection, electrical, 289
Insulators, 179-184
Intrinsic material, 327-339

K

k space, 95, 122, 128
Ket symbol, 42
Kinetics of photoconductivity, 454, 455, 490-505
$k \cdot P$ approximation, 127, 171
  in germanium, 192-194
Kronecker delta, 34
Kronig-Penney approximation, 163, 164, 307

L

Landau levels, 389
Lasers, 347, 349
Lattice, 74, 122
  cubic, 122, 130
  empty, 134
  reciprocal, 122, 128-141
  vectors, 122
Lifetime, 449, 490-505, 451-453
  minority-carrier, 508, 509
Linde's rule, 238
Lorentz force, 222, 360
Luminescence, 349

M

Magnetic field effects, 16, 357–396
  Az'bel–Kaner, 359, 390
  de Haas–van Alphen, 359, 394
  Hall, 357, 359–374, 377–380, 383–385
  high field, 387–394
  magnetoabsorption, 359, 392–394
  magnetoresistance, 357, 370, 374–377, 380–383, 385, 386
  magnetothermal, 357, 386, 387
Matrix elements, 142, 259, 265, 268, 319, 417, 432
  momentum, 193, 418
Matthiessen's rule, 238
Maxwell–Boltzmann distribution, 99, 100, 295
Maxwell's equations, 16
Metals
  electrical conductivity, 237–239
  one-dimensional, 97, 98
  three-dimensional, 83–96
  two-dimensional, 97
Mobility, 212
  conductivity, 235, 367
  drift, 367
  field dependence, 288–299
  Hall, 367
  microscopic, 367
  photoexcitation effects, 452
  ratio, 376
  temperature dependence, 254, 255, 279

N

n-p-n junction, 478, 479
n space, 88, 95
Néel switches, 352
Nernst effect, 387
Neutrality condition, 331
Nordheim rule, 239
Normalization, 33

O

Observable, 26
Occupational probability, 102
  of bands by holes, 184
Operator, 26
  equation, 28
  Hermitian, 35
  linear, 35
  Schroedinger, 35
Optical absorption, 401–445
  attenuation factor, 403
  constant, 404, 419
  degenerate material, 423
  direct band-to-band, 401, 413–427
  exciton, 422, 442–444
  free carrier, 403–412
  index, 403
  indirect band-to-band, 401, 427–442
  plasma, 411
  polarization effects, 426
  pressure dependence, 425, 426
  temperature dependence, 424–426
  virtual state, 428
OPW method, 168
Orthogonality, 34
Orthonormal set, 34
Oscillator, simple harmonic, 22, 23, 53–58
Oscillator strength, 420

P

p-n junctions, 347–349, 477
p state, 61
Pair formation, 345, 346
Paramagnetism, spin, 107
Particle-wave duality, 40
Pauli exclusion principle, 73, 79, 80, 99, 122, 158
Penetration depth, 404
Periodic boundary conditions, 74
Periodic potential 121 ff.
Periodic table, 73, 75, 79, 80, 202, 205
Permeability, 15, 403
Perturbation theory, 75, 141–145, 258–260, 430, 431
Phase velocity, 4
Phonon, 53
  absorption, 292, 427
  drag, 245, 264
  emission, 292, 345, 427
Photoconductivity, 347, 449–460, 471–505
  barrier, 479–484
  homogeneous materials, 484, 485

Index 523

lifetime, 449, 451–453, 490–505
n-p-n junction, 478, 479
one-center models, 490-502
p-n junction, 477
Shockley-Read model, 491-498
spectral response, 455–459
two-center models, 502–505
Photoelectronic effects, 449–510
Photomagnetoelectric effect, 450, 508, 509
Photon, 53
emission, 317, 449
Photosensitivity, 453
Photovoltaic effect, 347, 450, 507
Physical reality, 5, 31, 54, 70
Piezoelectric effect, 286
Plasma frequency, 411
Plasmon, 412
Poisson bracket, 26
Poisson's equation, 461, 471
Polarons, 200
Poynting vector, 420
Probability, 31
distribution, 100
Pseudopotentials, 168–171
Pseudowavefunction, 168–171

Q

Quantization, 21
Quantum efficiency, 453
Quantum mechanics, 20–43
postulates of, 26–31
simple systems, 47–70
Quantum number
angular momentum, 60
magnetic, 60
one-dimensional box, 50
oscillator, linear harmonic, 58
principal, 64
spin, 78
Quantum statistics, 98–106
Quenching of photoconductivity
optical, 503
thermal, 502

R

Reciprocal lattice, 122, 128–141
Recombination, 449

Auger, 488, 489
detection of, 349
nonradiative, 487, 488
radiative, 346, 349, 486, 487
surface, 457, 458
Rectifier, 347
Reflection coefficient, 252, 405
Refraction index, 15, 403–407
Relaxation time, 212, 254, 270, 279, 359
approximation, 211, 222–231, 359, 377
Reststrahlen, 11
Richardson's equation, 112
Righi–Leduc effect, 387
Rutherford scattering, 276

S

s state, 61
Scattering
center, 317
cross sections, 260, 270, 273, 277
Hall effect, in, 377–380
magnetoresistance, in, 380–383
normal, 263
probability, 269
Umklapp, 263, 268
Scattering processes, 211, 250–302
acoustic phonon, 231, 251–273, 288–299
charged impurities, 231, 273–285, 410
dislocations, 286
elastic, 223
inelastic, 223, 237
inhomogeneity, 287, 288
intervalley, 286, 287
neutral imperfections, 285
optical phonon, 245, 285, 297, 379, 410
piezoelectric, 286
Schroedinger equation, 23–26
Screening, 344
Secular equation, 142
Semiconductors, 179–184, 205, 206, 232–237, 288
Sensitizing centers, 318, 502
Shockley-Read model, 491–498
Skin effect, 408
Solar cell, 349
Sommerfeld model, 74, 83–90
Sound velocity, 2, 296
Space charge, 287

Space-charge-limited currents, 464, 466–471
  double injection, 469–471
  single injection, 466–469
Spherical harmonics, 60
Spin, 73, 77, 78
  paramagnetism, 106, 107
Spin-orbit interaction, 187
Square-integrable functions, 32
Stationary state, 25
Superconductivity, 237, 239
Superlinear photoconductivity, 502
Surface recombination velocity, 457, 458
Surface states, 305–308

T

Temperature, electron, 289
Thermal conductivity, 211
Thermal gradient, 211, 225–228
Thermionic emission, 110–114
Thermoelectric effect, 212, 244–246, 387
Thomas–Fermi approximation, 120
Transistor, 347
Transit time, 454
Transport, carrier, 211–248
Transport integrals, 240, 241
Traps, 317, 368, 441, 499–502
Tunnel diode, 349–351
Tunneling, 20, 114–117, 351

U

Ultraviolet transparency of metals, 411
Units, 514, 515

V

Valence electrons, 73, 79
Variational method, 75
Vector(s), *see* specific types
Vector potential, 388, 416
Velocity
  band, in, 174–179
  group, 4
  phase, 4

Virtual cathode, 465
Virtual state, 428

W

Wave(s), 1–16
  acoustic mode, 8
  approach to quantum mechanics, 20-26
  classical, 1–18
  diatomic crystal, in, 7, 8
  electromagnetic, 15–17
  equation, 5
  harmonic, 1
  longitudinal, 3
  monatomic crystal, in, 5, 7
  normal modes, 10
  optical modes, 9
  rod, in, 5, 6
  standing, 2
  string, in, 5, 6
  transverse, 3
  traveling, 2, 90
  velocity, 4
Wavefunction, 31–34
  antisymmetric, 81
  determinantal, 76, 81
  one-electron, 76ff.
  symmetric, 81
Wavenumber, 2
Wave packets, 211, 212–220, 274
  Bloch function, 217
  Gaussian, 213
  maximum of, 216
  width of, 216
Wavevector, 22, 122
  electron, 23–26
Well-behaved functions, 30
Wiedemann–Franz ratio, 243, 244
Wigner–Seitz cell, 124, 262
WKB approximation, 116
Work function, 110

Z

Zeeman effect, 143–145
Zero-point energy, 56